Student's Solutions Manual

Intermediate Algebra
Second Edition

Mark Dugopolski
Southeastern Louisiana University

ADDISON-WESLEY PUBLISHING COMPANY
Reading, Massachusetts • Menlo Park, California • New York
Don Mills, Ontario • Wokingham, England • Amsterdam • Bonn
Sydney • Singapore • Tokyo • Madrid • San Juan • Milan • Paris

Reproduced by Addison-Wesley from camera-ready copy supplied by the authors.

Copyright © 1996 Addison-Wesley Publishing Company, Inc.

ISBN   0-201-89515-3

6 7 8 9 10 CRS 9897

To the student

This manual contains a complete solution to every odd-numbered exercise in the accompanying text. Solutions are given for all exercises in the Warm-ups, Chapter Tests, and Tying It All Together sections. This manual should be used as a reference only after you have attempted to solve a problem on your own.

TABLE OF CONTENTS

1.1 WARM-UPS

1. False, since 5 is a counting number and $5 \notin A$.   **2.** False, since B has only 3 elements.
3. False.   **4.** False, since $1 \notin B$.   **5.** True, since $3 \in A$.   **6.** True.   **7.** True.
8. False, since $1 \in A$ but $1 \notin B$.   **9.** True, since $\emptyset$ is a subset of every set.   **10.** True, since $1 \in A$ and $1 \notin C$.

1.1 EXERCISES

1. False, 6 is not odd.

3. True, because $1 \in A$ but $1 \notin B$.

5. True.

7. False, because $5 \in A$.

9. False, because 0 is not a natural number.

11. False, because N is infinite and C is finite.

13. $A = \{1, 3, 5, 7, 9\}$ and $B = \{2, 4, 6, 8\}$ so $A \cap B = \emptyset$.

15. $A = \{1, 3, 5, 7, 9\}$ and $C = \{1, 2, 3, 4, 5\}$ so $A \cap C = \{1, 3, 5\}$.

17. The elements of B together with those of C give us $B \cup C = \{1, 2, 3, 4, 5, 6, 8\}$.

19. Since the empty set has no members, $A \cup \emptyset = A$.

21. Since $\emptyset$ has no members in common with A, $A \cap \emptyset = \emptyset$.

23. Since every member of A is also a member of N, $A \cap N = A = \{1, 3, 5, 7, 9\}$.

25. Since the members of A are odd and the members of B are even, they have no members in common. So $A \cap B = \emptyset$.

27. $A \cup B = \{1, 2, 3, 4, 5, 6, 7, 8, 9\}$ from problem 14.

29. Take the elements that B has in common with C to get $B \cap C = \{2, 4\}$.

31. $3 \notin A \cap B$ since $3 \notin B$.

33. Since 4 is in both sets, $4 \in B \cap C$.

35. True, since each member of A is a counting number.

37. True, since both 2 and 3 are members of C.

39. True, since $6 \in B$ but $6 \notin C$.

41. True, since $\emptyset$ is a subset of every set.

43. False, since $1 \in A$ but $1 \notin \emptyset$.

45. True, since $A \cap B = \emptyset$ and $\emptyset$ is a subset of any set.

47. Using all numbers that belong to D or to E yields $D \cup E = \{2, 3, 4, 5, 6, 7, 8\}$.

49. Using only numbers that belong to both D and F gives $D \cap F = \{3, 5\}$.

51. Using all numbers that belong to E or to F gives $E \cup F = \{1, 2, 3, 4, 5, 6, 8\}$.

53. Intersect $D \cup E$ from Exercise 47 with F to get $(D \cup E) \cap F = \{2, 3, 4, 5\}$.

55. Take $E \cap F = \{2, 4\}$ together with D to get $D \cup (E \cap F) = \{2, 3, 4, 5, 7\}$.

57. Take the union of $D \cap F = \{3, 5\}$ with $E \cap F = \{2, 4\}$ to get $\{2, 3, 4, 5\}$.

59. Intersect $D \cup E = \{2, 3, 4, 5, 6, 7, 8\}$ with $D \cup F = \{1, 2, 3, 4, 5, 7\}$ to get $\{2, 3, 4, 5, 7\}$.

61. Use $\subseteq$, since every element of D is an odd natural number.

63. Use $\in$, since 3 is a member of D.

65. Use $\cap$, since D and E have no elements in common.

67. Use $\subseteq$, since every member of $D \cap F$ is a member of F.

69. Use $\cap$, since $8 \in E$ but not in $E \cap F$.

71. Use $\cup$, since $D \cup F$ and $F \cup D$ have exactly the same members.

73. The set of even natural number less than 20 is {2, 4, 6, . . . , 18}.

75. The set of odd natural numbers greater than 11 is {13, 15, 17, . . .}.

77. The set of even natural numbers between 4 and 79 is {6, 8, 10, . . . , 78}.

79. {x | x is a natural number between 2 and 7}

81. {x | x is an odd natural number greater than 4}

83. {x | x is an even natural number between 5 and 83}

1.2 WARM-UPS

1. False, π is irrational. **2.** True. **3.** False, 0 is not irrational. **4.** False, the set of irrational numbers is a subset of the set of real numbers. **5.** True, since it is a repeating decimal. **6.** False, it is not repeating. **7.** True. **8.** True. **9.** False, it is a finite set. **10.** True.

1.2 EXERCISES

1. True, since −6 is rational.

3. False, since 0 is rational.

5. True, since a repeating decimal is rational.

7. True, since every natural number is rational.

9. {0, 1, 2, 3, 4, 5}

11. {−4, −3, −2, −1, 0, 1, . . .}

13. {1, 2, 3, 4}

15. {−2, −1, 0, 1, 2, 3, 4}

17. All of them

19. The whole numbers include 0. {0, 8/2}

21. {−3, −5/2, −0.025, 0, $3\frac{1}{2}$, 8/2}

23. True, since every rational number is also a real number.

25. False, since 0 is not irrational.

27. True, because the rational numbers together with the irrational numbers make up the set of real numbers.

29. False, since nonrepeating decimals are irrational.

31. False, since repeating decimals are rational.

33. False, since repeating decimals are rational.

35. True, since π is irrational.

37. N ⊆ W, since every natural number is a whole number.

39. J ⊄ N, since −9 ∈ J but −9 ∉ N.

41. Q ⊆ R, since every rational number is a real number.

43. ∅ ⊆ I, since ∅ is a subset of every set.

45. N ⊆ R, since every natural number is a real number.

47. 5 ∈ J, since 5 is an integer.

49. 7 ∈ Q, since 7 is a rational number.

51. $\sqrt{2}$ ∈ R, since $\sqrt{2}$ is a real number.

53. 0 ∉ I, since 0 is rational.

55. {2, 3} ⊆ Q, since both 2 and 3 are rational.

57. {3, $\sqrt{2}$} ⊆ R, since both numbers are real.

1.3 WARM-UPS

1. True, since −6 + 6 = 0. **2.** True.

3. False, |6| = 6. **4.** True, since b − a = b + (−a). **5.** True, because the product of two number with opposite signs is negative.

6. False, since 6 + (−4) = 2.

47. False, since −3 − (−6) = −3 + 6 = 3.

8. False, since 6 ÷ (−1/2) = 6(−2) = −12.

9. False, since division by zero is undefined.

10. True, because 0 divided by any nonzero number is 0.

1.3 EXERCISES

1. $|-34| = -(-34) = 34$

3. $|0| = 0$

5. $|-6| - |-6| = 6 - 6 = 0$

7. $-|-9| = -9$

9. $-(-9) = 9$

11. $-(-(-3)) = -(3) = -3$

13. $(-5) + 9 = 9 - 5 = 4$

15. $(-4) + (-3) = -(4 + 3) = -7$

17. $-6 + 4 = -(6 - 4) = -2$

19. $7 + (-17) = -(17 - 7) = -10$

21. $(-11) + (-15) = -(11 + 15) = -26$

23. $18 + (-20) = -(20 - 18) = -2$

25. $-14 + 9 = -(14 - 9) = -5$

27. $-4 + 4 = 0$

29. $-\frac{1}{10} + \frac{1}{5} = \frac{1}{5} - \frac{1}{10} = \frac{2}{10} - \frac{1}{10} = \frac{1}{10}$

31. $\frac{1}{2} + \left(-\frac{2}{3}\right) = \frac{3}{6} + \left(-\frac{4}{6}\right) = -\frac{1}{6}$

33. $-15 + 0.02 = -(15.00 - 0.02) = -14.98$

35. $-2.7 + (-0.01) = -(2.70 + 0.01) = -2.71$

37. $47.39 + (-44.587) = 2.803$

39. $0.2351 + (-0.5) = -0.2649$

41. $7 - 10 = -(10 - 7) = -3$

43. $-4 - 7 = -4 + (-7) = -11$

45. $7 - (-6) = 7 + 6 = 13$

47. $-1 - 5 = -1 + (-5) = -6$

49. $-12 - (-3) = -12 + 3 = -9$

51. $20 - (-3) = 20 + 3 = 23$

53. $\frac{9}{10} - \left(-\frac{1}{10}\right) = \frac{9}{10} + \frac{1}{10} = 1$

55. $1 - \frac{3}{2} = \frac{2}{2} - \frac{3}{2} = -\frac{1}{2}$

57. $2.00 - 0.03 = 1.97$

59. $5.3 - (-2) = 5.3 + 2 = 7.3$

61. $-2.44 - 48.29 = -50.73$

63. $-3.89 - (-5.16) = 1.27$

65. $(25)(-3) = -(25 \cdot 3) = -75$

67. $\left(-\frac{1}{3}\right)\left(-\frac{1}{2}\right) = \frac{1}{6}$

69. $(0.3)(-0.3) = -0.09$

71. $(-0.02)(-10) = 0.02(10) = 0.2$

73. The reciprocal of 20 is $\frac{1}{20}$ or 0.05.

75. The reciprocal of $-\frac{6}{5}$ is $-\frac{5}{6}$.

77. The reciprocal of -0.3 is $-\frac{1}{0.3}$ or $-\frac{10}{3}$.

79. $-6 \div 3 = -2$

81. $30 \div (-0.8) = -37.5$

83. $(-0.8)(0.1) = -0.08$

85. $(-0.1) \div (-0.4) = 0.25$

87. $9 \div \left(-\frac{3}{4}\right) = 9\left(-\frac{4}{3}\right) = -\frac{36}{3} = -12$

89. $-\frac{2}{3}\left(-\frac{9}{10}\right) = \frac{2}{3} \cdot \frac{9}{10} = \frac{18}{30} = \frac{3}{5}$

91. $(0.25)(-365) = -91.25$

93. $(-51) \div (-0.003) = 17,000$

95. $-62 + 13 = -(62 - 13) = -49$

97. $-32 - (-25) = -32 + 25 = -7$

99. $|-15| = -(-15) = 15$

101. $\frac{1}{2}(-684) = -342$

103. $\frac{1}{2} - \left(-\frac{1}{4}\right) = \frac{2}{4} + \frac{1}{4} = \frac{3}{4}$

105. $-\frac{1}{3}(-96) = \frac{96}{3} = 32$

107. $-0.2 + 19 = 19.0 - 0.2 = 18.8$

109. $(-0.3) + (-0.03) = -0.33$

111. $-57 \div 19 = -3$

113. $|-17| + |-3| = 17 + 3 = 20$

115. $0 \div (-0.15) = 0$

117. $27 \div (-0.15) = -180$

119. $-\frac{1}{3} + \frac{1}{6} = -\frac{2}{6} + \frac{1}{6} = -\frac{1}{6}$

121. $-63 + |8| = -63 + 8 = -55$

123. $-\frac{1}{2} + \left(-\frac{1}{2}\right) = -\frac{2}{2} = -1$

125. $-\frac{1}{2} - 19 = -\frac{1}{2} - \frac{38}{2} = -\frac{39}{2}$

127. $28 - 0.01 = 27.99$

129. $-29 - 0.3 = -29.3$

131. $(-2)(0.35) = -0.7$

133. $(-10)(-0.2) = 2$

135. $-45,000 + (-2,300) + (-1500) + 1200$
$+ 2(3500) + 20,000 = -20,600$

Net worth is $-\$20,600$.

137. $14° - (-6°) = 20°$

139. $-282 - (-1296) = 1014$ ft

1.4 WARM-UPS

1. False, $2^3 = 8$.　　**2.** True.

3. True, since $-2^2 = -(2^2) = -4$.

4. False, since $6 + 3 \cdot 2 = 6 + 6 = 12$.

5. False, since $(6 + 3) \cdot 2 = 9 \cdot 2 = 18$.

6. False, since $(6 + 3)^2 = 9^2 = 81$.

7. True, since $6 + 3^2 = 6 + 9 = 15$.

8. True, since $(-3)^3 = -27$ and $-3^3 = -27$.

9. False, since $|-3 - (-2)| = |-1| = 1$.

10. False, since $|7 - 8| = 1$ and $|7| - |8| = 7 - 8 = -1$.

1.4 EXERCISES

1. $(-3 \cdot 4) - (2 \cdot 5) = -12 - 10 = -22$

3. $4[5 - |3 - (2 \cdot 5)|] = 4[5 - 7] = -8$

5. $(6 - 8)(|2 - 3| + 6) = (-2)(7) = -14$

7. $2^5 = 2 \cdot 2 \cdot 2 \cdot 2 \cdot 2 = 32$

9. $(-1)^4 = (-1)(-1)(-1)(-1) = 1$

11. $(-4)(-4) = 16$

13. $\left(-\frac{3}{4}\right)^3 = \left(-\frac{3}{4}\right)\left(-\frac{3}{4}\right)\left(-\frac{3}{4}\right) = -\frac{27}{64}$

15. $(0.1)^3 = (0.1)(0.1)(0.1) = 0.001$

17. $4 - 6 \cdot 2 = 4 - 12 = -8$

19. $5 - 6(3 - 5) = 5 - 6(-2) = 5 + 12 = 17$

21. $\left(\frac{1}{3} - \frac{1}{2}\right)\left(\frac{1}{4} - \frac{1}{2}\right) = \left(-\frac{1}{6}\right)\left(-\frac{1}{4}\right) = \frac{1}{24}$

23. $7^2 + 9 = 49 + 9 = 58$

25. $-(2 + 3)^2 = -(5)^2 = -25$

27. $-5^2 \cdot 2^3 = -25 \cdot 8 = -200$

29. $5 \cdot 2^3 = 5 \cdot 8 = 40$

31. $-(3^2 - 4)^2 = -(9 - 4)^2 = -5^2 = -25$

33. $-60 \div 10 \cdot 3 \div 2 \cdot 5 \div 6 = -6 \cdot 3 \div 2 \cdot 5 \div 6$
$= -18 \div 2 \cdot 5 \div 6 = -9 \cdot 5 \div 6 = -45 \div 6 = -7.5$

35. $30 \cdot 2 - (-6) \div 3 \cdot 12 = 60 - (-2 \cdot 12)$
$= 60 - (-24) = 84$

37. $5.5 - 2.3^4 = -22.4841$

39. $(1.3 - 0.31)(2.9 - 4.88) = -1.9602$

41. $-388.8 \div (13.5)(9.6) = -276.48$

43. $\frac{2 - 6}{9 - 7} = \frac{-4}{2} = -2$

45. $\frac{-3 - 5}{6 - (-2)} = \frac{-8}{8} = -1$

47. $\frac{4 + 2 \cdot 7}{3 \cdot 2 - 9} = \frac{18}{-3} = -6$

49. $\frac{-3^2 - (-9)}{2 - 3^2} = \frac{-9 + 9}{2 - 9} = \frac{0}{-7} = 0$

51. $3^2 - 4(-1)(-4) = 9 - 16 = -7$

53. $\frac{-1 - 3}{-1 - (-4)} = \frac{-4}{-1 + 4} = -\frac{4}{3}$

55. $(-1 - 3)(-1 + 3) = (-4)(2) = -8$

57. $(-4)^2 - 2(-4) + 1 = 16 + 8 + 1 = 25$

59. $\dfrac{2}{-1} + \dfrac{3}{-4} - \dfrac{1}{-4} = -\dfrac{4}{2} - \dfrac{1}{2} = -\dfrac{5}{2}$

61. $|-1-3| = |-4| = 4$

63. $\dfrac{-6-4}{-7-2} = \dfrac{-10}{-9} = \dfrac{10}{9}$

65. $\dfrac{2-(-1)}{1-(-3)} = \dfrac{3}{4}$

67. $\dfrac{5.6-2.4}{4.7-5.9} = -2.67$

69. $(-2)^2 - 4(-1)(3) = 4 + 12 = 16$

71. $3^2 - 4(2)(-5) = 9 + 40 = 49$

73. $8^2 - 4(0.5)(2) = 64 - 4 = 60$

75. $(3.2)^2 - 4(-1.2)(5.6) = 37.12$

77. $-2^2 + 5(3)^2 = -4 + 45 = 41$

79. $(-2+5)3^2 = 3 \cdot 9 = 27$

81. $5^2 - 4(1)(6) = 25 - 24 = 1$

83. $[13 + 2(-5)]^2 = 3^2 = 9$

85. $\dfrac{4-(-1)}{-3-2} = \dfrac{5}{-5} = -1$

87. $3(-2)^2 - 5(-2) + 4 = 12 + 10 + 4 = 26$

89. $-4\left(\dfrac{1}{2}\right)^2 + 3\left(\dfrac{1}{2}\right) - 2 = -4\left(\dfrac{1}{4}\right) + \dfrac{3}{2} - 2$

$= -\dfrac{2}{2} + \dfrac{3}{2} - \dfrac{4}{2} = -\dfrac{3}{2}$

91. $-\dfrac{1}{2}|6-2| = -\dfrac{1}{2}(4) = -2$

93. $\dfrac{1}{2} - \dfrac{1}{3}\left|\dfrac{1}{4} - \dfrac{1}{2}\right| = \dfrac{1}{2} - \dfrac{1}{3} \cdot \dfrac{1}{4} = \dfrac{1}{2} - \dfrac{1}{12} = \dfrac{5}{12}$

95. $|6 - 3 \cdot 7| + |7 - 5| = |-15| + |2| = 17$

97. $3 - 7[4 - (2-5)] = 3 - 7[4+3] = 3 - 49$

$\qquad\qquad = -46$

99. $3 - 4(2 - |4-6|) = 3 - 4(2-2) = 3$

101. $4[2 - (5 - |-3|)^2] = 4[2 - (2^2)] = 4[-2]$

$\qquad\qquad = -8$

103. $0.65(220 - 25) - 0.65(220 - 65) = 26$

The target heart rate for a 25-yr old woman is 26 beats per minute larger than the target heart rate for a 65 yr-old woman. A woman's target heart rate is 115 at about 43 years of age.

105. The perimeter is $2(34) + 2(18)$ or 104 feet.

107. The investment amounts to $10,000(1 + 0.062)^{30}$ or \$60,776.47.

109. The actual amount owed when the payments start is $4000(1 + 0.08)^4$ or \$5441.96.

111. $29,930(1.08)^{15} - 62,222 = 32,721$

Using the formula in this exercise, a one-year stay costs \$32,721 more.

1.5 WARM-UPS

1. True. **2.** False, since $8 \div (4 \div 2) = 4$ and $(8 \div 4) \div 2 = 2$.

3. False, since $10 \div 2 = 5$ and $2 \div 10 = 0.2$.

4. False, since $5 - 3 = 2$ and $3 - 5 = -2$.

5. False, since $10 - (7-3) = 6$ and $(10-7) - 3 = 0$. **6.** False, since $4(6 \div 2) = 12$ and $(4 \cdot 6) \div (4 \cdot 2) = 3$.

7. True, since $(0.02)(50) = 1$.

8. True, because of Warm-up 2.

9. False, because if x = 0 we get 3 = 0.

10. True, because $\dfrac{1 \text{ car}}{0.04 \text{ hours}} = \dfrac{25 \text{ cars}}{1 \text{ hour}}$.

1.5 EXERCISES

1. $9 - 4 + 6 - 10 = 15 - 14 = 1$

3. $6 - 10 + 5 - 8 - 7 = 11 - 25 = -14$

5. $-4 - 11 + 6 - 8 + 13 - 20 = -43 + 19 = -24$

7. $-3.2 + 1.4 - 2.8 + 4.5 - 1.6$

$= -7.6 + 5.9 = -1.7$

9. $3.27 - 11.41 + 5.7 - 12.36 - 5 = 8.97 - 28.77$

$= -19.8$

11. $4(x - 6) = 4 \cdot x - 4 \cdot 6 = 4x - 24$

13. $2m + 10 = 2 \cdot m + 2 \cdot 5 = 2(m + 5)$

15. $a(3 + t) = 3a + at$

17. $-2(w - 5) = -2w - (-10) = -2w + 10$

19. $-2(3-y) = -6 - (-2y) = -6 + 2y$

21. $5x - 5 = 5 \cdot x - 5 \cdot 1 = 5(x-1)$

23. $-1(-2x - y) = (-1)(-2x) - (-1)y = 2x + y$

25. $-3(-2w - 3y) = -3(-2w) - (-3)3y$
$$= 6w + 9y$$

27. $3y - 15 = 3 \cdot y - 3 \cdot 5 = 3(y-5)$

29. $3a + 9 = 3 \cdot a + 3 \cdot 3 = 3(a+3)$

31. $\frac{1}{2}(4x + 8) = \frac{4}{2}x + \frac{8}{2} = 2x + 4$

33. $-\frac{1}{2}(2x - 4) = -\frac{2}{2}x + \frac{4}{2} = -x + 2$

35. The reciprocal of $\frac{1}{2}$ is 2 because $2 \cdot \frac{1}{2} = 1$.

37. The reciprocal of 1 is 1 because $1 \cdot 1 = 1$.

39. The reciprocal of 6 is $\frac{1}{6}$, because $6 \cdot \frac{1}{6} = 1$.

41. Since $0.25 = \frac{1}{4}$, its reciprocal is 4.

43. Since $-0.7 = -\frac{7}{10}$, its reciprocal is $-\frac{10}{7}$.

45. Since $-1.8 = -\frac{18}{10}$ or $-\frac{9}{5}$, the reciprocal of -1.8 is $-\frac{5}{9}$.

47. $\frac{1}{2.3} + \frac{1}{5.4} \approx 0.6200$

49. $\dfrac{\frac{1}{4.3}}{\frac{1}{5.6} + \frac{1}{7.2}} \approx 0.7326$

51. $\dfrac{1 \text{ mile}}{0.0006897 \text{ hours}} \approx 1450$ mph

53. $\frac{1}{0.01} + \frac{1}{0.02} + \frac{1}{0.015}$ buttons per hour

≈ 217 buttons per hour

55. Commutative

57. Distributive

59. Associative

61. Inverse

63. Commutative

65. Identity

67. Distributive

69. Inverse

71. Multiplication property of zero

73. Distributive

75. $5 + w = w + 5$

77. $5(xy) = (5x)y$

79. $\frac{1}{2}x - \frac{1}{2} = \frac{1}{2}x - \frac{1}{2} \cdot 1 = \frac{1}{2}(x-1)$

81. $6x + 9 = 3 \cdot 2x + 3 \cdot 3 = 3(2x + 3)$

83. Since 8 and 0.125 are reciprocals, $8(0.125) = 1$

85. $0 = 5(0)$

87. $0.25(4) = 1$

1.6 WARM-UPS

1. True by the distributive property.

2. False, because $-4x + 8 = -4(x - 2)$.

3. True, because multiplying by -1 is equivalent to finding the opposite.

4. True, by the distributive property.

5. False, $(2x)(5x) = 10x^2$ **6.** True.

7. False, $a + a = 2a$.   **8.** False, $b \cdot b = b^2$.

9. False, because 1 and $7x$ are not like terms.

10. True.

1.6 EXERCISES

1. $(45 \cdot 2) \cdot 100 = 90 \cdot 100 = 9000$

3. $\frac{4}{3}(0.75) = \frac{4}{3} \cdot \frac{3}{4} = 1$

5. $427 + (68 + 32) = 427 + 100 = 527$

7. $47 \cdot 4 + 47 \cdot 6 = 47(4 + 6) = 470$

9. $19 \cdot 2 \cdot 5 \cdot \frac{1}{5} = 19 \cdot 2 \cdot 1 = 19 \cdot 2 = 38$

11. $120 \cdot 4 \cdot 100 = 480 \cdot 100 = 48{,}000$

13. $13 \cdot 377 \cdot 0 = 0$

15. $348 + (5 + 45) = 348 + 50 = 398$

17. $\frac{2}{3} \cdot 1.5 = \frac{2}{3} \cdot \frac{3}{2} = 1$

19. $17 \cdot 101 - 17 \cdot 1 = 17(101 - 1) = 1700$

21. $354 + (7 + 3) + (8 + 2)$
$= 354 + 10 + 10 = 374$

23. $(567 + 874)(0) = 0$

25. $-4n + 6n = (-4 + 6)n = 2n$

27. $3w - (-4w) = 3w + 4w = 7w$

29. $4mw^2 - 15mw^2 = (4 - 15)mw^2 = -11mw^2$

31. $-5x - (-2x) = -5x + 2x = -3x$

33. Not like terms: $-4 - 7z$

35. $4t^2 + 5t^2 = (4 + 5)t^2 = 9t^2$

37. Not like terms: $-4ab + 3a^2b$

39. $9mn - mn = (9 - 1)mn = 8mn$

41. $x^3y - 3x^3y = 1x^3y - 3x^3y = -2x^3y$

43. $-kz^6 - kz^6 = -1kz^6 + (-1kz^6) = -2kz^6$

45. $4(7t) = (4 \cdot 7)t = 28t$

47. $(-2x)(-5x) = (-2)(-5)x \cdot x = 10x^2$

49. $(-h)(-h) = (-1)(-1)h \cdot h = h^2$

51. $7w(-4) = -4 \cdot 7w = -28w$

53. $-x(1 - x) = -x \cdot 1 - (-x)(x) = -x + x^2$

55. $5k \cdot 5k = 5 \cdot 5 \cdot k \cdot k = 25k^2$

57. $3 \cdot \frac{y}{3} = 3 \cdot \frac{1}{3} \cdot y = 1 \cdot y = y$

59. $9 \cdot \frac{2y}{9} = 9 \cdot \frac{1}{9} \cdot 2y = 1 \cdot 2y = 2y$

61. $\frac{6x^3}{2} = \frac{6}{2}x^3 = 3x^3$

63. $\frac{3x^2y + 15x}{3} = \frac{3x^2y}{3} + \frac{15x}{3} = x^2y + 5x$

65. $\frac{2x - 4}{-2} = \frac{2x}{-2} - \frac{4}{-2} = -x + 2$

67. $\frac{-xt + 10}{-2} = \frac{-xt}{-2} + \frac{10}{-2} = \frac{1}{2}xt - 5$

69. $a - (4a - 1) = a - 4a + 1 = -3a + 1$

71. $6 - (x - 4) = 6 - x + 4 = 10 - x$

73. $4m + 6 - m - 5 = 3m + 1$

75. $-5b + at - 7b = -12b + at$

77. $t^2 - 5w + 2w + t^2 = 2t^2 - 3w$

79. $x^2 - x^2 + y^2 + z = y^2 + z$

81. $(2x + 7x) + (3 + 5) = 9x + 8$

83. $(-3x + 4) + (5x - 6) = 2x - 2$

85. $4a^2 - 5c - 6a^2 + 7c = -2a^2 + 2c$

87. $5t^2 - 15w + 6w + 2t^2 = 7t^2 - 9w$

89. $-7m + 3m - 12 + 5m = m - 12$

91. $8 - 7k^3 - 21 - 4 = -7k^3 - 17$

93. $x - 0.04x - 0.04(50) = 0.96x - 2$

95. $0.10x + 0.5 - 0.04x - 2 = 0.06x - 1.5$

97. $3k + 5 - 6k + 8 - k + 3 = -4k + 16$

99. $5.7 - 4.5x + (4.5)(3.9) - 5.42$
$= -4.5x + 17.83$

101. $3 - 3xy - 2xy + 10 - 35 + xy$
$= -4xy - 22$

103. $3w^2 - 30w^2 - 2w^2 = -29w^2$

105. $3a^2w^2 - 5a^2w^2 - 4a^2w^2 = -6a^2w^2$

107. $\frac{1}{6} - \frac{1}{3}\left(-6x^2y - \frac{1}{2}\right) = \frac{1}{6} + 2x^2y + \frac{1}{6}$
$= 2x^2y + \frac{1}{3}$

109. $-\frac{1}{2}m\left(-\frac{1}{2}m\right) - \frac{1}{2}m - \frac{1}{2}m = \frac{1}{4}m^2 - m$

111. $\frac{-8t^3}{-2} - \frac{6t^2}{-2} + \frac{2}{-2} = 4t^3 + 3t^2 - 1$

113. $\frac{-6xyz}{-3} - \frac{3xy}{-3} + \frac{9z}{-3} = 2xyz + xy - 3z$

115. Perimeter is $s + s + 2 + s + 4$ or $3s + 6$ ft.

117. Perimeter is $2x + 2\left(x + \frac{1}{6}x\right)$ or $\frac{13}{3}x$ meters.

CHAPTER 1   REVIEW

1. True, because 3 is the only number common to A and B.

3. False, because $A \cup B = \{1, 2, 3, 4, 5\}$.

5. True, because $B \subseteq C$.

7. False, $A \cap \emptyset = \emptyset$.

9. True, because $A \cap B = \{3\}$.

11. True, because every member of B is also a member of C.

13. False, because $1 \in A$ and $1 \notin B$.

15. True, because 3 is a member of D.

17. False, because $0 \notin E$.

19. True, because $\emptyset$ is a subset of every set.

21. $\{0, 1, 31\}$

23. $\{-1, 0, 1, 31\}$

25. $\{-\sqrt{2}, \sqrt{3}, \pi\}$

27. False, because 0 is a whole number but not a natural number.

29. False, because -2, -1 and 0 are not natural numbers.

31. False, because 3.14 is not exactly π.

33. True, because any decimal number that repeats or terminates is a rational number.

35. False, 0 is rational only.

37. $9 - 4 = 5$

39. $25 - 37 = -12$

41. $(-4)(6) = -24$

43. $(-8) \div (-4) = 2$

45. $-\frac{3}{12} + \frac{1}{12} = -\frac{2}{12} = -\frac{1}{6}$

47. $\frac{-20}{-2} = 10$

49. $10.00 - 0.04 = 9.96$

51. $-6 - (-2) = -6 + 2 = -4$

53. $-0.5 + 0.5 = 0$

55. $3.2 \div (-0.8) = -4$

57. $0 \div (-0.3545) = 0$

59. $\frac{1}{4}(-12) = -\frac{12}{4} = -3$

61. $4 + 7 \cdot 5 = 4 + 35 = 39$

63. $(4 + 7)^2 = 11^2 = 121$

65. $5 + 3 \cdot |6 - 12| = 5 + 3 \cdot 6 = 5 + 18 = 23$

67. $(-2) - (-4) = -2 + 4 = 2$

69. $3 - 5(6 - 10) = 3 - 5(-4) = 23$

71. $25 - (11)^2 = 25 - 121 = -96$

73. $\frac{6}{2} + 2 = 3 + 2 = 5$

75. $\frac{-9}{7 + 2} = \frac{-9}{9} = -1$

77. $1.00 - 0.24 = 0.76$

79. $9 - 8 = 1$

81. $3 - 2 \, |-3| = 3 - 2 \cdot 3 = 3 - 6 = -3$

83. $\frac{-8}{-2} = 4$

85. $3^2 - 4(-2)(-1) = 9 - 8 = 1$

87. $(-1 - 3)(-1 + 3) = -4 \cdot 2 = -8$

89. $(-2)^2 + 2(-2)(3) + 3^2 = 4 - 12 + 9 = 1$

91. $(-2)^3 - 3^3 = -8 - 27 = -35$

93. $\frac{3 + (-1)}{-2 + 3} = \frac{2}{1} = 2$

95. $|-2 - 3| = |-5| = 5$

97. $(-2 + 3)(-1) = (1)(-1) = -1$

99. Commutative

101. Distributive

103. Associative

105. Identity

107. Inverse

109. Multiplication property of zero

111. Identity

113. Inverse

115. $3x - 3a = 3(x - a)$

117. $3 \cdot w + 3 \cdot 1 = 3w + 3$

119. $7x + 7 \cdot 1 = 7(x + 1)$

121. $5(x - 5) = 5x - 5 \cdot 5 = 5x - 25$

123. $-3 \cdot 2x - (-3)(5) = -6x + 15$

125. $p \cdot 1 - pt = p(1 - t)$

127. $3a + 7 + 4a - 5 = 7a + 2$

129. $5t - 20 - 6t + 18 = -t - 2$

131. $-a + 2 + 2 - a = -2a + 4$

133. $5 - 3x + 6 + 7x + 28 - 6 = 4x + 33$

135. $0.2x + 0.02 - x - 0.50 = -0.8x - 0.48$

137. $0.05x + 0.15 - 0.10x - 2 = -0.05x - 1.85$

139. $\frac{1}{2}x + 2 - \frac{1}{4}x + 2 = \frac{1}{4}x + 4$

141. $\frac{-9x^2}{3} - \frac{6x}{3} + \frac{3}{3} = -3x^2 - 2x + 1$

143. $32(4)(-6 + 6) = 32(4)(0) = 0$,

Inverse, Multiplication property of 0

145. $768(4) + 768(6) = 768(10) = 7680$,

Distributive

147. $12 \cdot 4 + (-6 + 6) = 48 + 0 = 48$,

Associative, Inverse

149. $752(-6) + 752(6) = 752(0) = 0$,

Distributive, inverse

151. $(47 \cdot 6)\frac{4}{24} = 47 \cdot 1 = 47$, Associative, Inverse

153. $(-6 \cdot 24)\frac{1}{6} = 24(-6 \cdot \frac{1}{6}) = -24$,

Commutative, Associative

155. $5(-6 + 6)(4 + 24) = 5 \cdot 0 \cdot 28 = 0$, Inverse,

Multiplication property of 0

157. $\frac{1}{0.2} + \frac{1}{0.5} = 7$ shingles per minute

159. $18,264(1.05)^6 \approx 24{,}476$

The car will cost \$24,476 in the year 2000. The first year in which the car is over \$21,000 will be 1997.

CHAPTER 1 TEST

1. List the elements in A or in B:

$A \cup B = \{2, 3, 4, 5, 6, 7, 8, 10\}$

2. Only 6 and 7 are in both B and C. So

$B \cap C = \{6, 7\}$

3. The intersection of $A = \{2, 4, 6, 8, 10\}$ with $B \cup C = \{3, 4, 5, 6, 7, 8, 9, 10\}$ is $\{4, 6, 8, 10\}$.

4. $\{0, 8\}$

5. $\{-4, 0, 8\}$ **6.** $\{-4, -\frac{1}{2}, 0, 1.65, 8\}$

7. $\{-\sqrt{3}, \sqrt{5}, \pi\}$ **8.** $6 + (-15) = -9$

9. $4 - 4(-10) = 4 + 40 = 44$

10. $-17 + 6 = -11$ **11.** $0.02 - 2 = -1.98$

12. $\frac{-3 + 7}{-2} = \frac{4}{-2} = -2$ **13.** $\frac{-8}{2} = -4$

14. $\left(\frac{2}{3} - 1\right)\left(\frac{1}{3} - \frac{1}{2}\right) = \left(-\frac{1}{3}\right)\left(-\frac{1}{6}\right) = \frac{1}{18}$

15. $-\frac{4}{7} - \frac{1}{2}\left(24 - \frac{8}{7}\right) = -\frac{4}{7} - 12 + \frac{4}{7} = -12$

16. $|3 - 10| = |-7| = 7$

17. $5 - 2|-4| = 5 - 2 \cdot 4 = 5 - 8 = -3$

18. $(452 + 695)[-8 + 8] = (452 + 695)[0] = 0$

19. $478(8 + 2) = 478(10) = 4780$

20. $-24 - 4(6 - 9 \cdot 8) = -24 - 4(-66) = 240$

21. $-3 + 5 - 18 - 4 + 8 - 18 = -43 + 13 = -30$

22. $(-4)^2 - 4(-3)(2) = 16 + 24 = 40$

23. $\frac{(-3)^2 - (-4)^2}{-4 - (-3)} = \frac{9 - 16}{-1} = 7$

24. $\frac{(-3)(-4) - 6(2)}{(-4)^2 - 2^2} = \frac{12 - 12}{16 - 4} = 0$

25. Distributive   **26.** Commutative
27. Associative   **28.** Inverse
29. Commutative
30. Multiplication property of 0
31. $3m - 15 + 8m + 12 = 11m - 3$
32. $x + 3 - 0.05x - 0.1 = 0.95x + 2.9$

33. $\frac{1}{2}x - \frac{4}{2} + \frac{1}{4}x + \frac{3}{4} = \frac{2}{4}x + \frac{1}{4}x - \frac{8}{4} + \frac{3}{4}$

$$= \frac{3}{4}x - \frac{5}{4}$$

34. $-3x^2 + 6y - 6y + 8x^2 = 5x^2$

35. $\frac{-6x^2}{-2} - \frac{4x}{-2} + \frac{2}{-2} = 3x^2 + 2x - 1$

36. $\frac{-8xy}{2} - \frac{6xy}{2} = -4xy - 3xy = -7xy$

37. $5x - 40 = 5x - 5 \cdot 8 = 5(x - 8)$

38. $7t - 7 = 7t - 7 \cdot 1 = 7(t - 1)$

39. $\frac{1 \text{ tree}}{0.0625 \text{ hours}} = 16$ trees/hour

40. The perimeter is $2x + 2(x - 4)$ or $4x - 8$ feet. If $x = 9$, the perimeter is $4 \cdot 9 - 8$ or 28 feet. The area is $x^2 - 4x$. If $x = 9$, then the area is $9^2 - 4 \cdot 9$, or 45 ft^2.

41. $6(1.03)^{25} \approx 12.6$ billion

2.1 WARM-UPS

1. False, it is equivalent to $-2x = 5$.
2. True.   3. False, multiply each side by $\frac{4}{3}$.
4. True.   5. True.   6. True.
7. True.   8. True, it is equivalent to $x = -3$.
9. True, it is equivalent to $0.8x = 0.8x$.
10. True, since it is equivalent to $3x - 12 = 0$.

2.1 EXERCISES

1. Yes, because $3 \cdot (-4) + 7 = -5$ is correct.

3. Yes, because $\frac{1}{2} \cdot 12 - 4 = \frac{1}{3} \cdot 12 - 2$.

5. No, because $0.2(200 - 50) = 30$ and

$20 - 0.05 \cdot 200 = 10$.

7. No because $0.1 \cdot 80 - 30 = -22$ and

$16 - 0.06 \cdot 80 = 11.2$.

9. $-72 - x + 72 = 15 + 72$
$\qquad -x = 87$
$\quad (-1)(-x) = -1 \cdot 87$
$\qquad\quad x = -87$
The solution set is $\{-87\}$.

11. $\quad -3x - 19 = 5 - 2x$
$\qquad\quad -3x = 24 - 2x$
$\quad -3x + 2x = 24$
$\qquad\qquad -x = 24$
$\qquad\qquad x = -24$
The solution set is $\{-24\}$.

13. $\quad 2x - 3 + 3 = 0 + 3$

$\qquad\qquad 2x = 3$

$\qquad\qquad \frac{2x}{2} = \frac{3}{2}$

$\qquad\qquad x = \frac{3}{2}$

The solution set is $\left\{\frac{3}{2}\right\}$.

15. $\quad -2x + 5 - 5 = 7 - 5$
$\qquad\qquad -2x = 2$

$\qquad\qquad \frac{-2x}{-2} = \frac{2}{-2}$

$\qquad\qquad x = -1$
The solution set is $\{-1\}$.

17. $\quad -12x - 15 + 15 = 21 + 15$
$\qquad\quad -12x \qquad = 36$

$\qquad\quad \frac{-12x}{-12} = \frac{36}{-12}$

$\qquad\quad x \qquad = -3$

The solution set is $\{-3\}$.

19. $\qquad 26 - 16 = 4x + 16 - 16$
$\qquad\qquad 10 = 4x$

$\qquad\qquad \frac{10}{4} = \frac{4x}{4}$

$\qquad\qquad \frac{5}{2} = x$

The solution set is $\left\{\frac{5}{2}\right\}$.

21. $\qquad -3(x - 16) = 12 - x$
$\qquad\quad -3x + 48 = 12 - x$
$\qquad\qquad -3x = -36 - x$
$\qquad\qquad -2x = -36$
$\qquad\qquad x = 18$
The solution set is $\{18\}$.

23. $\qquad 2x + 18 - x = 36$
$\qquad\qquad x + 18 = 36$
$\qquad x + 18 - 18 = 36 - 18$
$\qquad\qquad x \qquad = 18$
The solution set is $\{18\}$.

25. $\qquad 2 + 3x - 3 = x - 1$
$\qquad\qquad 3x - 1 = x - 1$
$\qquad\qquad 3x - x = -1 + 1$
$\qquad\qquad 2x = 0$
$\qquad\qquad x = 0$
The solution set is $\{0\}$.

27. $\qquad -\frac{7}{3}\left(-\frac{3}{7}x\right) = -\frac{7}{3}(4)$

$\qquad\qquad x \qquad = -\frac{28}{3}$

The solution set is $\left\{-\frac{28}{3}\right\}$.

29. $\qquad\qquad -\frac{5}{7}x = 4$

$\qquad -\frac{7}{5}\left(-\frac{5}{7}x\right) = -\frac{7}{5}(4)$

$\qquad\qquad x \quad = -\frac{28}{5}$

The solution set is $\left\{-\frac{28}{5}\right\}$.

31.
$$6\left(\frac{x}{3}+\frac{1}{2}\right)=6\cdot\frac{7}{6}$$

$$2x+3=7$$
$$2x=4$$
$$x=2$$
The solution set is $\{2\}$.

33.
$$3\left(\frac{2}{3}x+5\right)=3\left(-\frac{1}{3}x+17\right)$$

$$2x+15=-x+51$$
$$3x+15=51$$
$$3x=36$$
The solution set is $\{12\}$.

35.
$$4\left(\frac{1}{2}x+\frac{1}{4}\right)=4\cdot\frac{1}{4}(x-6)$$

$$2x+1=x-6$$
$$x+1=-6$$
$$x=-7$$
The solution set is $\{-7\}$.

37.
$$4\left(8-\frac{x-2}{2}\right)=4\left(\frac{x}{4}\right)$$

$$32-2(x-2)=x$$
$$32-2x+4=x$$
$$-3x=-36$$
$$x=12$$
The solution set is $\{12\}$.

39.
$$6\left(\frac{y-3}{3}\right)-6\left(\frac{y-2}{2}\right)=6(-1)$$

$$2y-6-3y+6=-6$$
$$-y=-6$$
$$y=6$$
The solution set is $\{6\}$.

41.
$$10(x-0.2x)=10(72)$$
$$10x-2x=720$$
$$8x=720$$
$$x=90$$
The solution set is $\{90\}$.

43.
$$0.03x+0.03(200)+0.05x=86$$
$$0.08x+6=86$$
$$0.08x=80$$
$$x=1000$$
The solution set is $\{1000\}$.

45.
$$0.1x+0.05x-0.05(300)=105$$
$$0.15x-15=105$$
$$0.15x=120$$
$$x=800$$
The solution set is $\{800\}$.

47.
$$2x+2=2x+6$$
$$2=6$$
The solution set is $\emptyset$. The equation is inconsistent.

49. $2x=2x$ The solution set is R. The equation is an identity.

51.
$$2x=2$$
$$x=1$$
Solution set: $\{1\}$ Equation is conditional.

53. $x=x$
Solution set: R   Equation is an identity.

55. $x^2=x^2$
Solution set: R   Equation is an identity.

57.
$$2x+6-7=25-5x+7x+7$$
$$2x-1=32+2x$$
$$-1=32$$
The solution set is $\emptyset$. The equation is inconsistent.

59.
$$x+3-\frac{7}{2}=\frac{3}{2}x+\frac{3}{2}-\frac{1}{2}x-2$$

$$2x+6-7=3x+3-x-4$$
$$2x-1=2x-1$$
Solution set: R   Equation is an identity.

61.
$$x+3-3.5=1.5x+1.5$$
$$x-0.5=1.5x+1.5$$
$$10x-5=15x+15$$
$$-20=5x$$
$$-4=x$$
The solution set is $\{-4\}$. Conditional

63.
$$4-12x+18+1=3+10-2x$$
$$-12x+23=13-2x$$
$$-10x=-10$$
$$x=1$$
The solution set is $\{1\}$.

65.
$$\frac{1}{2}y-\frac{1}{12}+\frac{2}{3}=\frac{5}{6}+\frac{1}{6}-y$$
$$6y-1+8=10+2-12y$$
$$6y+7=12-12y$$
$$18y=5$$
$$y=\frac{5}{18}$$
The solution set is $\left\{\frac{5}{18}\right\}$.

67.
$$8-40x+600=200x+3$$
$$-40x+608=200x+3$$
$$-240x=-605$$

$$x=\frac{605}{240}=\frac{121}{48}$$

The solution set is $\left\{\frac{121}{48}\right\}$.

69. $12 \cdot \dfrac{a-3}{4} - 12 \cdot \dfrac{2a-5}{2} = 12 \cdot \dfrac{a+1}{3} - 12 \cdot \dfrac{1}{6}$

$$3a - 9 - 12a + 30 = 4a + 4 - 2$$
$$-9a + 21 = 4a + 2$$
$$-13a = -19$$
$$a = \frac{19}{13}$$

The solution set is $\left\{\dfrac{19}{13}\right\}$.

71. $1.3 - 1.2 + 0.6x = 0.02x + 0.3$
$$130 - 120 + 60x = 2x + 30$$
$$10 + 60x = 2x + 30$$
$$58x = 20$$
$$x = \frac{20}{58} = \frac{10}{29}$$

The solution set is $\left\{\dfrac{10}{29}\right\}$.

73. $\quad 3x = 9$
$$x = 3$$
Solution set: $\{3\}$

75. $\quad 7 - z = -9$
$$16 = z$$
Solution set: $\{16\}$

77. $\quad \dfrac{2}{3}x = \dfrac{1}{2}$
$$x = \frac{3}{2} \cdot \frac{1}{2} = \frac{3}{4}$$
Solution set: $\left\{\dfrac{3}{4}\right\}$

79. $\quad -\dfrac{3}{5}y = 9$
$$y = -\frac{5}{3} \cdot 9 = -15$$
Solution set: $\{-15\}$

81. $\quad 3y + 5 = 4y - 1$
$$6 = y$$
The solution set is $\{6\}$.

83. $\quad 5x + 10x + 20 = 110$
$$15x = 90$$
$$x = 6$$
Solution set: $\{6\}$

85. $15\left(\dfrac{p+7}{3}\right) - 15\left(\dfrac{p-2}{5}\right) = 15\left(\dfrac{7}{3}\right) - 15\left(\dfrac{p}{15}\right)$
$$5p + 35 - 3p + 6 = 35 - p$$
$$2p + 41 = 35 - p$$
$$3p = -6$$
$$p = -2$$
Solution set: $\{-2\}$

87. $5 - [4 - t + 6] = 17 + 2[t + 3t - 18]$
$$5 - 4 + t - 6 = 17 + 2t + 6t - 36$$
$$t - 5 = 8t - 19$$
$$14 = 7t$$
$$2 = t$$
Solution set: $\{2\}$

89. $0.3 - [4 - t + 0.2] = [t - 2t - 6] + 3 - t$
$$0.3 - 4 + t - 0.2 = -t - 6 + 3 - t$$
$$t - 3.9 = -2t - 3$$
$$3t = 0.9$$
$$t = 0.3$$
Solution set: $\{0.3\}$

91. $\quad 0.94x = 50,000$
$$x = \frac{50,000}{0.94}$$
Solution set: $\{53,191.49\}$

93. $\quad 2.365x = 14.8095 - 3.694$
$$2.365x = 11.1155$$
$$x = \frac{11.1155}{2.365}$$
$$x = 4.7$$
Solution set: $\{4.7\}$

95. $5.39 - [4.21 - x + 3.52] = 17.6 + 2x + 27.162$

$$5.39 - 4.21 + x - 3.52 = 2x + 44.762$$
$$x - 2.34 = 2x + 44.762$$
$$-47.102 = x$$

Solution set: $\{-47.102\}$

97. The solution set to $2x + 1 = 7$ is $\{3\}$. If $x - 1 = b$ has the same solution set, then $3 - 1 = b$ or $b = 2$.

99. The solution set to $5 - 2x = -3$ is $\{4\}$. If $bx - 1 = 3b$ has the same solution set, then $b(4) - 1 = 3b$, $4b - 1 = 3b$, or $b = 1$.

101. The solution set to $x - 6 = 5x$ is $\{-3/2\}$.

If $2b - x = 4b$ has the same solution set, then

$2b - (-3/2) = 4b$, $2b = 3/2$, or $b = 3/4$.

103. $0.45(7) + 39.05 = 42.2$

Public school enrollment in 1992 was 42.2 million.

$$0.45x + 39.05 = 45$$
$$0.45x = 5.95$$
$$x \approx 13.2$$

Public school enrollment will reach 45 million in $1985 + 13$, or 1998.

105. The year in which the U.S. generated over 100 million tons of solid waste was 1964.

$0.576n + 3.78 = 0.13(3.14n + 87.1)$

$0.576n + 3.78 = 0.13 \cdot 3.14n + 0.13 \cdot 87.1$

$(0.576 - 0.13 \cdot 3.14)n = 0.13 \cdot 87.1 - 3.78$

$$n = \frac{0.13 \cdot 87.1 - 3.78}{0.576 - 0.13 \cdot 3.14}$$

$$n = 44.95$$

U.S. recovered 13% in $1960 + 45$, or 2005.

2.2 WARM-UPS

1. False, P is on both sides of the equation.
2. False, because we use the distributive property to factor out P on the right side.
3. False, because we divide each side by Pr to get $t = \frac{I}{Pr}$. **4.** True, $\frac{5 \cdot 6}{2} = 15$.
5. False, $P = 2L + 2W$. **6.** True, $V = LWH$.
7. False, $A = \frac{1}{2}h(b_1 + b_2)$. **8.** True.
9. True, because $-2(-3) - 4 = 6 - 4 = 2$.
10. False, perimeter is the distance around the outside edge.

2.2 EXERCISES

1. $\dfrac{I}{Pr} = \dfrac{Prt}{Pr}$

$\dfrac{I}{Pr} = t$

$t = \dfrac{I}{Pr}$

3. $F - 32 = \dfrac{9}{5}C$

$\dfrac{5}{9}(F - 32) = \dfrac{5}{9} \cdot \dfrac{9}{5}C$

$C = \dfrac{5}{9}(F - 32)$

5. $\dfrac{A}{L} = \dfrac{LW}{L}$

$\dfrac{A}{L} = W$

$W = \dfrac{A}{L}$

7. $2A = 2 \cdot \dfrac{1}{2}(b_1 + b_2)$

$2A = b_1 + b_2$

$2A - b_2 = b_1$

$b_1 = 2A - b_2$

9. $3y = -2x + 9$

$\dfrac{3y}{3} = \dfrac{-2x}{3} + \dfrac{9}{3}$

$y = -\dfrac{2}{3}x + 3$

11. $-y = -x + 4$

$\dfrac{-y}{-1} = \dfrac{-x}{-1} + \dfrac{4}{-1}$

$y = x - 4$

13. $-2y = -4x + 6$

$\dfrac{-2y}{-2} = \dfrac{-4x}{-2} + \dfrac{6}{-2}$

$y = 2x - 3$

15. $3y = -x - 6$

$\dfrac{3y}{3} = \dfrac{-x}{3} - \dfrac{6}{3}$

$y = -\dfrac{1}{3}x - 2$

17. $2y = x - 5$

$\dfrac{2y}{2} = \dfrac{x}{2} - \dfrac{5}{2}$

$y = \dfrac{1}{2}x - \dfrac{5}{2}$

19. $\dfrac{1}{2}y = x - 3$

$2\left(\dfrac{1}{2}y\right) = 2(x - 3)$

$y = 2x - 6$

21. $-\dfrac{1}{3}y = -\dfrac{1}{2}x + 2$

$-3\left(-\dfrac{1}{3}y\right) = -3\left(-\dfrac{1}{2}x + 2\right)$

$y = \dfrac{3}{2}x - 6$

23. $y - 2 = \dfrac{1}{2}x - \dfrac{3}{2}$

$y = \dfrac{1}{2}x + \dfrac{1}{2}$

25. $y + 5 = 4x - 24$

$y = 4x - 29$

27. $0.0575y = -0.1725x + 56.58$

$y = -3x + 984$

29. $A - P = Prt$

$$\frac{A - P}{Pr} = \frac{Prt}{Pr}$$

$$\frac{A - P}{Pr} = t$$

$$t = \frac{A - P}{Pr}$$

31. $a(b + 1) = 1$

$$\frac{a(b + 1)}{b + 1} = \frac{1}{b + 1}$$

$$a = \frac{1}{b + 1}$$

33. $xy - y = -5 - 7$

$$y(x - 1) = -12$$

$$y = \frac{-12(-1)}{(x - 1)(-1)}$$

$$y = \frac{12}{1 - x}$$

35. $xy^2 + xz^2 - xw^2 = -6$

$$x(y^2 + z^2 - w^2) = -6$$

$$x = \frac{-6}{y^2 + z^2 - w^2}$$

$$x = \frac{6}{w^2 - y^2 - z^2}$$

37. $RR_1R_2\left(\frac{1}{R}\right) = RR_1R_2\left(\frac{1}{R_1}\right) + RR_1R_2\left(\frac{1}{R_2}\right)$

$$R_1R_2 = RR_2 + RR_1$$

$$R_1R_2 - RR_1 = RR_2$$

$$R_1(R_2 - R) = RR_2$$

$$R_1 = \frac{RR_2}{R_2 - R}$$

39. $3.35x + 4.58x = 44.3 + 54.6$

$$x(3.35 + 4.58) = 44.3 + 54.6$$

$$x = \frac{44.3 + 54.6}{3.35 + 4.58}$$

$$x \approx 12.472$$

41. $4.59x - 66.7 = 3.2x - 3.2(5.67)$

$$x(4.59 - 3.2) = 66.7 - 3.2(5.67)$$

$$x = \frac{66.7 - 3.2(5.67)}{4.59 - 3.2}$$

$$x \approx 34.932$$

43. $\frac{x}{19} + \frac{3x}{7} = \frac{4}{31} + \frac{3}{23}$

$$x\left(\frac{1}{19} + \frac{3}{7}\right) = \frac{4}{31} + \frac{3}{23}$$

$$x = \frac{\frac{4}{31} + \frac{3}{23}}{\frac{1}{19} + \frac{3}{7}} \approx 0.539$$

45. $2(3) - 3y = 5$

$$-3y = -1$$

$$y = \frac{1}{3}$$

47. $-4(3) + 2y = 1$

$$2y = 13$$

$$y = \frac{13}{2}$$

49. $y = -2(3) + 5$
$y = -6 + 5$
$y = -1$

51. $-3 + 2y = 5$
$2y = 8$
$y = 4$

53. $y - 1.046 = 2.63(3 - 5.09)$
$y - 1.046 = -5.4967$
$y = -4.4507$

55. $4(2)x = 5$
$8x = 5$

$$x = \frac{5}{8}$$

57. $x + (-3)x = 7$

$$-2x = 7$$

$$x = -\frac{7}{2}$$

59. $4[x - (-3)] = 2(x - 4)$
$4x + 12 = 2x - 8$
$2x = -20$
$x = -10$

61. $4 = \frac{1}{2}x(-3)$

$$4 = -\frac{3}{2}x$$

$$-\frac{2}{3}(4) = -\frac{2}{3}\left(-\frac{3}{2}x\right)$$

$$-\frac{8}{3} = x$$

63. $\frac{1}{4} + \frac{1}{x} = \frac{1}{2}$

$4x \cdot \frac{1}{4} + 4x \cdot \frac{1}{x} = 4x \cdot \frac{1}{2}$

$x + 4 = 2x$

$4 = x$

65. $I = Prt$

$300 = 1000 \cdot 2 \cdot r$

$r = \frac{300}{2000} = 15\%$

67. $A = LW$

$23 = L(4)$

$L = \frac{23}{4} = 5.75$ yards

69. $V = LWH$

$36 = 2(2.5)H$

$36 = 5H$

$H = 7.2$ feet

71. $V = 900$ gal $\cdot \left(\frac{1 \ ft^3}{7.5 \ gal} \right) = 120 \ ft^3$

$V = LWH$

$120 = 4 \cdot 6 \cdot H$

$H = \frac{120}{24} = 5$ feet

73. $A = \frac{1}{2}bh$

$30 = \frac{1}{2}4h$

$30 = 2h$

$h = 15$ feet

75. $A = \frac{1}{2}h(b_1 + b_2)$

$300 = \frac{1}{2}(20)(16 + b_2)$

$300 = 10(16 + b_2)$

$30 = 16 + b_2$

$b_2 = 14$ inches

77. $P = 2L + 2W$

$600 = 2L + 2(132)$

$336 = 2L$

$L = 168$ feet

79. $C = 2\pi r$

$3\pi = 2\pi r$

$r = \frac{3\pi}{2\pi}$

$r = 1.5$ meters

81. $C = 2\pi r$

$r = \frac{C}{2\pi}$

$r = \frac{25000}{2\pi}$

$r \approx 3,979$ miles

83. $V = \pi r^2 h$

$30 = \pi(1)^2 h$

$30 = \pi h$

$h = \frac{30}{\pi} = 9.55$ inches

85. Let $W = 62$ lb/ft^3, $D = 32$ ft, and $A = 48$ ft^2 in $F = WDA$:

$F = 62$ lb/ft$^3 \cdot 32$ ft $\cdot 48$ ft$^2 = 95,232$ lb

87. a) $500,000 = 5000c - 525,000$

$1,025,000 = 5000c$

$c = 205$

b) $d = 5000(200) - 525,000 = 475,000$

$500,000 - 475,000 = 25,000$

There will be 25,000 fewer deaths in 2000.

c) Increases

89. $\$1000 \div \$2 = 500$ feet

$b_1 + b_2 = 500$

$A = \frac{1}{2}h(b_1 + b_2)$

$50,000 = \frac{1}{2}h \cdot 500$

$h = 200$ feet

91. $A = bh$

$60,000 = b \cdot 200$

$b = 300$ feet

There is 300 ft on each street. So 600 ft at \$2 each is a \$1200 assessment.

2.3 WARM-UPS

1. False, first identify what the variable stands for. **2.** True. **3.** False, you may have the wrong equation. **4.** False, odd integers differ by 2. **5.** True, $x + 6 - x = 6$. **6.** True, $x + 7 - x = 7$. **7.** False, $5x - 2 = 3(x + 20)$ since $5x$ is larger than $3(x + 20)$. **8.** True, because 8% of x is 0.08x. **9.** False, because 10% of \$88,000 is \$8,800 and you will not get \$80,000. **10.** False, because the acid in the mixture must be between 10% and 14%.

2.3 EXERCISES

1. Since two consecutive even integers differ by 2, we can use x and $x + 2$ to represent them.

3. The expressions x and $10 - x$ have a sum of 10, since $x + 10 - x = 10$.

5. If x is the selling price, then eighty-five percent of the selling price is $0.85x$.

7. Since $D = RT$, the distance is $3x$ miles.

9. Since the perimeter is twice the length plus twice the width, we can represent the perimeter by
$2(x + 5) + 2(x)$ or $4x + 10$.

11. Let $x =$ the first integer, $x + 1 =$ the second integer, and $x + 2 =$ the third integer. Their sum is 84:

$$x + x + 1 + x + 2 = 84$$
$$3x + 3 = 84$$
$$x = 27$$

If $x = 27$, then $x + 1 = 28$, and $x + 2 = 29$. The integers are 27, 28, and 29.

13. Let $x =$ the first even integer, $x + 2 =$ the second, and $x + 4 =$ the third. Their sum is 252:

$$x + x + 2 + x + 4 = 252$$
$$3x + 6 = 252$$
$$3x = 246$$
$$x = 82$$

If $x = 82$, then $x + 2 = 84$ and $x + 4 = 86$. The integers are 82, 84, and 86.

15. Let $x =$ the first odd integer and $x + 2 =$ the second. Their sum is 128:

$$x + x + 2 = 128$$
$$2x = 126$$
$$x = 63$$

If $x = 63$, then $x + 2 = 65$. The integers are 63 and 65.

17. Let $x =$ the width, and $2x + 1 =$ the length. Since $P = 2L + 2W$ we can write the following equation.

$$2(x) + 2(2x + 1) = 278$$
$$6x + 2 = 278$$
$$6x = 276$$
$$x = 46$$

If $x = 46$, then $2x + 1 = 93$. The width is 46 meters and the length is 93 meters.

19. Let $x =$ the length of the first side, $2x - 10$ = the length of the second side, and $x + 50$ = the length of the third side. Since the perimeter is 684, we have the following equation.

$$x + 2x - 10 + x + 50 = 684$$
$$4x + 40 = 684$$

$$4x = 644$$
$$x = 161$$

If $x = 161$ feet, then $2x - 10 = 312$ feet, and $x + 50 = 211$ feet.

21. Let $x =$ the width, and $2x + 5 =$ the length. To fence the 3 sides, we use 2 widths and 1 length:

$$2(x) + (2x + 5) = 50$$
$$4x = 45$$
$$x = 11.25$$
$$2x + 5 = 27.5$$

Width is 11.25 feet and the length is 27.5 feet.

23. Let $x =$ the amount invested at 6% and $x + 1000 =$ the amount invested at 10%. Since the interest on the investments is $0.06x$ and $0.10(x + 1000)$ we can write the equation

$$0.06x + 0.10(x + 1000) = 340$$
$$0.16x + 100 = 340$$
$$0.16x = 240$$
$$x = 1500$$
$$x + 1000 = 2500$$

He invested \$1,500 at 6% and \$2,500 at 10%.

25. Let $x =$ the amount of his inheritance. He invests $\frac{1}{2}x$ at 10% and $\frac{1}{4}x$ at 12%. His total income of \$6400 can be expressed as

$$0.10\left(\frac{1}{2}x\right) + 0.12\left(\frac{1}{4}x\right) = 6400$$
$$0.05x + 0.03x = 6400$$
$$0.08x = 6400$$
$$x = 80,000$$

His inheritance was \$80,000.

27. Let $x =$ the number of gallons of 5% solution. In the 5% solution there are $0.05x$ gallons of acid, and in the 20 gallons of 10% solution there are $0.10(20)$ gallons of acid. The final mixture consists of $x + 20$ gallons of which 8% is acid. The total acid in the mixture is the sum of the acid from each solution mixed together:

$$0.05x + 0.10(20) = 0.08(x + 20)$$
$$0.05x + 2 = 0.08x + 1.6$$
$$0.4 = 0.03x$$
$$x = \frac{0.4}{0.03} = \frac{40}{3}$$

Use $\frac{40}{3}$ gallons of 5% solution.

29. Let $x =$ the number of gallons of pure acid. After mixing we will have $1 + x$ gallons of 6% solution. The original gallon has $0.05(1)$ gallons of acid in it. The equation totals up the acid:
$$0.05(1) + x = 0.06(1 + x)$$
$$0.05 + x = 0.06 + 0.06x$$
$$0.94x = 0.01$$
$$x = 0.010638 \text{ gallons}$$
Use 1 gallon = 128 ounces, to get 1.36 ounces of pure acid.

31. Let $x =$ his speed in the fog and $x + 30 =$ his increased speed. Since $D = RT$, his distance in the fog was $3x$ and his distance later was $6(x + 30)$. The equation gives the total distance:
$$3x + 6(x + 30) = 540$$
$$9x + 180 = 540$$
$$9x = 360$$
$$x = 40$$
His speed in the fog was 40 mph.

33. Let $x =$ the speed of the commuter bus and $x + 25 =$ the speed of the express bus. Use $D = RT$ to get the distance traveled by each as $2x$ and $\frac{3}{4}(x + 25)$. The equation expresses the fact that they travel the same distance:
$$2x = \frac{3}{4}(x + 25)$$
$$4 \cdot 2x = 4 \cdot \frac{3}{4}(x + 25)$$
$$8x = 3x + 75$$
$$5x = 75$$
$$x = 15$$
The speed of the commuter bus was 15 mph.

35. If $x =$ the selling price, then the commission is $0.08x$. The owner gets the selling price minus the commission:
$$x - 0.08x = 80,000$$
$$0.92x = 80,000$$
$$x = \$86,957$$

37. If $x =$ the selling price, then $0.07x =$ the amount of sales tax. The selling price plus the sales tax is the total amount paid:

$$x + 0.07x = 9041.50$$
$$1.07x = 9041.50$$
$$x = \$8450$$

39. Let $x =$ the distance from the base line to the service line and $x - 3 =$ the distance from the service line to the net. The total is 39 feet:
$$x + x - 3 = 39$$
$$2x = 42$$
$$x = 21$$
The distance from service line to the net is 18 ft.

41. Let $x =$ the number of points scored by the Packers and $x - 25 =$ the number of points scored by the Chiefs.
$$x + x - 25 = 45$$
$$2x = 70$$
$$x = 35$$
The score was Packers 35, Chiefs 10.

43. Let $x =$ the price per pound for the blended coffee. The price of 0.75 lb of Brazilian coffee at \$10 per lb is \$7.50, and the price of 1.5 lb of Colombian coffee at \$8 per lb is \$12. The total price for 2.25 lb of blended coffee at x dollars per kg is $2.25x$ dollars. We write an equation expressing the total cost:
$$7.50 + 12 = 2.25x$$
$$19.50 = 2.25x$$
$$x = 8.67$$
The blended coffee should sell for \$8.67 per lb.

45. Let $x =$ the number of pounds of apricots. The total cost of 10 pounds of bananas at \$3.20 per pound is \$32, the total cost of x pounds of apricots at \$4 per pound is $4x$, and the total cost $x + 10$ pounds of mix at \$3.80 per pound is $3.80(x + 10)$. Write an equation expressing the total cost:
$$32 + 4x = 3.80(x + 10)$$
$$32 + 4x = 3.80x + 38$$
$$0.20x = 6$$
$$x = 30$$
The mix should contain 30 pounds of apricots.

47. Let $x =$ the number of quarts to be drained out. In the original 20 qt radiator there are $0.30(20)$ qts of antifreeze. In the x qts that are drained there are $0.30x$ qts of antifreeze, but in the x qts put back in there are x qts of antifreeze. The equation accounts for all of the antifreeze:
$$0.30(20) - 0.30x + x = 0.50(20)$$
$$6 + 0.70x = 10$$
$$0.7x = 4$$
$$x = \frac{4}{0.7} = \frac{40}{7}$$
The amount drained should be $\frac{40}{7}$ qts.

49. If x = the number of jobs in retail sales in 1990, then $x + 0.15x$ = the number of jobs in retail sales in 2005.

$$x + 0.15x = 4,180,000$$
$$1.15x = 4,180,000$$
$$x \approx 3,634,783$$

51. Let x = Brian's inheritance, $\frac{1}{2}x$ = Daniel's inheritance, and $\frac{1}{3}x - 1000$ = Raymond's inheritance. The sum of the three amounts is $25,400:

$$x + \frac{1}{2}x + \frac{1}{3}x - 1000 = 25400$$
$$\frac{11}{6}x = 26400$$
$$x = \frac{6}{11} \cdot 26400 = 14400$$
$$\frac{1}{2}x = 7,200$$
$$\frac{1}{3}x - 1000 = 3,800$$

Brian gets $14,400, Daniel $7,200, and Raymond $3,800.

53. Let x = the first integer and x + 1 = the second. Subtract the larger from twice the smaller to get 21:

$$2x - (x + 1) = 21$$
$$x - 1 = 21$$
$$x = 22$$
$$x + 1 = 23$$

The integers are 22 and 23.

55. Let x = Berenice's time and x − 2 = Jarrett's time. Berenice's distance is 50x and Jarrett's distance is $56(x - 2)$.

$$50x + 56(x - 2) = 683$$
$$106x - 112 = 683$$
$$106x = 795$$
$$x = 7.5$$

Berenice drove for 7.5 hours.

57. Let x = the length of a side of the square. She will use x meters of fencing for each of 3 sides, but only $\frac{1}{2}x$ meters for the side with the opening.

$$3x + \frac{1}{2}x = 70$$
$$\frac{7}{2}x = 70$$
$$x = \frac{2}{7} \cdot 70 = 20$$

The square will by 20 meters by 20 meters.

59. Let x = the amount invested at 8% and $3000 - x$ = the amount invested at 10%. Income on the first investment is 0.08x and income on the second is $0.10(3000 - x)$. The total income is $290:

$$0.08x + 0.10(3000 - x) = 290$$
$$0.08x + 300 - 0.10x = 290$$
$$-0.02x = -10$$
$$x = \frac{-10}{-0.02} = 500$$
$$3000 - x = 2500$$

She invested $500 at 8% and $2,500 at 10%.

61. Let x = the number of gallons of 5% alcohol and $5 - x$ = the number of gallons of 10% alcohol. The alcohol in the final 5 gallons is the sum of the alcohol in the two separate quantities that are mixed together:

$$0.05x + 0.10(5 - x) = 0.08(5)$$
$$0.05x + 0.5 - 0.10x = 0.4$$
$$-0.05x = -0.1$$
$$x = \frac{-0.1}{-0.05} = 2$$
$$5 - x = 3$$

Use 2 gallons of 5% solution and 3 gallons of 10% solution.

63. Let x = Darla's age now and $78 - x$ = Todd's age now. In 6 years Todd will be $78 - x + 6$ or $84 - x$, and 6 years ago Darla was $x - 6$. Todd's age in 6 years is twice what Darla's age was 6 years ago:

$$84 - x = 2(x - 6)$$
$$84 - x = 2x - 12$$
$$-3x = -96$$
$$x = 32$$
$$78 - x = 46$$

Todd is 46 now and Darla is 32 now.

2.4 WARM-UPS

1. False, $0 = 0$.   2. False, $-300 < -2$.
3. True, because $-60 = -60$.
4. False, since $6 < x$ is equivalent to $x > 6$.
5. False, $-2x < 10$ is equivalent to $x > -5$.
6. False, $3x \geq -12$ is equivalent to $x \geq -4$.
7. True, multiply each side by -1.   8. True.
9. True, because of the trichotomy property.
10. True, because $3 - 4(-2) \leq 11$ is correct.

2.4 EXERCISES

1. False, $-3 > -9$.

3. True, because $0 < 8$.

5. True, because $-60 > -120$.

7. True, because $9 - (-3) = 12$.

9. Yes, because $2(-3) - 4 < 8$ simplifies to $-10 < 8$.

11. No, because $2(5) - 3 \leq 3(5) - 9$ simplifies to $7 \leq 6$.

13. No, because $5 - (-1) < 4 - 2(-1)$ simplifies to $6 < 6$.

15. Shade the numbers to the left of -1, including -1.

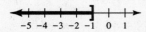

17. Shade numbers to the right of 20.

19. Since $3 \leq x$ is equivalent to $x \geq 3$, we shade the numbers to the right of 3, including 3.

21. Shade to the left of 2.3.

23. The set of all real numbers greater than 1 is expressed as $(1, \infty)$.

25. The set of all real numbers less than or equal to -3 is expressed as $(-\infty, -3]$.

27. The set of all real numbers less than 5 is expressed as $(-\infty, 5)$.

29. The set of all real numbers greater than or equal to -4 is expressed as $[-4, \infty)$.

31. $x + 5 > 12$ is equivalent to $x > 7$.

33. $-x < 6$ is equivalent to $x > -6$.

35. $-2x \geq 8$ is equivalent to $x \leq -4$.

37. $4 < x$ is equivalent to $x > 4$.

39. $\quad 7x > -14$
$\quad\quad x > -2$
Solution set is $(-2, \infty)$.

41. $\quad -3x \leq 12$
$\quad\quad\quad x \geq -4$
Solution set is $[-4, \infty)$.

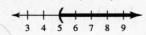

43. $\quad 2x - 3 > 7$
$\quad\quad\quad 2x > 10$
$\quad\quad\quad\quad x > 5$
Solution set is $(5, \infty)$.

45. $\quad 3 - 5x \leq 18$
$\quad\quad\quad -5x \leq 15$
$\quad\quad\quad\quad x \geq -3$
Solution set is $[-3, \infty)$.

47. $\quad \dfrac{x-3}{-5} < -2$
$\quad\quad x - 3 > 10$
$\quad\quad\quad x > 13$
Solution set is $(13, \infty)$.

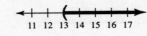

49. $\quad \dfrac{5 - 3x}{4} \leq 2$
$\quad\quad 5 - 3x \leq 8$
$\quad\quad\quad -3x \leq 3$
$\quad\quad\quad\quad x \geq -1$
Solution set is $[-1, \infty)$.

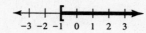

51. $\quad 3x - 8 > 0$
$\quad\quad\quad 3x > 8$
$\quad\quad\quad\quad x > \dfrac{8}{3}$
Solution set is $(8/3, \infty)$.

53. $\quad 4 - 3x < 0$
$\quad\quad\quad -3x < -4$
$\quad\quad\quad\quad x > \dfrac{4}{3}$
Solution set is $(4/3, \infty)$.

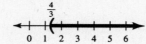

55.
$$-\frac{5}{6}x \geq -10$$
$$-\frac{6}{5}\left(-\frac{5}{6}x\right) \leq -\frac{6}{5}(-10)$$
$$x \leq 12$$

Solution set is $(-\infty, 12]$.

57.
$$4\left(3 - \frac{1}{4}x\right) \geq 4(2)$$
$$12 - x \geq 8$$
$$-x \geq -4$$
$$x \leq 4$$

Solution set is $(-\infty, 4]$.

59.
$$12\left(\frac{1}{4}x - \frac{1}{2}\right) < 12\left(\frac{1}{2}x - \frac{2}{3}\right)$$
$$3x - 6 < 6x - 8$$
$$-3x < -2$$
$$x > \frac{2}{3}$$

Solution set is $(2/3, \infty)$.

61.
$$4 \cdot \frac{y-3}{2} > 4 \cdot \frac{1}{2} - 4 \cdot \frac{y-5}{4}$$
$$2y - 6 > 2 - y + 5$$
$$3y > 13$$
$$y > \frac{13}{3}$$

Solution set is $(13/3, \infty)$.

63.
$$2x + 3 > 2x - 8$$
$$3 > -8$$

Solution set is $(-\infty, \infty)$.

65.
$$-8x + 20 \leq 12 - 8x$$
$$20 \leq 12$$

Solution set is $\emptyset$.

67.
$$-\frac{1}{2}x + 3 < \frac{1}{2}x + 2$$
$$-x < -1$$
$$x > 1$$

Solution set is $(1, \infty)$.

69.
$$-x + \frac{3}{2} + \frac{4}{3} - 2x \geq \frac{7}{4} - \frac{1}{2}x - 3$$
$$-3x + \frac{17}{6} \geq -\frac{1}{2}x - \frac{5}{4}$$
$$12\left(-3x + \frac{17}{6}\right) \geq 12\left(-\frac{1}{2}x - \frac{5}{4}\right)$$
$$-36x + 34 \geq -6x - 15$$
$$-30x \geq -49$$
$$x \leq \frac{49}{30}$$

Solution set is $(-\infty, 49/30]$.

71.
$$\frac{3}{10}x - \frac{1}{16} + 1 > \frac{3}{32}x + \frac{1}{80}$$
$$160\left(\frac{3}{10}x - \frac{1}{16} + 1\right) > 160\left(\frac{3}{32}x + \frac{1}{80}\right)$$
$$48x - 10 + 160 > 15x + 2$$
$$33x > -148$$
$$x > -\frac{148}{33}$$

Solution set is $(-148/33, \infty)$.

73.
$$4.273 + 2.8x \leq 10.985$$
$$2.8x \leq 6.712$$
$$x \leq 2.397$$

Solution set is $(-\infty, 2.397]$.

75.
$$3.25x - 27.39 > 4.06 + 5.1x$$
$$-1.85x > 31.45$$
$$x < -17$$

Solution set is $(-\infty, -17)$.

77. If x = Tony's height, then x > 6.

79. If s = Wilma's salary, then s < 80,000.

81. If v = speed of the Concorde, then v ≤ 1450.

83. If a = amount Julie can afford, then a ≤ 400.

85. If b = Burt's height, then b ≤ 5.

87. If t = Tina's hourly wage, then t ≤ 8.20.

89. Let x = the price of the car and 0.08x = the amount of tax. To spend less than $10,000 we must satisfy the inequality
$$x + 0.08x + 172 < 10,000$$
$$1.08x < 9828$$
$$x < 9100$$
The price range for the car is x < $9,100.

91. Let x = the price of the truck and 0.09x = the amount of sales tax. The total cost of at least $10,000 is expressed as
$$x + .09x + 80 \geq 10,000$$
$$1.09x \geq 9920$$
$$x \geq 9100.9174$$
The price range for the truck is x ≥ $9,100.92.

93. a) Increasing

b)
$$16.45n + 980.20 > 1,200$$
$$16.45n > 219.8$$
$$n > 13.362$$
Bachelor's degrees will exceed 1.2 million in the year 1985 + 14, or 1999.

95. Let x = the final exam score. One-third of the midterm plus two-thirds of the final must be at least 70:
$$\frac{1}{3}(56) + \frac{2}{3}x \geq 70$$

$$3\left(\frac{1}{3}(56) + \frac{2}{3}x\right) \geq 3(70)$$

$$56 + 2x \geq 210$$
$$2x \geq 154$$
$$x \geq 77$$
The final exam score must satisfy x ≥ 77.

97. Let x = the price of a pair of A-Mart jeans and x + 50 = the price of a pair of designer jeans. Four pairs of A-Mart jeans cost less than one pair of designer jeans is written as follows.
$$4x < x + 50$$
$$3x < 50$$
$$x < 16.6666$$
The price range for A-Mart jeans is x < $16.67.

2.5 WARM-UPS

1. True, because both inequalities are true.

2. True, because both inequalities are correct.

3. False, because 3 > 5 is incorrect.

4. True, because 3 ≤ 10 is correct.

5. True, because both inequalities are correct.

6. True, because both are correct.

7. False, because 0 < −2 is incorrect.

8. True, because only numbers larger than 8 are larger than 3 and larger than 8.

9. False, because (3, ∞) ∪ [8, ∞) = (3, ∞).

10. True, because the numbers greater than −2 and less than 9 are between −2 and 9.

2.5 EXERCISES

1. No, because −6 > −3 is incorrect.

3. Yes, because both inequalities are correct.

5. No, because both inequalities are incorrect.

7. No, because −4 > −3 is incorrect.

9. Yes, because −4 − 3 ≥ −7 is correct.

11. Yes, because 2(−4) − 1 < −7 is correct.

13. The set of numbers between −1 and 4:

15. Graph the union of the two solution sets:

17. The intersection of the two solution sets consists of all real numbers:

19. ∅, because no number is greater than 9 and less than or equal to 6.

21. The union of the two solution sets is graphed as follows:

23. ∅, there is no intersection to the two solution sets.

25.
$$x - 3 > 7 \quad \text{or} \quad 3 - x > 2$$
$$x > 10 \quad \text{or} \quad -x > -1$$
$$x > 10 \quad \text{or} \quad x < 1$$
$$(-\infty, 1) \cup (10, \infty)$$

21

27.
$$3 < x \text{ and } 1 + x > 10$$
$$x > 3 \text{ and } \qquad x > 9$$
$(9, \infty)$

29.
$$\tfrac{1}{2}x > 5 \quad \text{or} \quad -\tfrac{1}{3}x < 2$$
$$x > 10 \quad \text{or} \qquad x > -6$$
$(-6, \infty)$

31.
$$2x - 3 \le 5 \text{ and } x - 1 > 0$$
$$2x \le 8 \quad \text{and} \qquad x > 1$$
$$x \le 4 \quad \text{and} \qquad x > 1$$
$(1, 4]$

33.
$$\tfrac{1}{2}x - \tfrac{1}{3} \ge -\tfrac{1}{6} \quad \text{or} \quad \tfrac{2}{7}x \le \tfrac{1}{10}$$
$$3x - 2 \ge -1 \quad \text{or} \qquad x \le \tfrac{7}{2} \cdot \tfrac{1}{10}$$
$$x \ge \tfrac{1}{3} \quad \text{or} \qquad x \le \tfrac{7}{20}$$
$(-\infty, \infty)$

35.
$$0.5x < 2 \text{ and } -0.6x < -3$$
$$x < 4 \quad \text{and} \qquad x > 5$$
The solution set is $\emptyset$, because there are no numbers that are less than 4 and greater than 5.

37.
$$5 + 4x - 4 < 3 \text{ and } 1 + 3x - 3 < 1$$
$$4x < 2 \quad \text{and} \qquad 3x < 3$$
$$x < \tfrac{1}{2} \text{ and} \qquad x < 1$$
$(-\infty, 1/2)$

39.
$$0.2(x - 3) > 0.5 \quad \text{or} \quad 0.05x < 2$$
$$0.2x - 0.6 > 0.5 \quad \text{or} \qquad x < 40$$
$$0.2x > 1.1 \quad \text{or} \qquad x < 40$$
$$x > 5.5 \quad \text{or} \qquad x < 40$$
$(-\infty, \infty)$

41.
$$5 < 2x - 3 < 11$$
$$5 + 3 < 2x - 3 + 3 < 11 + 3$$
$$8 < 2x < 14$$
$$4 < x < 7$$
$(4, 7)$

43.
$$-1 < 5 - 3x \le 14$$
$$-6 < -3x \le 9$$
$$\frac{-6}{-3} > \frac{-3x}{-3} \ge \frac{9}{-3}$$
$$2 > x \ge -3$$
$[-3, 2)$

45.
$$2(-3) < 2 \cdot \frac{3m + 1}{2} \le 2 \cdot 5$$
$$-6 < 3m + 1 \le 10$$
$$-7 < 3m \le 9$$
$$-\tfrac{7}{3} < m \le 3$$
$(-7/3, 3]$

47.
$$-2(-2) > -2 \cdot \frac{1 - 3x}{-2} > -2(7)$$
$$4 > 1 - 3x > -14$$
$$3 > -3x > -15$$
$$-1 < x < 5$$
$(-1, 5)$

49.
$$3 \le 3 - 5x + 15 \le 8$$
$$3 \le 18 - 5x \le 8$$
$$-15 \le -5x \le -10$$
$$3 \ge x \ge 2$$
$[2, 3]$

51.
$$-1.2 \le 1 - 0.02x + 0.122 < 1.4$$
$$-1.2 \le 1.122 - 0.02x < 1.4$$
$$-2.322 \le -0.02x < 0.278$$
$$116.1 \ge x > -13.9$$
$(-13.9, 116.1]$

22

53. $(2, \infty) \cup (4, \infty) = (2, \infty)$
55. $(-\infty, 5) \cap (-\infty, 9) = (-\infty, 5)$
57. $(-\infty, 4] \cap [2, \infty) = [2, 4]$
59. $(-\infty, 5) \cup [-3, \infty) = (-\infty, \infty)$
61. $(3, \infty) \cap (-\infty, 3] = \emptyset$
63. $(3, 5) \cap [4, 8) = [4, 5)$
65. $[1, 4) \cup (2, 6] = [1, 6]$
67. The graph shows real numbers to the right of 2: $x > 2$
69. The graph shows the real numbers to the left of 3: $x < 3$
71. This graph is the union of the numbers greater than 2 with the numbers less than or equal to -1: $x > 2$ or $x \le -1$
73. This graph shows real numbers between -2 and 3: $-2 \le x < 3$
75. The graph shows real numbers greater than or equal to -3: $x \ge -3$
77. $\quad 2 < x < 7$ and $x > 5$
The solution set consists of the numbers between 2 and 7 that are also greater than 5: $(5, 7)$

79. $\quad -1 < 3x + 2 \le 5$ or $\frac{3}{2}x - 6 > 9$

$\quad -3 < 3x \le 3 \quad$ or $\quad 3x - 12 > 18$

$\quad -1 < x \le 1 \quad$ or $\quad\quad x > 10$

Solution set is $(-1, 1] \cup (10, \infty)$.

81. $\quad -6 < x - 1 < 10$ and $-2 < 1 - x < 4$

$\quad -5 < x < 11 \quad$ and $-3 < -x < 3$

$\quad -5 < x < 11 \quad$ and $\quad 3 > x > -3$

The solution set is $(-3, 3)$.

83. $\quad -0.16 \le 0.05x < 0.24$
$\quad\quad -3.2 \le x < 4.8$
or $-8.14 \le 5.3 - 2.1x < 1.39$
$\quad -13.44 \le -2.1x < -3.91$
$\quad\quad 6.4 \ge x > 1.86$
Solution set is $[-3.2, 6.4]$.

85. Let $x =$ the final exam score. We write an inequality expressing the fact that 1/3 of the midterm plus 2/3 of the final must be between 70 and 79 inclusive.

$$70 \le \frac{1}{3} \cdot 64 + \frac{2}{3}x \le 79$$

$$210 \le 64 + 2x \le 237$$
$$146 \le 2x \le 173$$
$$73 \le x \le 86.5$$

87. Let $x =$ the price of the truck. The total spent will be $x + 0.08x + 84$.

$$12,000 \le x + 0.08x + 84 \le 15,000$$
$$12,000 \le 1.08x + 84 \le 15,000$$
$$11,916 \le 1.08x \le 14,916$$
$$\$11,033 \le x \le \$13,811$$

89. Let $x =$ the number of cigarettes smoked on the run, giving the equivalent of $\frac{1}{2}x$ cigarettes smoked. Thus, she smokes $3 + \frac{1}{2}x$ whole cigarettes per day and this number is between 5 and 12:

$$5 \le 3 + \frac{1}{2}x \le 12$$
$$10 \le 6 + x \le 24$$
$$4 \le x \le 18$$

She smokes from 4 to 18 cigarettes on the run.

91. $16.45n + 980.20 > 1,300$

$\quad 16.45n > 319.8$

$\quad\quad n > 19.44$

$\quad 7.79n + 287.87 > 500$

$\quad\quad 7.79n > 212.13$

$\quad\quad n > 27.23$

Both happen in the year $1985 + 28$, or 2013.

Either happens in the year $1985 + 20$, or 2005.

2.6 WARM-UPS

1. True, because both 2 and -2 have absolute value 2. **2.** False, because $|x| = 0$ has only one solution. **3.** False, because it is equivalent to $2x - 3 = 7$ or $2x - 3 = -7$. **4.** True. **5.** False, because this equation has no solution. **6.** True, because only 3 satisfies the equation. **7.** False, because only inequalities that express x between two numbers are written this way. **8.** True, because x is between -7 and 7. **9.** True, subtract 2 from each side. **10.** False, because $|x| < -2$ has no solution.

2.6 EXERCISES

1. $a = 5$ or $a = -5$
Solution set: $\{-5, 5\}$

3. $x - 3 = 1$ or $x - 3 = -1$
$\qquad x = 4 \qquad$ or $\qquad x = 2$
Solution set: $\{2, 4\}$

5. $3 - x = 6$ or $3 - x = -6$
$\quad -x = 3$ or $\quad -x = -9$
$\quad\ x = -3\ $ or $\ x = 9$
Solution set: $\{-3, 9\}$

7. $3x - 4 = 12 \quad$ or $\quad 3x - 4 = -12$
$\quad 3x = 16 \quad$ or $\qquad 3x = -8$
$$x = \frac{16}{3} \quad \text{or} \quad x = -\frac{8}{3}$$
Solution set: $\left\{ -\dfrac{8}{3}, \dfrac{16}{3} \right\}$

9. $\dfrac{2}{3}x - 8 = 0$
$\qquad \dfrac{2}{3}x = 8$
$\qquad\quad x = 12$
Solution set: $\{12\}$

11. $6 - 0.2x = 10 \quad$ or $\quad 6 - 0.2x = -10$
$\quad -0.2x = 4 \quad$ or $\qquad -0.2x = -16$
$\qquad\ x = -20$ or $\qquad\quad x = 80$
Solution set: $\{-20, 80\}$

13. Since absolute value is nonnegative, the solution set is $\emptyset$.

15. $2(x - 4) + 3 = 5$ or $2(x - 4) + 3 = -5$
$\quad 2x - 5 = 5 \quad$ or $\qquad 2x - 5 = -5$
$\qquad 2x = 10 \quad$ or $\qquad\quad 2x = 0$
$\qquad\ x = 5 \quad$ or $\qquad\qquad x = 0$
Solution set: $\{0, 5\}$

17. $7.3x - 5.26 = 4.215$
$\qquad 7.3x = 9.475$
$\qquad\quad x = 1.298$

or $\quad 7.3x - 5.26 = -4.215$
$\qquad\qquad 7.3x = 1.045$
$\qquad\qquad\quad x = 0.143$
Solution set: $\{0.143, 1.298\}$

19. $3 + |x| = 5$
$\qquad |x| = 2$
$\quad x = 2 \quad$ or $\quad x = -2$
Solution set: $\{-2, 2\}$

21. $2 - |x + 3| = -6$
$\quad -|x + 3| = -8$
$\qquad |x + 3| = 8$
$x + 3 = 8 \quad$ or $\quad x + 3 = -8$
$\quad x = 5 \quad$ or $\qquad x = -11$
Solution set: $\{-11, 5\}$

23. $15 - |3 - 2x| = 12$
$\quad -|3 - 2x| = -3$
$\qquad |3 - 2x| = 3$
$3 - 2x = 3 \qquad$ or $\qquad 3 - 2x = -3$
$\quad -2x = 0 \qquad$ or $\qquad\quad -2x = -6$
$\qquad x = 0 \qquad$ or $\qquad\qquad x = 3$
The solution set is $\{0, 3\}$.

25. $x - 5 = 2x + 1 \quad$ or $\quad x - 5 = -(2x + 1)$
$\quad -6 = x \qquad$ or $\quad x - 5 = -2x - 1$
$\qquad\qquad\qquad\qquad\qquad 3x = 4$
$\qquad\qquad\qquad\qquad\qquad\ x = 4/3$
Solution set: $\left\{ -6, \dfrac{4}{3} \right\}$

27. $\dfrac{5}{2} - x = 2 - \dfrac{x}{2} \quad$ or $\quad \dfrac{5}{2} - x = -\left(2 - \dfrac{x}{2}\right)$
$5 - 2x = 4 - x \quad$ or $\quad \dfrac{5}{2} - x = -2 + \dfrac{x}{2}$
$\qquad 1 = x \quad$ or $\quad 5 - 2x = -4 + x$
$\qquad\qquad\qquad\qquad\qquad -3x = -9$
$\qquad\qquad\qquad\qquad\qquad\ \ x = 3$
Solution set: $\{1, 3\}$

29. $x - 3 = 3 - x$ or $x - 3 = -(3 - x)$
$\quad 2x = 6 \qquad$ or $x - 3 = -3 + x$
$\quad\ x = 3 \qquad$ or $x - 3 = x - 3$
The second equation is an identity. So the solution set is the set of real numbers, $(-\infty, \infty)$.

31. $.5w + 3 = .5w - 3$ or $.5w + 3 = -(.5w - 3)$
$\quad 3 = -3 \qquad\qquad$ or $\quad w = 0$
The first equation is inconsistent, but 0 is a solution to the second. The solution set is $\{0\}$.

33. The graph shows the real numbers between -2 and 2: $|x| < 2$

35. The graph shows numbers greater than 3 or less than -3: $\quad |x| > 3$

37. The graph shows numbers between -1 and 1 inclusive: $|x| \leq 1$

39. The graph shows numbers two or more units away from 0: $|x| \geq 2$

41. No, because $|x| < 3$ is equivalent to $-3 < x < 3$.

43. Yes.

45. No, because $|x - 3| \geq 1$ is equivalent to $x - 3 \geq 1$ or $x - 3 \leq -1$.

47. Yes, because the following compound inequalities are equivalent.

$$|4 - x| < 1$$
$$4 - x < 1 \quad \text{and} \quad 4 - x > -1$$
$$4 - x < 1 \quad \text{and} \quad -(4 - x) < 1$$

49. No, because $|4 - x| \leq 1$ is equivalent to $-1 \leq 4 - x \leq 1$.

51.
$$|x| > 6$$
$$x > 6 \quad \text{or} \quad x < -6$$
$$(-\infty, -6) \cup (6, \infty)$$

53.
$$|2a| < 6$$
$$-6 < 2a < 6$$
$$-3 < a < 3$$
$$(-3, 3)$$

55.
$$x - 2 \geq 3 \quad \text{or} \quad x - 2 \leq -3$$
$$x \geq 5 \quad \text{or} \quad x \leq -1$$
$$(-\infty, -1] \cup [5, \infty)$$

57.
$$|2x - 4| < 5$$
$$-5 < 2x - 4 < 5$$
$$-1 < 2x < 9$$
$$-\frac{1}{2} < x < \frac{9}{2}$$
$$\left(-\frac{1}{2}, \frac{9}{2}\right)$$

59.
$$|5 - x| \leq 7$$
$$-7 \leq 5 - x \leq 7$$
$$-12 \leq -x \leq 2$$
$$12 \geq x \geq -2$$
$$[-2, 12]$$

61.
$$2|3 - 2x| \geq 24$$
$$|3 - 2x| \geq 12$$
$$3 - 2x \geq 12 \quad \text{or} \quad 3 - 2x \leq -12$$
$$-2x \geq 9 \quad \text{or} \quad -2x \leq -15$$
$$x \leq -\frac{9}{2} \quad \text{or} \quad x \geq \frac{15}{2}$$
$$(-\infty, -\frac{9}{2}] \cup [\frac{15}{2}, \infty)$$

63.
$$\frac{2w + 7}{3} > 1 \quad \text{or} \quad \frac{2w + 7}{3} < -1$$
$$2w + 7 > 3 \quad \text{or} \quad 2w + 7 < -3$$
$$2w > -4 \quad \text{or} \quad 2w < -10$$
$$w > -2 \quad \text{or} \quad w < -5$$
$$(-\infty, -5) \cup (-2, \infty)$$

65. $|x - 2| > 0$, except when $x = 2$.

$$(-\infty, 2) \cup (2, \infty)$$

67. Since the absolute value of any quantity is greater than or equal to zero, the solution set is R. $(-\infty, \infty)$

69. $-2|3x - 7| > 6$ is equivalent to
$$|3x - 7| < -3.$$
Absolute value of an expression cannot be less than a negative number. Solution set is $\emptyset$.

71. $|2x + 3| > -6$
Since the absolute value of any expression is greater than or equal to zero, it is greater than any negative number. The solution set is R.

73.
$$|x + 2| > 1$$
$$x + 2 > 1 \quad \text{or} \quad x + 2 < -1$$
$$x > -1 \quad \text{or} \quad x < -3$$
$$(-\infty, -3) \cup (-1, \infty)$$

75.
$$|x| + 1 < 5$$
$$|x| < 4$$
$$-4 < x < 4$$
$$(-4, 4)$$

77. $3 - 5|x| > -2$

$-5|x| > -5$

$|x| < 1$

$-1 < x < 1$

$(-1, 1)$

79. $-1.68 < 5.67x - 3.124 < 1.68$

$1.444 < 5.67x < 4.804$

$0.255 < x < 0.847$

$(0.255, 0.847)$

81. $-3 < 2x - 1 < 3$ and $2x - 3 > 2$

$-2 < 2x < 4$ and $2x > 5$

$-1 < x < 2$ and $x > 2.5$

$\emptyset$

83. $-3 < x - 2 < 3$ and $-3 < x - 7 < 3$

$-1 < x < 5$ and $4 < x < 10$

$(4, 5)$

85. $t - 5 = t$ or $t - 5 = -t$

$-5 = 0$ or $2t = 5$

$t = \frac{5}{2}$

Solution set: $\left\{ \frac{5}{2} \right\}$

87. $7 - w = w + 5$ or $7 - w = -(w + 5)$

$-2w = -2$ or $7 - w = -w - 5$

$w = 1$ or $7 = -5$

Solution set: $\{1\}$

89. $|x| = 3 - x$

$x = 3 - x$ or $x = -(3 - x)$

$2x = 3$ or $x = -3 + x$

$x = \frac{3}{2}$ or $0 = -3$

Solution set: $\left\{ \frac{3}{2} \right\}$

91. For $x \geq 0$, $|x| = x$. For $x < 0$, $|x| = -x$.

For $x < 0$, $-x > x$. Thus, $|x| > x$ for $x < 0$.

$(-\infty, 0)$

93. Case 1. If $x \geq 5$, then $x - 5 \geq 0$ and $|x - 5| = x - 5$.

Solve: $x - 5 < 2x + 1$

$-6 < x$

Thus the inequality is satisfied if $x \geq 5$ and $x > -6$, equivalently $x \geq 5$.

Case 2. If $x < 5$, then $x - 5 < 0$ and $|x - 5| = -(x - 5)$.

Solve: $-(x - 5) < 2x + 1$

$-x + 5 < 2x + 1$

$-3x < -4$

$x > \frac{4}{3}$

Thus the inequality is also satisfied if $x < 5$ and $x > \frac{4}{3}$, equivalently $\frac{4}{3} < x < 5$.

The union of these two solutions is the set of real numbers greater than $\frac{4}{3}$.

$\left(\frac{4}{3}, \infty \right)$

95. Let $x =$ the year of the battle of Orleans. Since the difference between x and 1415 is 14 years, we have $|x - 1415| = 14$.

$x - 1415 = 14$ or $x - 1415 = -14$

$x = 1429$ or $x = 1401$

The battle Agincourt was either in 1401 or 1429.

97. Let $x =$ the weight of Kathy. The difference between their weights is less than 6 pounds is expressed by the absolute value inequality

$|x - 127| < 6$

$-6 < x - 127 < 6$

$121 < x < 133$

Kathy weighs between 121 and 133 pounds.

99. Height of first ball is $S = -16t^2 + 50t$ and height of second ball is $S = -16t^2 + 40t + 10$. The difference between the heights is less than 5 feet when

$|-16t^2 + 50t - (-16t^2 + 40t + 10)| < 5$

$|10t - 10| < 5$

$-5 < 10t - 10 < 5$

$5 < 10t < 15$

$0.5 < t < 1.5$

CHAPTER 2 REVIEW

1. $2x = 16$

$x = 8$

Solution set: $\{8\}$

3. $5 - 4x = 11$

$-4x = 6$

$x = \dfrac{6}{-4} = -\dfrac{3}{2}$

Solution set: $\{-\dfrac{3}{2}\}$

5. $x - 6 - x + 6 = 0$

$0 = 0$

Solution set: $\mathbb{R}$

7. $2x - 6 - 5 = 5 - 3 + 2x$

$2x - 11 = 2 + 2x$

$-11 = 2$

Solution set: $\emptyset$

9. $\dfrac{3}{17}x = 0$

$x = 0$

Solution set: $\{0\}$

11. $20\left(\dfrac{1}{4}x - \dfrac{1}{5}\right) = 20\left(\dfrac{1}{5}x + \dfrac{4}{5}\right)$

$5x - 4 = 4x + 16$

$x = 20$

Solution set: $\{20\}$

13. $6\left(\dfrac{t}{2}\right) - 6\left(\dfrac{t-2}{3}\right) = 6\left(\dfrac{3}{2}\right)$

$3t - 2t + 4 = 9$

$t + 4 = 9$

$t = 5$

Solution set: $\{5\}$

15. $1 - 0.4x + 1.6 + 0.6x - 4.2 = -0.6$

$0.2x - 1.6 = -0.6$

$0.2x = 1$

$x = 5$

The solution set is $\{5\}$.

17. $ax + b = 0$

$ax = -b$

$x = -\dfrac{b}{a}$

19. $ax + 2 = cx$

$ax - cx = -2$

$x(a - c) = -2$

$\dfrac{x(a-c)}{a-c} = \dfrac{-2}{a-c}$

$x = \dfrac{2}{c-a}$

21. $\dfrac{mwx}{mw} = \dfrac{P}{mw}$

$x = \dfrac{P}{mw}$

23. $2x\left(\dfrac{1}{x}\right) + 2x\left(\dfrac{1}{2}\right) = 2x(w)$

$2 + x = 2xw$

$x - 2xw = -2$

$x(1 - 2w) = -2$

$x = \dfrac{-2}{1 - 2w}$

$x = \dfrac{2}{2w - 1}$

25. $3x - 2y = -6$

$-2y = -3x - 6$

$y = \dfrac{-3x - 6}{-2}$

$y = \dfrac{3}{2}x + 3$

27. $y - 2 = -\dfrac{1}{3}x + 2$

$y = -\dfrac{1}{3}x + 4$

29. $4\left(\dfrac{1}{2}x - \dfrac{1}{4}y\right) = 4(5)$

$2x - y = 20$

$-y = -2x + 20$

$y = 2x - 20$

31. Let $W = $ the width and $W + 5.5 = $ the length. Since $2L + 2W = P$, we have

$2(W + 5.5) + 2W = 45$

$2W + 11 + 2W = 45$

$4W = 34$

$W = 8.5$

Length is 14 inches and width is 8.5 inches.

33. Let $x = $ the wife's income and $x + 8000 = $ Roy's income. Roy saves $0.10(x + 8000)$ and his wife saves $0.08x$. Since the total saved is $5,660, we can write

$0.10(x + 8000) + 0.08x = 5660$

$0.10x + 800 + 0.08x = 5660$

$0.18x = 4860$

$x = 27000$

Roy's wife earns $27,000 and Roy earns $35,000.

35. Let $x =$ the list price and $0.20x =$ the discount. The list price minus the discount is equal to \$7,600.

$$x - 0.20x = 7600$$
$$0.80x = 7600$$
$$x = 9500$$

The list price was \$9,500.

37. Let $x =$ the number of nickels and $15 - x =$ the number of dimes. The value of x nickels is $5x$ cents and the value of $15 - x$ dimes is $10(15 - x)$ cents. Since she has a total of 95 cents, we can write the equation

$$5x + 10(15 - x) = 95$$
$$5x + 150 - 10x = 95$$
$$-5x = -55$$
$$x = 11$$
$$15 - x = 4$$

She has 11 nickels and 4 dimes.

39. Let $x =$ her walking speed and $x + 9 =$ her riding speed. Since she walked for 3 hours, $3x$ is the number of miles that she walked. Since she rode for 5 hours, $5(x + 9)$ is the number of miles that she rode. Her total distance was 85 miles.

$$3x + 5(x + 9) = 85$$
$$3x + 5x + 45 = 85$$
$$8x = 40$$
$$x = 5$$

She walked 5 miles per hour for 3 hours, so she walked 15 miles.

41.
$$3 - 4x < 15$$
$$-4x < 12$$
$$x > -3$$

$(-3, \infty)$

43.
$$2x - 6 > -6$$
$$2x > 0$$
$$x > 0$$

$(0, \infty)$

45.
$$-\frac{3}{4}x \geq 6$$
$$-3x \geq 24$$
$$x \leq -8$$

$(-\infty, -8]$

47.
$$3x + 6 > 5x - 5$$
$$-2x > -11$$
$$x < \frac{11}{2}$$

$(-\infty, 11/2)$

49.
$$4\left(\frac{1}{2}x + 7\right) \leq 4\left(\frac{3}{4}x - 5\right)$$
$$2x + 28 \leq 3x - 20$$
$$48 \leq x$$

$[48, \infty)$

51.
$$x + 2 > 3 \quad \text{or} \quad x - 6 < -10$$
$$x > 1 \quad \text{or} \quad x < -4$$

$(-\infty, -4) \cup (1, \infty)$

53.
$$x > 0 \quad \text{and} \quad x < 9$$

$(0, 9)$

55.
$$-x < -3 \quad \text{or} \quad -x < 0$$
$$x > 3 \quad \text{or} \quad x > 0$$

$(0, \infty)$

57.
$$2x < 8 \quad \text{and} \quad 2x - 6 < 6$$
$$x < 4 \quad \text{and} \quad 2x < 12$$
$$x < 4 \quad \text{and} \quad x < 6$$

$(-\infty, 4)$

59.
$$x - 6 > 2 \quad \text{and} \quad 6 - x > 0$$
$$x > 8 \quad \text{and} \quad -x > -6$$
$$x > 8 \quad \text{and} \quad x < 6$$

No number is greater than 8 and less than 6. $\emptyset$

61.
$$0.5x > 10 \quad \text{or} \quad 0.1x < 3$$
$$x > 20 \quad \text{or} \quad x < 30$$

Every number is either greater than 20 or less than 30.   Solution set is R or $(-\infty, \infty)$

63. $\quad 10(-2) \leq 10 \cdot \dfrac{2x-3}{10} \leq 10(1)$

$$-20 \leq 2x - 3 \leq 10$$
$$-17 \leq 2x \leq 13$$
$$-\dfrac{17}{2} \leq x \leq \dfrac{13}{2}$$

$\left[-\dfrac{17}{2}, \dfrac{13}{2}\right]$

65. $[1, 4) \cup (2, \infty) = [1, \infty)$
67. $(3, 6) \cap [2, 8] = (3, 6)$
69. $(-\infty, 5) \cup [5, \infty) = (-\infty, \infty)$
71. $(-3, -1] \cap [-2, 5] = [-2, -1]$
73. $\quad 2x \geq 8 \quad \text{or} \quad 2x \leq -8$
$\qquad\quad x \geq 4 \quad \text{or} \quad x \leq -4$

75. Since $|x| \geq 0$ for any real number, the solution set is $\emptyset$.

77. $\quad 1 - \dfrac{x}{5} > \dfrac{9}{5} \quad \text{or} \quad 1 - \dfrac{x}{5} < -\dfrac{9}{5}$
$\qquad 5 - x > 9 \quad \text{or} \quad 5 - x < -9$
$\qquad\quad -x > 4 \quad \text{or} \quad -x < -14$
$\qquad\qquad\quad x < -4 \quad \text{or} \quad x > 14$

79. Since $|x - 3| \geq 0$ for any value of x, the solution set is $\emptyset$.

81. $\qquad\qquad \left|\dfrac{x}{2}\right| - 5 = -1$
$\qquad\qquad\qquad \left|\dfrac{x}{2}\right| = 4$
$\qquad\quad \dfrac{x}{2} = 4 \quad \text{or} \quad \dfrac{x}{2} = -4$
$\qquad\qquad x = 8 \quad \text{or} \quad x = -8$

83. Since $|x + 4| \geq 0$ for any value of x, $|x + 4| \geq -1$ for any x. Solution set: $\mathbb{R}$

85. $\qquad 2 - 3|x - 2| < -1$
$\qquad\qquad -3|x - 2| < -3$
$\qquad\qquad\quad |x - 2| > 1$
$\qquad x - 2 > 1 \quad \text{or} \quad x - 2 < -1$
$\qquad\quad x > 3 \quad\quad \text{or} \quad x < 1$

87. Let x = the rental price, $0.45x$ = the overhead per tape, and $x - 0.45x$ or $0.55x$ = the profit per tape. The rental price must be less than or equal to \$5 and satisfy the inequality
$$0.55x \geq 1.65$$
$$x \geq 3$$
The range of the rental price is \$3 $\leq x \leq$ \$5.

89. Since $150 < h < 180$, we have
$$150 < 60.089 + 2.238F < 180$$
$$89.911 < 2.238F < 119.911$$
$$40.2 < F < 53.6$$
The length of the femur is in $(40.2, 53.6)$.

91. Let x = the number on the mile marker where Dane was picked up. We can write the absolute value equation
$$|x - 86| = 5$$
$$x - 86 = 5 \quad \text{or} \quad x - 86 = -5$$
$$x = 91 \quad \text{or} \quad x = 81$$
He was either at 81 or 91.

93. $\qquad\qquad b = 0.20(300,000 - b)$
$\qquad\qquad b = 60,000 - 0.20b$
$\qquad 1.2b = 60,000$
$\qquad\qquad b = 50,000$
Bonus is \$50,000 according to the accountant, and \$60,000 according the employees.

95. Let x = the number of cows in Washington County and $3600 - x$ = the number of cows in Cade County. We have the following equation:
$$0.30x + 0.60(3600 - x) = 0.50(3600)$$
$$-0.30x + 2160 = 1800$$
$$-0.30x = -360$$
$$x = 1200$$
$$3600 - x = 2400$$
There are 1200 cows in Washington County and 2400 in Cade County.

97. The numbers to the right of 1 are described by the inequality $x > 1$.

99. The number 2 satisfies the equation $x = 2$.

101. The numbers 3 and -3 both satisfy the equation $|x| = 3$.

103. The numbers to the left of and including -1 satisfy $x \leq -1$.

105. The numbers between -2 and 2 including the endpoints satisfy $|x| \leq 2$.

107. $x \leq 2$ or $x \geq 7$

109. The numbers greater than 3 or less than -3 satisfy $|x| > 3$.

111. $5 < x < 7$

113. Every number except 0 has a positive absolute value and satisfies $|x| > 0$.

CHAPTER 2   TEST

1.
$$-6x - 5 = -4x + 3$$
$$-2x = 8$$
$$x = -4$$

Solution set: $\{-4\}$

2.
$$6\left(\frac{y}{2}\right) - 6\left(\frac{y-3}{3}\right) = 6\left(\frac{y+6}{6}\right)$$
$$3y - 2y + 6 = y + 6$$
$$y + 6 = y + 6$$

The equation is an identity.   Solution set: R

3.
$$|w| + 3 = 9$$
$$|w| = 6$$
$$w = 6 \quad \text{or} \quad w = -6$$

Solution set: $\{-6, 6\}$

4.
$$|3 - 2(5 - x)| = 3$$
$$|-7 + 2x| = 3$$
$$-7 + 2x = 3 \quad \text{or} \quad -7 + 2x = -3$$
$$2x = 10 \quad \text{or} \qquad 2x = 4$$
$$x = 5 \quad \text{or} \qquad x = 2$$

Solution set: $\{2, 5\}$

5.
$$2x - 5y = 20$$
$$-5y = -2x + 20$$
$$y = \frac{2}{5}x - 4$$

6.
$$a = axt + S$$
$$a - axt = S$$
$$a(1 - xt) = S$$
$$a = \frac{S}{1 - xt}$$

7.
$$-2 \leq m - 6 \leq 2$$
$$4 \leq m \leq 8$$

$[4, 8]$

8.
$$2|x - 3| > 20$$
$$|x - 3| > 10$$
$$x - 3 > 10 \quad \text{or} \quad x - 3 < -10$$
$$x > 13 \quad \text{or} \qquad x < -7$$

$(-\infty, -7) \cup (13, \infty)$

9.
$$2 - 3(w - 1) < -2w$$
$$2 - 3w + 3 < -2w$$
$$-w < -5$$
$$w > 5$$

$(5, \infty)$

10.
$$3(2) < 3\left(\frac{5 - 2x}{3}\right) < 3(7)$$
$$6 < 5 - 2x < 21$$
$$1 < -2x < 16$$
$$-\frac{1}{2} > x > -8$$

$\left(-8, -\frac{1}{2}\right)$

11.
$$3x - 2 < 7 \quad \text{and} \quad -3x \leq 15$$
$$3x < 9 \quad \text{and} \qquad x \geq -5$$
$$x < 3 \quad \text{and} \qquad x \geq -5$$

$[-5, 3)$

12.
$$\frac{3}{2}\left(\frac{2}{3}y\right) < \frac{3}{2}(4) \quad \text{or} \quad y - 3 < 12$$
$$y < 6 \qquad \text{or} \qquad y < 15$$

$(-\infty, 15)$

13.
$$|2x - 7| = 3$$
$$2x - 7 = 3 \quad \text{or} \quad 2x - 7 = -3$$
$$2x = 10 \quad \text{or} \qquad 2x = 4$$
$$x = 5 \quad \text{or} \qquad x = 2$$

$\{2, 5\}$

14. $x > 5 \quad \text{or} \quad x < 12$
$(-\infty, \infty)$

15. $x < 0 \quad \text{and} \quad x > 7$

No real number is both less than 0 and greater than 7. Solution set: $\emptyset$

16. Since no real number satisfies $|2x - 5| < 0$, we need only solve $|2x - 5| = 0$, which is equivalent to $2x - 5 = 0$, or $x = 2.5$.

Solution set: $\{2.5\}$

17. Since no real number satisfies $|x - 3| < 0$, the solution set is $\emptyset$.

18. Since $x + 3x = 4x$ is equivalent to $4x = 4x$, the solution set is R.

19.
$$2x + 14 = 2x + 9$$
$$14 = 9$$

Solution set is ∅.

20. Since $|x - 6| \geq 0$ for any real number x, $|x - 6| \geq -6$ for any real number x. The solution set is R.

21.
$$x - 0.04x + 0.4 = 96.4$$
$$0.96x = 96$$
$$x = 100$$

{100}

22. Let W = the width and $W + 16$ = the length. Use $2W + 2L = P$ to write the equation

$$2W + 2(W + 16) = 84$$
$$4W + 32 = 84$$
$$4W = 52$$
$$W = 13$$

The width is 13 meters.

23. Let h = the height. Use the formula $A = \frac{1}{2}bh$ to write the following equation.

$$\frac{1}{2} \cdot 3h = 21$$
$$3h = 42$$
$$h = 14$$

The height is 14 inches.

24. Let x = the original price and $0.30x$ = the amount of discount. The original price minus the discount is equal to the price she paid.

$$x - 0.30x = 210$$
$$0.70x = 210$$
$$x = 300$$

The original price was $300.

25. Let x = the number of liters of 11% alcohol. The mixture will be $x + 60$ liters of 7% alcohol. There are $0.11x$ liters of alcohol in the 11% solution, $0.05(60)$ liters of alcohol in the 5% solution, and $0.07(x + 60)$ liters of alcohol in the 7% solution.

$$0.11x + 0.05(60) = 0.07(x + 60)$$
$$0.11x + 3 = 0.07x + 4.2$$
$$0.04x = 1.2$$
$$x = 30$$

Use 30 liters of 11% alcohol solution.

26. Let b = Brenda's salary.
$$|b - 28,000| > 3000$$
$$b - 28,000 > 3000 \quad \text{or} \quad b - 28,000 < -3000$$
$$b > 31,000 \quad \text{or} \quad b < 25,000$$

Brenda's salary is either greater than $31,000 or less than $25,000.

Tying It All Together Chapters 1-2

1. $5x + 6x = 11x$

2. $5x \cdot 6x = 30x^2$

3. $\frac{6x + 2}{2} = \frac{6x}{2} + \frac{2}{2} = 3x + 1$

4. $5 - 4(2 - x) = 5 - 8 + 4x = 4x - 3$

5. $(30 - 1)(30 + 1) = 29 \cdot 31 = 899$

6. $(30 + 1)^2 = 31^2 = 961$

7. $(30 - 1)^2 = 29^2 = 841$

8. $(2 + 3)^2 = 5^2 = 25$

9. $2^2 + 3^2 = 4 + 9 = 13$

10. $(8 - 3)(3 - 8) = 5(-5) = -25$

11. $(-1)(3 - 8) = -1(-5) = 5$

12. $-2^2 = -(2^2) = -4$

13. $3x + 8 - 5(x - 1) = 3x + 8 - 5x + 5$
$= -2x + 13$

14. $(-6)^2 - 4(-3)2 = 36 + 24 = 60$

15. $3^2 \cdot 2^3 = 9 \cdot 8 = 72$

16. $4(-6) - (-5)(3) = -24 + 14 = -9$

17. $-3x \cdot x \cdot x = -3x^3$

18. $(-1)^6 = 1$

19.
$$5x + 6x = 8x$$
$$11x = 8x$$
$$3x = 0$$
$$x = 0$$

Solution set: {0}

20.
$$5x + 6x = 11x$$
$$11x = 11x$$

This equation is an identity. Solution set: R

31

21.
$$5x + 6x = 0$$
$$11x = 0$$
$$x = 0$$
Solution set: $\{0\}$

22.
$$5x + 6 = 11x$$
$$-6x = -6$$
$$x = 1$$
Solution set: $\{1\}$

23.
$$3x + 1 = 0$$
$$3x = -1$$
$$x = -\frac{1}{3}$$
Solution set: $\left\{-\frac{1}{3}\right\}$

24.
$$5 - 4(2 - x) = 1$$
$$5 - 8 + 4x = 1$$
$$4x = 4$$
$$x = 1$$
Solution set: $\{1\}$

25.
$$3x + 6 = 3(x + 2)$$
$$3x + 6 = 3x + 6$$
This equation is an identity. Solution set: R

26.
$$x - 0.01x = 990$$
$$0.99x = 990$$
$$x = 1000$$
Solution set: $\{1000\}$

27.
$$|5x + 6| = 11$$
$$5x + 6 = 11 \quad \text{or} \quad 5x + 6 = -11$$
$$5x = 5 \quad \text{or} \quad 5x = -17$$
$$x = 1 \quad \text{or} \quad x = -17/5$$

Solution set: $\{-17/5, 1\}$

28. If x = the number of copies made in 5 years, then the cost for renting is $60(75) + 0.06x$ dollars. The cost if the copier is purchased is $8000 + 0.02x$ dollars.

$$60(75) + 0.06x = 8000 + 0.02x$$
$$0.04x = 3500$$
$$x = 87,500$$

Five-year cost is same for 87,500 copies. If 90,000 copies are made, then renting cost is $9900 and buying cost is $9800. So over 87,500 copies it is cheaper to buy.

3.1 WARM-UPS

1. True, because $3^5 \cdot 3^4 = 3^{5+4} = 3^9$.

2. False, because $3 \cdot 3 \cdot 3^{-1} = 9 \cdot \frac{1}{3} = 3$.

3. False, because $10^{-3} = \frac{1}{1000} = 0.001$.

4. True, because the reciprocal of 1 is 1.

5. True, because $\frac{2^5}{2^{-2}} = 2^{5-(-2)} = 2^7$.

6. False, because $2^3 \cdot 5^2 = 8 \cdot 25 = 200$.

7. True, because $-2^{-2} = -\frac{1}{2^2} = -\frac{1}{4}$.

8. True, $46.7 \times 10^5 = 4.67 \times 10^1 \times 10^5$.

9. True, $0.512 \times 10^{-3} = 5.12 \times 10^{-1} \times 10^{-3}$.

10. False, because $\frac{8 \times 10^{30}}{2 \times 10^{-5}} = 4 \times 10^{35}$.

3.1 EXERCISES

1. $4^2 = 4 \cdot 4 = 16$

3. $4^{-2} = \frac{1}{4^2} = \frac{1}{16}$

5. $\left(\frac{1}{4}\right)^{-2} = 4^2 = 16$

7. $\left(-\frac{3}{4}\right)^{-2} = \left(-\frac{4}{3}\right)^2 = \frac{16}{9}$

9. $-2^{-1}(-6)^2 = -\frac{1}{2} \cdot 36 = -18$

11. $3 \cdot 10^{-3} = 3 \cdot \frac{1}{1000} = 0.003$

13. $\frac{1}{2^{-4}} = 2^4 = 16$

15. $2^5 \cdot 2^{12} = 2^{5+12} = 2^{17}$

17. $2y^{3+2-7} = 2y^{-2} = \frac{2}{y^2}$

19. $5 \cdot 6 \cdot x^{-3}x^{-4} = 30x^{-7} = \frac{30}{x^7}$

21. $-8^0 - 8^0 = -1 - 1 = -2$

23. $(x - y)^0 = 1$

25. $2w^{-3+7-4} = 2w^0 = 2 \cdot 1 = 2$

27. $\frac{2}{4^{-2}} = 2 \cdot 4^2 = 2 \cdot 16 = 32$

29. $\frac{3^{-1}}{10^{-2}} = \frac{10^2}{3^1} = \frac{100}{3}$

31. $\frac{2x^{-3}(4x)}{5y^{-2}} = \frac{8x^{-2}}{5y^{-2}} = \frac{8y^2}{5x^2}$

33. $\frac{4^{-2}x^3x^{-6}}{3x^{-3}x^2} = \frac{x^{-3}}{16 \cdot 3x^{-1}} = \frac{x^{-2}}{48} = \frac{1}{48x^2}$

35. $x^{5-3} = x^2$

37. $3^{6-(-2)} = 3^8$

39. $\frac{4a^{-5-(-2)}}{12} = \frac{a^{-3}}{3} = \frac{1}{3a^3}$

41. $-3w^{-5-3} = -3w^{-8} = \frac{-3}{w^8}$

43. $\frac{3^3w^3}{3^{-5}w^{-3}} = 3^8w^6$

45. $\frac{3x^{-4}y}{6x^{-5}} = \frac{x^{-4-(-5)}y}{2} = \frac{xy}{2}$

47. $3^{-1} \cdot \left(\frac{1}{3}\right)^{-3} = \frac{1}{3} \cdot 3^3 = 9$

49. $-2^4 + \left(\frac{1}{2}\right)^{-1} = -16 + 2 = -14$

51. $-(-2)^{-3} \cdot 2^{-1} = -\left(-\frac{1}{8}\right)\frac{1}{2} = \frac{1}{16}$

53. $\left(1 + \frac{1}{2}\right)^{-2} = \left(\frac{3}{2}\right)^{-2} = \left(\frac{2}{3}\right)^2 = \frac{4}{9}$

55. $-5a^2 \cdot 3a^2 = -15a^4$

57. $5a \cdot a^2 + 3a^{-2} \cdot a^5 = 5a^3 + 3a^3 = 8a^3$

59. $\frac{-3a^5(-2a^{-1})}{6a^3} = \frac{6a^4}{6a^3} = a$

61. $\frac{(-3x^3y^2)(-2xy^{-3})}{-9x^2y^{-5}} = \frac{6x^4y^{-1}}{-9x^2y^{-5}} = \frac{-2x^2y^4}{3}$

63. $\frac{2^{-1} + 3^{-1}}{2^{-1}} = \left(\frac{1}{2} + \frac{1}{3}\right) \cdot \frac{2}{1} = \frac{5}{6} \cdot \frac{2}{1} = \frac{5}{3}$

65. $\frac{(2+3)^{-1}}{2^{-1} + 3^{-1}} = \frac{5^{-1}}{\frac{1}{2} + \frac{1}{3}} = \frac{1}{5} \div \frac{5}{6} = \frac{6}{25}$

67. $8 = 2^3$

69. $\frac{1}{4} = 2^{-2}$

71. $16 = \left(\frac{1}{2}\right)^{-4}$

73. $10^{-3} = 0.001$

75. Since the exponent in 4.86×10^8 is positive, we move the decimal point 8 places to the right: 486,000,000

77. Since the exponent in 2.37×10^{-6} is negative, we move the decimal point 6 places to the left: 0.00000237

79. Since the exponent in 4×10^6 is positive, we move the decimal point 6 places to the right: 4,000,000

81. Since the exponent in 5×10^{-6} is negative, we move the decimal point 6 places to the left: 0.000005

83. The decimal point in 320,000 must be moved 5 places to the left for scientific notation. Since the original number is larger than 10 the exponent is positive: 3.2×10^5

85. The decimal point in 0.00000071 must be moved 7 places to the right to get scientific notation. Since the original number is smaller than 1, the exponent is negative: 7.1×10^{-7}

87. Move the decimal point 5 places to the right: 7×10^{-5}

89. $235 \times 10^5 = 2.35 \times 10^2 \times 10^5 = 2.35 \times 10^7$

91. $\frac{5 \times 10^6 \cdot 3 \times 10^{-4}}{2 \times 10^3} = \frac{7.5 \times 10^2}{10^3}$

$$= 7.5 \times 10^{-1} = 0.75$$

93. By the quotient rule, we subtract exponents when we divide exponential expressions: 3×10^{22}

95. $\frac{(-4 \times 10^5)(6 \times 10^{-9})}{2 \times 10^{-16}}$

$$= -12 \times 10^{5 + (-9) - (-16)} = -12 \times 10^{12}$$

$$= -1.2 \times 10^1 \times 10^{12} = -1.2 \times 10^{13}$$

97. Enter the numbers into a calculator using scientific notation. The calculator will give answers to the computations in scientific notation. 1.578×10^5

99. 9.187×10^{-5}

33

101. Multiply 9.27 and 6.43 on a calculator to get 59.606. Since most calculators do not handle scientific notation with exponents higher than 99, we must add the exponents without entering the numbers as scientific notation. We get
$59.606 \times 10^{156} = 5.961 \times 10^1 \times 10^{156}$
$= 5.961 \times 10^{157}$

103. 3.828×10^{30}

105. $9.3 \times 10^7 \text{ miles} \times \dfrac{5280 \text{ feet}}{1 \text{ mile}} = 4.910 \times 10^{11} \text{ ft}$

107. Since $T = \dfrac{D}{R}$, $T = \dfrac{4.6 \times 10^{12} \text{ km}}{1.2 \times 10^5 \text{ kps}}$

$= 3.83 \times 10^7 \text{ seconds}$

109. $\dfrac{8.71 \times 10^7 \text{ ton} \times \dfrac{2000 \text{ lb}}{1 \text{ ton}}}{1.80863 \times 10^8 \text{ people}} \div 365 \text{ da} =$

2.6 lb/person/day

3.2 WARM-UPS

1. False, because $(2^2)^3 = 2^6$.

2. True, because $(2^{-3})^{-1} = 2^3 = 8$.

3. True, because we $3(-3) = -9$.

4. False, because $(2^3)^3 = 2^9$.

5. False, because $(2x)^3 = 2^3 x^3 = 8x^3$.

6. False, because $(-3y^3)^2 = 9y^6$.

7. True, the reciprocal of $\frac{2}{3}$ is $\frac{3}{2}$.

8. True, because $\left(\frac{2}{3}\right)^3 = \frac{2^3}{3^3} = \frac{8}{27}$.

9. True, because $\left(\frac{x^2}{2}\right)^3 = \frac{x^6}{2^3} = \frac{x^6}{8}$.

10. True, because $\left(\frac{2}{x}\right)^{-2} = \left(\frac{x}{2}\right)^2 = \frac{x^2}{4}$.

3.2 EXERCISES

1. $2^6 = 64$

3. $y^{2 \cdot 5} = y^{10}$

5. $(x^2)^{-4} = x^{(2)(-4)} = x^{-8} = \frac{1}{x^8}$

7. $(m^{-3})^{-6} = m^{(-3)(-6)} = m^{18}$

9. $x^6 x^{-6} = x^0 = 1$

11. $\dfrac{x^{-12}}{x^{-10}} = x^{-2} = \dfrac{1}{x^2}$

13. $9^2 y^2 = 81 y^2$

15. $(-5w^3)^2 = (-5)^2 w^6 = 25 w^6$

17. $x^9 y^{-6} = \dfrac{x^9}{y^6}$

19. $3^{-2} a^{-2} b^2 = \dfrac{b^2}{9a^2}$

21. $\dfrac{2xy^{-2}}{3^{-1} x^{-2} y^{-1}} = 2 \cdot 3 x^3 y^{-1} = \dfrac{6x^3}{y}$

23. $\dfrac{2^{-2} a^{-2} b^{-2}}{2ab^2} = 2^{-3} a^{-3} b^{-4} = \dfrac{1}{8a^3 b^4}$

25. $\dfrac{w^3}{2^3} = \dfrac{w^3}{8}$

27. $\dfrac{(-3)^3 a^3}{4^3} = -\dfrac{27a^3}{64}$

29. $\dfrac{2^{-2} x^2}{y^{-2}} = \dfrac{x^2 y^2}{4}$

31. $\left(\dfrac{-3x^3}{y}\right)^{-2} = \dfrac{(-3)^{-2}(x^3)^{-2}}{y^{-2}} = \dfrac{x^{-6} y^2}{(-3)^2} = \dfrac{y^2}{9x^6}$

33. $\left(\dfrac{2}{5}\right)^{-2} = \left(\dfrac{5}{2}\right)^2 = \dfrac{25}{4}$

35. $\left(-\dfrac{1}{2}\right)^{-2} = (-2)^2 = 4$

37. $\left(-\dfrac{3}{2x}\right)^3 = -\dfrac{3^3}{2^3 x^3} = -\dfrac{27}{8x^3}$

39. $\left(\dfrac{3y}{2x^2}\right)^3 = \dfrac{3^3 y^3}{2^3 (x^2)^3} = \dfrac{27y^3}{8x^6}$

41. $5^{2t} \cdot 5^{4t} = 5^{6t}$

43. $\left(2^{-3w}\right)^{-2w} = 2^{6w^2}$

45. $\dfrac{7^{2m+6}}{7^{m+3}} = 7^{2m+6-m-3} = 7^{m+3}$

47. $8^{2a-1}\left(8^{a+4}\right)^3 = 8^{2a-1} \cdot 8^{3a+12}$

$= 8^{5a+11}$

49. $\dfrac{2^{3m-1} 2^{10-2m}}{2^{3-3m} 2^{-m-1}} = \dfrac{2^{m+9}}{2^{-4m+2}} = 2^{m+9-(-4m+2)}$

$= 2^{5m+7}$

51. Add the exponents: $6x^9$

53. Multiply the exponents: $-8x^6$

55. Eliminate negative exponents: $\dfrac{3z}{x^2y}$

57. The reciprocal of $-\dfrac{2}{3}$ is $-\dfrac{3}{2}$.

59. Multiply the exponents: $\dfrac{4x^6}{9}$

61. $\left(-2x^{-2}\right)^{-1} = (-2)^{-1}x^2 = -\dfrac{x^2}{2}$

63. $\left(\dfrac{2x^2y}{xy^2}\right)^{-3} = \left(\dfrac{xy^2}{2x^2y}\right)^3 = \dfrac{x^3y^6}{8x^6y^3} = \dfrac{y^3}{8x^3}$

65. $\dfrac{5^3a^{-3}b^6}{5^4a^4b^{-8}} = 5^{-1}a^{-7}b^{14} = \dfrac{b^{14}}{5a^7}$

67. $\dfrac{2^{-3}x^6y^{-3}}{2^2x^2y^{-2}} \cdot 2x^2y^{-7} = 2^{-4}x^6y^{-8} = \dfrac{x^6}{16y^8}$

69. $\dfrac{6^{-2}a^4b^{-6}}{2^{-2}c^{-8}} \cdot 3^3a^{-3}b^6 = 3a^1b^0c^8 = 3ac^8$

71. $\dfrac{3^4a^{-8}b^{12}}{a^{-12}b^{20}} \cdot \dfrac{9^{-3}a^{15}b^{-12}a^{-6}}{b^{18}a^{-15}} = \dfrac{81a}{9^3a^{-27}b^{38}}$

$= \dfrac{a^{28}}{9b^{38}}$

73. $\dfrac{(-2)^{-3}x^9y^{-3}z^{-3}x^2y^{-2}z^43^{-2}x^2y^2z^4}{2^{-3}x^6y^33^2x^{-1}y(-x^{-3})z^{-6}}$

$= \dfrac{x^{13}y^{-3}z^5}{3^4x^2y^4z^{-6}} = \dfrac{x^{11}z^{11}}{81y^7}$

75. $32 \cdot 64 = 2^5 \cdot 2^6 = 2^{11}$

77. $81 \cdot 6^{-4} = 3^4(2 \cdot 3)^{-4} = 3^42^{-4}3^{-4} = 2^{-4}$

79. $4^{3n} = (2^2)^{3n} = 2^{6n}$

81. $\left(\dfrac{1}{(2^4)^{-m}}\right)^m = \left(\dfrac{1}{2^{-4m}}\right)^m = \left(2^{4m}\right)^m = 2^{4m^2}$

83. $\dfrac{1}{5^{-2}} = 25$

85. $2^{-1} + 2^{-2} = 0.75$

87. $(0.036)^{-2} + (4.29)^3 \approx 850.559$

89. $\dfrac{(5.73)^{-1} + (4.29)^{-1}}{(3.762)^{-1}} \approx 1.533$

91. $40{,}000(1 + 0.12)^3 = 40{,}000(1.12)^3$

$= \$56{,}197.12$

93. $10{,}000(1 + 0.07)^{-18} = \$2{,}958.64$

95. $L = 67.0166(1.00308)^{20} \approx 71.3$ yr

$L = 67.0166(1.00308)^{70} \approx 83.1$ yr

3.3 WARM-UPS

1. False, because it has a negative exponent.
2. False, because the coefficient of x is -5.
3. False, because the highest power of x is 3.
4. True, because $5^2 - 3 = 22$.
5. False, because $30(0) + 10 = 10$. **6.** True, because the two trinomials are added correctly.
7. False, because $(x^2 - 5x) - (x^2 - 3x) = -2x$ for any value of x. **8.** True, because the monomial and binomial are correctly multiplied.
9. True, because $-(a - b) = b - a$ for any values of a and b. **10.** False, because $-(y + 5)$ $= -y - 5$ for any value of y.

3.3 EXERCISES

1. Yes, it is a monomial.
3. No, because polynomials do not have negative exponents.
5. Yes, it is a trinomial.
7. No, because $\frac{1}{x}$ is not allowed in a polynomial.
9. The degree is 4 because 4 is the highest power. The coefficient of x^3 is -8. It is a binomial because it has two terms.
11. The degree of -8 is 0 and the coefficient of x^3 is 0. It is a monomial.
13. The degree is 7 because 7 is the highest power. The coefficient of x^3 is 0, because x^3 is missing. It is a monomial because it has one term.
15. The degree is 6 because 6 is the highest power. The coefficient of x^3 is 1. It is a trinomial because it has three terms.
17. Replace x by 3 in $P(x) = x^4 - 1$.

$P(3) = 3^4 - 1 = 81 - 1 = 80$

19. Replace x by -2:

$M(-2) = -3(-2)^2 + 4(-2) - 9 = -29$

21. Replace x by 1:

$R(1) = 1^5 - 1^4 + 1^3 - 1^2 + 1 - 1 = 0$

23. $2a + a - 3 + 5 = 3a + 2$

25. $7xy + 30 - 2xy - 5 = 5xy + 25$

27. $x^2 - x^2 - 3x + 5x - 9 = 2x - 9$

29. $2x^3 - x^2 - 4x + 2x - 3 - 5$
$= 2x^3 - x^2 - 2x - 8$

31. Add the like terms to get $11x^2 - 2x - 9$.

33. $5x - 4x + 2 - (-3) = x + 5$.

35. Change the sign of every term on the bottom and add it to the appropriate like term on the top to get $-6x^2 + 5x + 2$.

37. Add the like terms to get $2x$.

39. Add exponents to get $-15x^6$.

41. $-1(3x - 2) = -3x + 2$

43. $5x^2y^3(3x^2y - 4x) = 15x^4y^4 - 20x^3y^3$

45. $(x - 2)(x + 2) = (x - 2)x + (x - 2)2$
$= x^2 - 2x + 2x - 4 = x^2 - 4$

47. $(x^2 + x + 2)2x + (x^2 + x + 2)(-3)$
$= 2x^3 + 2x^2 + 4x - 3x^2 - 3x - 6$
$= 2x^3 - x^2 + x - 6$

49. Multiply $-5x$ by $2x - 3$ to get $-10x^2 + 15x$.

51.

$$\begin{array}{r} x + 5 \\ x + 5 \\ \hline 5x + 25 \\ x^2 + 5x \\ \hline x^2 + 10x + 25 \end{array}$$

53.

$$\begin{array}{r} x + 6 \\ 2x - 3 \\ \hline -3x - 18 \\ 2x^2 + 12x \\ \hline 2x^2 + 9x - 18 \end{array}$$

55.

$$\begin{array}{r} x^2 + xy + y^2 \\ x - y \\ \hline -x^2y - xy^2 - y^3 \\ x^3 + x^2y + xy^2 \\ \hline x^3 - y^3 \end{array}$$

57. Add like terms to get $2x - 5$.

59. Add like terms: $4a^2 - 11a - 4$

61. $w^2 - 7w - 2 - w + 3w^2 - 5$
$= 4w^2 - 8w - 7$

63. $(x - 2)x^2 + (x - 2)2x + (x - 2)4$
$= x^3 - 2x^2 + 2x^2 - 4x + 4x - 8$
$= x^3 - 8$

65. $(x - w)z + (x - w)2w$
$= xz - wz + 2xw - 2w^2$

67. $(2xy - 1)3xy + (2xy - 1)5$
$= 6x^2y^2 - 3xy + 10xy - 5$
$= 6x^2y^2 + 7xy - 5$

69. $(2.31x - 5.4)6.25x + (2.31x - 5.4)1.8$
$= 14.4375x^2 - 29.592x - 9.72$

71. $(3.759 + 11.61)x^2 - 4.71x + 6.59x + 2.85$
$- 3.716 = 15.369x^2 + 1.88x - 0.866$

73. $\frac{2}{4}x + \frac{4}{2} + \frac{1}{4}x - \frac{1}{2} = \frac{3}{4}x + \frac{3}{2}$

75. $\frac{1}{2}x^2 + \frac{1}{3}x - \frac{1}{5} - \frac{2}{2}x^2 + \frac{2}{3}x + \frac{1}{5}$
$= -\frac{1}{2}x^2 + x$

77. Combine like terms within the first set of brackets:

$x^2 - 3 - x^2 - 5x + 4 = -5x + 1$

Combine like terms within the second set of brackets:

$x - 3x^2 + 15x = -3x^2 + 16x$

Now subtract these results:

$[-5x + 1] - [-3x^2 + 16x]$
$= -5x + 1 + 3x^2 - 16x$
$= 3x^2 - 21x + 1$

79. $[x + 12][-4x - 14]$
$= [x + 12](-4x) + [x + 12](-14)$
$= -4x^2 - 48x - 14x - 168$
$= -4x^2 - 62x - 168$

81. $[x^2 - m - 2][x^2 + m + 2]$
$= x^4 - mx^2 - 2x^2 + mx^2 - m^2 - 2m$
$ + 2x^2 - 2m - 4$
$= x^4 - m^2 - 4m - 4$

83. $10x^2 - 8x - 3x[2x^2 + 21x]$

$\quad = 10x^2 - 8x - 6x^3 - 63x^2$

$\quad = -6x^3 - 53x^2 - 8x$

85. $(a^{2m} + 3a^m - 3) + (-5a^{2m} - 7a^m + 8)$

$\quad = -4a^{2m} - 4a^m + 5$

87. $(x^n - 1)(x^n + 3) = (x^n - 1)x^n + (x^n - 1)3$

$\quad = x^{2n} - x^n + 3x^n - 3 = x^{2n} + 2x^n - 3$

89. $z^{3w} - z^{2w}(z^{1-w} - 4z^w)$

$\quad = z^{3w} - z^{1+w} + 4z^{3w} = 5z^{3w} - z^{1+w}$

91. $(x^{2r} + y)x^{4r} + (x^{2r} + y)(-x^{2r}y) + (x^{2r} + y)y^2$

$\quad = x^{6r} + x^{4r}y - x^{4r}y - x^{2r}y^2 + x^{2r}y^2 + y^3$

$\quad = x^{6r} + y^3$

93. $C(3) = 20(3) + 15 = \$75$

95. $C(3) = 50 \cdot 3 - 0.01(3)^4 = 149.19$

$C(2) = 50 \cdot 2 - 0.01(2)^4 = 99.84$

$C(3) - C(2) = 49.35$

Marginal cost of 3rd window is \$49.35.

$C(10) - C(9) = 400 - 384.39$

Marginal cost of 10th window is \$15.61.

97. $L(1990) = 71.4$, $L(1960) = 66.7$

Life expectancy for male born in 1990 is 4.76 years more than a male born in 1960.

3.4 WARM-UPS

1. True, because the binomials are multiplied correctly.   **2.** True, because the binomials are multiplied correctly.   **3.** False, $(2+3)^2 = 5^2 = 25$ and $2^2 + 3^2 = 4 + 9 = 13$.
4. True, because the binomial is squared correctly.   **5.** False, $(8-3)^2 = 5^2 = 25$ and $64 - 9 = 55$.   **6.** True.
7. True, $(60-1)(60+1) = 60^2 - 1^2 = 3600 - 1$.
8. True, because the binomial is squared correctly.   **9.** False, $(x-3)^2 = x^2 - 6x + 9$ for any value of x.
10. False, it is a product of two monomials.

3.4 EXERCISES

1. $x^2 - 2x + 4x - 8 = x^2 + 2x - 8$

3. $2x^2 + x + 6x + 3 = 2x^2 + 7x + 3$

5. $2a^2 + 3a - 10a - 15 = 2a^2 - 7a - 15$

7. $(2x^2 - 7)(2x^2 + 7) = 4x^2 - 14x^2 + 14x^2 - 49$

$\quad = 4x^4 - 49$

9. $(2x^3 - 1)(x^3 + 4) = 2x^6 - x^3 + 8x^3 - 4$

$\quad = 2x^6 + 7x^3 - 4$

11. $6zw + w^2 - 6z^2 - wz = w^2 + 5wz - 6z^2$

13. $(3k - 2t)(4t + 3k) = 12kt - 8t^2 + 9k^2 - 6tk$

$\quad = 9k^2 + 6kt - 8t^2$

15. $xy - 3y + xw - 3w$

17. $\left(\frac{1}{2} + t\right)\left(\frac{1}{3} + t\right) = \frac{1}{6} + \frac{5}{6}t + t^2$

19. $\left(\frac{1}{2}x^2 - 4\right)\left(\frac{1}{2}x^2 + 6\right) = \frac{1}{4}x^4 + x^2 - 24$

21. $m^2 + 2(3)(m) + 3^2 = m^2 + 6m + 9$

23. $a^2 - 2(4)(a) + 4^2 = a^2 - 8a + 16$

25. $(2w)^2 + 2(2w)(1) + 1^2 = 4w^2 + 4w + 1$

27. $(3t)^2 - 2(3t)(5u) + (5u)^2$

$\qquad\qquad = 9t^2 - 30tu + 25u^2$

29. $x^2 - 2(-x)(1) + 1^2 = x^2 + 2x + 1$

31. $a^2 - 2(a)(3y^3) + (3y^3)^2 = a^2 - 6ay^3 + 9y^6$

33. $\left(\frac{1}{2}x - 1\right)^2 = \frac{1}{4}x^2 - 2 \cdot \frac{1}{2}x \cdot 1 + 1 = \frac{1}{4}x^2 - x + 1$

35. $\frac{9}{16}a^2 + 2 \cdot \frac{3}{4}a \cdot \frac{1}{3} + \frac{1}{9} = \frac{9}{16}a^2 + \frac{1}{2}a + \frac{1}{9}$

37. $(-2)^2 + 2(-2)(gt) + (gt)^2 = g^2t^2 - 4gt + 4$

39. $(8y^2z^4)^2 + 2(8y^2z^4)(3) + 3^2$

$\quad = 64y^4z^8 + 48y^2z^4 + 9$

41. $w^2 - 9^2 = w^2 - 81$

43. $(w^3)^2 - y^2 = w^6 - y^2$

45. $(2x)^2 - 7^2 = 4x^2 - 49$

47. $(3x^2)^2 - 2^2 = 9x^4 - 4$

49. $[(m+t) + 5][(m+t) - 5] = (m+t)^2 - 25$

$\quad = m^2 + 2mt + t^2 - 25$

51. $[y - (r+5)][y + (r+5)] = y^2 - (r+5)^2$

$\quad = y^2 - r^2 - 10r - 25$

53. $[(2y-t) + 3]^2 = (2y-t)^2 + 6(2y-t) + 9$

$\quad = 4y^2 - 4yt + t^2 + 12y - 6t + 9$

55. $[3h + (k-1)]^2 = 9h^2 + 6h(k-1) + (k-1)^2$

$\quad = 9h^2 + 6hk - 6h + k^2 - 2k + 1$

57. $x^2 - 6x + 9x - 54 = x^2 + 3x - 54$

59. $5^2 - x^2 = 25 - x^2$

61. $6x^2 - 8ax + 15ax - 20a^2$

$\qquad\qquad = 6x^2 + 7ax - 20a^2$

63. $2t^2 + 2tw - 3t - 3w$

65. $(3x^2)^2 + 2(3x^2)(2y^3) + (2y^3)^2$
$$= 9x^4 + 12x^2y^3 + 4y^6$$

67. $6y^2 + 6y - 10y - 10 = 6y^2 - 4y - 10$

69. $50^2 - 2^2 = 2500 - 4 = 2496$

71. $3^2 + 2(3)(7x) + (7x)^2 = 49x^2 + 42x + 9$

73. $4y\left(3y + \frac{1}{2}\right)^2 = 4y\left((3y)^2 + 2(3y)\frac{1}{2} + \left(\frac{1}{2}\right)^2\right)$
$$= 4y\left(9y^2 + 3y + \frac{1}{4}\right) = 36y^3 + 12y^2 + y$$

75. $(a + h)^2 - a^2 = a^2 + 2ah + h^2 - a^2$
$$= 2ah + h^2$$

77. $(x + 2)(x + 2)^2 = (x + 2)(x^2 + 4x + 4)$
$$= x(x^2 + 4x + 4) + 2(x^2 + 4x + 4)$$
$$= x^3 + 6x^2 + 12x + 8$$

79. $(y - 3)^3 = (y - 3)(y^2 - 6y + 9)$
$$= y(y^2 - 6y + 9) - 3(y^2 - 6y + 9)$$
$$= y^3 - 9y^2 + 27y - 27$$

81. $(3.2)(5.1)x^2 - (4.5)(5.1)x + (3.9)(3.2)x$
$$- (4.5)(3.9)$$
$$= 16.32x^2 - 10.47x - 17.55$$

83. $(3.6y)^2 + 2(3.6)(4.4)y + (4.4)^2$
$$= 12.96y^2 + 31.68y + 19.36$$

85. $x^m x^{2m} + 2x^{2m} + 3x^m + 6$
$$= x^{3m} + 2x^{2m} + 3x^m + 6$$

87. $a^{n+1+2n} + a^{n+1+n} - 3a^{n+1}$
$$= a^{3n+1} + a^{2n+1} - 3a^{n+1}$$

89. $(a^m)^2 + 2a^m a^n + (a^n)^2$
$$= a^{2m} + 2a^{m+n} + a^{2n}$$

91. $5y^m 3y^{2m} + 3 \cdot 8y^{2m}z^k + 4 \cdot 5y^m z^{3-k}$
$$+ 32z^{k+3-k}$$
$$= 15y^{3m} + 24y^{2m}z^k + 20\,y^m z^{3-k} + 32z^3$$

93. Use $L = x + 3$ and $W = x + 1$ in the formula for the area of a rectangle, $A = LW$.
$A = (x + 3)(x + 1) = x^2 + 4x + 3$

95. $A = (8 - 2x)(10 - 2x) = 4x^2 - 36x + 80$
If $x = 0.4$, then $A = 4(0.4)^2 - 36(0.4) + 80$
$= 66.24$ km^2

97. $V = x(4 - 2x)(6 - 2x) = x(4x^2 - 20x + 24)$
$= 4x^3 - 20x^2 + 24x$

If $x = 4$ in. $= \frac{1}{3}$ ft, then
$V = 4\left(\frac{1}{3}\right)^3 - 20\left(\frac{1}{3}\right)^2 + 24\left(\frac{1}{3}\right) = 5\frac{25}{27}$ ft$^3 \approx 5.9$ ft^3

3.5 WARM-UPS

1. False, c is the quotient. **2.** False, (quotient)(divisor) + remainder = dividend.
3. True, because $(x + 2)(x + 3) = x^2 + 5x + 6$ and $(x + 2)(x + 3) + 1 = x^2 + 5x + 7$.
4. True. From warm-up number 3 we can see that if $x + 3$ is the divisor then $x + 2$ is the quotient.
5. True. From warm-up number 2 we can see that if $x + 2$ is the divisor then $x + 3$ is the quotient and 1 is the remainder.
6. False, to divide by $x - c$ with synthetic division we use c. **7.** False, synthetic division is used only for dividing by $x - c$. **8.** True, when dividing by $x - c$ the degree of the quotient is 1 less than the degree of the dividend.
9. True, because if remainder is zero, then (quotient)(divisor) = dividend.
10. True, because if remainder is zero, then (quotient)(divisor) = dividend.

3.5 EXERCISES

1. $\frac{36x^7}{3x^3} = 12x^4$

3. $\frac{16x^2}{-8x^2} = -2$

5. $\frac{6b - 9}{3} = \frac{6b}{3} - \frac{9}{3} = 2b - 3$

7. $\frac{3x^2 + 6x}{3x} = \frac{3x^2}{3x} + \frac{6x}{3x} = x + 2$

9. $\frac{10x^4 - 8x^3 + 6x^2}{-2x^2} = \frac{10x^4}{-2x^2} - \frac{8x^3}{-2x^2} + \frac{6x^2}{-2x^2}$
$$= -5x^2 + 4x - 3$$

11. $\frac{7x^3 - 4x^2}{2x} = \frac{7x^3}{2x} - \frac{4x^2}{2x} = \frac{7}{2}x^2 - 2x$

13.
$$
\begin{array}{r}
x + 5 \\
x + 3 \overline{\smash{\big)}\ x^2 + 8x + 13} \\
\underline{x^2 + 3x} \\
5x + 13 \\
\underline{5x + 15} \\
-2
\end{array}
$$

Quotient: $x + 5$ Remainder: -2

15.

$$
\require{enclose}
\begin{array}{r}
x-4 \\
x+2 \enclose{longdiv}{\,x^2-2x} \\
\underline{x^2+2x} \\
-4x \\
\underline{-4x-8} \\
8
\end{array}
$$

Quotient: $x-4$ Remainder: 8

17.

$$
\begin{array}{r}
x^2-2x+4 \\
x+2 \enclose{longdiv}{\,x^3+0x^2+0x+8} \\
\underline{x^3+2x^2} \\
-2x^2+0x \\
\underline{-2x^2-4x} \\
4x+8 \\
\underline{4x+8} \\
0
\end{array}
$$

Quotient: x^2-2x+4 Remainder: 0

19.

$$
\begin{array}{r}
a^2+2a+8 \\
a-2 \enclose{longdiv}{\,a^3+0a^2+4a-5} \\
\underline{a^3-2a^2} \\
2a^2+4a \\
\underline{2a^2-4a} \\
8a-5 \\
\underline{8a-16} \\
11
\end{array}
$$

Quotient: a^2+2a+8 Remainder: 11

21.

$$
\begin{array}{r}
x^2-2x+3 \\
x+1 \enclose{longdiv}{\,x^3-x^2+x-3} \\
\underline{x^3+x^2} \\
-2x^2+x \\
\underline{-2x^2-2x} \\
3x-3 \\
\underline{3x+3} \\
-6
\end{array}
$$

Quotient: $x^2 - 2x + 3$ Remainder: -6

23.

$$
\begin{array}{r}
x^3+3x^2+6x+11 \\
x-2 \enclose{longdiv}{\,x^4+x^3+0x^2-x-1} \\
\underline{x^4-2x^3} \\
3x^3+0x^2 \\
\underline{3x^3-6x^2} \\
6x^2-x \\
\underline{6x^2-12x} \\
11x-1 \\
\underline{11x-22} \\
21
\end{array}
$$

Quotient: $x^3+3x^2+6x+11$ Remainder: 21

25.

$$
\begin{array}{r}
-3x^2-1 \\
x^2-2 \enclose{longdiv}{\,-3x^4+0x^3+5x^2+x-2} \\
\underline{-3x^4+6x^2} \\
-x^2+x-2 \\
\underline{-x^2+2} \\
x-4
\end{array}
$$

Quotient: $-3x^2-1$ Remainder: $x-4$

27.

$$
\begin{array}{r}
\tfrac{1}{2}x-\tfrac{5}{4} \\
2x-3 \enclose{longdiv}{\,x^2-4x+2} \\
\underline{x^2-\tfrac{3}{2}x} \\
-\tfrac{5}{2}x+2 \\
\underline{-\tfrac{5}{2}x+\tfrac{15}{4}} \\
-\tfrac{7}{4}
\end{array}
$$

Quotient: $\frac{1}{2}x-\frac{5}{4}$ Remainder: $-\frac{7}{4}$

29.

$$
\begin{array}{r}
\tfrac{2}{3}x+\tfrac{1}{9} \\
3x-2 \enclose{longdiv}{\,2x^2-x+6} \\
\underline{2x^2-\tfrac{4}{3}x} \\
\tfrac{1}{3}x+6 \\
\underline{\tfrac{1}{3}x-\tfrac{2}{9}} \\
\tfrac{56}{9}
\end{array}
$$

Quotient: $\frac{2}{3}x+\frac{1}{9}$ Remainder: $\frac{56}{9}$

31.

$$\begin{array}{r} 2 \\ x-5 \overline{)\, 2x+0\,} \\ 2x-10 \\ \hline 10 \end{array}$$

$$\frac{2x}{x-5} = 2 + \frac{10}{x-5}$$

33.

$$\begin{array}{r} x-1 \\ x+1 \overline{)\, x^2+0x+0\,} \\ x^2+\ x \\ \hline -x+0 \\ -x-1 \\ \hline 1 \end{array}$$

$$\frac{x^2}{x+1} = x-1 + \frac{1}{x+1}$$

35.

$$\begin{array}{r} x^2-2x+4 \\ x+2 \overline{)\, x^3+0x^2+0x+0\,} \\ x^3+2x^2 \\ \hline -2x^2+0x \\ -2x^2\ -4x \\ \hline 4x+0 \\ 4x+8 \\ \hline -8 \end{array}$$

$$\frac{x^3}{x+2} = x^2-2x+4 + \frac{-8}{x+2}$$

37. $\dfrac{x^3+\ 2x}{x^2} = \dfrac{x^3}{x^2} + \dfrac{2x}{x^2} = x + \dfrac{2}{x}$

39.

$$\begin{array}{r} x-6 \\ x+2 \overline{)\, x^2-4x+9\,} \\ x^2+2x \\ \hline -6x+9 \\ -6x-12 \\ \hline 21 \end{array}$$

$$\frac{x^2-4x+9}{x+2} = x-6 + \frac{21}{x+2}$$

41.

$$\begin{array}{r} 3x^2-\ x-1 \\ x-1 \overline{)\, 3x^3-4x^2+0x+7\,} \\ 3x^3-3x^2 \\ \hline -x^2+0x \\ -x^2+\ x \\ \hline -x+7 \\ -x+1 \\ \hline 6 \end{array}$$

$$\frac{3x^3-4x^2+7}{x-1} = 3x^2-x-1 + \frac{6}{x-1}$$

43.

$$\begin{array}{r|rrrr} 2 & 1 & -5 & 6 & -3 \\ & & 2 & -6 & 0 \\ \hline & 1 & -3 & 0 & -3 \end{array}$$

The quotient is x^2-3x and the remainder is -3.

45.

$$\begin{array}{r|rrr} -1 & 2 & -4 & 5 \\ & & -2 & 6 \\ \hline & 2 & -6 & 11 \end{array}$$

The quotient is $2x-6$ and the remainder is 11.

47.

$$\begin{array}{r|rrrrr} 3 & 3 & 0 & -15 & 7 & -9 \\ & & 9 & 27 & 36 & 129 \\ \hline & 3 & 9 & 12 & 43 & 120 \end{array}$$

The quotient is $3x^3 + 9x^2 + 12x + 43$ and the remainder is 120.

49.

$$\begin{array}{r|rrrrrr} 1 & 1 & 0 & 0 & 0 & 0 & -1 \\ & & 1 & 1 & 1 & 1 & 1 \\ \hline & 1 & 1 & 1 & 1 & 1 & 0 \end{array}$$

The quotient is $x^4 + x^3 + x^2 + x + 1$ and the remainder is 0.

51.

$$\begin{array}{r|rrrr} -2 & 1 & 0 & -5 & 6 \\ & & -2 & 4 & 2 \\ \hline & 1 & -2 & -1 & 8 \end{array}$$

The quotient is $x^2 - 2x - 1$ and the remainder is 8.

53.

$$\begin{array}{r|rrr} 0.32 & 2.3 & -0.14 & 0.6 \\ & & 0.736 & 0.19072 \\ \hline & 2.3 & 0.596 & 0.79072 \end{array}$$

The quotient is $2.3x + 0.596$ and the remainder is 0.79072.

55. If we divide the polynomials we get a quotient of $x^2 - 3x + 1$ and remainder 4. So the first polynomial is not a factor of the second.

57. If we divide the polynomials we get a quotient of $x^2 + 4x + 3$ and remainder 0. So the first polynomial is a factor of the second.

59. If we divide the polynomials we get a quotient of $x^2 - 2$ and remainder 0. So the first polynomial is a factor of the second.

61. Divide the polynomials to get a quotient of $9w^2 - 3w + 1$ and remainder 0. So the first polynomial is a factor of the second.

40

63. If we divide the polynomials we get a quotient of $a^2 + 5a + 25$ and remainder 0. So the first polynomial is a factor of the second.

65. If we divide the polynomials we get a quotient of $x^2 + 3x + 2$ and remainder 0. So the first polynomial is a factor of the second.

67. Divide $x^2 - 6x + 8$ by $x - 4$:

$$
\begin{array}{r|rrr}
4 & 1 & -6 & 8 \\
  &   & 4 & -8 \\
\hline
  & 1 & -2 & 0
\end{array}
$$

So $x^2 - 6x + 8 = (x - 4)(x - 2)$.

69. Divide $w^3 - 27$ by $w - 3$:

$$
\begin{array}{r|rrrr}
3 & 1 & 0 & 0 & -27 \\
  &   & 3 & 9 & 27 \\
\hline
  & 1 & 3 & 9 & 0
\end{array}
$$

So $w^3 - 27 = (w - 3)(w^2 + 3w + 9)$.

71. Divide $x^3 - 4x^2 + 6x - 4$ by $x - 2$:

$$
\begin{array}{r|rrrr}
2 & 1 & -4 & 6 & -4 \\
  &   & 2 & -4 & 4 \\
\hline
  & 1 & -2 & 2 & 0
\end{array}
$$

So $x^3 - 4x^2 + 6x - 4 = (x^2 - 2x + 2)(x - 2)$.

73. Divide $z^2 + 6z + 9$ by $z + 3$:

$$
\begin{array}{r|rrr}
-3 & 1 & 6 & 9 \\
   &   & -3 & -9 \\
\hline
   & 1 & 3 & 0
\end{array}
$$

So $z^2 + 6z + 9 = (z + 3)(z + 3)$.

75. Divide $6y^2 + 5y + 1$ by $2y + 1$:

$$
\begin{array}{r}
3y + 1 \\
2y + 1 \overline{\smash{)}\, 6y^2 + 5y + 1} \\
\underline{6y^2 + 3y} \\
2y + 1 \\
\underline{2y + 1} \\
0
\end{array}
$$

So $6y^2 + 5y + 1 = (2y + 1)(3y + 1)$.

77. Divide $0.03x^2 + 300x$ by x to get $AC(x) = 0.03x + 300$.

$AC(12) = 0.03(12) + 300 = \$300.36$

79. Divide $x^2 - 1$ by $x + 1$ to get $x - 1$. The width is $x - 1$ ft.

81. $V = \dfrac{10(30^3 - 20^3)}{3(30 - 20)} \approx 6{,}333.3$ m^3

3.6 WARM-UPS

1. True, $3xy(x - 2y) = -3xy(-x + 2y)$.

2. True.   **3.** True, because $-2(2 - x)$ $= -4 + 2x = 2x - 4$ for any value of x.

4. True, $a^2 - b^2 = (a + b)(a - b)$.

5. False, because $x^2 + 12x + 36$ is a perfect square trinomial.   **6.** False, because $y^2 + 8y + 16$ is a perfect square trinomial.

7. False, because $(3x + 7)^2 = 9x^2 + 42x + 49$.

8. True, $x^3 + 1 = (x + 1)(x^2 - x + 1)$.

9. False, $x^3 - 27 = (x - 3)(x^2 + 3x + 9)$.

10. False, $x^3 - 8 = (x - 2)(x^2 + 2x + 4)$.

3.6 EXERCISES

1. Since $48 = 2 \cdot 2 \cdot 2 \cdot 2 \cdot 3$ and $36x = 2 \cdot 2 \cdot 3 \cdot 3x$, the greatest common factor is $2 \cdot 2 \cdot 3 = 12$.

3. The gcf for 9, 21, and 15 is 3. Since there are no variables in common to all three terms, the gcf is 3.

5. Since $24 = 2 \cdot 2 \cdot 2 \cdot 3$, $42 = 2 \cdot 3 \cdot 7$, and $66 = 2 \cdot 3 \cdot 11$, the GCF of 24, 42, and 66 is $2 \cdot 3$ or 6. Using each common variable with the lowest power, we get the GCF 6xy.

7. $x^3 - 5x = x \cdot x^2 - x \cdot 5 = x(x^2 - 5)$

9. $48wx + 36wy \quad = 12w \cdot 4x + 12w \cdot 3y$
$$= 12w(4x + 3y)$$

11. $2x \cdot x^2 - 2x \cdot 2x + 2x \cdot 3 = 2x(x^2 - 2x + 3)$

13. $36a^3b^6 - 24a^4b^2 + 60a^5b^3$
$$= 12a^3b^2(3b^4 - 2a + 5a^2b)$$

15. Factor out the quantity $x - 6$:
$$(x - 6)a + (x - 6)b = (x - 6)(a + b)$$

17. Factor out $(y - 1)^2$:
$$(y - 1)^2 y + (y - 1)^2 z = (y - 1)^2(y + z)$$

19. $2 \cdot x - 2 \cdot y = 2(x - y)$
$$-2(-x) - 2 \cdot y = -2(-x + y)$$

21. $6x^2 - 3x = 3x(2x - 1)$
$$= -3x(-2x + 1)$$

23. $w^2 \cdot (-w) + w^2(3) = w^2(-w + 3)$
$$-w^2 \cdot w - (-w^2) \cdot 3 = -w^2(w - 3)$$

25. $-a^3 + a^2 - 7a = a(-a^2 + a - 7)$
$$= -a(a^2 - a + 7)$$

27. The polynomial is a difference of two squares and so it factors as the product of a sum and a difference: $(x - 10)(x + 10)$

29. $4y^2 - 49 \qquad = (2y)^2 - 7^2$
$$= (2y - 7)(2y + 7)$$

31. $9x^2 - 25a^2 \qquad = (3x)^2 - (5a)^2$
$$= (3x - 5a)(3x + 5a)$$

33. $(12wz)^2 - 1^2 = (12wz - 1)(12wz + 1)$

35. This polynomial is a perfect square trinomial. The middle term is twice the product of the square roots of the first term and last term. $x^2 - 2(x)(10) + 10^2 = (x - 10)^2$

37. $4m^2 - 4m + 1 = (2m)^2 - 2 \cdot 2m + 1^2$
$$= (2m - 1)^2$$

39. $w^2 - 2wt + t^2 = (w - t)^2$

41. $a^3 - 1^3 = (a - 1)(a^2 + a + 1)$

43. $w^3 + 3^3 = (w + 3)(w^2 - 3w + 9)$

45. $(2x)^3 - 1^3 = (2x - 1)(4x^2 + 2x + 1)$

47. $a^3 + 2^3 = (a + 2)(a^2 - 2a + 4)$

49. $2(x^2 - 4) = 2(x + 2)(x - 2)$

51. $x(x^2 + 10x + 25) = x(x + 5)^2$

53. $(2x)^2 + 2(2x)(1) + 1^2 = (2x + 1)^2$

55. $(x + 3)(x + 7)$

57. $3y \cdot 2y + 3y \cdot 1 = 3y(2y + 1)$

59. $(2x)^2 - 2(2x)(5) + 5^2 = (2x - 5)^2$

61. $2m^4 - 2mn^3 = 2m(m^3 - n^3)$
$$= 2m(m - n)(m^2 + mn + n^2)$$

63. $(2x - 3)(x - 2)$

65. $a(9a^2 - w^2) = a(3a + w)(3a - w)$

67. $-5(a^2 - 6a + 9) = -5(a - 3)^2$

69. $16 - 54x^3 = 2(8 - 27x^3)$
$$= 2(2 - 3x)(4 + 6x + 9x^2)$$

71. $-3y(y^2 + 6y + 9) = -3y(y + 3)^2$

73. $-7(a^2b^2 - 1) = -7(ab + 1)(ab - 1)$

75. If $a = x^5$ then
$$x^{10} - 9 = (x^5)^2 - 3^2 = a^2 - 3^2$$
$$= (a - 3)(a + 3) = (x^5 - 3)(x^5 + 3)$$

77. $z^{12} - 6z^6 + 9 = (z^6 - 3)^2$

79. $2x^7 + 8x^4 + 8x = 2x(x^6 + 4x^3 + 4)$
$$= 2x(x^3 + 2)^2$$

81. $4x^5 + 4x^3 + x = x(4x^4 + 4x^2 + 1)$
$$= x(2x^2 + 1)^2$$

83. $x^6 - 8 = (x^2)^3 - 2^3$
$$= (x^2 - 2)(x^4 + 2x^2 + 4)$$

85. $2x^9 + 16 = 2(x^9 + 8) = 2\big((x^3)^3 + 2^3\big)$
$$= 2(x^3 + 2)(x^6 - 2x^3 + 4)$$

87. $a^{2n} - 1 = (a^n)^2 - 1^2 = (a^n - 1)(a^n + 1)$

89. $a^{2r} + 6a^r + 9 = (a^r)^2 + 6a^r + 3^2$
$$= (a^r + 3)^2$$

91. $x^{3m} - 8 = (x^m)^3 - 2^3$
$$= (x^m - 2)(x^{2m} + 2x^m + 4)$$

93. $a^{3m} - b^3 = (a^m)^3 - b^3$
$$= (a^m - b)(a^{2m} + a^m b + b^2)$$

95. $k^{2w + 1} - 10k^{w + 1} + 25k$
$$= k(k^{2w} - 10k^w + 25) = k(k^w - 5)^2$$

97. $uv^{6k} - 2u^2v^{4k} + u^3v^{2k}$
$$= uv^{2k}(v^{4k} - 2uv^{2k} + u^2) = uv^{2k}(v^{2k} - u)^2$$

99. If we choose $k = 9$, then $x^2 + 6x + k$ will be the perfect square $x^2 + 6x + 9$.

101. If we choose $k = 20$, then $4a^2 - ka + 25$ will be the perfect square $4a^2 - 20a + 25$.

103. If we choose $k = 16$, then $km^2 - 24m + 9$ will be the perfect square $16m^2 - 24m + 9$.

105. Choose $k = 100$ so that $81y^2 - 180y + k$ is the perfect square $81y^2 - 180y + 100$

107. $16b^3 + 24b^2 + 9b = b(16b^2 + 24b + 9)$
$$= b(4b + 3)^2$$

Since $V = LWH$ and the height is b, the sides of the square bottom are each $4b + 3$ cm.

3.7 WARM-UPS

1. True. Check by multiplying $(x + 3)(x + 6)$.
2. False, $y^2 + 2y - 35 = (y + 7)(y - 5)$.
3. False, $x^2 + 4$ cannot be factored. It is a sum of two squares.
4. False, $x^2 - 5x - 6 = (x - 6)(x + 1)$.
5. True. Check by multiplying $(x - 6)(x + 2)$.
6. False, $x^2 + 15x + 36 = (x + 12)(x + 3)$.
7. False, $3x^2 + 4x - 15 = (3x - 5)(x + 3)$.
8. False, $4x^2 + 4x - 3 = (2x + 3)(2x - 1)$.
9. True. Check by multiplying.
10. True. Check by multiplying.

3.7 EXERCISES

1. Two numbers that have a product of 3 and a sum of 4 are 1 and 3: $(x + 1)(x + 3)$
3. Two numbers that have a product of 50 and a sum of 15 are 5 and 10: $(a + 10)(a + 5)$
5. Two numbers that have a product of -14 and a sum of -5 are -7 and 2: $(y - 7)(y + 2)$
7. Two numbers that have a product of 8 and a sum of -6 are -2 and -4: $(x - 2)(x - 4)$
9. Two numbers with product 27 and sum -12 are -9 and -3: $a^2 - 12a + 27 = (a - 9)(a - 3)$
11. Two numbers that have a product of -30 and a sum of 7 are -3 and 10: $(a + 10)(a - 3)$
13. Two numbers with a product of $6 \cdot 1 = 6$ and a sum of 5 are 2 and 3:
$$
\begin{aligned}
6w^2 + 5w + 1 &= 6w^2 + 3w + 2w + 1 \\
&= 3w(2w + 1) + 1(2w + 1) \\
&= (3w + 1)(2w + 1)
\end{aligned}
$$

15. Using the AC method we find two numbers that have a product of $2(-3) = -6$ and a sum of -5. These numbers are -6 and 1:
$$
\begin{aligned}
2x^2 - 5x - 3 &= 2x^2 - 6x + 1x - 3 \\
&= 2x(x - 3) + 1(x - 3) \\
&= (2x + 1)(x - 3)
\end{aligned}
$$
17. Two numbers with a product of $4(15) = 60$ and a sum of 16 are 6 and 10:
$$
\begin{aligned}
4x^2 + 16x + 15 &= 4x^2 + 6x + 10x + 15 \\
&= 2x(2x + 3) + 5(2x + 3) \\
&= (2x + 5)(2x + 3)
\end{aligned}
$$
19. Two numbers with a product of $6 \cdot 1 = 6$ and a sum of -5 are -2 and -3:
$$
\begin{aligned}
6x^2 - 5x + 1 &= 6x^2 - 2x - 3x + 1 \\
&= 2x(3x - 1) - 1(3x - 1) \\
&= (2x - 1)(3x - 1)
\end{aligned}
$$

21. Two numbers with a product of $12(-1) = -12$ and a sum of 1 are 4 and -3:
$$
\begin{aligned}
12y^2 + y - 1 &= 12y^2 - 3y + 4y - 1 \\
&= 3y(4y - 1) + 1(4y - 1) \\
&= (3y + 1)(4y - 1)
\end{aligned}
$$

23. Two numbers with a product of $-5(6) = -30$ and a sum of 1 are 6 and -5:
$$
\begin{aligned}
6a^2 + a - 5 &= 6a^2 + 6a - 5a - 5 \\
&= 6a(a + 1) - 5(a + 1) \\
&= (6a - 5)(a + 1)
\end{aligned}
$$

25. $2x^2 + 15x - 8 = (2x - 1)(x + 8)$
27. $3b^2 - 16b - 35 = (3b + 5)(b - 7)$
29. $6w^2 - 35w + 36 = (3w - 4)(2w - 9)$
31. $4x^2 - 5x + 1 = (4x - 1)(x - 1)$
33. $5m^2 + 13m - 6 = (5m - 2)(m + 3)$
35. $6y^2 - 7y - 20 = (3y + 4)(2y - 5)$
37. $x^6 - 2x^3 - 35 = (x^3 + 5)(x^3 - 7)$
39. $a^{20} - 20a^{10} + 100 = (a^{10} - 10)^2$
41. $-12a^5 - 10a^3 - 2a = -2a(6a^4 + 5a^2 + 1)$
$= -2a(3a^2 + 1)(2a^2 + 1)$
43. Two numbers with a product of -15 and a sum of 2 are 5 and -3:
$$
x^{2a} + 2x^a - 15 = (x^a + 5)(x^a - 3)
$$
45. $x^{2a} - y^{2b} = (x^a)^2 - (y^b)^2$
$$
= (x^a - y^b)(x^a + y^b)
$$
47. $x^8 - x^4 - 6 = (x^4 - 3)(x^4 + 2)$
49. $x^a(x^2 - 1) = x^a(x - 1)(x + 1)$
51. $(x^a)^2 + 2 \cdot x^a \cdot 3 + 3^2 = (x^a + 3)^2$
53. $4y^3z^6 + 5z^3y^2 - 6y = y(4y^2z^6 + 5z^3y - 6)$
$= y(4yz^3 - 3)(yz^3 + 2)$
55. $2x^2 + 20x + 50 = 2(x^2 + 10x + 25)$
$$
= 2(x + 5)^2
$$
57. $a^3 - 36a = a(a^2 - 36) = a(a - 6)(a + 6)$
59. $5(a^2 + 5a - 6) = 5(a + 6)(a - 1)$
61. $2(x^2 - 64y^2) = 2(x + 8y)(x - 8y)$
63. $-3(x^2 - x - 12) = -3(x + 3)(x - 4)$
65. $m^3(m^2 + 20m + 100) = m^3(m + 10)^2$
67. Two numbers with a product of $6 \cdot 20 = 120$ and a sum of 23 are 8 and 15:
$$
\begin{aligned}
6x^2 + 23x + 20 &= 6x^2 + 15x + 8x + 20 \\
&= 3x(2x + 5) + 4(2x + 5) \\
&= (3x + 4)(2x + 5)
\end{aligned}
$$

69. $y^2 - 2 \cdot 6y + 6^2 = (y - 6)^2$
71. $(3m)^2 - (5n)^2 = (3m - 5n)(3m + 5n)$

73. $5(a^2 + 4a - 12) = 5(a + 6)(a - 2)$

75. $-2(w^2 - 9w - 10) = -2(w - 10)(w + 1)$

77. $x^2(w^2 - 100) = x^2(w + 10)(w - 10)$

79. $(3x)^2 - 1^2 = (3x + 1)(3x - 1)$

81. Two numbers with a product of $-15 \cdot 8 = -120$ and a sum of -2 are -12 and 10:

$$8x^2 - 2x - 15 \quad = 8x^2 - 12x + 10x - 15$$
$$= 4x(2x - 3) + 5(2x - 3)$$
$$= (4x + 5)(2x - 3)$$

83. $(2x)^2 - 2 \cdot 10x + 5^2 = (2x - 5)^2$

85. $(3a)^2 + 2 \cdot 3a \cdot 10 + 10^2 = (3a + 10)^2$

87. $4(a^2 + 6a + 8) = 4(a + 2)(a + 4)$

89. $3m^4 - 24m = 3m(m^3 - 8)$

$$= 3m(m - 2)(m^2 + 2m + 4)$$

3.8 WARM-UPS

1. False, $x^2 - 9 = (x - 3)(x + 3)$ for any value of x. **2.** True, because $4x^2 + 12x + 9 = (2x + 3)^2$.
3. True. **4.** False, because $x^2 - 4$ is not a prime polynomial. **5.** False, because $y^3 - 27 = (y - 3)(y^2 + 3y + 9)$ for any value of y.
6. True, because $y^6 - 1 = (y^3)^2 - 1^2$.
7. False, because $2x - 4$ is not prime.
8. True, because it cannot be factored.
9. True, because $a^6 - 1 = (a^2)^3 - 1^3$.
10. False, because if we factor x out of the first two terms and a out of the last two terms we get $x^2 + 3x - ax + 3a = x(x + 3) + a(-x + 3)$ and we cannot finish. If we factor x out of the first two terms and $-a$ out of the last two terms we get $x(x + 3) - a(x - 3)$ and again we cannot finish the factoring.

3.8 EXERCISES

1. Prime, because it is a sum of two squares.

3. Not prime, because $-9w^2 - 9 = -9(w^2 + 1)$.

5. Not prime, $x^2 - 2x - 3 = (x - 3)(x + 1)$.

7. Prime, because no two integers have a product of 3 and a sum of 2.

9. Prime, because no two numbers have a product of -3 and a sum of -4.

11. Prime, because no two numbers have a product of $6 \cdot (-4) = -24$ and a sum of 3.

13. Let $y = a^2$ in the polynomial:

$$a^4 - 10a^2 + 25 = y^2 - 10y + 25$$
$$= (y - 5)^2 = (a^2 - 5)^2$$

15. Let $y = x^2$ in the polynomial:

$$x^4 - 6x^2 + 8 = y^2 - 6y + 8 = (y - 2)(y - 4)$$
$$= (x^2 - 2)(x^2 - 4) = (x^2 - 2)(x - 2)(x + 2)$$

17. Let $y = 3x - 5$ in the polynomial:

$$(3x - 5)^2 - 1 = y^2 - 1 = (y - 1)(y + 1)$$
$$= (3x - 5 - 1)(3x - 5 + 1) = (3x - 6)(3x - 4)$$
$$= 3(x - 2)(3x - 4)$$

19. If we let $a = y^3$, then $a^2 = (y^3)^2 = y^6$.

$$2y^6 - 128 = 2(y^6 - 64) = 2(a^2 - 64)$$
$$= 2(a - 8)(a + 8) = 2(y^3 - 8)(y^3 + 8)$$
$$= 2(y - 2)(y^2 + 2y + 4)(y + 2)(y^2 - 2y + 4)$$

21. $32a^4 - 18 = 2(16a^4 - 9)$

$$= 2(4a^2 + 3)(4a^2 - 3)$$

23. Let $a = x^2$ and $b = x - 6$:

$$x^4 - (x - 6)^2 = a^2 - b^2 = (a + b)(a - b)$$
$$= (x^2 + x - 6)(x^2 - [x - 6])$$
$$= (x + 3)(x - 2)(x^2 - x + 6)$$

25. Let $y = m + 2$ in the polynomial:

$$(m + 2)^2 + 2(m + 2) - 3 = y^2 + 2y - 3$$
$$= (y + 3)(y - 1) = (m + 2 + 3)(m + 2 - 1)$$
$$= (m + 5)(m + 1)$$

27. Let $w = y - 1$ in the polynomial:

$$3(y - 1)^2 + 11(y - 1) - 20 = 3w^2 + 11w - 20$$
$$= (3w - 4)(w + 5) = [3(y - 1) - 4][y - 1 + 5]$$
$$= (3y - 7)(y + 4)$$

29. Let $a = y^2 - 3$:

$$(y^2 - 3)^2 - 4(y^2 - 3) - 12 = a^2 - 4a - 12$$
$$= (a - 6)(a + 2) = (y^2 - 3 - 6)(y^2 - 3 + 2)$$
$$= (y^2 - 9)(y^2 - 1) = (y - 3)(y + 3)(y - 1)(y + 1)$$

31. $ax + ay + bx + by = a(x+y) + b(x+y)$
$$= (a+b)(x+y)$$

33. $x^3 + x^2 - 9x - 9 = x^2(x+1) - 9(x+1)$
$$= (x^2 - 9)(x+1) = (x-3)(x+3)(x+1)$$

35. $aw - bw - 3a + 3b = w(a-b) - 3(a-b)$
$$= (w-3)(a-b)$$

37. $a^4 + 3a^3 + 27a + 81 = a^3(a+3) + 27(a+3)$
$$= (a^3 + 27)(a+3) = (a+3)(a^2 - 3a + 9)(a+3)$$
$$= (a+3)^2(a^2 - 3a + 9)$$

39. $y^4 - 5y^3 + 8y - 40 = y^3(y-5) + 8(y-5)$
$$= (y-5)(y^3 + 8) = (y-5)(y+2)(y^2 - 2y + 4)$$

41. $ady + d - w - awy = d(ay+1) - w(ay+1)$
$$= (d-w)(ay+1)$$

43. $x^2 y - y + ax^2 - a = y(x^2 - 1) + a(x^2 - 1)$
$$= (a+y)(x^2 - 1) = (a+y)(x-1)(x+1)$$

45. $y^4 + y + by^3 + b = y(y^3 + 1) + b(y^3 + 1)$
$$= (y+b)(y^3 + 1) = (y+b)(y+1)(y^2 - y + 1)$$

47. This is a perfect square trinomial.
$$9x^2 - 24x + 16 = (3x-4)^2$$

49. Two numbers whose product is 36 and whose sum is -13 are -4 and -9.
$$12x^2 - 13x + 3 = 12x^2 - 4x - 9x + 3$$
$$= 4x(3x-1) - 3(3x-1) = (3x-1)(4x-3)$$

51. $3a^4 + 81a = 3a(a^3 + 27)$
$$= 3a(a+3)(a^2 - 3a + 9)$$

53. $32 + 2x^2 = 2(x^2 + 16)$

55. Prime, because there are no two numbers that have a product of 72 and a sum of -5.

57. Let $a = x + y$: $(x+y)^2 - 1 = a^2 - 1$
$$= (a-1)(a+1) = (x+y-1)(x+y+1)$$

59. $a^3 b - ab^3 = ab(a^2 - b^2) = ab(a-b)(a+b)$

61. $x^4 + 2x^3 - 8x - 16 = x^3(x+2) - 8(x+2)$
$$= (x+2)(x^3 - 8) = (x+2)(x-2)(x^2 + 2x + 4)$$

63. $m^2 n + 2mn^2 + n^3 = n(m^2 + 2mn + n^2)$
$$= n(m+n)^2$$

65. $2m + 2n + wm + wn = 2(m+n) + w(m+n)$
$$= (2+w)(m+n)$$

67. Two numbers with a product of -12 and a sum of 4 are 6 and -2:
$$4w^2 + 4w - 3 = 4w^2 - 2w + 6w - 3$$
$$= 2w(2w-1) + 3(2w-1) = (2w+3)(2w-1)$$

69. Let $a = t^2$: $t^4 + 4t^2 - 21 = a^2 + 4a - 21$
$$= (a+7)(a-3) = (t^2 + 7)(t^2 - 3)$$

71. $-a^3 - 7a^2 + 30a = -a(a^2 + 7a - 30)$
$$= -a(a+10)(a-3)$$

73. Let $a = y + 5$: $(y+5)^2 - 2(y+5) - 3$
$$= a^2 - 2a - 3 = (a-3)(a+1)$$
$$= (y+5-3)(y+5+1) = (y+2)(y+6)$$

75. $-2w^4 + 1250 = -2(w^4 - 625)$
$$= -2(w^2 - 25)(w^2 + 25)$$
$$= -2(w-5)(w+5)(w^2 + 25)$$

77. $8a^3 + 8a = 8a(a^2 + 1)$

79. Let $y = w + 5$: $(w+5)^2 - 9 = y^2 - 9$
$$= (y-3)(y+3) = (w+5-3)(w+5+3)$$
$$= (w+2)(w+8)$$

81. $4aw^2 - 12aw + 9a = a(4w^2 - 12w + 9)$
$$= a(2w-3)^2$$

83. $x^2 - 6x + 9 = (x-3)^2$

85. $3x^4 - 75x^2 = 3x^2(x^2 - 25)$
$$= 3x^2(x-5)(x+5)$$

87. $m^3 n - n = n(m^3 - 1)$
$$= n(m-1)(m^2 + m + 1)$$

89. $12x^2 + 2x - 30 = 2(6x^2 + x - 15)$
$$= 2[6x^2 - 9x + 10x - 15]$$
$$= 2[3x(2x-3) + 5(2x-3)] = 2(3x+5)(2x-3)$$

91. $2a^3 - 32 = 2(a^3 - 16)$

93. $a^{3m} - 1 = (a^m)^3 - 1^3$
$$= (a^m - 1)(a^{2m} + a^m + 1)$$

95. $a^{3w} - b^{6n} = (a^w)^3 - (b^{2n})^3$

$\qquad = (a^w - b^{2n})(a^{2w} + a^w b^{2n} + b^{4n})$

97. $t^{8n} - 16 = (t^{4n})^2 - 4^2 = (t^{4n} - 4)(t^{4n} + 4)$

$= (t^{2n} - 2)(t^{2n} + 2)(t^{4n} + 4)$

99. $a^{2n+1} - 2a^{n+1} - 15a = a(a^{2n} - 2a^n - 15)$

$\qquad = a(a^n - 5)(a^n + 3)$

101. $a^{2n} - 3a^n + a^n b - 3b = a^n(a^n - 3) + b(a^n - 3)$

$\qquad = (a^n - 3)(a^n + b)$

3.9 WARM-UPS

1. False, because 4 is a solution to
$x - 1 = 3$, but $(4-1)(4+3) \neq 12$.
2. True, both 2 and 3 satisfy $(x-2)(x-3) = 0$.
3. True, because of the zero factor property.
4. False, $|x^2 + 4| = 5$ is equivalent to

$x^2 + 4 = 5$ or $x^2 + 4 = -5$.
5. True, because of the zero factor property.
6. False, the Pythagorean theorem applies only to right triangles.
7. True, because the sum of the length and width of a rectangle is one-half of the perimeter.
8. True, $x + (8 - x) = 8$ for any number x.
9. False, the solution set also includes 0 because of the factor x.
10. False, because 3 is not a solution.

3.9 EXERCISES

1. $(x-5)(x+4) = 0$
$\qquad x - 5 = 0$ or $x + 4 = 0$
$\qquad x = 5$ or $\qquad x = -4$
Solution set: $\{-4, 5\}$

3. $(2x-5)(3x+4) = 0$
$\qquad 2x - 5 = 0$ or $3x + 4 = 0$
$\qquad 2x = 5$ or $\qquad 3x = -4$
$\qquad x = \frac{5}{2}$ or $\qquad x = -\frac{4}{3}$
Solution set: $\left\{ \frac{5}{2}, -\frac{4}{3} \right\}$

5. $w^2 + 5w - 14 = 0$
$\qquad (w+7)(w-2) = 0$
$\qquad w + 7 = 0$ or $w - 2 = 0$
$\qquad w = -7$ or $\qquad w = 2$
Solution set: $\{-7, 2\}$

7. $m^2 - 7m = 0$
$\qquad m(m-7) = 0$
$\qquad m = 0$ or $m - 7 = 0$
$\qquad m = 0$ or $\qquad m = 7$
Solution set: $\{0, 7\}$

9. $a^2 - a = 20$
$\qquad a^2 - a - 20 = 0$
$\qquad (a-5)(a+4) = 0$
$\qquad a - 5 = 0$ or $a + 4 = 0$
$\qquad a = 5$ or $\qquad a = -4$
Solution set: $\{-4, 5\}$

11. $3(x^2 - x - 12) = 0$
$\qquad 3(x-4)(x+3) = 0$
$\qquad x - 4 = 0$ or $x + 3 = 0$
$\qquad x = 4$ or $\qquad x = -3$
Solution set: $\{-3, 4\}$

13. $2\left(z^2 + \frac{3}{2}z\right) = 2(10)$

$\qquad 2z^2 + 3z = 20$
$\qquad 2z^2 + 3z - 20 = 0$
$\qquad (2z-5)(z+4) = 0$
$\qquad 2z - 5 = 0$ or $z + 4 = 0$
$\qquad 2z = 5$ or $\qquad z = -4$
$\qquad z = \frac{5}{2}$

Solution set: $\left\{ -4, \frac{5}{2} \right\}$

15. $x^3 - 4x = 0$
$\qquad x(x^2 - 4) = 0$
$\qquad x(x-2)(x+2) = 0$
$x = 0$ or $x - 2 = 0$ or $x + 2 = 0$
$x = 0$ or $\qquad x = 2$ or $\qquad x = -2$
Solution set: $\{-2, 0, 2\}$

17. $w^3 + 4w^2 - 25w - 100 = 0$
$\qquad w^2(w+4) - 25(w+4) = 0$
$\qquad (w^2 - 25)(w+4) = 0$
$\qquad (w-5)(w+5)(w+4) = 0$
$w - 5 = 0$ or $w + 5 = 0$ or $w + 4 = 0$
$\qquad w = 5$ or $\qquad w = -5$ or $\qquad w = -4$
Solution set: $\{-5, -4, 5\}$

19. $n^3 - 2n^2 - n + 2 = 0$
$\qquad n^2(n-2) - 1(n-2) = 0$
$\qquad (n^2 - 1)(n-2) = 0$
$\qquad (n-1)(n+1)(n-2) = 0$
$n - 1 = 0$ or $n + 1 = 0$ or $n - 2 = 0$
$\qquad n = 1$ or $\qquad n = -1$ or $\qquad n = 2$
Solution set: $\{-1, 1, 2\}$

21. $|x^2 - 5| = 4$

$x^2 - 5 = 4$ or $x^2 - 5 = -4$

$x^2 - 9 = 0$ or $x^2 - 1 = 0$

$(x-3)(x+3) = 0$ or $(x-1)(x+1) = 0$

$x - 3 = 0$ or $x + 3 = 0$ or $x - 1 = 0$ or $x + 1 = 0$

$x = 3$ or $x = -3$ or $x = 1$ or $x = -1$

Solution set: $\{-3, -1, 1, 3\}$

23. $|x^2 + 2x - 36| = 12$

$x^2 + 2x - 36 = 12$ or $x^2 + 2x - 36 = -12$

$x^2 + 2x - 48 = 0$ or $x^2 + 2x - 24 = 0$

$(x+8)(x-6) = 0$ or $(x+6)(x-4) = 0$

$x = -8$ or $x = 6$ or $x = -6$ or $x = 4$

Solution set: $\{-8, -6, 4, 6\}$

25. $|x^2 + 4x + 2| = 2$

$x^2 + 4x + 2 = 2$ or $x^2 + 4x + 2 = -2$

$x^2 + 4x = 0$ or $x^2 + 4x + 4 = 0$

$x(x+4) = 0$ or $(x+2)^2 = 0$

$x = 0$ or $x + 4 = 0$ or $x + 2 = 0$

$x = 0$ or $x = -4$ or $x = -2$

Solution set: $\{-4, 0, -2\}$

27. $|x^2 + 6x + 1| = 8$

$x^2 + 6x + 1 = 8$ or $x^2 + 6x + 1 = -8$

$x^2 + 6x - 7 = 0$ or $x^2 + 6x + 9 = 0$

$(x+7)(x-1) = 0$ or $(x+3)^2 = 0$

$x + 7 = 0$ or $x - 1 = 0$ or $x + 3 = 0$

$x = -7$ or $x = 1$ or $x = -3$

Solution set: $\{-7, -3, 1\}$

29. $2x^2 - x = 6$

$2x^2 - x - 6 = 0$

$(2x+3)(x-2) = 0$

$2x + 3 = 0$ or $x - 2 = 0$

$x = -\frac{3}{2}$ or $x = 2$

Solution set: $\{-\frac{3}{2}, 2\}$

31. $|x^2 + 5x| = 6$

$x^2 + 5x = 6$ or $x^2 + 5x = -6$

$x^2 + 5x - 6 = 0$ or $x^2 + 5x + 6 = 0$

$(x+6)(x-1) = 0$ or $(x+2)(x+3) = 0$

$x + 6 = 0$ or $x - 1 = 0$ or $x + 2 = 0$ or $x + 3 = 0$

$x = -6$ or $x = 1$ or $x = -2$ or $x = -3$

Solution set: $\{-6, -3, -2, 1\}$

33. $x^2 + 5x = 6$

$x^2 + 5x - 6 = 0$

$(x+6)(x-1) = 0$

$x + 6 = 0$ or $x - 1 = 0$

$x = -6$ or $x = 1$

Solution set: $\{-6, 1\}$

35. $(x+2)(x+1) = 12$

$x^2 + 3x + 2 = 12$

$x^2 + 3x - 10 = 0$

$(x+5)(x-2) = 0$

$x + 5 = 0$ or $x - 2 = 0$

$x = -5$ or $x = 2$

The solution set is $\{-5, 2\}$.

37. $y^3 + 9y^2 + 20y = 0$

$y(y^2 + 9y + 20) = 0$

$y(y+4)(y+5) = 0$

$y = 0$ or $y + 4 = 0$ or $y + 5 = 0$

$y = 0$ or $y = -4$ or $y = -5$

Solution set: $\{-5, -4, 0\}$

39. $5a^3 = 45a$

$5a^3 - 45a = 0$

$5a(a^2 - 9) = 0$

$5a(a-3)(a+3) = 0$

$a = 0$ or $a - 3 = 0$ or $a + 3 = 0$

$a = 0$ or $a = 3$ or $a = -3$

Solution set: $\{-3, 0, 3\}$

41. $(2x-1)(x^2 - 9) = 0$

$(2x-1)(x-3)(x+3) = 0$

$2x - 1 = 0$ or $x - 3 = 0$ or $x + 3 = 0$

$x = \frac{1}{2}$ or $x = 3$ or $x = -3$

The solution set is $\left\{-3, \frac{1}{2}, 3\right\}$.

43. $4x^2 - 12x + 9 = 0$

$(2x-3)^2 = 0$

$2x - 3 = 0$

$x = \frac{3}{2}$

The solution set is $\left\{\frac{3}{2}\right\}$.

45. $y^2 + by = 0$

$y(y+b) = 0$

$y = 0$ or $y + b = 0$

$y = 0$ or $y = -b$

Solution set: $\{0, -b\}$

47. $a^2y^2 - b^2 = 0$

$(ay - b)(ay + b) = 0$

$ay - b = 0$ or $ay + b = 0$

$ay = b$ or $ay = -b$

$y = \frac{b}{a}$ or $y = -\frac{b}{a}$

Solution set: $\left\{-\frac{b}{a}, \frac{b}{a}\right\}$

49.
$$4y^2 + 4by + b^2 = 0$$
$$(2y + b)^2 = 0$$
$$2y + b = 0$$
$$2y = -b$$
$$y = -\frac{b}{2}$$

Solution set: $\{-\frac{b}{2}\}$

51.
$$ay^2 + 3y - ay = 3$$
$$ay^2 + 3y - ay - 3 = 0$$
$$y(ay + 3) \; -1(ay + 3) = 0$$
$$(y - 1)(ay + 3) = 0$$
$$y - 1 = 0 \quad \text{or} \quad ay + 3 = 0$$
$$y = 1 \quad \text{or} \quad y = -\frac{3}{a}$$

Solution set: $\{-\frac{3}{a}, 1\}$

53. Let x = the width and $x + 2$ = the length. Since $A = LW$, we can write the equation
$$x(x + 2) = 24$$
$$x^2 + 2x = 24$$
$$x^2 + 2x - 24 = 0$$
$$(x + 6)(x - 4) = 0$$
$$x + 6 = 0 \quad \text{or} \quad x - 4 = 0$$
$$x = -6 \quad \text{or} \quad x = 4$$
$$x + 2 = 6$$

Since the width cannot be -6, we have a width of 4 inches and a length of 6 inches.

55. Let x = one number and $13 - x$ = the other. Their product is 36:
$$x(13 - x) = 36$$
$$-x^2 + 13x - 36 = 0$$
$$x^2 - 13x + 36 = 0$$
$$(x - 4)(x - 9) = 0$$
$$x - 4 = 0 \quad \text{or} \quad x - 9 = 0$$
$$x = 4 \quad \text{or} \quad x = 9$$
$$13 - x = 9 \quad \text{or} \quad 13 - x = 4$$
The numbers are 4 and 9.

57. Let x = the width and $x + 21$ = the length.
$$x(x + 21) = 946$$
$$x^2 + 21x - 946 = 0$$
$$(x + 43)(x - 22) = 0$$
$$x + 43 = 0 \quad \text{or} \quad x - 22 = 0$$
$$x = -43 \quad \text{or} \quad x = 22$$
$$x + 21 = 43$$

The length is 43 in. and the width is 22 in.

59. To find the time the ball is in the air we need the values of t that give D a value of 0. We must solve the equation
$$-16t^2 + 64t = 0$$
$$-16t(t - 4) = 0$$
$$t = 0 \quad \text{or} \quad t - 4 = 0$$
$$t = 0 \quad \text{or} \quad t = 4$$
The ball is in the air for 4 seconds.

61. Let x = the width and $2x + 2$ = the length. Using the Pythagorean theorem we can write
$$x^2 + (2x + 2)^2 = 13^2$$
$$x^2 + 4x^2 + 8x + 4 = 169$$
$$5x^2 + 8x - 165 = 0$$
$$(5x + 33)(x - 5) = 0$$
$$5x + 33 = 0 \quad \text{or} \quad x - 5 = 0$$
$$x = -\frac{33}{5} \quad \text{or} \quad x = 5$$
$$2x + 2 = 12$$

Since the lengths must be positive numbers, the width is 5 feet and the length is 12 feet.

63. Since the perimeter is 34 feet the sum of the length and width is 17 feet. Let x = the width and $17 - x$ = the length. Using the Pythagorean theorem we can write
$$x^2 + (17 - x)^2 = 13^2$$
$$x^2 + 289 - 34x + x^2 = 169$$
$$2x^2 - 34x + 120 = 0$$
$$x^2 - 17x + 60 = 0$$
$$(x - 5)(x - 12) = 0$$
$$x - 5 = 0 \quad \text{or} \quad x - 12 = 0$$
$$x = 5 \quad \text{or} \quad x = 12$$
$$17 - x = 12 \quad \text{or} \quad 17 - x = 5$$
The width is 5 feet and the length is 12 feet.

65. Let x = the first integer and $x + 1$ = the second integer. The sum of their squares is 25:
$$x^2 + (x + 1)^2 = 25$$
$$x^2 + x^2 + 2x + 1 = 25$$
$$2x^2 + 2x - 24 = 0$$
$$x^2 + x - 12 = 0$$
$$(x + 4)(x - 3) = 0$$
$$x + 4 = 0 \quad \text{or} \quad x - 3 = 0$$
$$x = -4 \quad \text{or} \quad x = 3$$
$$x + 1 = -3 \quad \text{or} \quad x + 1 = 4$$
The integers are 3 and 4, or -4 and -3.

67. Let x = the original length. Since the length times the width is 240 square feet, the width is $240/x$. The new length is $x - 4$ and the new width is $(240/x) + 3$. Since the area is still

240 square feet, we can write the following equation.

$$(x-4)\left(\frac{240}{x}+3\right)=240$$

$$x\cdot\frac{240}{x}-4\cdot\frac{240}{x}+3x-12=240$$

$$240-\frac{960}{x}+3x-252=0$$

$$-\frac{960}{x}+3x-12=0$$

$$x\left(-\frac{960}{x}+3x-12\right)=x\cdot0$$

$$-960+3x^2-12x=0$$

$$3x^2-12x-960=0$$

$$x^2-4x-320=0$$

$$(x-20)(x+16)=0$$

$$x-20=0 \text{ or } x+16=0$$

$$x=20 \text{ or } \quad x=-16$$

$$\frac{240}{x}=12$$

The length is 20 feet and the width is 12 feet.

CHAPTER 3 REVIEW

1. $2\cdot2\cdot2^{-1}=4\cdot\frac{1}{2}=2$

3. $2^2\cdot3^2=4\cdot9=36$

5. $(-3)^{-3}=-\frac{1}{3^3}=-\frac{1}{27}$

7. $-(-1)^{-3}=-\frac{1}{(-1)^3}=-\frac{1}{-1}=1$

9. $2x^3\cdot4x^{-6}=8x^{-3}=\frac{8}{x^3}$

11. $\frac{y^{-5}}{y^{-3}}=y^{-2}=\frac{1}{y^2}$

13. $\frac{a^5a^{-2}}{a^{-4}}=\frac{a^3}{a^{-4}}=a^7$

15. $\frac{6x^{-2}}{3x^2}=2x^{-4}=\frac{2}{x^4}$

17. Move the decimal 6 places to the right: $8.36\times10^6=8,360,000$

19. Move the decimal point 4 places to the left: $5.7\times10^{-4}=0.00057$

21. Move the decimal point 6 places to the left: $8,070,000=8.07\times10^6$

23. Move the decimal point 4 places to the right: $0.000709=7.09\times10^{-4}$

25. $\frac{(4\times10^9)(6\times10^{-7})}{(1.2\times10^{-5})(2\times10^6)}=\frac{24\times10^2}{2.4\times10^1}$

$$=10\times10^1=1\times10^2$$

27. $(a^{-3})^{-2}\cdot a^{-7}=a^6\cdot a^{-7}=a^{-1}=\frac{1}{a}$

29. $(m^2n^3)^{-2}(m^{-3}n^2)^4=m^{-4}n^{-6}m^{-12}n^8$

$$=m^{-16}n^2=\frac{n^2}{m^{16}}$$

31. $\left(\frac{2}{3}\right)^{-4}=\left(\frac{3}{2}\right)^4=\frac{3^4}{2^4}=\frac{81}{16}$

33. $\left(\frac{1}{2}+\frac{1}{3}\right)^2=\left(\frac{5}{6}\right)^2=\frac{25}{36}$

35. $\left(-\frac{3a}{4b^{-1}}\right)^{-1}=-\frac{4b^{-1}}{3a}=-\frac{4}{3ab}$

37. $\frac{(a^{-3}b)^4}{(ab^2)^{-5}}=\frac{a^{-12}b^4}{a^{-5}b^{-10}}=\frac{b^{14}}{a^7}$

39. $5^{2w}\cdot5^{4w}\cdot5^{-1}=5^{6w-1}$

41. $\left(\frac{7^{3a}}{7^8}\right)^5=\frac{7^{15a}}{7^{40}}=7^{15a-40}$

43. $(2w-3)+(6w+5)=8w+2$

45. $(x^2-3x-4)-(x^2+3x-7)$

$$=x^2-3x-4-x^2-3x+7=-6x+3$$

47. $(x^2-2x+4)(x)-(x^2-2x+4)(2)$

$$=x^3-2x^2+4x-2x^2+4x-8$$

$$=x^3-4x^2+8x-8$$

49. $xy+7z-5(xy-3z)$

$$=xy+7z-5xy+15z=-4xy+22z$$

51. $m^2(5m^3-m+2)=5m^5-m^3+2m^2$

53. $(x-3)(x+7)=x^2-3x+7x-21$

$$=x^2+4x-21$$

55. $(z-5y)(z+5y)=z^2-(5y)^2=z^2-25y^2$

57. $(m+8)^2=m^2+2\cdot8m+8^2$

$$=m^2+16m+64$$

59. $(w-6x)(w-4x)=w^2-6xw-4xw+24x^2$

$$=w^2-10xw+24x^2$$

61. $(k-3)^2=k^2-2\cdot3\cdot k+9=k^2-6k+9$

63. $(m^2-5)(m^2+5)=m^4-5m^2+5m^2-25$

$$=m^4-25$$

65.

$$\begin{array}{r} x^2 + 3x - 5 \\ x - 2 \overline{\smash{)}\ x^3 + x^2 - 11x + 10} \\ \underline{x^3 - 2x^2} \\ 3x^2 - 11x \\ \underline{3x^2 - 6x} \\ -5x + 10 \\ \underline{-5x + 10} \\ 0 \end{array}$$

Quotient: $x^2 + 3x - 5$ Remainder: 0

67.

$$\begin{array}{r} m^3 - m^2 + m - 1 \\ m + 1 \overline{\smash{)}\ m^4 + 0m^3 + 0m^2 + 0m - 1} \\ \underline{m^4 + m^3} \\ -m^3 + 0m^2 \\ \underline{-m^3 - m^2} \\ m^2 + 0m \\ \underline{m^2 + m} \\ -m - 1 \\ \underline{-m - 1} \\ 0 \end{array}$$

Quotient: $m^3 - m^2 + m - 1$ Remainder: 0

69.

$$\begin{array}{r} a^6 + 2a^3 + 4 \\ a^3 - 2 \overline{\smash{)}\ a^9 + 0a^6 + 0a^3 - 8} \\ \underline{a^9 - 2a^6} \\ 2a^6 + 0a^3 \\ \underline{2a^6 - 4a^3} \\ 4a^3 - 8 \\ \underline{4a^3 - 8} \\ 0 \end{array}$$

Quotient: $a^6 + 2a^3 + 4$ Remainder: 0

71. $\dfrac{3m^3 + 6m^2 - 18m}{3m} = \dfrac{3m^3}{3m} + \dfrac{6m^2}{3m} - \dfrac{18m}{3m}$

$= m^2 + 2m - 6$

Quotient: $m^2 + 2m - 6$ Remainder: 0

73.

$$\begin{array}{r} x + 1 \\ x - 1 \overline{\smash{)}\ x^2 + 0x - 5} \\ \underline{x^2 - x} \\ x - 5 \\ \underline{x - 1} \\ -4 \end{array}$$

$\dfrac{x^2 - 5}{x - 1} = x + 1 + \dfrac{-4}{x - 1}$

75.

$$\begin{array}{r} 3 \\ x - 2 \overline{\smash{)}\ 3x + 0} \\ \underline{3x - 6} \\ 6 \end{array}$$

$\dfrac{3x}{x - 2} = 3 + \dfrac{6}{x - 2}$

77. When $x^3 - 2x^2 + 3x + 22$ is divided by $x + 2$ the quotient is $x^2 - 4x + 11$ and the remainder is 0. So the first polynomial is a factor of the second.

79. If $x^3 - x - 120$ is divided by $x - 5$ the quotient is $x^2 + 5x + 24$ and remainder is 0. So the first polynomial is a factor of the second.

81. When $x^3 + x^2 - 3$ is divided by $x - 1$ the quotient is $x^2 + 2x + 2$ and the remainder is -1. So the first polynomial is not a factor of the second.

83. When $x^4 + x^3 + 5x^2 + 2x + 6$ is divided by $x^2 + 2$ the quotient is $x^2 + x + 3$ and the remainder is 0. So the first polynomial is a factor of the second.

85. $3x - 6 = 3(x - 2)$

87. $4a - 20 = -4(-a + 5)$

89. $3w - w^2 = -w(-3 + w) = -w(w - 3)$

91. $y^2 - 81 = (y - 9)(y + 9)$

93. $4x^2 + 28x + 49$

$= (2z)^2 + 2 \cdot 2x \cdot 7 + 7^2 = (2x + 7)^2$

95. $t^2 - 18t + 81 = t^2 - 2 \cdot 9t + 9^2 = (t - 9)^2$

97. $t^3 - 125 = (t - 5)(t^2 + 5t + 25)$

99. Two numbers with a product of -30 and a sum of -7 are -10 and 3.
$x^2 - 7x - 30 = (x - 10)(x + 3)$

101. Two numbers with a product of -28 and a sum of -3 are -7 and 4.

$w^2 - 3w - 28 = (w - 7)(w + 4)$

103. Two numbers with a product of -14 and a sum of 5 are -2 and 7.

$2m^2 + 5m - 7 = 2m^2 - 2m + 7m - 7$

$= 2m(m - 1) + 7(m - 1) = (2m + 7)(m - 1)$

105. $m^7 - 3m^4 - 10m = m(m^6 - 3m^3 - 10)$

$= m(m^3 - 5)(m^3 + 2)$

107. $5x^3 + 40 = 5(x^3 + 8)$

$$= 5(x + 2)(x^2 - 2x + 4)$$

109. Two numbers with a product of 18 and a sum of 9 are 6 and 3.

$9x^2 + 9x + 2 = 9x^2 + 3x + 6x + 2$

$= 3x(3x + 1) + 2(3x + 1) = (3x + 2)(3x + 1)$

111. $x^3 + x^2 - x - 1 = x^2(x + 1) - 1(x + 1)$

$= (x^2 - 1)(x + 1) = (x - 1)(x + 1)(x + 1)$

$$= (x - 1)(x + 1)^2$$

113. $-x^2y + 16y = -y(x^2 - 16)$

$$= -y(x - 4)(x + 4)$$

115. $-a^3b^2 + 2a^2b^2 - ab^2 = -ab^2(a^2 - 2a + 1)$

$$= -ab^2(a - 1)^2$$

117. $x^3 - x^2 + 9x - 9 = x^2(x - 1) + 9(x - 1)$

$$= (x - 1)(x^2 + 9)$$

119. $x^4 - x^2 - 12 = (x^2 - 4)(x^2 + 3)$

$$= (x - 2)(x + 2)(x^2 + 3)$$

121. $a^6 - a^3 = a^3(a^3 - 1)$

$$= a^3(a - 1)(a^2 + a + 1)$$

123. $-8m^2 - 24m - 18 = -2(4m^2 + 12m + 9)$

$$= -2(2m + 3)^2$$

125. Let $y = 2x - 3$: $(2x - 3)^2 - 16 = y^2 - 16$

$= (y - 4)(y + 4) = (2x - 3 - 4)(2x - 3 + 4)$

$$= (2x - 7)(2x + 1)$$

127. $x^6 + 7x^3 - 8 = (x^3 + 8)(x^3 - 1)$

$= (x + 2)(x^2 - 2x + 4)(x - 1)(x^2 + x + 1)$

129. Let $y = a^2 - 9$: $(a^2 - 9)^2 - 5(a^2 - 9) + 6$

$y^2 - 5y + 6 = (y - 2)(y - 3)$

$= (a^2 - 9 - 2)(a^2 - 9 - 3) = (a^2 - 11)(a^2 - 12)$

131. $x^{2k} - 49 = (x^k - 7)(x^k + 7)$

133. $m^{2a} - 2m^a - 3 = (m^a - 3)(m^a + 1)$

135. $9z^{2k} - 12z^k + 4 = (3z^k - 2)^2$

137. $y^{2a} - by^a + cy^a - bc$

$= y^a(y^a - b) + c(y^a - b) = (y^a - b)(y^a + c)$

139.
$$x^3 - 5x^2 = 0$$
$$x^2(x - 5) = 0$$
$$x^2 = 0 \quad \text{or} \quad x - 5 = 0$$
$$x = 0 \quad \text{or} \quad x = 5$$
Solution set: $\{0, 5\}$

141.
$$(a - 2)(a - 3) = 6$$
$$a^2 - 5a + 6 = 6$$
$$a^2 - 5a = 0$$
$$a(a - 5) = 0$$
$$a = 0 \quad \text{or} \quad a - 5 = 0$$
$$a = 0 \quad \text{or} \quad a = 5$$
Solution set: $\{0, 5\}$

143.
$$2m^2 - 9m - 5 = 0$$
$$(2m + 1)(m - 5) = 0$$
$$2m + 1 = 0 \quad \text{or} \quad m - 5 = 0$$
$$m = -1/2 \quad \text{or} \quad m = 5$$
Solution set: $\{-1/2, 5\}$

145.
$$w^3 + 5w^2 - w - 5 = 0$$
$$w^2(w + 5) - 1(w + 5) = 0$$
$$(w^2 - 1)(w + 5) = 0$$
$$(w - 1)(w + 1)(w + 5) = 0$$
$$w - 1 = 0 \quad \text{or} \quad w + 1 = 0 \quad \text{or} \quad w + 5 = 0$$
$$w = 1 \quad \text{or} \quad w = -1 \quad \text{or} \quad w = -5$$
Solution set: $\{-5, -1, 1\}$

147.
$$|x^2 - 5| = 4$$
$$x^2 - 5 = 4 \quad \text{or} \quad x^2 - 5 = -4$$
$$x^2 - 9 = 0 \quad \text{or} \quad x^2 - 1 = 0$$
$$(x - 3)(x + 3) = 0 \quad \text{or} \quad (x - 1)(x + 1) = 0$$
$$x = 3 \text{ or } x = -3 \quad \text{or} \quad x = 1 \text{ or } x = -1$$
Solution set: $\{-3, -1, 1, 3\}$

149. Let x = the distance from the bottom of the ladder and the cactus and $x + 2$ = the distance from the top of the ladder to the ground. Since the ladder is the hypotenuse of a right triangle, we can write

$$x^2 + (x + 2)^2 = 10^2$$
$$x^2 + x^2 + 4x + 4 = 100$$
$$2x^2 + 4x - 96 = 0$$
$$x^2 + 2x - 48 = 0$$
$$(x + 8)(x - 6) = 0$$
$$x + 8 = 0 \quad \text{or} \quad x - 6 = 0$$
$$x = -8 \quad \text{or} \quad x = 6$$

The ladder should be placed 6 feet from the cactus.

151. $A = 9a^2 + 6a + 1 = (3a + 1)^2$
So the side of the square is $3a + 1$ and its perimeter is $4(3a + 1)$ or $12a + 4$ km.
153. $L = 59.4022(1.00459)^{20} \approx 65.1$ yr
From Exercise 95 Section 3.2, a 20-yr old white male is expected to live 71.3 yr. The White male is expected to live 6.2 years longer.

155. $172,000 = R \cdot \dfrac{(1 + 0.07)^{20} - 1}{0.07}$

$R = 172,000 \div \dfrac{(1 + 0.07)^{20} - 1}{0.07} \approx \$4195.58$

CHAPTER 3   TEST

1. $3^{-2} = \dfrac{1}{3^2} = \dfrac{1}{9}$ **2.** $\dfrac{1}{6^{-2}} = 6^2 = 36$

3. $\left(\dfrac{1}{2}\right)^{-3} = 2^3 = 8$ **4.** $3 \cdot 4x^4 x^3 = 12x^7$

5. $\dfrac{8y^9}{2y^{-3}} = 4y^{9-(-3)} = 4y^{12}$

6. $(4a^2 b)^3 = 4^3 (a^2)^3 b^3 = 64a^6 b^3$

7. $\left(\dfrac{x^2}{3}\right)^{-3} = \left(\dfrac{3}{x^2}\right)^3 = \dfrac{27}{x^6}$

8. $\dfrac{(2^{-1})^{-3}(a^2)^{-3} b^{-3}}{4a^{-9}} = \dfrac{2^3 a^{-6} b^{-3}}{4a^{-9}} = \dfrac{8a^3}{4b^3} = \dfrac{2a^3}{b^3}$

9. $3.24 \times 10^9 = 3,240,000,000$

10. $8.673 \times 10^{-4} = 0.0008673$

11. $\dfrac{(8 \times 10^4)(6 \times 10^{-4})}{2 \times 10^6} = 24 \times 10^{-6}$

$= 2.4 \times 10^1 \times 10^{-6} = 2.4 \times 10^{-5}$

12. $\dfrac{(6 \times 10^{-5})^2 (5 \times 10^2)}{(3 \times 10^4)^2 (1 \times 10^{-2})} = \dfrac{36 \times 10^{-10} \cdot 5 \times 10^2}{9 \times 10^8 \cdot 1 \times 10^{-2}}$

$= 20 \times 10^{-14} = 2 \times 10^{-13}$

13. Add like terms to get $3x^3 + 3x^2 - 2x + 3$.

14. $(x^2 - 6x - 7) - (3x^2 + 2x - 4)$
$= x^2 - 6x - 7 - 3x^2 - 2x + 4 = -2x^2 - 8x - 3$

15. $(x^2 - 3x + 7)(x) - (x^2 - 3x + 7)(2)$
$= x^3 - 3x^2 + 7x - 2x^2 + 6x - 14$
$= x^3 - 5x^2 + 13x - 14$

16.
$$\begin{array}{r} x^2 + 4x - 5 \\ x+3 \overline{)\ x^3 + 7x^2 + 7x - 15} \\ \underline{x^3 + 3x^2} \\ 4x^2 + 7x \\ \underline{4x^2 + 12x} \\ -5x - 15 \\ \underline{-5x - 15} \\ 0 \end{array}$$

Quotient is $x^2 + 4x - 5$.

17. $(x - 2)^3 = (x - 2)^2 (x - 2)$
$= (x^2 - 4x + 4)(x - 2)$
$= (x^2 - 4x + 4)(x) - (x^2 - 4x + 4)(2)$
$= x^3 - 4x^2 + 4x - 2x^2 + 8x - 8$
$= x^3 - 6x^2 + 12x - 8$

18. $\dfrac{x - 3}{3 - x} = \dfrac{(-1)(3 - x)}{3 - x} = -1$

19. $(x - 7)(2x + 3) = 2x^2 - 14x + 3x - 21$
$\qquad\qquad\qquad\ = 2x^2 - 11x - 21$

20. $(x - 6)^2 = x^2 - 2 \cdot 6x + 36 = x^2 - 12x + 36$

21. $(2x + 5)^2 = 4x^2 + 20x + 25$

22. $(3y^2 - 5)(3y^2 + 5) = 9y^4 - 25$

23.
$$\begin{array}{r} 5 \\ x+3 \overline{)\ 5x + 0} \\ \underline{5x + 15} \\ -15 \end{array}$$

$\dfrac{5x}{x + 3} = 5 + \dfrac{-15}{x + 3}$

24.
$$\begin{array}{r} x + 5 \\ x-2 \overline{)\ x^2 + 3x - 6} \\ \underline{x^2 - 2x} \\ 5x - 6 \\ \underline{5x - 10} \\ 4 \end{array}$$

$\dfrac{x^2 + 3x - 6}{x - 2} = x + 5 + \dfrac{4}{x - 2}$

25. Two numbers with a product of -24 and a sum of -2 are -6 and 4.
$a^2 - 2a - 24 = (a - 6)(a + 4)$

26. $4x^2 + 28x + 49$
$= (2x)^2 + 2 \cdot 2x \cdot 7 + 7^2 = (2x + 7)^2$

27. $3m^3 - 24 = 3(m^3 - 8)$
$\qquad\qquad\ = 3(m - 2)(m^2 + 2m + 4)$

28. $2x^2 y - 32y = 2y(x^2 - 16)$
$\qquad\qquad\ = 2y(x - 4)(x + 4)$

29. $2xa + 3a - 10x - 15$
$= a(2x + 3) - 5(2x + 3) = (a - 5)(2x + 3)$

30. $x^4 + 3x^2 - 4 = (x^2 - 1)(x^2 + 4)$
$\qquad = (x - 1)(x + 1)(x^2 + 4)$

31. $\qquad 2m^2 + 7m - 15 = 0$
$\qquad (2m - 3)(m + 5) = 0$
$\qquad 2m - 3 = 0 \quad \text{or} \quad m + 5 = 0$
$\qquad m = \frac{3}{2} \quad \text{or} \quad m = -5$

Solution set: $\left\{ -5, \frac{3}{2} \right\}$

32. $\qquad\qquad x^3 - 4x = 0$
$\qquad\qquad x(x^2 - 4) = 0$
$\qquad\qquad x(x - 2)(x + 2) = 0$
$x = 0 \quad \text{or} \quad x - 2 = 0 \quad \text{or} \quad x + 2 = 0$
$x = 0 \quad \text{or} \quad x = 2 \quad \text{or} \quad x = -2$
Solution set: $\{-2, 0, 2\}$

33. $\qquad\qquad |x^2 + x - 9| = 3$

$x^2 + x - 9 = 3 \quad \text{or} \quad x^2 + x - 9 = -3$
$x^2 + x - 12 = 0 \quad \text{or} \quad x^2 + x - 6 = 0$
$(x + 4)(x - 3) = 0 \quad \text{or} \quad (x + 3)(x - 2) = 0$
$x = -4 \text{ or } x = 3 \quad \text{or} \quad x = -3 \text{ or } x = 2$
Solution set: $\{-4, -3, 2, 3\}$

34. Let x = the height and $x + 2$ = the width.
Using the Pythagorean theorem we can write
$\qquad x^2 + (x + 2)^2 = 10^2$
$\qquad x^2 + x^2 + 4x + 4 = 100$
$\qquad 2x^2 + 4x - 96 = 0$
$\qquad x^2 + 2x - 48 = 0$
$\qquad (x + 8)(x - 6) = 0$
$\qquad x = -8 \text{ or } \qquad x = 6$
$\qquad x + 2 = -6 \text{ or } x + 2 = 8$
The width is 8 inches and the height is 6 inches.

35. $d = (1.2527 \times 10^{69})(1.08243)^{-1900} \approx 5473$

$d = (1.2527 \times 10^{69})(1.08243)^{-1950} \approx 104$

$d = (1.2527 \times 10^{69})(1.08243)^{-1990} \approx 4$

Tying It All Together Chapters 1-3.

1. $4^2 = 4 \cdot 4 = 16$ **2.** $4(-2) = -8$

3. $4^{-2} = \frac{1}{4^2} = \frac{1}{16}$ **4.** $2^3 \cdot 4^{-1} = 8 \cdot \frac{1}{4} = 2$

5. $2^{-1} + 2^{-1} = \frac{1}{2} + \frac{1}{2} = 1$

6. $2^{-1} \cdot 3^{-1} = \frac{1}{2} \cdot \frac{1}{3} = \frac{1}{6}$

7. $3^{-1} - 2^{-2} = \frac{1}{3} - \frac{1}{4} = \frac{4}{12} - \frac{3}{12} = \frac{1}{12}$

8. $3^2 - 4(5)(-2) = 9 - (-40) = 49$

9. $2^7 - 2^6 = 128 - 64 = 64$

10. $0.08(32) + 0.08(68) = 0.08(32 + 68)$
$\qquad\qquad\qquad = 0.08(100) = 8$

11. $3 - 2|5 - 7 \cdot 3| = 3 - 2|-16|$
$\qquad = 3 - 2 \cdot 16 = 3 - 32 = -29$

12. $5^{-1} + 6^{-1} = \frac{1}{5} + \frac{1}{6} = \frac{6}{30} + \frac{5}{30} = \frac{11}{30}$

13. $\qquad 0.05a - 0.04(a - 50) = 4$
$\qquad 0.05a - 0.04a + 2 = 4$
$\qquad\qquad 0.01a = 2$
$\qquad\qquad\qquad a = 200$

Solution set: $\{200\}$

14. $\qquad\qquad 15b - 27 = 0$
$\qquad\qquad\qquad 15b = 27$
$\qquad\qquad\qquad b = \frac{27}{15} = \frac{9}{5}$

Solution set: $\left\{ \frac{9}{5} \right\}$

15. $\qquad\qquad 2c^2 + 15c - 27 = 0$
$\qquad (2c - 3)(c + 9) = 0$
$\qquad 2c - 3 = 0 \quad \text{or} \quad c + 9 = 0$
$\qquad c = \frac{3}{2} \quad \text{or} \quad c = -9$
Solution set: $\left\{ -9, \frac{3}{2} \right\}$

16. $\qquad\qquad 2t^2 + 15t = 0$

$\qquad t(2t + 15) = 0$

$\qquad t = 0 \quad \text{or} \quad 2t + 15 = 0$

$\qquad t = 0 \quad \text{or} \quad t = -\frac{15}{2}$
Solution set: $\left\{ -\frac{15}{2}, 0 \right\}$

17. $\qquad\qquad |15v - 27| = 3$
$15v - 27 = 3 \quad \text{or} \quad 15v - 27 = -3$
$\qquad 15v = 30 \quad \text{or} \qquad 15v = 24$
$\qquad v = 2 \quad \text{or} \qquad v = \frac{8}{5}$

Solution set: $\left\{ 2, \frac{8}{5} \right\}$

18. $\qquad\qquad |15v - 27| = 0$
$\qquad\qquad 15v - 27 = 0$
$\qquad\qquad\qquad 15v = 27$
$\qquad\qquad\qquad v = \frac{27}{15} = \frac{9}{5}$

Solution set: $\left\{ \frac{9}{5} \right\}$

19. Absolute value of any quantity is greater than or equal to 0. So the solution set is ∅.

20.
$$|x^2 + x - 4| = 2$$
$$x^2 + x - 4 = 2 \text{ or } x^2 + x - 4 = -2$$
$$x^2 + x - 6 = 0 \text{ or } x^2 + x - 2 = 0$$
$$(x+3)(x-2) = 0 \text{ or } (x+2)(x-1) = 0$$
$$x = -3 \text{ or } x = 2 \text{ or } x = -2 \text{ or } x = 1$$
Solution set: $\{-3, -2, 1, 2\}$

21.
$$(2x-1)(x+5) = 0$$
$$2x - 1 = 0 \text{ or } x + 5 = 0$$
$$x = \frac{1}{2} \text{ or } x = -5$$
Solution set: $\left\{-5, \frac{1}{2}\right\}$

22.
$$|3x - 1| + 6 = 9$$
$$|3x - 1| = 3$$
$$3x - 1 = 3 \text{ or } 3x - 1 = -3$$
$$3x = 4 \text{ or } 3x = -2$$
$$x = \frac{4}{3} \text{ or } x = -\frac{2}{3}$$
Solution set: $\left\{-\frac{2}{3}, \frac{4}{3}\right\}$

23. $(1.5 \times 10^{-4})w = 7 \times 10^6 + 5 \times 10^5$
$$w = \frac{7 \times 10^6 + 5 \times 10^5}{1.5 \times 10^{-4}} = 5 \times 10^{10}$$

24. $y - 5 \times 10^3 = \dfrac{6 \times 10^{12}}{3 \times 10^7}$
$$y = \frac{6 \times 10^{12}}{3 \times 10^7} + 5 \times 10^3$$
$$y = 2.05 \times 10^5$$

25. $7000 + 500(64 - 62) = \$8,000$

$$10,000 + 800(a - 67) = 11,600$$
$$800(a - 67) = 1,600$$
$$a - 67 = 2$$
$$a = 69$$
The age of a person who gets $11,600 in 2005 is 69.

4.1 WARM-UPS
1. True, because a number is a monomial.
2. True, because it is a ratio of two binomials.
3. False, because the domain is any number except 2.
4. True, because 9 and $-\frac{1}{2}$ both give a denominator of 0. **5.** False, because the numerator may be zero in a rational expression.
6. False, because 5 is not a factor of the numerator. **7.** False, we would multiply the numerator and denominator by $x + 1$.
8. True, because we can multiply the numerator and denominator by -1. **9.** True, because it is a correctly reduced rational expression.
10. False, it reduces to $x + y$.

4.1 EXERCISES
1. If $x = 1$, then $x - 1 = 0$. So the domain is $\{x \mid x \neq 1\}$.
3. If $z = 0$ then $7z = 0$. So the domain is $\{z \mid z \neq 0\}$.
5. If $y = -2$, then $y^2 - 4 = 0$. So the domain is $\{y \mid y \neq -2 \text{ and } y \neq 2\}$.
7. To find the domain solve
$$a^2 + 5a + 6 = 0$$
$$(a+2)(a+3) = 0$$
$$a = -2 \text{ or } a = -3$$
The domain is $\{a \mid a \neq -2 \text{ and } a \neq -3\}$.
9. To find the domain solve
$$x^2 + 4x = 0$$
$$x(x+4) = 0$$
$$x = 0 \text{ or } x = -4$$
The domain is $\{x \mid x \neq -4 \text{ and } x \neq 0\}$.
11. Solve the equation
$$x^3 + x^2 - 6x = 0$$
$$x(x+3)(x-2) = 0$$
$$x = 0 \text{ or } x = -3 \text{ or } x = 2$$
So the domain is $\{x \mid x \neq -3 \text{ and } x \neq 0 \text{ and } x \neq 2\}$.
13. $\dfrac{6}{57} = \dfrac{3 \cdot 2}{3 \cdot 19} = \dfrac{2}{19}$
15. $\dfrac{42}{210} = \dfrac{42 \cdot 1}{42 \cdot 5} = \dfrac{1}{5}$
17. $\dfrac{2x+2}{4} = \dfrac{2(x+1)}{2 \cdot 2} = \dfrac{x+1}{2}$
19. $\dfrac{3x - 6y}{10y - 5x} = \dfrac{3(x - 2y)}{-5(x - 2y)} = -\dfrac{3}{5}$

21. $\dfrac{ab^2}{a^3b} = \dfrac{b}{a^2}$ **23.** $\dfrac{-2w^2x^3y}{6wx^5y^2} = \dfrac{-w}{3x^2y}$

25. $\dfrac{a^3b^2}{a^3 + a^4} = \dfrac{a^3 \cdot b^2}{a^3(1+a)} = \dfrac{b^2}{1+a}$

27. $\dfrac{a-b}{2b-2a} = \dfrac{a-b}{-2(a-b)} = -\dfrac{1}{2}$

29. $\dfrac{3x+6}{3x} = \dfrac{3(x+2)}{3x} = \dfrac{x+2}{x}$

31. $\dfrac{a^3 - b^3}{a-b} = \dfrac{(a-b)(a^2 + ab + b^2)}{a-b}$

$$= a^2 + ab + b^2$$

33. $\dfrac{4x^2 - 4}{4x^2 + 4} = \dfrac{4(x^2 - 1)}{4(x^2 + 1)} = \dfrac{x^2 - 1}{x^2 + 1}$

35. $\dfrac{2(x+3)(x-2)}{4(x-3)(x+3)} = \dfrac{x-2}{2x-6}$

37. $\dfrac{x^3 + 7x^2 - 4x}{x^3 - 16x} = \dfrac{x(x^2 + 7x - 4)}{x(x-4)(x+4)}$

$$= \dfrac{x^2 + 7x - 4}{x^2 - 16}$$

39. $\dfrac{ab + 3a - by - 3y}{a^2 - y^2} = \dfrac{(a-y)(b+3)}{(a-y)(a+y)} = \dfrac{b+3}{a+y}$

41. $\dfrac{1}{5} = \dfrac{1 \cdot 10}{5 \cdot 10} = \dfrac{10}{50}$ **43.** $\dfrac{1}{x} = \dfrac{1 \cdot 3x}{x \cdot 3x} = \dfrac{3x}{3x^2}$

45. $\dfrac{5}{x-1} = \dfrac{5(x-1)}{(x-1)(x-1)} = \dfrac{5x-5}{x^2 - 2x + 1}$

47. $\dfrac{1}{2x+2} = \dfrac{1(-3)}{(2x+2)(-3)} = \dfrac{-3}{-6x-6}$

49. $5 = \dfrac{5 \cdot a}{1 \cdot a} = \dfrac{5a}{a}$

51. $\dfrac{x+2}{x+3} = \dfrac{(x+2)(x-1)}{(x+3)(x-1)} = \dfrac{x^2 + x - 2}{x^2 + 2x - 3}$

53. $\dfrac{7}{x-1} = \dfrac{7(-1)}{(x-1)(-1)} = \dfrac{-7}{1-x}$

55. $\dfrac{3}{x+2} = \dfrac{3(x^2 - 2x + 4)}{(x+2)(x^2 - 2x + 4)} = \dfrac{3x^2 - 6x + 12}{x^3 + 8}$

57. $\dfrac{x+2}{3x-1} = \dfrac{(x+2)(2x+5)}{(3x-1)(2x+5)} = \dfrac{2x^2 + 9x + 10}{6x^2 + 13x - 5}$

59. $R(3) = \dfrac{3 \cdot 3 - 5}{3 + 4} = \dfrac{4}{7}$

61. $H(-2) = \dfrac{(-2)^2 - 5}{3(-2) - 4} = \dfrac{-1}{-10} = \dfrac{1}{10}$

63. $W(-2) = \dfrac{4(-2)^3 - 1}{(-2)^2 - (-2) - 6} = \dfrac{-33}{0}$

$W(-2)$ is undefined.

65. $\dfrac{1}{3} = \dfrac{1 \cdot 7}{3 \cdot 7} = \dfrac{7}{21}$

67. $5 = \dfrac{5 \cdot 2}{1 \cdot 2} = \dfrac{10}{2}$

69. $\dfrac{3}{a} = \dfrac{3a}{a \cdot a} = \dfrac{3a}{a^2}$

71. $\dfrac{2}{a-b} = \dfrac{2(-1)}{(a-b)(-1)} = \dfrac{-2}{b-a}$

73. $\dfrac{2}{x-1} = \dfrac{2(x+1)}{(x-1)(x+1)} = \dfrac{2x+2}{x^2 - 1}$

75. $\dfrac{2}{w-3} = \dfrac{2(-1)}{(w-3)(-1)} = \dfrac{-2}{3-w}$

77. $\dfrac{2x+4}{6} = \dfrac{2(x+2)}{2 \cdot 3} = \dfrac{x+2}{3}$

79. $\dfrac{x+4}{x^2 - 16} = \dfrac{x+4}{(x-4)(x+4)} = \dfrac{1}{x-4}$

81. $\dfrac{3a+3}{3a} = \dfrac{3(a+1)}{3a} = \dfrac{a+1}{a}$

83. $\dfrac{1}{x-1} = \dfrac{1(x^2 + x + 1)}{(x-1)(x^2 + x + 1)} = \dfrac{x^2 + x + 1}{x^3 - 1}$

85. $\dfrac{x^{2a} - 4}{x^a + 2} = \dfrac{(x^a - 2)(x^a + 2)}{x^a + 2} = x^a - 2$

87. $\dfrac{x^a + m + wx^a + wm}{x^{2a} - m^2} = \dfrac{(x^a + m)(1 + w)}{(x^a - m)(x^a + m)}$

$$= \dfrac{1+w}{x^a - m}$$

89. $\dfrac{x^{3b+1} - x}{x^{2b+1} - x} = \dfrac{x(x^b - 1)(x^{2b} + x^b + 1)}{x(x^b - 1)(x^b + 1)}$

$$= \dfrac{x^{2b} + x^b + 1}{x^b + 1}$$

91. Since $D = RT$, his rate is $\dfrac{500}{2x}$ or $\dfrac{250}{x}$ mph.

93. Average cost per invitation is

$A(n) = \dfrac{0.50n + 45}{n}$ dollars.

$A(200) = \dfrac{0.50(200) + 45}{200} = \$0.725$

$A(300) = \dfrac{0.50(300) + 45}{300} = \$0.65$

It costs 7.5 cents more per invitation to print 200 rather than 300 invitations.

95. In 1990, about 12% of solid waste was recovered.

$p(n) = \dfrac{0.576n + 3.78}{3.14n + 87.1}$

$p(0) = \dfrac{0.576(0) + 3.78}{3.14(0) + 87.1} = 4.3\%$

$p(30) = \dfrac{0.576(30) + 3.78}{3.14(30) + 87.1} = 11.6\%$

$p(60) = \dfrac{0.576(60) + 3.78}{3.14(60) + 87.1} = 13.9\%$

4.2 WARM-UPS

1. False, because we can multiply any two fractions. **2.** False, $\dfrac{2}{7} \cdot \dfrac{3}{7} = \dfrac{6}{49}$.

3. True. **4.** False, $a \div b = a \cdot \dfrac{1}{b}$.

5. True, because the expressions are correctly multiplied.

6. True, $\dfrac{1}{2} \cdot \dfrac{1}{3} = \dfrac{1}{6}$. **7.** True, $\dfrac{1}{3} \div \dfrac{1}{2} = \dfrac{1}{3} \cdot \dfrac{2}{1} = \dfrac{2}{3}$.

8. True, $\dfrac{w - z}{z - w} = -1$.

9. True, $\dfrac{x}{3} \div 2 = \dfrac{x}{3} \cdot \dfrac{1}{2} = \dfrac{x}{6}$.

10. False, $\dfrac{a}{b} \div \dfrac{b}{a} = \dfrac{a}{b} \cdot \dfrac{a}{b} = \dfrac{a^2}{b^2}$.

4.2 EXERCISES

1. $\dfrac{12}{42} \cdot \dfrac{35}{22} = \dfrac{2 \cdot 2 \cdot 3}{2 \cdot 3 \cdot 7} \cdot \dfrac{5 \cdot 7}{2 \cdot 11} = \dfrac{5}{11}$

3. $\dfrac{3a}{2 \cdot 5b} \cdot \dfrac{5b^2}{2 \cdot 3} = \dfrac{ab}{4}$

5. $\dfrac{3(x - 1)}{2 \cdot 3} \cdot \dfrac{x}{x(x - 1)} = \dfrac{1}{2}$

7. $\dfrac{5(2x + 1)}{5(x^2 + 1)} \cdot \dfrac{(2x - 1)(x + 1)}{(2x - 1)(2x + 1)} = \dfrac{x + 1}{x^2 + 1}$

9. $\dfrac{(a + b)(x + w)}{(x - w)(x + w)} \cdot \dfrac{x - w}{(a - b)(a + b)} = \dfrac{1}{a - b}$

11. $\dfrac{a^2 - 2a + 4}{(a + 2)(a^2 - 2a + 4)} \cdot \dfrac{(a + 2)^3}{2(a + 2)} = \dfrac{a + 2}{2}$

13. $\dfrac{x - 9}{4 \cdot 3y} \cdot \dfrac{4 \cdot 2y}{-1(x - 9)} = \dfrac{2}{-3} = -\dfrac{2}{3}$

15. $(a^2 - 4) \cdot \dfrac{7}{2 - a} = (a - 2)(a + 2) \cdot \dfrac{7}{-1(a - 2)}$

$\qquad = -7(a + 2) = -7a - 14$

17. $\dfrac{15}{17} \div \dfrac{10}{17} = \dfrac{3 \cdot 5}{17} \cdot \dfrac{17}{2 \cdot 5} = \dfrac{3}{2}$

19. $\dfrac{36x}{5y} \div \dfrac{20x}{35y} = \dfrac{2^2 3^2 x}{5y} \cdot \dfrac{5 \cdot 7y}{2^2 5x} = \dfrac{3^2 7}{5} = \dfrac{63}{5}$

21. $\dfrac{2^3 \cdot 3a^5 b^2}{5c^3} \cdot \dfrac{1}{2^2 a^5 bc^5} = \dfrac{6b}{5c^8}$

23. $(w + 1) \div \dfrac{w^2 - 1}{w} = (w + 1) \cdot \dfrac{w}{w^2 - 1}$

$\qquad = (w + 1) \cdot \dfrac{w}{(w - 1)(w + 1)} = \dfrac{w}{w - 1}$

25. $\dfrac{x - y}{5} \div \dfrac{x^2 - 2xy + y^2}{10} = \dfrac{x - y}{5} \cdot \dfrac{2 \cdot 5}{(x - y)^2}$

$= \dfrac{2}{x - y}$

27. $\dfrac{2(2x - 1)}{x(x - 5)} \cdot \dfrac{(x - 5)(x + 5)}{(2x - 1)(x + 5)} = \dfrac{2}{x}$

29. $\dfrac{x - y}{3} \cdot \dfrac{6}{1} = 2x - 2y$

31. $\dfrac{(x - 5)(x + 5)}{3} \cdot \dfrac{2 \cdot 3}{x - 5} = 2x + 10$

33. $\dfrac{a - b}{2} \cdot \dfrac{1}{3} = \dfrac{a - b}{6}$

35. $(a - b)(a + b) \cdot \dfrac{3}{a + b} = 3a - 3b$

37. $\dfrac{5x}{2} \cdot \dfrac{1}{3} = \dfrac{5x}{6}$ **39.** $\dfrac{3}{4} \cdot \dfrac{4}{1} = 3$

41. $\dfrac{1}{2} \cdot \dfrac{1}{6} = \dfrac{1}{12}$ **43.** $\dfrac{1}{2} \cdot \dfrac{4x}{3} = \dfrac{2x}{3}$

45. $\dfrac{a - b}{b - a} = -1$

47. $\dfrac{x-y}{3} \cdot \dfrac{2 \cdot 3}{y-x} = -2$

49. $\dfrac{2(a+b)}{a} \cdot \dfrac{1}{2} = \dfrac{a+b}{a}$

51. $(a+b) \cdot 2 = 2a + 2b$

53. $\dfrac{3x}{5} \cdot \dfrac{1}{y} = \dfrac{3x}{5y}$ **55.** $\dfrac{3a}{5b} \cdot \dfrac{1}{2} = \dfrac{3a}{10b}$

57. $\dfrac{(3x-2)(x+5)}{x} \cdot \dfrac{x^3}{(3x-2)(3x+2)} \cdot \dfrac{7(x-5)}{(x-5)(x+5)}$

$$= \dfrac{7x^2}{3x+2}$$

59. $\dfrac{(a^2b^3c)^2}{(-2ab^2c)^3} \cdot \dfrac{(a^3b^2c)^3}{(abc)^4} = \dfrac{a^4b^6c^2a^9b^6c^3}{-8a^3b^6c^3a^4b^4c^4}$

$$= -\dfrac{a^6b^2}{8c^2}$$

61. $\dfrac{(2mn)^3}{6mn^2} \cdot \dfrac{(m^2n)^4}{2m^2n^3} = \dfrac{8m^3n^3m^8n^4}{12m^3n^5} = \dfrac{2m^8n^2}{3}$

63. $\dfrac{(2x-3)(x+5)}{4(x-5)(x+5)} \cdot \dfrac{(2x+1)(x-5)}{(2x-1)(2x+1)} = \dfrac{2x-3}{8x-4}$

65. We can factor $k^2 - 2km + m^2$ as

$(k-m)^2$ or $(m-k)^2$.

$\dfrac{(k+m)^2}{(m-k)^2} \cdot \dfrac{(m-k)(m+3)}{(m+3)(m+k)} = \dfrac{k+m}{m-k}$

67.

$\dfrac{(5w-2t)(3w+t)}{(3w+t)(9w^2-3wt+t^2)} \cdot \dfrac{3(9w^2-3wt+t^2)}{(5w-2t)(5w+2t)} \cdot \dfrac{(5w+2t)^2}{2(5w+2t)}$

$$= \dfrac{3}{2}$$

69. $\dfrac{x^a}{y^2} \cdot \dfrac{y^{b+2}}{x^{2a}} = \dfrac{y^b}{x^a}$

71. $\dfrac{x^{2a}+x^a-6}{x^{2a}+6x^a+9} \div \dfrac{x^{2a}-4}{x^{2a}+2x^a-3}$

$= \dfrac{(x^a+3)(x^a-2)}{(x^a+3)^2} \cdot \dfrac{(x^a+3)(x^a-1)}{(x^a-2)(x^a+2)}$

$= \dfrac{x^a-1}{x^a+2}$

73.

$\dfrac{m^k v^k + 3v^k - 2m^k - 6}{m^{2k} - 9} \cdot \dfrac{m^{2k} - 2m^k - 3}{v^k m^k - 2m^k + 2v^k - 4}$

$= \dfrac{(m^k+3)(v^k-2)}{(m^k-3)(m^k+3)} \cdot \dfrac{(m^k-3)(m^k+1)}{(v^k-2)(m^k+2)}$

$= \dfrac{m^k+1}{m^k+2}$

75. $\dfrac{1}{40} \div \dfrac{9}{25} = \dfrac{1}{40} \cdot \dfrac{25}{9} = \dfrac{5}{72} \approx 6.9\%$

77. Her rate is $\dfrac{100}{x}$ miles/hour. Since $D = RT$,

in $\dfrac{3}{4}$ hr she traveled $\dfrac{100}{x} \cdot \dfrac{3}{4}$ or $\dfrac{75}{x}$ miles.

4.3 WARM-UPS

1. False, the LCM is 30.   **2.** False, the LCM is $24a^2b^3$.   **3.** True, because $x-1$ is a factor of x^2-1.   **4.** False, the LCD is $x(x+1)$.

5. False, $\dfrac{1}{2} + \dfrac{2}{3} = \dfrac{7}{6}$.

6.   False, $5 + \dfrac{1}{x} = \dfrac{5x}{x} + \dfrac{1}{x} = \dfrac{5x+1}{x}$ for   any nonzero x.   **7.** True, because the expressions are added correctly.   **8.** True, because the expressions are subtracted correctly.

9. True, because $\dfrac{8}{12} + \dfrac{9}{12} = \dfrac{17}{12}$.

10. False, he uses x reams per day

4.3 EXERCISES

1. $\dfrac{3x}{2} + \dfrac{5x}{2} = \dfrac{8x}{2} = 4x$

3. $\dfrac{x-3}{2x} - \dfrac{3x-5}{2x} = \dfrac{x-3-3x+5}{2x} = \dfrac{-2x+2}{2x}$

$= \dfrac{2(-x+1)}{2x} = \dfrac{-x+1}{x}$

5. $\dfrac{3x-4}{2x-4} + \dfrac{2x-6}{2x-4} = \dfrac{5x-10}{2x-4} = \dfrac{5(x-2)}{2(x-2)} = \dfrac{5}{2}$

7. $\dfrac{x^2+4x-6}{x^2-9} - \dfrac{x^2+2x-12}{x^2-9} = \dfrac{2x+6}{x^2-9}$

$$= \dfrac{2(x+3)}{(x-3)(x+3)} = \dfrac{2}{x-3}$$

9. Since $24 = 2^3 \cdot 3$ and $20 = 2^2 \cdot 5$, the LCM $= 2^3 \cdot 3 \cdot 5 = 120$.

11. The LCM for 10 and 15 is 30. So the LCM is $30x^3y$.

13. The highest power of a is 3, the highest power of b is 5, and the highest power of c is 2. So the LCM $= a^3b^5c^2$.

15. Since the three polynomials have no common factors, the LCM is the product of the polynomials: $x(x+2)(x-2)$

17. $4a + 8 = 4(a+2)$, $6a + 12 = 6(a+2)$

LCM $= 12(a+2) = 12a + 24$

19. $x^2 - 1 = (x-1)(x+1)$

$x^2 + 2x + 1 = (x+1)^2$

LCM $= (x-1)(x+1)^2$

21. $x^2 - 4x = x(x-4)$

$x^2 - 16 = (x-4)(x+4)$

$x^2 + 6x + 8 = (x+4)(x+2)$

LCM $= x(x-4)(x+4)(x+2)$

23. $\dfrac{1 \cdot 5}{28 \cdot 5} + \dfrac{3 \cdot 4}{35 \cdot 4} = \dfrac{5}{140} + \dfrac{12}{140} = \dfrac{17}{140}$

25. $\dfrac{7 \cdot 5}{24 \cdot 5} - \dfrac{4 \cdot 8}{15 \cdot 8} = \dfrac{35}{120} - \dfrac{32}{120} = \dfrac{3}{120} = \dfrac{1}{40}$

27. $\dfrac{3}{wz^2} + \dfrac{5}{w^2z} = \dfrac{3 \cdot w}{wz^2 \cdot w} + \dfrac{5 \cdot z}{w^2z \cdot z}$

$= \dfrac{3w}{w^2z^2} + \dfrac{5z}{w^2z^2} = \dfrac{3w + 5z}{w^2z^2}$

29. $\dfrac{2x-3}{8} - \dfrac{x-2}{6} = \dfrac{(2x-3)3}{8 \cdot 3} - \dfrac{(x-2)4}{6 \cdot 4}$

$= \dfrac{6x-9}{24} - \dfrac{4x-8}{24} = \dfrac{2x-1}{24}$

31. $\dfrac{x \cdot 5}{2a \cdot 5} + \dfrac{3x \cdot 2}{5a \cdot 2} = \dfrac{5x}{10a} + \dfrac{6x}{10a} = \dfrac{11x}{10a}$

33. $\dfrac{9}{4y} - x = \dfrac{9}{4y} - \dfrac{x \cdot 4y}{4y} = \dfrac{9}{4y} - \dfrac{4xy}{4y} = \dfrac{9 - 4xy}{4y}$

35. $\dfrac{5}{a+2} - \dfrac{7}{a} = \dfrac{5 \cdot a}{(a+2)a} - \dfrac{7(a+2)}{a(a+2)}$

$= \dfrac{5a}{a(a+2)} - \dfrac{7a+14}{a(a+2)}$

$= \dfrac{5a - (7a+14)}{a(a+2)} = \dfrac{-2a - 14}{a(a+2)}$

37. $\dfrac{1}{a-b} + \dfrac{2}{a+b}$

$= \dfrac{1(a+b)}{(a-b)(a+b)} + \dfrac{2(a-b)}{(a+b)(a-b)}$

$= \dfrac{a+b}{(a-b)(a+b)} + \dfrac{2a-2b}{(a-b)(a+b)}$

$= \dfrac{3a-b}{(a-b)(a+b)}$

39. $\dfrac{x}{(x+3)(x-3)} + \dfrac{3(x+3)}{(x-3)(x+3)}$

$= \dfrac{4x+9}{(x+3)(x-3)}$

41. $\dfrac{1}{a-b} + \dfrac{1(-1)}{(b-a)(-1)} = \dfrac{1}{a-b} + \dfrac{-1}{a-b}$

$= \dfrac{0}{a-b} = 0$

43. $\dfrac{5}{2(x-2)} - \dfrac{3(-2)}{(2-x)(-2)} = \dfrac{5}{2x-4} - \dfrac{-6}{2x-4}$

$= \dfrac{11}{2x-4}$

45. $\dfrac{5(x+3)}{(x+2)(x-1)(x+3)} - \dfrac{6(x+2)}{(x+3)(x-1)(x+2)}$

$= \dfrac{5x+15}{(x-1)(x+2)(x+3)} - \dfrac{6x+12}{(x-1)(x+2)(x+3)}$

$= \dfrac{-x+3}{(x-1)(x+2)(x+3)}$

47. $\dfrac{x(x+1)}{(x-3)(x+3)(x+1)} + \dfrac{6(x-3)}{(x+3)(x+1)(x-3)}$

$= \dfrac{x^2+x}{(x+1)(x+3)(x-3)} + \dfrac{6x-18}{(x+1)(x+3)(x-3)}$

$= \dfrac{x^2+7x-18}{(x+1)(x+3)(x-3)}$

49. $\dfrac{1(x-1)(x+2)}{x(x-1)(x+2)} + \dfrac{2x(x+2)}{(x-1)x(x+2)} - \dfrac{3x(x-1)}{(x+2)x(x-1)}$

$= \dfrac{x^2+x-2}{x(x-1)(x+2)} + \dfrac{2x^2+4x}{x(x-1)(x+2)} - \dfrac{3x^2+3x}{x(x-1)(x+2)}$

$= \dfrac{8x-2}{x(x-1)(x+2)}$

51. $\dfrac{1}{3} + \dfrac{1}{4} = \dfrac{1 \cdot 4 + 1 \cdot 3}{12} = \dfrac{7}{12}$

53. $\dfrac{1}{8} - \dfrac{3}{5} = \dfrac{1 \cdot 5 - 8 \cdot 3}{40} = -\dfrac{19}{40}$

55. $\dfrac{x}{3} + \dfrac{x}{2} = \dfrac{x \cdot 2 + x \cdot 3}{6} = \dfrac{5x}{6}$

57. $\dfrac{a}{b} - \dfrac{2}{3} = \dfrac{3 \cdot a - 2 \cdot b}{3b} = \dfrac{3a - 2b}{3b}$

59. $a + \dfrac{2}{3} = \dfrac{a \cdot 3 + 1 \cdot 2}{3 \cdot 1} = \dfrac{3a + 2}{3}$

61. $\dfrac{3}{a} + 1 = \dfrac{a \cdot 1 + 3 \cdot 1}{a \cdot 1} = \dfrac{a + 3}{a}$

63. $\dfrac{3+x}{x} - 1 = \dfrac{3 + x - x}{x} = \dfrac{3}{x}$

65. $\dfrac{2}{3} + \dfrac{1}{4x} = \dfrac{2 \cdot 4x + 1 \cdot 3}{12x} = \dfrac{8x + 3}{12x}$

67. $\dfrac{w^2 - 3w + 6}{w - 5} + \dfrac{9 - w^2}{w - 5} = \dfrac{-3w + 15}{w - 5}$

$= \dfrac{-3(w - 5)}{w - 5} = -3$

69. $\dfrac{1}{x+2} - \dfrac{2}{x+3} = \dfrac{1(x+3) - 2(x+2)}{(x+2)(x+3)}$

$= \dfrac{-x - 1}{(x+2)(x+3)}$

71. $\dfrac{1}{a^3 - 1} - \dfrac{1}{a^3 + 1} = \dfrac{a^3 + 1 - (a^3 - 1)}{(a^3 - 1)(a^3 + 1)}$

$= \dfrac{2}{(a^3 - 1)(a^3 + 1)}$

73. $\dfrac{a^8 b^{12}}{a^3 b^{12}} \cdot \dfrac{a^3 b^3}{a^8 b^2} = \dfrac{a^{11} b^{15}}{a^{11} b^{14}} = b$

75. $\dfrac{x^2 - 3x}{(x-1)(x^2+x+1)} + \dfrac{4(x^2+x+1)}{(x-1)(x^2+x+1)}$

$= \dfrac{x^2 - 3x}{(x-1)(x^2+x+1)} + \dfrac{4x^2 + 4x + 4}{(x-1)(x^2+x+1)}$

$= \dfrac{5x^2 + x + 4}{(x-1)(x^2 + x + 1)}$

77. $\dfrac{x^2 + 25}{(x-5)(x+5)} \cdot \dfrac{(x+5)^2}{x(x+5)} = \dfrac{x^2 + 25}{x(x-5)}$

79. $\dfrac{(w^2 - 3)}{3(w+3)(w^2 - 3w + 9)} - \dfrac{1(w^2 - 3w + 9)}{3(w+3)(w^2 - 3w + 9)}$

$- \dfrac{3(w-4)(w+3)}{3(w+3)(w^2 - 3w + 9)}$

$= \dfrac{w^2 - 3 - w^2 + 3w - 9 - 3(w^2 - w - 12)}{3(w+3)(w^2 - 3w + 9)}$

$= \dfrac{-w^2 + 2w + 8}{(w+3)(w^2 - 3w + 9)}$

81. $\dfrac{(a-3)^2}{(a-2)(a^2 + 2a + 4)} \cdot \dfrac{(a-2)(a+2)}{(a-3)(a+2)}$

$= \dfrac{a - 3}{a^2 + 2a + 4}$

83. $\dfrac{(w^2 + 3)(w + 2)}{(w-2)(w^2 + 2w + 4)(w + 2)}$

$- \dfrac{2w(w^2 + 2w + 4)}{(w-2)(w+2)(w^2 + 2w + 4)}$

$= \dfrac{w^3 + 3w + 2w^2 + 6 - (2w^3 + 4w^2 + 8w)}{(w-2)(w+2)(w^2 + 2w + 4)}$

$= \dfrac{-w^3 - 2w^2 - 5w + 6}{(w-2)(w+2)(w^2 + 2w + 4)}$

85. $\dfrac{x + 1}{(x-1)(x^2 + x + 1)(x + 1)}$

$- \dfrac{x^2 + x + 1}{(x-1)(x+1)(x^2 + x + 1)} + \dfrac{(x+1)(x^2 + x + 1)}{(x-1)(x+1)(x^2 + x + 1)}$

$= \dfrac{x + 1 - x^2 - x - 1 + x^3 + x^2 + x + x^2 + x + 1}{(x-1)(x+1)(x^2 + x + 1)}$

$= \dfrac{x^3 + x^2 + 2x + 1}{(x-1)(x+1)(x^2 + x + 1)}$

87. Joe processes $\dfrac{1}{x}$ claims/hr while Ellen processes $\dfrac{1}{x+1}$ claims/hr. Together they process $\dfrac{1}{x} + \dfrac{1}{x+1}$ or $\dfrac{2x+1}{x(x+1)}$ claims/hr. In 8 hours they process $\dfrac{8(2x+1)}{x(x+1)}$ or $\dfrac{16x+8}{x^2+x}$ claims.

89. George's rate is $\dfrac{1}{20}$ mag/min while Theresa's rate is $\dfrac{1}{x}$ mag/min. Their rate together is $\dfrac{1}{x} + \dfrac{1}{20}$ or $\dfrac{x+20}{20x}$ mag/min in 60 min they will sell $60 \cdot \dfrac{x+20}{20x}$ or $\dfrac{3x+60}{x}$ magazines.

91. Let x be Joan's original speed and $x + 5$ be her increased speed. Since $T = \dfrac{D}{R}$, here total travel time is $\dfrac{100}{x} + \dfrac{200}{x+5}$ or $\dfrac{300x + 500}{x^2 + 5x}$ hours.

4.4 WARM-UPS

1. False, the LCM is $6x^2$.   2. True, since $2b - 2a = -2(a-b)$.   3. True.

4. True, because $\frac{1}{2} + \frac{1}{3} = \frac{5}{6}$ and $1 + \frac{1}{2} = \frac{3}{2}$.

5. False, because $2^{-1} + 3^{-1} = \frac{1}{2} + \frac{1}{3} = \frac{5}{6}$ and $(2+3)^{-1} = \frac{1}{5}$.

6. False, because the left side is the reciprocal of $\frac{5}{6}$ and the right side is 5.

7. False, because $2 + 3^{-1} = \frac{7}{3}$ and $5^{-1} = \frac{1}{5}$.

8. False, because $x + 2^{-1} = x + \frac{1}{2} = \frac{2x+1}{2}$ for any value of x.

9. True, because ab is the LCD.

10. True, because multiplying by a^5b^2 will eliminate all negative exponents.

4.4 EXERCISES

1. $\dfrac{\left(\frac{1}{2} - \frac{1}{3}\right)60}{\left(\frac{1}{4} - \frac{1}{5}\right)60} = \dfrac{30 - 20}{15 - 12} = \dfrac{10}{3}$

3. $\dfrac{24 \cdot \frac{2}{3} + 24 \cdot \frac{5}{6} - 24 \cdot \frac{1}{2}}{24 \cdot \frac{1}{8} - 24 \cdot \frac{1}{3} + 24 \cdot \frac{1}{12}} = \dfrac{16 + 20 - 12}{3 - 8 + 2}$

$= -8$

5. $\dfrac{ab\left(\frac{a+b}{b}\right)}{ab\left(\frac{a-b}{ab}\right)} = \dfrac{a^2 + ab}{a - b}$

7. $\dfrac{ab \cdot a + ab \cdot \frac{3}{b}}{ab \cdot \frac{b}{a} + ab \cdot \frac{1}{b}} = \dfrac{a^2b + 3a}{a + b^2}$

9. $\dfrac{xy\left(\frac{x-3y}{xy}\right)}{xy \cdot \frac{1}{x} + xy \cdot \frac{1}{y}} = \dfrac{x - 3y}{x + y}$

11. $\dfrac{18m \cdot 3 - 18m \cdot \frac{m-2}{6}}{18m \cdot \frac{4}{9} + 18m \cdot \frac{2}{m}} = \dfrac{54m - 3m^2 + 6m}{8m + 36}$

$= \dfrac{60m - 3m^2}{8m + 36}$

13. $\dfrac{a^3b^3\left(\frac{a^2 - b^2}{a^2b^3}\right)}{a^3b^3\left(\frac{a+b}{a^3b}\right)} = \dfrac{a(a^2 - b^2)}{b^2(a+b)}$

$= \dfrac{a(a-b)(a+b)}{b^2(a+b)} = \dfrac{a^2 - ab}{b^2}$

15. $\dfrac{\frac{1}{x^2y^2} + \frac{1}{xy^3}}{\frac{1}{x^3y} - \frac{1}{xy}} = \dfrac{\left(\frac{1}{x^2y^2} + \frac{1}{xy^3}\right)x^3y^3}{\left(\frac{1}{x^3y} - \frac{1}{xy}\right)x^3y^3}$

$= \dfrac{xy + x^2}{y^2 - x^2y^2}$

17. $\dfrac{(x+4)x + (x+4) \cdot \frac{4}{x+4}}{(x+4)x - (x+4) \cdot \frac{4x+4}{x+4}}$

$= \dfrac{x^2 + 4x + 4}{x^2 + 4x - 4x - 4} = \dfrac{(x+2)^2}{(x-2)(x+2)} = \dfrac{x+2}{x-2}$

19. $\dfrac{(y-1)(y+1)\left(1 - \frac{1}{y-1}\right)}{(y-1)(y+1)\left(3 + \frac{1}{y+1}\right)}$

$= \dfrac{y^2 - 1 - (y+1)}{3(y^2 - 1) + y - 1} = \dfrac{y^2 - y - 2}{3y^2 + y - 4}$

$= \dfrac{y^2 - y - 2}{(y-1)(3y+4)}$

21. $\dfrac{(x-3) \cdot \frac{2}{3-x} - (x-3)4}{(x-3) \cdot \frac{1}{x-3} - (x-3)1}$

$= \dfrac{-2 - 4x + 12}{1 - x + 3} = \dfrac{-4x + 10}{-x + 4} = \dfrac{4x - 10}{x - 4}$

23. $\dfrac{w(w-1)\left(\frac{w+2}{w-1} - \frac{w-3}{w}\right)}{w(w-1)\left(\frac{w+4}{w} + \frac{w-2}{w-1}\right)}$

$= \dfrac{w^2 + 2w - (w-1)(w-3)}{(w-1)(w+4) + w^2 - 2w}$

$= \dfrac{w^2 + 2w - w^2 + 4w - 3}{w^2 + 3w - 4 + w^2 - 2w} = \dfrac{6w - 3}{2w^2 + w - 4}$

25. $\dfrac{(a-b)(a+b)\left(\dfrac{1}{a-b}-\dfrac{3}{a+b}\right)}{(a-b)(a+b)\left(\dfrac{2}{b-a}+\dfrac{4}{a+b}\right)}$

$\quad = \dfrac{a+b-3(a-b)}{-2(a+b)+4(a-b)} = \dfrac{-2a+4b}{2a-6b}$

$\quad = \dfrac{2(2b-a)}{2(a-3b)} = \dfrac{2b-a}{a-3b}$

27. $\dfrac{(a-1)3-(a-1)\cdot\dfrac{4}{a-1}}{(a-1)5-(a-1)\cdot\dfrac{3}{1-a}}$

$\quad\quad = \dfrac{3a-3-4}{5a-5+3} = \dfrac{3a-7}{5a-2}$

29. $\dfrac{m(m-3)(m-2)\left(\dfrac{2}{m-3}+\dfrac{4}{m}\right)}{m(m-3)(m-2)\left(\dfrac{3}{m-2}+\dfrac{1}{m}\right)}$

$\quad = \dfrac{2m(m-2)+4(m-3)(m-2)}{3m(m-3)+(m-3)(m-2)}$

$\quad = \dfrac{2m^2-4m+4m^2-20m+24}{3m^2-9m+m^2-5m+6}$

$\quad = \dfrac{6m^2-24m+24}{4m^2-14m+6} = \dfrac{3m^2-12m+24}{2m^2-7m+3}$

$\quad = \dfrac{3m^2-12m+12}{(m-3)(2m-1)}$

31. $\dfrac{(x-1)(x+1)(x^2+x+1)\left(\dfrac{3}{x^2-1}-\dfrac{x-2}{x^3-1}\right)}{(x-1)(x+1)(x^2+x+1)\left(\dfrac{3}{x^2+x+1}+\dfrac{x-3}{x^3-1}\right)}$

$\quad = \dfrac{3x^2+3x+3-(x+1)(x-2)}{3x^2-3+(x+1)(x-3)}$

$\quad = \dfrac{3x^2+3x+3-x^2+x+2}{3x^2-3+x^2-2x-3}$

$\quad = \dfrac{2x^2+4x+5}{4x^2-2x-6}$

33. $\dfrac{wyz(w^{-1}+y^{-1})}{wyz(z^{-1}+y^{-1})} = \dfrac{yz+wz}{wy+wz}$

35. $\dfrac{x^2(1-x^{-1})}{x^2(1-x^{-2})} = \dfrac{x^2-x}{x^2-1} = \dfrac{x(x-1)}{(x-1)(x+1)}$

$\quad\quad\quad\quad = \dfrac{x}{x+1}$

37. $\dfrac{a^2b^2(a^{-2}+b^{-2})}{a^2b^2(a^{-1}b)} = \dfrac{a^2+b^2}{ab^3}$

39. $\dfrac{a(1-a^{-1})}{a} = \dfrac{a-1}{a}$

41. $\dfrac{x^2(x^{-1}+x^{-2})}{x^2(x+x^{-2})} = \dfrac{x+1}{x^3+1}$

$\quad = \dfrac{x+1}{(x+1)(x^2-x+1)} = \dfrac{1}{x^2-x+1}$

43. $\dfrac{m^2(2m^{-1}-3m^{-2})}{m^2(m^{-2})} = 2m-3$

45. $\dfrac{ab(a^{-1}-b^{-1})}{ab(a-b)} = \dfrac{b-a}{-ab(b-a)} = -\dfrac{1}{ab}$

47. $\dfrac{x^3-y^3}{x^{-3}-y^{-3}} = \dfrac{(x^3-y^3)x^3y^3}{(x^{-3}-y^{-3})x^3y^3}$

$\quad = \dfrac{(x^3-y^3)x^3y^3}{y^3-x^3} = (-1)x^3y^3 = -x^3y^3$

49. $\dfrac{x^3(1-8x^{-3})}{x^3(x^{-1}+2x^{-2}+4x^{-3})}$

$\quad = \dfrac{x^3-8}{x^2+2x+4} = \dfrac{(x-2)(x^2+2x+4)}{x^2+2x+4}$

$\quad\quad\quad\quad = x-2$

51. $(x^{-1}+y^{-1})^{-1} = \dfrac{1}{x^{-1}+y^{-1}}$

$\quad\quad = \dfrac{xy\cdot 1}{xy(x^{-1}+y^{-1})} = \dfrac{xy}{x+y}$

53. $\dfrac{\dfrac{5}{3}-\dfrac{4}{5}}{\dfrac{1}{3}-\dfrac{5}{6}} \approx -1.7333$

55. $\dfrac{4^{-1}-9^{-1}}{2^{-1}+3^{-1}} \approx 0.1667$

57. If x = the number of students at Central, then $\frac{1}{2}$x = the number at Northside and $\frac{2}{3}$x = the number at Southside. To find the

61

percentage of black students among the city's elementary students we divide the number of black students by the total number of students:

$$\frac{\frac{1}{3}\cdot\frac{1}{2}x+\frac{3}{4}\cdot x+\frac{1}{6}\cdot\frac{2}{3}x}{\frac{1}{2}x+x+\frac{2}{3}x}=\frac{\frac{1}{6}x+\frac{3}{4}x+\frac{1}{9}x}{\frac{1}{2}x+x+\frac{2}{3}x}=$$

$$=\frac{36\left(\frac{1}{6}x+\frac{3}{4}x+\frac{1}{9}x\right)}{36\left(\frac{1}{2}x+x+\frac{2}{3}x\right)}=\frac{6x+27x+4x}{18x+36x+24x}$$

$$=\frac{37x}{78x}=\frac{37}{78}=47.4\%$$

59. Let $x=$ the distance and $\frac{x}{45}=$ the time from Clarksville to Leesville. We can also say that $x=$ the distance and $\frac{x}{55}=$ the time for the return trip. To find the average speed for any trip we divide the total distance by the total time. In this case the total distance is $2x$ and the total time is $\frac{x}{45}+\frac{x}{55}$.

$$\frac{2x}{\frac{x}{45}+\frac{x}{55}}=\frac{2x\cdot5\cdot9\cdot11}{\left(\frac{x}{45}+\frac{x}{55}\right)5\cdot9\cdot11}=\frac{990x}{11x+9x}$$

$$=\frac{990x}{20x}=\frac{99}{2}=49.5$$

Her average speed for the trip is 49.5 mph.

4.5 WARM-UPS

1. True, because that will eliminate all denominators. **2.** False, we should multiply each side by 6x. **3.** False, extraneous roots are real numbers that are not roots to the equation. **4.** True, x^2-4 is the LCD. **5.** False, $-\frac{1}{2}$ cannot be a solution because it causes $2x+1$ to be 0. **6.** True, because the equation is equivalent to $2x=15$. **7.** False, because the extremes-means property is only applied to equations of the type $\frac{a}{b}=\frac{c}{d}$. **8.** False, because $x^2=x$ has two solutions, 0 and 1. **9.** False, the solution set is $\left\{\frac{3}{2},-\frac{4}{3}\right\}$. **10.** True, because of the extremes-means property.

4.5 EXERCISES

1.
$$24x\left(\frac{1}{x}+\frac{1}{6}\right)=24x\left(\frac{1}{8}\right)$$
$$24+4x=3x$$
$$x=-24$$
Solution set: $\{-24\}$

3.
$$30x\left(\frac{2}{3x}+\frac{1}{15x}\right)=30x\left(\frac{1}{2}\right)$$
$$20+2=15x$$
$$22=15x$$
$$\frac{22}{15}=x$$
Solution set: $\left\{\frac{22}{15}\right\}$

5.
$$x(x-2)\left(\frac{3}{x-2}+\frac{5}{x}\right)=x(x-2)\left(\frac{10}{x}\right)$$
$$3x+5x-10=10x-20$$
$$8x-10=10x-20$$
$$10=2x$$
$$5=x$$
Solution set: $\{5\}$

7.
$$x(x-2)\left(\frac{x}{x-2}+\frac{3}{x}\right)=x(x-2)2$$
$$x^2+3x-6=2x^2-4x$$
$$-x^2+7x-6=0$$
$$x^2-7x+6=0$$
$$(x-6)(x-1)=0$$
$$x-6=0 \quad\text{or}\quad x-1=0$$
$$x=6 \quad\text{or}\quad x=1$$
Solution set: $\{1, 6\}$

9.
$$x(x+5)\left(\frac{100}{x}\right)=x(x+5)\left(\frac{150}{x+5}-1\right)$$
$$100x+500=150x-x(x+5)$$
$$100x+500=150x-x^2-5x$$
$$x^2-45x+500=0$$
$$(x-25)(x-20)=0$$
$$x=25 \quad\text{or}\quad x=20$$
Solution set: $\{20, 25\}$

11.
$$(x-1)\left(\frac{3x-5}{x-1}\right)=(x-1)\left(2-\frac{2x}{x-1}\right)$$
$$3x-5=2x-2-2x$$
$$3x-5=-2$$
$$3x=3$$
$$x=1$$
Since replacing x by 1 gives 0 in the denominator, 1 is an extraneous root. The solution set is $\emptyset$.

13.
$$x + 1 + \frac{2x-5}{x-5} = \frac{x}{x-5}$$
$$(x-5)\left(x + 1 + \frac{2x-5}{x-5}\right) = (x-5)\frac{x}{x-5}$$
$$x^2 - 4x - 5 + 2x - 5 = x$$
$$x^2 - 3x - 10 = 0$$
$$(x-5)(x+2) = 0$$
$$x = 5 \quad \text{or} \quad x = -2$$

Since $x - 5$ is in the denominator, 5 is an extraneous root. The solution set is $\{-2\}$

15.
$$\frac{2}{x} = \frac{3}{4}$$
$$3x = 8$$
$$x = \frac{8}{3}$$
Solution set: $\left\{\frac{8}{3}\right\}$

17.
$$\frac{a}{3} = \frac{-1}{4}$$
$$4a = -3$$
$$a = -\frac{3}{4}$$
Solution set: $\left\{-\frac{3}{4}\right\}$

19.
$$\frac{-5}{7} = \frac{2}{x}$$
$$-5x = 14$$
$$x = -\frac{14}{5}$$
Solution set: $\left\{-\frac{14}{5}\right\}$

21.
$$\frac{10}{x} = \frac{20}{x+20}$$
$$10x + 200 = 20x$$
$$-10x = -200$$
$$x = 20$$
Solution set: $\{20\}$

23.
$$\frac{2}{x+1} = \frac{x-1}{4}$$
$$x^2 - 1 = 8$$
$$x^2 - 9 = 0$$
$$(x-3)(x+3) = 0$$
$$x = 3 \quad \text{or} \quad x = -3$$
Solution set: $\{-3, 3\}$

25.
$$\frac{x}{6} = \frac{5}{x-1}$$
$$x^2 - x = 30$$
$$x^2 - x - 30 = 0$$
$$(x-6)(x+5) = 0$$
$$x = 6 \quad \text{or} \quad x = -5$$
Solution set: $\{-5, 6\}$

27.
$$\frac{x}{x-3} = \frac{x+2}{x}$$
$$x^2 = (x+2)(x-3)$$
$$x^2 = x^2 - x - 6$$
$$x = -6$$
Solution set: $\{-6\}$

29.
$$\frac{x-2}{x-3} = \frac{x+5}{x+2}$$
$$(x-2)(x+2) = (x-3)(x+5)$$
$$x^2 - 4 = x^2 + 2x - 15$$
$$-2x = -11$$
$$x = \frac{11}{2}$$
Solution set: $\left\{\frac{11}{2}\right\}$

31.
$$\frac{a}{9} = \frac{4}{a}$$
$$a^2 = 36$$
$$a^2 - 36 = 0$$
$$(a-6)(a+6) = 0$$
$$a = 6 \quad \text{or} \quad a = -6$$
Solution set: $\{-6, 6\}$

33. $4(x-2)\left(\frac{1}{2(x-2)} + \frac{1}{x-2}\right) = 4(x-2)\cdot\frac{1}{4}$
$$2 + 4 = x - 2$$
$$8 = x$$
Solution set: $\{8\}$

35.
$$\frac{x-2}{4} = \frac{x-2}{x}$$
$$x^2 - 2x = 4x - 8$$
$$x^2 - 6x + 8 = 0$$
$$(x-2)(x-4) = 0$$
$$x = 2 \quad \text{or} \quad x = 4$$
Solution set: $\{2, 4\}$

37. $2(x+2)(x-1)\left(\frac{5}{2(x+2)} - \frac{1}{x-1}\right)$
$$= 2(x+2)(x-1)\left(\frac{3}{x+2}\right)$$
$$5x - 5 - 2(x+2) = 6(x-1)$$
$$5x - 5 - 2x - 4 = 6x - 6$$
$$3x - 9 = 6x - 6$$
$$-3x = 3$$
$$x = -1$$
Solution set: $\{-1\}$

39.
$$\frac{5}{x-3} = \frac{x}{x-3}$$
$$5x - 15 = x^2 - 3x$$
$$-x^2 + 8x - 15 = 0$$
$$x^2 - 8x + 15 = 0$$
$$(x-3)(x-5) = 0$$
$$x = 3 \quad \text{or} \quad x = 5$$

Since replacing x by 3 causes 0 to appear in a denominator, 3 is an extraneous root. The solution set is $\{5\}$.

41.
$$\frac{w}{6} = \frac{3}{2w}$$
$$2w^2 = 18$$
$$w^2 = 9$$
$$w^2 - 9 = 0$$
$$(w-3)(w+3) = 0$$
$$w = 3 \quad \text{or} \quad w = -3$$

Solution set: $\{-3, 3\}$

43. $6(2x-1)(x+2)\left(\frac{5}{2(2x-1)} - \frac{-1}{2x-1}\right)$
$$= 6(2x-1)(x+2)\left(\frac{7}{3(x+2)}\right)$$

$$15(x+2) + 6(x+2) = 14(2x-1)$$
$$15x + 30 + 6x + 12 = 28x - 14$$
$$21x + 42 = 28x - 14$$
$$-7x = -56$$
$$x = 8$$

Solution set: $\{8\}$

45.
$$\frac{5}{x} = \frac{2}{5}$$
$$2x = 25$$
$$x = \frac{25}{2}$$

Solution set: $\left\{\frac{25}{2}\right\}$

47. $(x-3)(x+3)\left(\frac{5}{x^2-9} + \frac{2}{x+3}\right)$
$$= (x-3)(x+3)\left(\frac{1}{x-3}\right)$$
$$5 + 2x - 6 = x + 3$$
$$x = 4$$

Solution set: $\{4\}$

49. $(x-1)(x^2+x+1)\left(\frac{9}{x^3-1} - \frac{1}{x-1}\right)$
$$= (x-1)(x^2+x+1)\left(\frac{2}{x^2+x+1}\right)$$
$$9 - (x^2+x+1) = 2(x-1)$$

$$9 - x^2 - x - 1 = 2x - 2$$
$$-x^2 - 3x + 10 = 0$$
$$x^2 + 3x - 10 = 0$$
$$(x+5)(x-2) = 0$$
$$x = -5 \quad \text{or} \quad x = 2$$

Solution set: $\{-5, 2\}$

51.
$$\frac{300,000}{250,000} = \frac{200,000}{a}$$
$$300,000a = 200,000 \cdot 250,000$$
$$a = \frac{200,000 \cdot 250,000}{300,000}$$
$$a = \$166,666.67$$

53.
$$\frac{2}{23} = \frac{12}{p}$$
$$2p = 12 \cdot 23$$
$$p = \frac{12 \cdot 23}{2} = 138$$

55. Let w = the width and w + 22 = the length.
$$\frac{7}{6} = \frac{w+22}{w}$$
$$7w = 6w + 132$$
$$w = 132$$
$$w + 22 = 154$$

The width is 132 cm and the length is 154 cm.

57.
$$1,000,000 = \frac{4,000,000p}{100-p}$$

$$100,000,000 - 1,000,000p = 4,000,000p$$
$$100,000,000 = 5,000,000p$$
$$p = 20$$

For \$1 million, 20% of the pollution can be cleaned up. For \$100 million, 96% of the pollution can be cleaned up.

59. Let s = the amount invested in stocks and s − 20,000 = the amount invested in bonds. For a wealth-building portfolio, the ratio of stocks to bonds should be 65 to 30:
$$\frac{65}{30} = \frac{s}{s-20,000}$$
$$65s - 1,300,000 = 30s$$
$$35s = 1,300,000$$
$$s = \$37,142.86$$
$$s - 20,000 = \$17,142.86$$

She invested \$17,142.86 in bonds and her annual bonus was \$54,285.72

4.6 WARM-UPS

1. False, t must not appear on both sides of the formula.   2. True, pqs is the LCD.

3. False, the price is $\frac{x}{50}$ dollars per pound.

4. True, rate is distance divided by time.
5. True, this is his rate.
6. True, this is the rate at which he is working.
7. False, the correct equation is $y - 1 = x$.
8. True, multiply by B and divide by m.

9. False, if $a = \frac{x}{y}$ then $y = \frac{x}{a}$.

10. False, the correct equation is $x - 3 = y$.

4.6 EXERCISES

1. $(x-2)\left(\dfrac{y-3}{x-2}\right) = (x-2)(5)$

$$y - 3 = 5x - 10$$
$$y = 5x - 7$$

3. $(x-6)\left(\dfrac{y+1}{x-6}\right) = -\dfrac{1}{3}(x-6)$

$$y + 1 = -\dfrac{1}{3}x + 2$$

$$y = -\dfrac{1}{3}x + 1$$

5. $(x-b)\left(\dfrac{y-a}{x-b}\right) = m(x-b)$

$$y - a = mx \ - mb$$
$$y = mx - bm + a$$

7. $(x+5)\left(\dfrac{y-2}{x+5}\right) = -\dfrac{7}{3}(x+5)$

$$y - 2 = -\dfrac{7}{3}x - \dfrac{35}{3}$$

$$y = -\dfrac{7}{3} - \dfrac{35}{3} + \dfrac{6}{3}$$

$$y = -\dfrac{7}{3}x - \dfrac{29}{3}$$

9. $f \cdot M = f \cdot \dfrac{F}{f}$

$$fM = F$$

$$f = \dfrac{F}{M}$$

11. $4 \cdot A = 4 \cdot \dfrac{\pi}{4} \cdot D^2$

$$4A = \pi D^2$$

$$D^2 = \dfrac{4A}{\pi}$$

13. $Fr^2 = km_1m_2$

$$m_1 = \dfrac{Fr^2}{km_2}$$

15. $pqf\left(\dfrac{1}{p} + \dfrac{1}{q}\right) = pqf \cdot \dfrac{1}{f}$

$$qf + pf = pq$$

$$qf - pq = -pf$$

$$q(f - p) = -pf$$

$$q = \dfrac{-pf}{f - p}$$

$$q = \dfrac{pf}{p - f}$$

17. $a^2(e^2) = a^2\left(1 - \dfrac{b^2}{a^2}\right)$

$$a^2e^2 = a^2 - b^2$$

$$a^2e^2 - a^2 = -b^2$$

$$a^2(e^2 - 1) = -b^2$$

$$a^2 = \dfrac{-b^2}{e^2 - 1} = \dfrac{b^2}{1 - e^2}$$

19. $T_1T_2\left(\dfrac{P_1V_1}{T_1}\right) = T_1T_2\left(\dfrac{P_2V_2}{T_2}\right)$

$$T_2P_1V_1 = T_1P_2V_2$$

$$\dfrac{T_2P_1V_1}{P_2V_2} = \dfrac{T_1P_2V_2}{P_2V_2}$$

$$T_1 = \dfrac{P_1V_1T_2}{P_2V_2}$$

21. $3V = 3 \cdot \dfrac{4}{3}\pi r^2 h$

$$3V = 4\pi r^2 h$$

$$\dfrac{3V}{4\pi r^2} = \dfrac{4\pi r^2 h}{4\pi r^2}$$

$$h = \dfrac{3V}{4\pi r^2}$$

23. $10 = \dfrac{5}{f}$

$$10f = 5$$

$$f = \dfrac{1}{2}$$

25.
$$6\pi = \frac{\pi}{4}D^2$$
$$\frac{4}{\pi} \cdot 6\pi = \frac{4}{\pi} \cdot \frac{\pi}{4}D^2$$
$$24 = D^2$$

27.
$$32 = k\frac{6 \cdot 8}{4^2}$$
$$32 = 3k$$
$$k = \frac{32}{3}$$

29.
$$\frac{1}{p} + \frac{1}{1.7} = \frac{1}{2.3}$$
$$\frac{1}{p} = -0.15345$$
$$p = \frac{1}{-0.15345} = -6.517$$

31.
$$(0.62)^2 = 1 - \frac{(3.5)^2}{a^2}$$
$$-0.6156 = -\frac{12.25}{a^2}$$
$$a^2 = \frac{12.25}{0.6156} = 19.899$$

33.
$$25.6 = \frac{4}{3}\pi r^2(3.2)$$
$$25.6 = 13.404r^2$$
$$r^2 = \frac{25.6}{13.404} = 1.910$$

35. Let $x =$ her walking speed and $x + 10 =$ her riding speed. Since $T = D/R$, her time to school is $7/(x + 10)$ and her time to the post office is $2/x$. Since the times are equal, we can write
$$\frac{2}{x} = \frac{7}{x + 10}$$
$$7x = 2x + 20$$
$$5x = 20$$
$$x = 4$$
She walks 4 mph.

37. Let $x =$ the speed of each. Since $T = \frac{D}{R}$, Patrick's time is $\frac{40}{x}$ and Guy's time is $\frac{60}{x}$. Since Guy's time is $\frac{1}{5}$ hr longer than Patrick's we can write the equation
$$\frac{60}{x} - \frac{1}{5} = \frac{40}{x}$$
$$5x\left(\frac{60}{x} - \frac{1}{5}\right) = 5x \cdot \frac{40}{x}$$
$$300 - x = 200$$

$$100 = x$$
They are both driving 100 mph. Patrick takes $\frac{40}{100} = \frac{2}{5}$ hr = 24 minutes to get to work and Guy takes 12 minutes longer or 36 minutes.

39. Let $x =$ his walking speed and $x + 6 =$ his running speed. His time walking was $\frac{1}{x}$ hours and his time running was $\frac{5}{x + 6}$ hours. Since his total time was 3/4 of an hour, we can write the equation
$$\frac{1}{x} + \frac{5}{x + 6} = \frac{3}{4}$$
$$4x(x+6)\left(\frac{1}{x} + \frac{5}{x+6}\right) = 4x(x+6)\frac{3}{4}$$
$$4x + 24 + 20x = 3x^2 + 18x$$
$$-3x^2 + 6x + 24 = 0$$
$$x^2 - 2x - 8 = 0$$
$$(x - 4)(x + 2) = 0$$
$$x = 4 \quad \text{or} \quad x = -2$$
Disregard the negative solution. Use $x = 4$ to find that his running speed was 10 mph.

41. Let $x =$ the number of hours for the smaller pump to drain the pool working alone. In one hour of operation, the larger pump drains $\frac{1}{3}$ of the pool, the smaller pump drains $\frac{1}{x}$ of the pool, and together $\frac{1}{2}$ of the pool is emptied.
$$\frac{1}{3} + \frac{1}{x} = \frac{1}{2}$$
$$6x\left(\frac{1}{3} + \frac{1}{x}\right) = 6x\left(\frac{1}{2}\right)$$
$$2x + 6 = 3x$$
$$6 = x$$
The smaller pump would take 6 hours to drain the pool by itself.

43. Let $x =$ the number of minutes to fill the tub with the drain left open. In one minute, the faucet fills $\frac{1}{10}$ of the tub, the drain takes $\frac{1}{12}$ of the tub, but the tub is $\frac{1}{x}$ full.

$$\frac{1}{10} - \frac{1}{12} = \frac{1}{x}$$
$$60x\left(\frac{1}{10} - \frac{1}{12}\right) = 60x \cdot \frac{1}{x}$$
$$6x - 5x = 60$$
$$x = 60$$
The tub is filled in 60 minutes.

45. Let $x =$ their time working together. Since Gina takes 90 minutes and Hilda works twice as fast, Hilda can do the job alone in 45 minutes. Hilda does $\frac{1}{45}$ of the job per minute and Gina does $\frac{1}{90}$ or the job per minute. Since together they do $\frac{1}{x}$ of the job per minute, we can write the following equation.

$$\frac{1}{45} + \frac{1}{90} = \frac{1}{x}$$
$$90x\left(\frac{1}{45} + \frac{1}{90}\right) = 90x \cdot \frac{1}{x}$$
$$2x + x = 90$$
$$3x = 90$$
$$x = 30$$

It will take them 30 minutes working together.

47. Let $x =$ the number of pounds of apples and $x + 2 =$ the number of pounds of oranges. The apples sell for $\frac{8.80}{x}$ dollars per pound and the oranges sell for $\frac{5.28}{x+2}$. The apples cost twice as much per pound is expressed as

$$\frac{8.80}{x} = 2\left(\frac{5.28}{x+2}\right)$$
$$\frac{8.80}{x} = \frac{10.56}{x+2}$$
$$8.80x + 17.60 = 10.56x$$
$$17.60 = 1.76x$$
$$10 = x$$

She bought 10 pounds of apples and 12 pounds of oranges.

49.
$$\frac{1}{2} = \frac{1}{3} + \frac{1}{R_2}$$
$$6R_2\left(\frac{1}{2}\right) = 6R_2\left(\frac{1}{3} + \frac{1}{R_2}\right)$$
$$3R_2 = 2R_2 + 6$$
$$R_2 = 6 \text{ ohms}$$

51. Let $x =$ the number in the original group. Their cost is $\frac{24,000}{x}$ per person. For 40 more people, the cost is $\frac{24,000}{x+40}$. We have

$$\frac{24,000}{x} = \frac{24,000}{x+40} + 100$$
$$x(x+40)\frac{24,000}{x} = x(x+40)\left(\frac{24,000}{x+40} + 100\right)$$
$$24,000x + 960,000 = 24,000x + 100x^2 + 4000x$$
$$100x^2 + 4000x - 960,000 = 0$$
$$x^2 + 40x - 9600 = 0$$

$$(x - 80)(x + 120) = 0$$
$$x = 80 \quad \text{or} \quad x = -120$$

There are 80 people in the initial group.

53. Let $x =$ the number of days for both dogs to eat a 50 pound bag together. Muffy eats at a rate of $\frac{25}{28}$ pounds per day, Missy eats at the rate of $\frac{25}{23}$ pounds per day, and together they eat at a rate of $\frac{50}{x}$ pounds per day.

$$\frac{25}{28} + \frac{25}{23} = \frac{50}{x}$$
$$1.9798 = \frac{50}{x}$$
$$x = \frac{50}{1.9798} = 25.255$$

It would take 25.255 days for them to eat 50 pounds of dog food together.

CHAPTER 4   REVIEW

1. If $x = 1$, then $3x - 3 = 0$. So the domain is $\{x \mid x \neq 1\}$.

3. Solve $x^2 - x - 2 = 0$
$$(x - 2)(x + 1) = 0$$
$$x - 2 = 0 \quad \text{or} \quad x + 1 = 0$$
$$x = 2 \quad \text{or} \quad x = -1$$
The domain is $\{x \mid x \neq -1 \text{ and } x \neq 2\}$.

5. $\dfrac{a^3bc^3}{a^5b^2c} = \dfrac{c^2}{a^2b}$

7. $\dfrac{2 \cdot 2 \cdot 17x^3}{3 \cdot 17xy} = \dfrac{4x^2}{3y}$

9. $\dfrac{a^3b^2b(a-b)}{b^3a(-a)(a-b)} = -a$

11. $\dfrac{w-4}{3w} \cdot \dfrac{3 \cdot 3w}{2(w-4)} = \dfrac{3}{2}$

13. Since $6x = 2 \cdot 3x$, $3x - 6 = 3(x - 2)$, and $x^2 - 2x = x(x-2)$, the LCM $= 6x(x-2)$.

15. Since $6ab^3 = 2 \cdot 3b^3$ and $4a^5b^2 = 2 \cdot 2a^5b^2$, the LCM $= 2 \cdot 2 \cdot 3a^5b^3 = 12a^5b^3$.

17. $\dfrac{3(x+3)}{2(x-3)(x+3)} + \dfrac{1 \cdot 2}{(x-3)(x+3)2}$

$$= \dfrac{3x+9}{2(x-3)(x+3)} + \dfrac{2}{2(x-3)(x+3)}$$

$$= \dfrac{3x+11}{2(x-3)(x+3)}$$

19. $\dfrac{w \cdot a}{ab^2 \cdot a} - \dfrac{5 \cdot b}{a^2b \cdot b} = \dfrac{aw - 5b}{a^2b^2}$

21. $\dfrac{30x\left(\frac{3}{2x}-\frac{4}{5x}\right)}{30x\left(\frac{1}{3}-\frac{2}{x}\right)}=\dfrac{45-24}{10x-60}=\dfrac{21}{10x-60}$

23. $\dfrac{(y-2)\left(\frac{1}{y-2}-3\right)}{(y-2)\left(\frac{5}{y-2}+4\right)}=\dfrac{1-3(y-2)}{5+4(y-2)}$

$$=\dfrac{7-3y}{4y-3}$$

25. $\dfrac{a^2b^3(a^{-2}-b^{-3})}{a^2b^3(a^{-1}b^{-2})}=\dfrac{b^3-a^2}{ab}$

27.
$$\dfrac{-3}{8}=\dfrac{2}{x}$$
$$-3x=16$$
$$x=-\dfrac{16}{3}$$
Solution set: $\left\{-\dfrac{16}{3}\right\}$

29. $(a-5)(a+5)\left(\dfrac{15}{a^2-25}+\dfrac{1}{a-5}\right)$
$$=(a-5)(a+5)\left(\dfrac{6}{a+5}\right)$$
$$15+a+5=6a-30$$
$$-5a=-50$$
$$a=10$$
Solution set: $\{10\}$

31.
$$m\left(\dfrac{y-b}{m}\right)=mx$$
$$y-b=mx$$
$$y=mx+b$$

33.
$$r\cdot F=r\left(\dfrac{mv^2}{r}\right)$$
$$rF=mv^2$$
$$m=\dfrac{Fr}{v^2}$$

35.
$$3\cdot A=3\cdot\dfrac{2}{3}\pi rh$$
$$3A=2\pi rh$$
$$\dfrac{3A}{2\pi h}=\dfrac{2\pi rh}{2\pi h}$$
$$r=\dfrac{3A}{2\pi h}$$

37.
$$(x-7)\left(\dfrac{y+3}{x-7}\right)=(x-7)2$$
$$y+3=2x-14$$
$$y=2x-17$$

39. $\dfrac{1}{x}+\dfrac{1}{3x}=\dfrac{3}{3x}+\dfrac{1}{3x}=\dfrac{4}{3x}$

41. $\dfrac{5}{3xy}+\dfrac{7}{6x}=\dfrac{5\cdot2}{3xy\cdot2}+\dfrac{7\cdot y}{6x\cdot y}=\dfrac{10+7y}{6xy}$

43. $\dfrac{5}{a-5}-\dfrac{3}{-1(a+5)}=\dfrac{5}{a-5}+\dfrac{3}{a+5}$

$$=\dfrac{5(a+5)}{(a-5)(a+5)}+\dfrac{3(a-5)}{(a+5)(a-5)}$$

$$=\dfrac{8a+10}{(a-5)(a+5)}$$

45. $15(x-2)(x+2)\left(\dfrac{1}{x-2}-\dfrac{1}{x+2}\right)$
$$=15(x-2)(x+2)\left(\dfrac{1}{15}\right)$$
$$15x+30-15(x-2)=x^2-4$$
$$60=x^2-4$$
$$x^2-64=0$$
$$(x-8)(x+8)=0$$
$$x=8 \quad\text{or}\quad x=-8$$
Solution set: $\{-8,8\}$

47. $\dfrac{-3}{x+2}\cdot\dfrac{5(x+2)}{2\cdot5}=-\dfrac{3}{2}$

49.
$$\dfrac{x}{-3}=\dfrac{-27}{x}$$
$$x^2=81$$
$$x^2-81=0$$
$$(x-9)(x+9)=0$$
$$x=9 \quad\text{or}\quad x=-9$$
Solution set: $\{-9,9\}$

51. $\dfrac{(w+3)(x+m)}{(w-3)(w+3)}\cdot\dfrac{w-3}{(x-m)(x+m)}=\dfrac{1}{x-m}$

53. $\dfrac{5(a+1)}{(a-5)(a+5)(a+1)}+\dfrac{3(a+5)}{(a-5)(a+1)(a+5)}$
$$=\dfrac{8a+20}{(a-5)(a+5)(a+1)}$$

55. $\dfrac{-7(a+2)}{2(a-3)(a+3)(a+2)}-\dfrac{4\cdot2(a-3)}{(a+2)(a+3)2(a-3)}$
$$=\dfrac{-7a-14-(8a-24)}{2(a+2)(a+3)(a-3)}$$
$$=\dfrac{-15a+10}{2(a+2)(a+3)(a-3)}$$

57. $(a-1)(a+1)\left(\dfrac{7}{a^2-1}+\dfrac{2}{1-a}\right)$
$$=(a-1)(a+1)\left(\dfrac{1}{a+1}\right)$$
$$7-2(a+1)=a-1$$
$$-2a+5=a-1$$
$$-3a=-6$$
$$a=2$$
Solution set: $\{2\}$

59. $(x-2)(x-3)\left(\dfrac{2x}{x-3}+\dfrac{3}{x-2}\right)$

$$= (x-2)(x-3)\dfrac{6}{(x-2)(x-3)}$$

$2x(x-2)+3(x-3)=6$

$\qquad 2x^2-x-15=0$

$\qquad (2x+5)(x-3)=0$

$\qquad 2x+5=0 \quad\text{or}\quad x-3=0$

$\qquad\qquad x=-\dfrac{5}{2} \quad\text{or}\quad x=3$

Since 3 is an extraneous root, the solution set is $\left\{-\dfrac{5}{2}\right\}$.

61. $\dfrac{x-2}{2\cdot 3}\cdot\dfrac{2}{2-x}=-\dfrac{1}{3}$

63. $\dfrac{x-3}{(x+2)(x+1)}\cdot\dfrac{(x-2)(x+2)}{3(x-3)}=\dfrac{x-2}{3x+3}$

65. $\dfrac{a+4}{(a-2)(a^2+2a+4)}-\dfrac{-3}{a-2}$

$=\dfrac{a+4}{(a-2)(a^2+2a+4)}-\dfrac{-3(a^2+2a+4)}{(a-2)(a^2+2a+4)}$

$=\dfrac{a+4+3a^2+6a+12}{a^3-8}=\dfrac{3a^2+7a+16}{a^3-8}$

67. $\dfrac{x(x-3)(x+3)}{(1-x)(1+x)}\cdot\dfrac{x-1}{x(x+3)^2}=\dfrac{3-x}{(x+1)(x+3)}$

69. $\dfrac{(a+w)(a+3)}{(a+2)(a+4)}\cdot\dfrac{(a+2)(a-w)}{(a-w)(a+3)}=\dfrac{a+w}{a+4}$

71. $5x(x+2)\left(\dfrac{5}{x}-\dfrac{4}{x+2}\right)=5x(x+2)\left(\dfrac{1}{5}+\dfrac{1}{5x}\right)$

$\qquad 25x+50-20x=x^2+2x+x+2$

$\qquad\qquad 5x+50=x^2+3x+2$

$\qquad\qquad -x^2+2x+48=0$

$\qquad\qquad x^2-2x-48=0$

$\qquad\qquad (x-8)(x+6)=0$

$\qquad\qquad x=8 \quad\text{or}\quad x=-6$

Solution set: $\{-6,\ 8\}$

73. $\dfrac{6}{x}=\dfrac{6\cdot 3}{x\cdot 3}=\dfrac{18}{3x}$

75. $\dfrac{3}{a-b}=\dfrac{3(-1)}{(a-b)(-1)}=\dfrac{-3}{b-a}$

77. $4=4\cdot\dfrac{x}{x}=\dfrac{4x}{x}$

79. $5x\div\dfrac{1}{2}=5x\cdot 2=10x$

81. $4a\div\dfrac{1}{3}=4a\cdot 3=12a$

83. $\dfrac{a-3}{a^2-9}=\dfrac{1(a-3)}{(a-3)(a+3)}=\dfrac{1}{a+3}$

85. $\dfrac{1}{2}-\dfrac{1}{5}=\dfrac{5}{10}-\dfrac{2}{10}=\dfrac{3}{10}$

87. $\dfrac{a}{3}+\dfrac{a}{2}=\dfrac{2a}{6}+\dfrac{3a}{6}=\dfrac{5a}{6}$

89. $\dfrac{1}{a}-\dfrac{1}{b}=\dfrac{b}{ab}-\dfrac{a}{ab}=\dfrac{b-a}{ab}$

91. $\dfrac{a}{3}-1=\dfrac{a}{3}-\dfrac{3}{3}=\dfrac{a-3}{3}$

93. $2+\dfrac{1}{a}=\dfrac{2a}{a}+\dfrac{1}{a}=\dfrac{2a+1}{a}$

95. $\dfrac{a}{5}-1=\dfrac{a}{5}-\dfrac{5}{5}=\dfrac{a-5}{5}$

97. $(x-1)(-1)=1-x$

99. $(m-2)\div(2-m)=\dfrac{-1(2-m)}{2-m}=-1$

101. $\dfrac{\frac{b}{3a}}{2}=\dfrac{b}{3a}\cdot\dfrac{1}{2}=\dfrac{b}{6a}$

103. $\dfrac{a-6}{5}\cdot\dfrac{2\cdot 5}{6-a}=-2$

105. $\dfrac{2}{x^a-2}+\dfrac{1}{x^a+3}$

$=\dfrac{2(x^a+3)}{(x^a-2)(x^a+3)}+\dfrac{1(x^a-2)}{(x^a+3)(x^a-2)}$

$=\dfrac{3x^a+4}{(x^a-2)(x^a+3)}$

107. $\dfrac{x^{2k}-9}{3x^{k+3}+9x^3}\cdot\dfrac{6x^{k+5}}{x^{k+2}-3x^2}$

$=\dfrac{(x^k-3)(x^k+3)}{3x^3(x^k+3)}\cdot\dfrac{6x^{k+5}}{x^2(x^k-3)}$

$=2x^k$

109. Let $x=$ the number of deaths in the 50-59 age group and $x+19{,}024=$ the number of deaths in the 40-49 age group.

$$\dfrac{2.875}{1}=\dfrac{x+19{,}024}{x}$$

$\qquad 2.875x=x+19{,}024$

$\qquad 1.875x=19{,}024$

$\qquad\qquad x\approx 10{,}146$

There were 10,146 deaths in the 50-59 age group.

111. If x = the time east of Louisville, then we also have x = the time west of Louisville. His speed east of Louisville was $\frac{310}{x}$ and his speed west of Louisville was $\frac{360}{x}$. We can write an equation expressing the fact that his speed west of Louisville was 10 mph greater than east of Louisville:

$$\frac{360}{x} - 10 = \frac{310}{x}$$
$$x\left(\frac{360}{x} - 10\right) = x\left(\frac{310}{x}\right)$$
$$360 - 10x = 310$$
$$-10x = -50$$
$$x = 5$$

His journey took a total of 10 hours.

113. Let x = the time for all three to make the quilt working together. Debbie makes $\frac{1}{2000}$ of the quilt per hour, Pat makes $\frac{1}{1000}$ of the quilt per hour, and Cheryl makes $\frac{1}{1000}$ of the quilt per hour. Working together, $\frac{1}{x}$ of the quilt gets done per hour:

$$\frac{1}{1000} + \frac{1}{1000} + \frac{1}{2000} = \frac{1}{x}$$
$$2000x\left(\frac{1}{1000} + \frac{1}{1000} + \frac{1}{2000}\right) = 2000x \cdot \frac{1}{x}$$

$$2x + 2x + x = 2000$$
$$5x = 2000$$
$$x = 400$$

It will take them 400 hours to make the quilt together.

CHAPTER 4   TEST

1. The solution to $x^2 - 9 = 0$ is $\{-3, 3\}$. So the domain of the rational expression is $\{x \mid x \neq 3 \text{ and } x \neq -3\}$.

2. The solution to $4 - 3x = 0$ is $\{4/3\}$. So the domain of the rational expression is $\left\{x \mid x \neq \frac{4}{3}\right\}$.

3. $\dfrac{12a^9b^8}{(2a^2b^3)^3} = \dfrac{12a^9b^8}{8a^6b^9} = \dfrac{3a^3}{2b}$

4. $\dfrac{(y-x)(y+x)}{2(x-y)^2} = \dfrac{-1(y+x)}{2(x-y)} = \dfrac{-x-y}{2x-2y}$

5. $\dfrac{5}{12} - \dfrac{4}{9} = \dfrac{15}{36} - \dfrac{16}{36} = -\dfrac{1}{36}$

6. $\dfrac{3}{y} + 7y = \dfrac{3}{y} + \dfrac{7y^2}{y} = \dfrac{7y^2 + 3}{y}$

7. $\dfrac{4}{a-9} - \dfrac{1(-1)}{(9-a)(-1)} = \dfrac{5}{a-9}$

8. $\dfrac{1}{6ab^2} + \dfrac{1}{8a^2b} = \dfrac{1 \cdot 4a}{6ab^2 \cdot 4a} + \dfrac{1 \cdot 3b}{8a^2b \cdot 3b}$

$$= \dfrac{4a}{24a^2b^2} + \dfrac{3b}{24a^2b^2} = \dfrac{4a + 3b}{24a^2b^2}$$

9. $\dfrac{3a^3b}{2 \cdot 2 \cdot 5ab} \cdot \dfrac{2a^2b}{3 \cdot 3ab^3} = \dfrac{a^3}{30b^2}$

10. $\dfrac{a-b}{7} \cdot \dfrac{3 \cdot 7}{(b-a)(b+a)} = -\dfrac{3}{a+b}$

11. $\dfrac{x-3}{x-1} \cdot \dfrac{1}{(x-3)(x+1)} = \dfrac{1}{x^2-1}$

12. $\dfrac{2}{(x-2)(x+2)} - \dfrac{6}{(x-5)(x+2)}$

$$= \dfrac{2(x-5)}{(x-2)(x+2)(x-5)} - \dfrac{6(x-2)}{(x-5)(x+2)(x-2)}$$

$$= \dfrac{2x - 10 - (6x - 12)}{(x+2)(x-2)(x-5)}$$

$$= \dfrac{-4x + 2}{(x+2)(x-2)(x-5)}$$

13. $\dfrac{(m-1)(m^2+m+1)}{(m-1)^2} \cdot \dfrac{(m-1)(m+1)}{3(m^2+m+1)}$

$$= \dfrac{m+1}{3}$$

14.
$$\frac{3}{x} = \frac{7}{4}$$
$$7x = 12$$
$$x = \frac{12}{7}$$

Solution set: $\left\{\frac{12}{7}\right\}$

15. $4x(x-2)\left(\dfrac{x}{x-2} - \dfrac{5}{x}\right) = 4x(x-2)\left(\dfrac{3}{4}\right)$
$$4x^2 - 5(4x - 8) = 3x^2 - 6x$$
$$x^2 - 14x + 40 = 0$$
$$(x-4)(x-10) = 0$$
$$x = 4 \quad \text{or} \quad x = 10$$

Solution set: $\{4, 10\}$

16.
$$\frac{3m}{2} = \frac{6}{m}$$
$$3m^2 = 12$$
$$m^2 = 4$$
$$m^2 - 4 = 0$$
$$(m-4)(m+4) = 0$$
$$m = 2 \quad \text{or} \quad m = -2$$

Solution set: $\{-2, 2\}$

17.
$$W = \frac{a^2}{t}$$
$$tW = a^2$$
$$t = \frac{a^2}{W}$$

18.
$$2ab\left(\frac{1}{a} + \frac{1}{b}\right) = 2ab \cdot \frac{1}{2}$$
$$2b + 2a = ab$$
$$2b - ab = -2a$$
$$b(2 - a) = -2a$$
$$b = \frac{-2a}{2 - a}$$
$$b = \frac{2a}{a - 2}$$

19. $\dfrac{12x\left(\frac{1}{x} + \frac{1}{3x}\right)}{12x\left(\frac{3}{4x} - \frac{1}{2}\right)} = \dfrac{12 + 4}{9 - 6x} = \dfrac{16}{9 - 6x}$

20. $\dfrac{m^2w^2(m^{-2} - w^{-2})}{m^2w^2(m^{-2}w^{-1} + m^{-1}w^{-2})}$

$= \dfrac{w^2 - m^2}{w + m} = \dfrac{(w - m)(w + m)}{w + m} = w - m$

21. $\dfrac{a^2b^3}{2 \cdot 2a} \cdot \dfrac{2 \cdot 3a^2}{ab^3} = \dfrac{3a^2}{2}$

22. Let $x =$ the number of minutes to fill the leaky pool. In one minute, the hose is supposed to fill $\frac{1}{6}$ of the pool, the leak removes $\frac{1}{8}$ of the pool, but together $\frac{1}{x}$ of the pool actually gets filled.

$$\frac{1}{6} - \frac{1}{8} = \frac{1}{x}$$
$$24x\left(\frac{1}{6} - \frac{1}{8}\right) = 24x \cdot \frac{1}{x}$$
$$4x - 3x = 24$$
$$x = 24$$

The leaky pool will be filled in 24 minutes.

23. Let $x =$ the number of miles hiked in one day. Milton's time is $\frac{x}{4}$ and Bonnie's time is $\frac{x}{3}$ hours. Since Milton's time is 2.5 hours less than Bonnie's, we can write the equation

$$\frac{x}{4} + \frac{5}{2} = \frac{x}{3}$$
$$12\left(\frac{x}{4} + \frac{5}{2}\right) = 12 \cdot \frac{x}{3}$$
$$3x + 30 = 4x$$
$$30 = x$$

They hiked 30 miles that day.

24. Let $x =$ the number of sailors in the original group and $x + 3 =$ the number in the larger group. The cost is $\frac{72,000}{x}$ dollars per person for x people and $\frac{72,000}{x + 3}$ dollars per person for $x + 3$ people.

$$\frac{72,000}{x} = \frac{72,000}{x + 3} + 2000$$
$$x(x + 3)\frac{72,000}{x} = x(x + 3)\left(\frac{72,000}{x + 3} + 2000\right)$$
$$72,000x + 216,000 = 72,000x + 2000x^2 + 6000x$$
$$2000x^2 + 6000x - 216,000 = 0$$
$$x^2 + 3x - 108 = 0$$
$$(x - 9)(x + 12) = 0$$
$$x = 9 \quad \text{or} \quad x = -12$$

There are 9 sailors in the original group.

Tying It All Together Chapters 1 - 4

1.
$$\frac{3}{x} = \frac{4}{5}$$
$$4x = 15$$
$$x = \frac{15}{4}$$

Solution set: $\left\{\frac{15}{4}\right\}$

2.
$$\frac{2}{x} = \frac{x}{8}$$
$$x^2 = 16$$
$$x^2 - 16 = 0$$
$$(x - 4)(x + 4) = 0$$
$$x = 4 \quad \text{or} \quad x = -4$$

Solution set: $\{-4, 4\}$

3.
$$\frac{x}{3} = \frac{4}{5}$$
$$5x = 12$$
$$x = \frac{12}{5}$$
Solution set: $\left\{\frac{12}{5}\right\}$

4.
$$\frac{3}{x} = \frac{x+3}{6}$$
$$x^2 + 3x = 18$$
$$x^2 + 3x - 18 = 0$$
$$(x+6)(x-3) = 0$$
$$x = -6 \quad \text{or} \quad x = 3$$
Solution set: $\{-6, 3\}$

5.
$$\frac{1}{x} = 4$$
$$4x = 1$$
$$x = \frac{1}{4}$$
Solution set: $\left\{\frac{1}{4}\right\}$

6.
$$\frac{2}{3}x = 4$$
$$2x = 12$$
$$x = 6$$
Solution set: $\{6\}$

7.
$$2x + 3 = 4$$
$$2x = 1$$
$$x = \frac{1}{2}$$
Solution set: $\left\{\frac{1}{2}\right\}$

8.
$$2x + 3 = 4x$$
$$3 = 2x$$
$$\frac{3}{2} = x$$
Solution set: $\left\{\frac{3}{2}\right\}$

9.
$$\frac{2a}{3} = \frac{6}{a}$$
$$2a^2 = 18$$
$$a^2 = 9$$
$$a^2 - 9 = 0$$
$$(a-3)(a+3) = 0$$
$$a = 3 \quad \text{or} \quad a = -3$$
Solution set: $\{-3, 3\}$

10. $2x(x+1)\left(\frac{12}{x} - \frac{14}{x+1}\right) = 2x(x+1) \cdot \frac{1}{2}$
$$24x + 24 - 28x = x^2 + x$$
$$-x^2 - 5x + 24 = 0$$
$$x^2 + 5x - 24 = 0$$
$$(x+8)(x-3) = 0$$
$$x = -8 \quad \text{or} \quad x = 3$$
Solution set: $\{-8, 3\}$

11.
$$|6x - 3| = 1$$
$$6x - 3 = 1 \quad \text{or} \quad 6x - 3 = -1$$
$$6x = 4 \quad \text{or} \quad 6x = 2$$
$$x = \frac{2}{3} \quad \text{or} \quad x = \frac{1}{3}$$
Solution set: $\left\{\frac{1}{3}, \frac{2}{3}\right\}$

12.
$$\frac{x}{2x+9} = \frac{3}{x}$$
$$x^2 = 6x + 27$$
$$x^2 - 6x - 27 = 0$$
$$(x-9)(x+3) = 0$$
$$x = 9 \quad \text{or} \quad x = -3$$
Solution set: $\{-3, 9\}$

13.
$$4(6x-3)(2x+9) = 0$$
$$6x - 3 = 0 \quad \text{or} \quad 2x + 9 = 0$$
$$6x = 3 \quad \text{or} \quad 2x = -9$$
$$x = \frac{1}{2} \quad \text{or} \quad x = -\frac{9}{2}$$
Solution set: $\left\{-\frac{9}{2}, \frac{1}{2}\right\}$

14. $5(x+2)\left(\frac{x-1}{x+2} - \frac{1}{5(x+2)}\right) = 5(x+2) \cdot 1$
$$5x - 5 - 1 = 5x + 10$$
$$-6 = 10$$
Solution set: $\emptyset$

15.
$$Ax + By = C$$
$$By = C - Ax$$
$$y = \frac{C - Ax}{B}$$

16.
$$\frac{y-3}{x+5} = -\frac{1}{3}$$
$$y - 3 = -\frac{1}{3}(x+5)$$
$$y - 3 = -\frac{1}{3}x - \frac{5}{3}$$
$$y = -\frac{1}{3}x - \frac{5}{3} + \frac{9}{3}$$
$$y = -\frac{1}{3}x + \frac{4}{3}$$

17.
$$Ay = By + C$$
$$Ay - By = C$$
$$y(A - B) = C$$
$$y = \frac{C}{A - B}$$

18.
$$\frac{A}{y} = \frac{y}{A}$$
$$y^2 = A^2$$
$$y^2 - A^2 = 0$$
$$(y - A)(y + A) = 0$$
$$y - A = 0 \quad \text{or} \quad y + A = 0$$
$$y = A \quad \text{or} \quad y = -A$$

19.
$$2y\left(\frac{A}{y} - \frac{1}{2}\right) = 2y \cdot \frac{B}{y}$$
$$2A - y = 2B$$
$$2A - 2B = y$$
$$y = 2A - 2B$$

20.
$$2Cy\left(\frac{A}{y} - \frac{1}{2}\right) = 2Cy \cdot \frac{B}{C}$$
$$2AC - Cy = 2yB$$
$$2AC = Cy + 2yB$$
$$2AC = y(C + 2B)$$
$$\frac{2AC}{C + 2B} = y$$
$$y = \frac{2AC}{2B + C}$$

21.
$$3x - 4y = 6$$
$$-4y = -3x + 6$$
$$y = \frac{-3x + 6}{-4}$$
$$y = \frac{3}{4}x - \frac{3}{2}$$

22.
$$y^2 - 2y - Ay + 2A = 0$$
$$y(y - 2) - A(y - 2) = 0$$
$$(y - A)(y - 2) = 0$$
$$y - A = 0 \quad \text{or} \quad y - 2 = 0$$
$$y = A \quad \text{or} \quad y = 2$$

23.
$$A = \tfrac{1}{2}B(C + y)$$
$$2A = B(C + y)$$
$$2A = BC + By$$
$$2A - BC = By$$
$$\frac{2A - BC}{B} = y$$
$$y = \frac{2A - BC}{B}$$

24.
$$y^2 + Cy = BC + By$$
$$y^2 + Cy - BC - By = 0$$
$$y(y + C) - B(y + C) = 0$$
$$(y - B)(y + C) = 0$$
$$y - B = 0 \quad \text{or} \quad y + C = 0$$
$$y = B \quad \text{or} \quad y = -C$$

25. $3x^5 \cdot 4x^8 = 12x^{13}$

26. $3x^2(x^3 + 5x^6) = 3x^5 + 15x^8$

27. $(5x^6)^2 = 25x^{12}$

28. $(3a^3b^2)^3 = 27a^9b^6$

29. $\dfrac{12a^9b^4}{-3a^3b^{-2}} = -4a^6b^6$

30. $\left(\dfrac{x^{-2}}{2}\right)^5 = \dfrac{x^{-10}}{2^5} = \dfrac{1}{32x^{10}}$

31. $\left(\dfrac{2x^{-4}}{3y^5}\right)^{-3} = \dfrac{2^{-3}x^{12}}{3^{-3}y^{-15}} = \dfrac{27x^{12}y^{15}}{8}$

32. $(-2a^{-1}b^3c)^{-2} = 2^{-2}a^2b^{-6}c^{-2} = \dfrac{a^2}{4b^6c^2}$

33. $\dfrac{a^{-1} + b^3}{a^{-2} + b^{-1}} = \dfrac{a^2b(a^{-1} + b^3)}{a^2b(a^{-2} + b^{-1})} = \dfrac{ab + a^2b^4}{b + a^2}$

34. $\dfrac{(a + b)^{-1}}{(a + b)^{-2}} = a + b$

35. For E = \$100,000,
$$D = 0.75(100{,}000) + 5000 = \$80{,}000$$
The tax is $100{,}000 - 80{,}000$ or \$20,000.

For E = 10,000,
$$D = 0.75(10{,}000) + 5000 = 12{,}500$$
The person receives $12{,}500 - 10{,}000$ or \$2500.

5.1 WARM-UPS

1. True, because $4^{-1/2} = \frac{1}{4^{1/2}} = \frac{1}{2}$.

2. False, because $16^{1/2} = 4$.

3. True, because $(3^{2/3})^3 = 3^{6/3} = 3^2 = 9$.

4. False, because $8^{-2/3} = \frac{1}{2^2} = \frac{1}{4}$.

5. True, $2^{1/2} \cdot 2^{1/2} = 2^{1/2 + 1/2} = 2^1 = 2$.

6. True, because the square root of $\frac{1}{4}$ is $\frac{1}{2}$.

7. True, $\frac{3^1}{3^{1/2}} = 3^{1 - 1/2} = 3^{1/2}$

8. False, because $(2^9)^{1/2} = 2^{9/2}$.

9. False, because $3^{1/3}6^{1/3} = (3 \cdot 6)^{1/3} = 18^{1/3}$.

10. False, because $2^{3/4} \cdot 2^{1/4} = 2^{4/4} = 2^1 = 2$.

5.1 EXERCISES

1. Because $10^2 = 100$, $100^{1/2} = 10$.

3. Because $3^4 = 81$, $81^{1/4} = 3$.

5. $-9^{1/2} = -(9^{1/2}) = -3$

7. $\left(\frac{1}{64}\right)^{1/6} = \frac{1}{2}$

9. Square root of a negative number is not real.

11. The cube root of 1000 is 10, because $10^3 = 1000$.

13. $(-64)^{1/3} = -4$, because $(-4)^3 = -64$.

15. $-1^{1/5} = -1$

17. Take the fifth root and then cube it: $32^{3/5} = 2^3 = 8$.

19. $(-27)^{2/3} = (-3)^2 = 9$

21. Take the square root of 25, cube it, and then take the opposite: $-25^{3/2} = -5^3 = -125$

23. The expression is not a real number because it is an even root of a negative number.

25. Take the square root and then find the reciprocal: $4^{-1/2} = \frac{1}{2}$.

27. $8^{-4/3} = \frac{1}{8^{4/3}} = \frac{1}{2^4} = \frac{1}{16}$

29. Take the fifth root, cube it, and then find the reciprocal: $(-32)^{-3/5} = \frac{1}{(-2)^3} = -\frac{1}{8}$.

31. The square root of -9 is not real.

33. $3^{1/3} \cdot 3^{1/4} = 3^{\frac{1}{3} + \frac{1}{4}} = 3^{7/12}$

35. $3^{1/3} \cdot 3^{-1/3} = 3^0 = 1$

37. $\frac{8^{1/3}}{8^{2/3}} = 8^{1/3 - 2/3} = 8^{-1/3} = \frac{1}{8^{1/3}} = \frac{1}{2}$

39. $4^{3/4} \div 4^{1/4} = 4^{3/4 - 1/4} = 4^{1/2} = 2$

41. $18^{1/2} \cdot 2^{1/2} = 36^{1/2} = 6$

43. $(2^6)^{1/3} = 2^{6/3} = 2^2 = 4$

45. $(3^8)^{1/2} = 3^{8/2} = 3^4 = 81$

47. $(2^{-4})^{1/2} = 2^{-2} = \frac{1}{4}$

49. $\left(\frac{3^4}{2^6}\right)^{1/2} = \frac{3^2}{2^3} = \frac{9}{8}$

51. $\left(\frac{2^{-9}}{3^{-6}}\right)^{-1/3} = \frac{2^3}{3^2} = \frac{8}{9}$

53. $(x^4)^{1/4} = |x|$

55. $(a^8)^{1/2} = a^4$

57. $(y^3)^{1/3} = y$

59. $(9x^6y^2)^{1/2} = |3x^3y|$

61. $\left(\frac{81x^{12}}{y^{20}}\right)^{1/4} = \left|\frac{3x^3}{y^5}\right|$

63. $x^{1/2}x^{1/4} = x^{2/4 + 1/4} = x^{3/4}$

65. $(x^{1/2}y)(x^{-3/4}y^{1/2}) = x^{-1/4}y^{3/2} = \frac{y^{3/2}}{x^{1/4}}$

67. $\frac{w^{1/3}}{w^3} = w^{\frac{1}{3} - 3} = w^{-8/3} = \frac{1}{w^{8/3}}$

74

69. $\dfrac{x^{1/2}y}{x} = x^{\frac{1}{2}-1}y = x^{-1/2}y = \dfrac{y}{x^{1/2}}$

71. $(144x^{16})^{1/2} = 12x^8$

73. $(4x^{-1/2}yz^{1/2})^{-1/2} = 4^{-1/2}x^{1/4}y^{-1/2}z^{-1/4}$

$$= \dfrac{x^{1/4}}{2y^{1/2}z^{1/4}}$$

75. $\left(\dfrac{a^{-1/2}}{b^{-1/4}}\right)^{-4} = \dfrac{a^{4/2}}{b^{4/4}} = \dfrac{a^2}{b}$

77. $\left(\dfrac{9x^{10}}{w^{12}}\right)^{-1/2} = \dfrac{9^{-1/2}x^{-5}}{w^{-6}} = \dfrac{w^6}{3x^5}$

79. $(9^2)^{1/2} = 9$

81. $-16^{-3/4} = -\dfrac{1}{2^3} = -\dfrac{1}{8}$

83. $125^{-4/3} = \dfrac{1}{5^4} = \dfrac{1}{625}$

85. $2^{1/2} \cdot 2^{-1/4} = 2^{2/4-1/4} = 2^{1/4}$

87. $3^{0.26}3^{0.74} = 3^{0.26+0.74} = 3^1 = 3$

89. $3^{1/4} \cdot 27^{1/4} = (3 \cdot 27)^{1/4} = 81^{1/4} = 3$

91. $\left(-\dfrac{8}{27}\right)^{2/3} = \dfrac{(-8)^{2/3}}{27^{2/3}} = \dfrac{4}{9}$

93. Not a real number, because the fourth root of $-1/16$ is not real.

95. $(9x^9)^{1/2} = 3x^{9/2}$

97. $(3a^{-2/3})^{-3} = 3^{-3}a^2 = \dfrac{a^2}{27}$

99. $(a^{1/2}b)^{1/2}(ab^{1/2}) = a^{1/4}b^{1/2}a^1b^{1/2}$

$$= a^{5/4}b$$

101. $(km^{1/2})^3(k^3m^5)^{1/2} = k^3m^{3/2}k^{3/2}m^{5/2}$

$$= k^{9/2}m^4$$

103. $\left(\dfrac{w^{-3/4}a^{-2/3}}{w^2a^{-1/2}}\right)^6 \left(\dfrac{w^{1/2}a^{-1/2}}{w^{-1/2}a^3}\right)^{-1/2}$

$= \dfrac{w^{-18/4}a^{-4}}{w^{12}a^{-3}} \cdot \dfrac{w^{-1/4}a^{1/4}}{w^{1/4}a^{-3/2}} = \dfrac{w^{-19/4}a^{-15/4}}{w^{49/4}a^{-9/2}}$

$$= w^{-68/4}a^{3/4} = \dfrac{a^{3/4}}{w^{17}}$$

105. $2^{1/3} = 2^{0.33333333} = 1.2599$

107. $-2^{1/2} = -(2^{0.5}) = -1.4142$

109. $1024^{1/10} = 2$ **111.** $8^{0.33} = 1.9862$

113. $\left(\dfrac{64}{15,625}\right)^{-1/6} = 2.5$

115. $\left(\dfrac{5^{-2/3}}{6^{1/4}}\right)^{-1/2} = 5^{1/3}6^{1/8} = 2.1392$

117. $a^{m/2} \cdot a^{m/4} = a^{m/2+m/4} = a^{3m/4}$

119. $\dfrac{a^{-m/5}}{a^{-m/3}} = a^{-m/5+m/3} = a^{2m/15}$

121. $\left(a^{-1/m}b^{-1/n}\right)^{-mn} = a^nb^m$

123. $\left(\dfrac{a^{-3m}b^{-6n}}{a^{9m}}\right)^{-1/3} = \dfrac{a^mb^{2n}}{a^{-3m}} = a^{4m}b^{2n}$

125. $D = (12^2 + 4^2 + 3^2)^{1/2}$

$= (144 + 16 + 9)^{1/2}$

$= 169^{1/2} = 13$ inches

127. $S = (13.0368 + 7.84(18.42)^{1/3} - 0.8(21.45))^2$

$S = 274.96$ m^2

129. $r = \left(\dfrac{41,200}{10,000}\right)^{1/10} - 1 = 0.152 = 15.2\%$

131. $r = \left(\dfrac{141,600,000,000}{450,000}\right)^{1/213} - 1$

$= 0.061 = 6.1\%$

5.2 WARM-UPS

1. True, because of the definition of the fractional exponent 1/2.

2. False, $3^{1/3} = \sqrt[3]{3}$.

3. True, $2^{2/3} = \sqrt[3]{2^2} = \sqrt[3]{4}$.

4. False, $\sqrt{81} = 9$.

5. True, $\sqrt{417^2} = (417^2)^{1/2} = 417$.

6. False, $\sqrt[3]{a^{27}} = (a^{27})^{1/3} = a^9$.

7. True, because $\dfrac{1 \cdot \sqrt{2}}{\sqrt{2} \cdot \sqrt{2}} = \dfrac{\sqrt{2}}{2}$.

8. False, $\dfrac{\sqrt{10}}{\sqrt{2}} = \sqrt{5}$.

9. True, $\sqrt{2^{-4}} = (2^{-4})^{1/2} = 2^{-2} = \dfrac{1}{4}$.

10. True, because $\sqrt{2} \cdot \sqrt{3} = \sqrt{6}$.

5.2 EXERCISES

1. $\sqrt{27} = 27^{1/2}$

3. The square root of x^5 is written $x^{5/2}$.

5. The cube root of a^{12} is written $a^{12/3}$.

7. The opposite of the cube root of 5: $-\sqrt[3]{5}$.

9. The fifth root of 2^2 is written $\sqrt[5]{2^2}$.

11. The cube root of x^{-2} is written $\sqrt[3]{x^{-2}}$.

13. Since $11^2 = 121$, $\sqrt{121} = 11$.

15. Since $(-10)^3 = -1000$, $\sqrt[3]{-1000} = -10$.

17. The expression is not a real number since it is an even root of a negative number.

19. $\sqrt{a^{16}} = a^{16/2} = a^8$

21. $\sqrt{4^{16}} = 4^{16/2} = 4^8$

23. $\sqrt[5]{w^{30}} = w^{30/5} = w^6$

25. $\sqrt{9w} = \sqrt{9}\sqrt{w} = 3\sqrt{w}$

27. $\sqrt{20} = \sqrt{4}\sqrt{5} = 2\sqrt{5}$

29. $\sqrt{45w} = \sqrt{9}\sqrt{5w} = 3\sqrt{5w}$

31. $\sqrt{288} = \sqrt{144}\sqrt{2} = 12\sqrt{2}$

33. $\sqrt[3]{54} = \sqrt[3]{27} \cdot \sqrt[3]{2} = 3 \cdot \sqrt[3]{2}$

35. $\sqrt[4]{32a} = \sqrt[4]{16} \cdot \sqrt[4]{2a} = 2 \cdot \sqrt[4]{2a}$

37. $\sqrt{\dfrac{9}{100}} = \dfrac{\sqrt{9}}{\sqrt{100}} = \dfrac{3}{10}$

39. $\sqrt{\dfrac{50}{9}} = \dfrac{\sqrt{50}}{\sqrt{9}} = \dfrac{\sqrt{25}\sqrt{2}}{3} = \dfrac{5\sqrt{2}}{3}$

41. $\sqrt[3]{-\dfrac{125}{8}} = \dfrac{\sqrt[3]{-125}}{\sqrt[3]{8}} = \dfrac{-5}{2} = -\dfrac{5}{2}$

43. $\sqrt[3]{\dfrac{16x}{27}} = \dfrac{\sqrt[3]{16x}}{\sqrt[3]{27}} = \dfrac{\sqrt[3]{8} \cdot \sqrt[3]{2x}}{3} = \dfrac{2 \cdot \sqrt[3]{2x}}{3}$

45. $\dfrac{2}{\sqrt{5}} = \dfrac{2\sqrt{5}}{\sqrt{5}\sqrt{5}} = \dfrac{2\sqrt{5}}{5}$

47. $\dfrac{\sqrt{3}}{\sqrt{7}} = \dfrac{\sqrt{3}\sqrt{7}}{\sqrt{7}\sqrt{7}} = \dfrac{\sqrt{21}}{\sqrt{49}} = \dfrac{\sqrt{21}}{7}$

49. $\dfrac{1}{\sqrt[3]{4}} = \dfrac{1 \cdot \sqrt[3]{2}}{\sqrt[3]{4} \cdot \sqrt[3]{2}} = \dfrac{\sqrt[3]{2}}{\sqrt[3]{8}} = \dfrac{\sqrt[3]{2}}{2}$

51. $\dfrac{\sqrt[3]{6}}{\sqrt[3]{5}} = \dfrac{\sqrt[3]{6} \cdot \sqrt[3]{25}}{\sqrt[3]{5} \cdot \sqrt[3]{25}} = \dfrac{\sqrt[3]{150}}{\sqrt[3]{125}} = \dfrac{\sqrt[3]{150}}{5}$

53. $\dfrac{\sqrt{5}}{\sqrt{12}} = \dfrac{\sqrt{5}\sqrt{3}}{\sqrt{12}\sqrt{3}} = \dfrac{\sqrt{15}}{\sqrt{36}} = \dfrac{\sqrt{15}}{6}$

55. $\dfrac{\sqrt{3}}{\sqrt{12}} = \dfrac{\sqrt{3}}{\sqrt{4}\sqrt{3}} = \dfrac{1}{\sqrt{4}} = \dfrac{1}{2}$

57. $\sqrt{\dfrac{1}{2}} = \dfrac{1}{\sqrt{2}} = \dfrac{1 \cdot \sqrt{2}}{\sqrt{2}\sqrt{2}} = \dfrac{\sqrt{2}}{2}$

59. $\sqrt[3]{\dfrac{7}{4}} = \dfrac{\sqrt[3]{7} \cdot \sqrt[3]{2}}{\sqrt[3]{4} \cdot \sqrt[3]{2}} = \dfrac{\sqrt[3]{14}}{2}$

61. $\sqrt{12x^8} = \sqrt{4x^8}\sqrt{3} = 2x^4 \cdot \sqrt{3}$

63. $\sqrt{60a^9b^3} = \sqrt{4a^8b^2}\sqrt{15ab} = 2a^4b \cdot \sqrt{15ab}$

65. $\sqrt{\dfrac{x}{y}} = \dfrac{\sqrt{x}\sqrt{y}}{\sqrt{y}\sqrt{y}} = \dfrac{\sqrt{xy}}{y}$

67. $\dfrac{\sqrt{a^3}}{\sqrt{b^7}} = \dfrac{\sqrt{a^2}\sqrt{a}}{\sqrt{b^6}\sqrt{b}} = \dfrac{a\sqrt{a}\sqrt{b}}{b^3\sqrt{b}\sqrt{b}} = \dfrac{a\sqrt{ab}}{b^4}$

69. $\sqrt[3]{16x^{13}} = \sqrt[3]{8x^{12}} \cdot \sqrt[3]{2x} = 2x^4 \cdot \sqrt[3]{2x}$

71. $\sqrt[4]{x^9y^6} = \sqrt[4]{x^8y^4} \cdot \sqrt[4]{xy^2} = x^2y \cdot \sqrt[4]{xy^2}$

73. $\sqrt[5]{64x^{22}} = \sqrt[5]{32x^{20}} \cdot \sqrt[5]{2x^2} = 2x^4 \cdot \sqrt[5]{2x^2}$

75. $\sqrt[3]{\dfrac{a}{b}} = \dfrac{\sqrt[3]{a}}{\sqrt[3]{b}} = \dfrac{\sqrt[3]{a} \cdot \sqrt[3]{b^2}}{\sqrt[3]{b} \cdot \sqrt[3]{b^2}} = \dfrac{\sqrt[3]{ab^2}}{\sqrt[3]{b^3}} = \dfrac{\sqrt[3]{ab^2}}{b}$

77. $\sqrt[4]{3^{12}} = 3^{12/4} = 3^3 = 27$

79. $\sqrt{10^{-2}} = 10^{-2/2} = 10^{-1} = \dfrac{1}{10}$

81. $\sqrt{\dfrac{8x}{49}} = \dfrac{\sqrt{8x}}{\sqrt{49}} = \dfrac{\sqrt{4}\sqrt{2x}}{7} = \dfrac{2\sqrt{2x}}{7}$

83. $\sqrt[4]{\dfrac{32a}{81}} = \dfrac{\sqrt[4]{32a}}{\sqrt[4]{81}} = \dfrac{\sqrt[4]{16} \cdot \sqrt[4]{2a}}{3} = \dfrac{2 \cdot \sqrt[4]{2a}}{3}$

85. $\sqrt[3]{-27x^9y^8} = \sqrt[3]{-27x^9y^6} \cdot \sqrt[3]{y^2}$

$$= -3x^3y^2 \cdot \sqrt[3]{y^2}$$

87. $\dfrac{\sqrt{ab^3}}{\sqrt{a^3b^2}} = \dfrac{\sqrt{ab^3}\sqrt{a}}{\sqrt{a^3b^2}\sqrt{a}} = \dfrac{\sqrt{a^2b^3}}{\sqrt{a^4b^2}}$

$$= \dfrac{\sqrt{a^2b^2}\sqrt{b}}{a^2b} = \dfrac{ab\sqrt{b}}{a^2b} = \dfrac{\sqrt{b}}{a}$$

89. $\dfrac{\sqrt[3]{a^2b}}{\sqrt[3]{4ab^2}\cdot\sqrt[3]{3ab^5}} = \dfrac{\sqrt[3]{a^2b}}{\sqrt[3]{12a^2b^7}}$

$$= \dfrac{\sqrt[3]{a^2b}\cdot\sqrt[3]{18ab^2}}{\sqrt[3]{12a^2b^7}\cdot\sqrt[3]{18ab^2}} = \dfrac{\sqrt[3]{18a^3b^3}}{\sqrt[3]{216a^3b^9}}$$

$$= \dfrac{ab\cdot\sqrt[3]{18}}{6ab^3} = \dfrac{\sqrt[3]{18}}{6b^2}$$

91. $\dfrac{6m\cdot\sqrt[5]{5mn^2}}{\sqrt[5]{12m^3n}} = \dfrac{6m\cdot\sqrt[5]{5mn^2}\cdot\sqrt[5]{648m^2n^4}}{\sqrt[5]{12m^3n}\cdot\sqrt[5]{648m^2n^4}}$

$$= \dfrac{6m\cdot\sqrt[5]{3240m^3n^6}}{\sqrt[5]{7776m^5n^5}} = \dfrac{6m\cdot\sqrt[5]{n^5}\cdot\sqrt[5]{3240m^3n}}{6mn}$$

$$= \dfrac{6mn\cdot\sqrt[5]{3240m^3n}}{6mn} = \sqrt[5]{3240m^3n}$$

93. $\dfrac{5}{\sqrt{3}} = 2.887$ **95.** $\sqrt[3]{\dfrac{1}{3}} = 0.693$

97. $\dfrac{\sqrt[3]{9}}{\sqrt[3]{4}} = 1.310$ **99.** $\dfrac{\sqrt[6]{16}}{\sqrt[3]{4}} = 1$

101. $w = 91.4$
$$-\dfrac{(10.5 + 6.7\sqrt{20} - 0.45\cdot 20)(457 - 5\cdot 25)}{110}$$

$w = -4°\text{F}$

If the air temp is 25°F and wind is 30mph, then

$w = -10°\text{F}$

103. $t = \sqrt{\dfrac{S}{16}} = \dfrac{\sqrt{S}}{4}$ $t = \dfrac{\sqrt{30}}{4} \approx 1.4$ sec

105. $M = 1.3\sqrt{20} \approx 5.8$ knots

107. $V = \sqrt{\dfrac{841\cdot 8700}{2.81\cdot 200}} \approx 114.1$ ft/sec

$114.1\,\dfrac{\text{ft}}{\text{sec}}\cdot\dfrac{1\text{ mi}}{5280\text{ ft}}\cdot\dfrac{3600\text{ sec}}{1\text{ hr}} = 77.8$ mph

5.3 WARM-UPS

1. False, because $\sqrt{3} + \sqrt{3} = 2\sqrt{3}$.
2. True, because $\sqrt{8} + \sqrt{2} = 2\sqrt{2} + \sqrt{2} = 3\sqrt{2}$.
3. False, because $2\sqrt{3}\cdot 3\sqrt{3} = 6\cdot 3 = 18$.
4. False, because $\sqrt[3]{2}\cdot\sqrt[3]{2} = \sqrt[3]{4}$.
5. True, because $\sqrt{5}\cdot\sqrt{2} = \sqrt{10}$.
6. False, because $2\sqrt{5} + 3\sqrt{5} = 5\sqrt{5}$.
7. True, because $\sqrt{2}\sqrt{3} = \sqrt{6}$ and $\sqrt{2}\sqrt{2} = 2$.
8. False, because $\sqrt{12} = \sqrt{4}\sqrt{3} = 2\sqrt{3}$.
9. False, because $(\sqrt{2} + \sqrt{3})^2 = 2 + 2\sqrt{2}\sqrt{3} + 3$
$$= 5 + 2\sqrt{6}.$$
10. True, $(\sqrt{3} - \sqrt{2})(\sqrt{3} + \sqrt{2}) = 3 - 2 = 1$.

5.3 EXERCISES

1. $1\sqrt{3} - 2\sqrt{3} = -1\sqrt{3} = -\sqrt{3}$
3. $5\sqrt{7x} + 4\sqrt{7x} = (5+4)\sqrt{7x} = 9\sqrt{7x}$
5. $2\cdot\sqrt[3]{2} + 3\cdot\sqrt[3]{2} = (2+3)\cdot\sqrt[3]{2} = 5\cdot\sqrt[3]{2}$
7. $\sqrt{3} + 3\sqrt{3} - \sqrt{5} - \sqrt{5} = 4\sqrt{3} - 2\sqrt{5}$
9. $\sqrt[3]{2} - \sqrt[3]{2} + \sqrt[3]{x} + 4\cdot\sqrt[3]{x} = 5\cdot\sqrt[3]{x}$
11. $\sqrt[3]{x} + \sqrt[3]{x} - \sqrt{2x} = 2\cdot\sqrt[3]{x} - \sqrt{2x}$
13. $\sqrt{8} + \sqrt{28} = \sqrt{4}\sqrt{2} + \sqrt{4}\sqrt{7} = 2\sqrt{2} + 2\sqrt{7}$
15. $\sqrt{2} - \sqrt{8} = \sqrt{2} - 2\sqrt{2} = -\sqrt{2}$
17. $\dfrac{\sqrt{2}}{2} + \sqrt{2} = \dfrac{\sqrt{2}}{2} + \dfrac{2\sqrt{2}}{2} = \dfrac{3\sqrt{2}}{2}$
19. $\sqrt{80} + \sqrt{\dfrac{1}{5}} = \sqrt{16}\sqrt{5} + \dfrac{1\sqrt{5}}{\sqrt{5}\sqrt{5}}$
$$= 4\sqrt{5} + \dfrac{\sqrt{5}}{5} = \dfrac{20\sqrt{5}}{5} + \dfrac{\sqrt{5}}{5} = \dfrac{21\sqrt{5}}{5}$$
21. $\sqrt{45x^3} - \sqrt{18x^2} + \sqrt{50x^2} - \sqrt{20x^3}$
$$= 3x\sqrt{5x} - 3x\sqrt{2} + 5x\sqrt{2} - 2x\sqrt{5x}$$
$$= x\sqrt{5x} + 2x\sqrt{2}$$
23. $\sqrt[3]{24} + \sqrt[3]{81} = \sqrt[3]{8}\cdot\sqrt[3]{3} + \sqrt[3]{27}\cdot\sqrt[3]{3}$
$$= 2\cdot\sqrt[3]{3} + 3\cdot\sqrt[3]{3} = 5\cdot\sqrt[3]{3}$$
25. $\sqrt[4]{16\cdot 3} - \sqrt[4]{81\cdot 3} = 2\cdot\sqrt[4]{3} - 3\cdot\sqrt[4]{3}$
$$= -\sqrt[4]{3}$$

27. $\sqrt[3]{54t^4y^3} - \sqrt[3]{16t^4y^3}$

$= \sqrt[3]{27t^3y^3} \cdot \sqrt[3]{2t} - \sqrt[3]{8t^3y^3} \cdot \sqrt[3]{2t}$

$= 3ty \cdot \sqrt[3]{2t} - 2ty \cdot \sqrt[3]{2t} = ty\sqrt[3]{2t}$

29. $\sqrt{3}\sqrt{5} = \sqrt{15}$

31. $(2\sqrt{5})(3\sqrt{10}) = 6\sqrt{50} = 6\sqrt{25}\sqrt{2}$
$$= 6 \cdot 5\sqrt{2} = 30\sqrt{2}$$

33. $(2\sqrt{7a})(3\sqrt{2a}) = 6\sqrt{14a^2} = 6\sqrt{a^2}\sqrt{14} = 6a\sqrt{14}$

35. $(\sqrt[4]{9})(\sqrt[4]{27}) = \sqrt[4]{243} = \sqrt[4]{81} \cdot \sqrt[4]{3} = 3 \cdot \sqrt[4]{3}$

37. $(2\sqrt{3})^2 = 4 \cdot 3 = 12$

39. $\sqrt[3]{\dfrac{4x^2}{3}} \cdot \sqrt[3]{\dfrac{2x^2}{3}} = \sqrt[3]{\dfrac{8x^4}{9}}$

$= \sqrt[3]{8x^3} \cdot \sqrt[3]{\dfrac{x}{9}} = 2x \cdot \dfrac{\sqrt[3]{x} \cdot \sqrt[3]{3}}{\sqrt[3]{9} \cdot \sqrt[3]{3}} = \dfrac{2x\sqrt[3]{3x}}{3}$

41. $2\sqrt{3}(\sqrt{6} + 3\sqrt{3}) = 2\sqrt{18} + 18$
$$= 2\sqrt{9}\sqrt{2} + 18 = 6\sqrt{2} + 18$$

43. $\sqrt{5}(\sqrt{10} - 2) = \sqrt{50} - 2\sqrt{5}$
$$= \sqrt{25}\sqrt{2} - 2\sqrt{5} = 5\sqrt{2} - 2\sqrt{5}$$

45. $\sqrt[3]{3t}(\sqrt[3]{9t} - \sqrt[3]{t^2}) = \sqrt[3]{27t^2} - \sqrt[3]{3t^3}$
$= 3\sqrt[3]{t^2} - t\sqrt[3]{3}$

47. $(\sqrt{3} + 2)(\sqrt{3} - 5) = 3 + 2\sqrt{3} - 5\sqrt{3} - 10$
$$= -7 - 3\sqrt{3}$$

49. $(\sqrt{11} - 3)(\sqrt{11} + 3) = 11 - 9 = 2$

51. $(2\sqrt{5} - 7)(2\sqrt{5} + 4) = 20 - 14\sqrt{5} + 8\sqrt{5} - 28$
$= -8 - 6\sqrt{5}$

53. $(2\sqrt{3} - \sqrt{6})(\sqrt{3} + 2\sqrt{6}) = 6 - \sqrt{18} + 4\sqrt{18} - 12$
$= -6 - 3\sqrt{2} + 4 \cdot 3\sqrt{2} = -6 + 9\sqrt{2}$

55. $(\sqrt{2} + 3)^2 = 2 + 2 \cdot 3\sqrt{2} + 9 = 11 + 6\sqrt{2}$

57. $(5 + \sqrt{x - 25})^2 = 25 + 10\sqrt{x - 25} + x - 25$
$$= x + 10\sqrt{x - 25}$$

59. $\sqrt[3]{3} \cdot \sqrt{3} = 3^{1/3}3^{1/2} = 3^{5/6} = \sqrt[6]{3^5}$

61. $\sqrt[3]{5} \cdot \sqrt[4]{5} = 5^{1/3}5^{1/4} = 5^{7/12} = \sqrt[12]{5^7}$

63. $\sqrt[3]{2} \cdot \sqrt{5} = 2^{1/3}5^{1/2} = 2^{2/6}5^{3/6}$

$= \sqrt[6]{2^2 5^3} = \sqrt[6]{500}$

65. $\sqrt[3]{2} \cdot \sqrt[4]{3} = 2^{1/3}3^{1/4} = 2^{4/12}3^{3/12}$

$= \sqrt[12]{2^4 3^3} = \sqrt[12]{432}$

67. $(\sqrt{3} - 2)(\sqrt{3} + 2) = 3 - 4 = -1$

69. $(\sqrt{5} + \sqrt{2})(\sqrt{5} - \sqrt{2}) = 5 - 2 = 3$

71. $(2\sqrt{5} + 1)(2\sqrt{5} - 1) = 4 \cdot 5 - 1 = 19$

73. $(3\sqrt{2} + \sqrt{5})(3\sqrt{2} - \sqrt{5}) = 9 \cdot 2 - 5 = 13$

75. $(5 - 3\sqrt{x})(5 + 3\sqrt{x}) = 25 - 9x$

77. $\sqrt{300} + \sqrt{3} = 10\sqrt{3} + \sqrt{3} = 11\sqrt{3}$

79. $2\sqrt{5} \cdot 5\sqrt{6} = 10\sqrt{30}$

81. $(3 + 2\sqrt{7})(\sqrt{7} - 2) = 3\sqrt{7} - 6 + 2 \cdot 7 - 4\sqrt{7}$

$= 8 - \sqrt{7}$

83. $4\sqrt{w} \cdot 4\sqrt{w} = 16(\sqrt{w})^2 = 16w$

85. $\sqrt{3x^3} \cdot \sqrt{6x^2} = \sqrt{18x^5} = \sqrt{9x^4} \cdot \sqrt{2x} = 3x^2 \cdot \sqrt{2x}$

87. $\dfrac{1}{\sqrt{2}} - \dfrac{1}{\sqrt{8}} + \dfrac{1}{\sqrt{18}} = \dfrac{\sqrt{2}}{2} - \dfrac{\sqrt{8}}{8} + \dfrac{\sqrt{2}}{6}$

$= \dfrac{\sqrt{2}}{2} - \dfrac{2\sqrt{2}}{8} + \dfrac{\sqrt{2}}{6} = \sqrt{2}\left(\dfrac{1}{2} - \dfrac{1}{4} + \dfrac{1}{6}\right) = \dfrac{5\sqrt{2}}{12}$

89. $(2\sqrt{5} + \sqrt{2})(3\sqrt{5} - \sqrt{2})$

$= 30 + 3\sqrt{10} - 2\sqrt{10} - 2 = 28 + \sqrt{10}$

91. $\dfrac{\sqrt{2}}{3} + \dfrac{\sqrt{2}}{5} = \dfrac{5\sqrt{2}}{5 \cdot 3} + \dfrac{3\sqrt{2}}{3 \cdot 5} = \dfrac{8\sqrt{2}}{15}$

93. $(5 + 2\sqrt{2})(5 - 2\sqrt{2}) = 25 - 4 \cdot 2 = 17$

95. $(3 + \sqrt{x})^2 = 9 + 2 \cdot 3\sqrt{x} + x = 9 + 6\sqrt{x} + x$

97. $(5\sqrt{x} - 3)^2 = 25x - 2 \cdot 3 \cdot 5\sqrt{x} + 9$

$$= 25x - 30\sqrt{x} + 9$$

99. $(1 + \sqrt{x + 2})^2 = 1 + 2\sqrt{x + 2} + x + 2$

$$= x + 3 + 2\sqrt{x + 2}$$

101. $\sqrt{4w} - \sqrt{9w} = 2\sqrt{w} - 3\sqrt{w} = -\sqrt{w}$

103. $2\sqrt{a^3} + 3\sqrt{a^3} - 2a\sqrt{4a} = 2a\sqrt{a} + 3a\sqrt{a} - 4a\sqrt{a}$

$= a\sqrt{a}$

105. $\sqrt{25 \cdot 2a} + \sqrt{9 \cdot 2a} - \sqrt{2a}$

$= 5\sqrt{2a} + 3\sqrt{2a} - \sqrt{2a} = 7\sqrt{2a}$

107. $\sqrt{x^5} + 2x\sqrt{x^3} = \sqrt{x^4}\sqrt{x} + 2x\sqrt{x^2}\sqrt{x}$

$\qquad\qquad = x^2 \cdot \sqrt{x} + 2x^2 \cdot \sqrt{x} = 3x^2 \cdot \sqrt{x}$

109. $(\sqrt{a} + a^3)(\sqrt{a} - a^3) = (\sqrt{a})^2 - (a^3)^2 = a - a^6$

111. $\sqrt[3]{-16x^4} + 5x\sqrt[3]{54x}$

$= -2x \cdot \sqrt[3]{2x} + 5x \cdot 3 \cdot \sqrt[3]{2x} = 13x \cdot \sqrt[3]{2x}$

113. $\sqrt[3]{\dfrac{y^7}{4x}} = \dfrac{y^2 \cdot \sqrt[3]{y}}{\sqrt[3]{4x}} = \dfrac{y^2 \cdot \sqrt[3]{y} \cdot \sqrt[3]{2x^2}}{\sqrt[3]{4x} \cdot \sqrt[3]{2x^2}}$

$= \dfrac{y^2 \cdot \sqrt[3]{2x^2y}}{\sqrt[3]{8x^3}} = \dfrac{y^2 \cdot \sqrt[3]{2x^2y}}{2x}$

115. $\sqrt[3]{\dfrac{x}{5}} \cdot \sqrt[3]{\dfrac{x^5}{5}} = \sqrt[3]{\dfrac{x^6}{25}} = \dfrac{x^2 \cdot \sqrt[3]{5}}{\sqrt[3]{25} \cdot \sqrt[3]{5}} = \dfrac{x^2 \cdot \sqrt[3]{5}}{5}$

117. $\sqrt[3]{2x} \cdot \sqrt{2x} = (2x)^{1/3}(2x)^{1/2} = (2x)^{5/6}$

$= \sqrt[6]{32x^5}$

119. $A = \sqrt{6} \cdot \sqrt{3} = \sqrt{18} = 3\sqrt{2}$ ft^2

121. $A = \frac{1}{2}\sqrt{6}(\sqrt{3} + \sqrt{12}) = \frac{1}{2}\sqrt{6}(\sqrt{3} + 2\sqrt{3})$

$= \frac{1}{2}\sqrt{6} \cdot 3\sqrt{3} = \dfrac{3\sqrt{18}}{2} = \dfrac{9\sqrt{2}}{2}$ ft^2

5.4 WARM-UPS

1. True, because $\sqrt{3}\sqrt{2} = \sqrt{6}$.

2. True, because $\dfrac{2}{\sqrt{2}} = \dfrac{2\sqrt{2}}{\sqrt{2}\sqrt{2}} = \dfrac{2\sqrt{2}}{2} = \sqrt{2}$.

3. False, because $\dfrac{4 - \sqrt{10}}{2} = 2 - \dfrac{\sqrt{10}}{2}$.

4. True, because $\dfrac{1}{\sqrt{3}} = \dfrac{1 \cdot \sqrt{3}}{\sqrt{3}\sqrt{3}} = \dfrac{\sqrt{3}}{3}$.

5. False, because $\dfrac{8\sqrt{7}}{2\sqrt{7}} = 4$.

6. True, because $(2 - \sqrt{3})(2 + \sqrt{3}) = 4 - 3 = 1$.

7. False, because $\dfrac{\sqrt{12}}{3} = \dfrac{2\sqrt{3}}{3}$.

8. True, because $\dfrac{\sqrt{20}}{\sqrt{5}} = \dfrac{\sqrt{4}\sqrt{5}}{\sqrt{5}} = \sqrt{4} = 2$.

9. True, because $(2\sqrt{4})^2 = 4 \cdot 4 = 16$.

10. True, because $(3\sqrt{5})^3 = 3^3 \cdot \sqrt{5^3} = 27\sqrt{125}$.

5.4 EXERCISES

1. $\sqrt{15} \div \sqrt{5} = \sqrt{\dfrac{15}{5}} = \sqrt{3}$

3. $\sqrt{3} \div \sqrt{5} = \dfrac{\sqrt{3}}{\sqrt{5}} = \dfrac{\sqrt{3}\sqrt{5}}{\sqrt{5}\sqrt{5}} = \dfrac{\sqrt{15}}{5}$

5. $\dfrac{3\sqrt{3}}{5\sqrt{6}} = \dfrac{3}{5\sqrt{2}} = \dfrac{3\sqrt{2}}{5\sqrt{2}\sqrt{2}} = \dfrac{3\sqrt{2}}{10}$

7. $\dfrac{2\sqrt{3}}{3\sqrt{6}} = \dfrac{2}{3\sqrt{2}} = \dfrac{2\sqrt{2}}{3\sqrt{2}\sqrt{2}} = \dfrac{2\sqrt{2}}{3 \cdot 2} = \dfrac{\sqrt{2}}{3}$

9. $\dfrac{\sqrt[3]{20}}{\sqrt[3]{2}} = \sqrt[3]{\dfrac{20}{2}} = \sqrt[3]{10}$

11. $\dfrac{\sqrt[3]{8x^7}}{\sqrt[3]{2x}} = \sqrt[3]{4x^6} = x^2 \cdot \sqrt[3]{4}$

13. $\dfrac{6 + \sqrt{9 \cdot 5}}{3} = \dfrac{6 + 3\sqrt{5}}{3} = 2 + \sqrt{5}$

15. $\dfrac{-2 + \sqrt{4 \cdot 3}}{-2} = \dfrac{-2 + 2\sqrt{3}}{-2} = 1 - \sqrt{3}$

17. $\dfrac{1 + \sqrt{2}}{\sqrt{3} - 1} = \dfrac{(1 + \sqrt{2})(\sqrt{3} + 1)}{(\sqrt{3} - 1)(\sqrt{3} + 1)}$

$= \dfrac{1 + \sqrt{6} + \sqrt{2} + \sqrt{3}}{2}$

19. $\dfrac{\sqrt{2}}{\sqrt{6}+\sqrt{3}} = \dfrac{\sqrt{2}(\sqrt{6}-\sqrt{3})}{(\sqrt{6}+\sqrt{3})(\sqrt{6}-\sqrt{3})} = \dfrac{\sqrt{12}-\sqrt{6}}{6-3}$

$\qquad\qquad = \dfrac{2\sqrt{3}-\sqrt{6}}{3}$

21. $\dfrac{2\sqrt{3}(3\sqrt{2}+\sqrt{5})}{(3\sqrt{2}-\sqrt{5})(3\sqrt{2}+\sqrt{5})} = \dfrac{6\sqrt{6}+2\sqrt{15}}{18-5}$

$\qquad\qquad\qquad = \dfrac{6\sqrt{6}+2\sqrt{15}}{13}$

23. $\dfrac{(1+3\sqrt{2})(2\sqrt{6}-3\sqrt{10})}{(2\sqrt{6}+3\sqrt{10})(2\sqrt{6}-3\sqrt{10})}$

$\qquad = \dfrac{2\sqrt{6}+6\sqrt{12}-3\sqrt{10}-9\sqrt{20}}{24-90}$

$\qquad = \dfrac{2\sqrt{6}+12\sqrt{3}-3\sqrt{10}-18\sqrt{5}}{-66}$

$\qquad = \dfrac{18\sqrt{5}+3\sqrt{10}-2\sqrt{6}-12\sqrt{3}}{66}$

25. $(2\sqrt{2})^5 = 2^5 \cdot \sqrt{2^5} = 32\sqrt{16}\sqrt{2} = 128\sqrt{2}$

27. $(\sqrt{x})^5 = \sqrt{x^5} = \sqrt{x^4}\sqrt{x} = x^2 \cdot \sqrt{x}$

29. $(-3\sqrt{x^3})^3 = -27\sqrt{x^9} = -27\sqrt{x^8}\sqrt{x}$

$\qquad = -27x^4 \cdot \sqrt{x}$

31. $(2x \cdot \sqrt[3]{x^2})^3 = 8x^3 \cdot \sqrt[3]{x^6} = 8x^3 \cdot x^2 = 8x^5$

33. $(-2 \cdot \sqrt[3]{5})^2 = (-2)^2 \cdot \sqrt[3]{5^2} = 4 \cdot \sqrt[3]{25}$

35. $(\sqrt[3]{x^2})^6 = (x^2)^{6/3} = (x^2)^2 = x^4$

37. $\dfrac{\sqrt{3}}{\sqrt{2}} + \dfrac{2}{\sqrt{2}} = \dfrac{\sqrt{3}\sqrt{2}}{\sqrt{2}\sqrt{2}} + \dfrac{2\sqrt{2}}{\sqrt{2}\sqrt{2}} = \dfrac{\sqrt{6}}{2} + \dfrac{2\sqrt{2}}{2}$

$\qquad\qquad\qquad\qquad\qquad = \dfrac{\sqrt{6}+2\sqrt{2}}{2}$

39. $\dfrac{\sqrt{3}}{\sqrt{2}} + \dfrac{3\sqrt{6}}{2} = \dfrac{\sqrt{3}\sqrt{2}}{\sqrt{2}\sqrt{2}} + \dfrac{3\sqrt{6}}{2} = \dfrac{\sqrt{6}}{2} + \dfrac{3\sqrt{6}}{2} = \dfrac{4\sqrt{6}}{2}$

$\qquad\qquad\qquad\qquad\qquad = 2\sqrt{6}$

41. $\dfrac{\sqrt{6}}{2} \cdot \dfrac{1}{\sqrt{3}} = \dfrac{\sqrt{2}\sqrt{3}}{2\sqrt{3}} = \dfrac{\sqrt{2}}{2}$

43. $\dfrac{2\sqrt{w}}{3\sqrt{w}} = \dfrac{2}{3}$

45. $\dfrac{8-\sqrt{16 \cdot 2}}{20} = \dfrac{8-4\sqrt{2}}{20} = \dfrac{4(2-\sqrt{2})}{4 \cdot 5} = \dfrac{2-\sqrt{2}}{5}$

47. $\dfrac{5+5\sqrt{3}}{10} = \dfrac{5(1+\sqrt{3})}{5 \cdot 2} = \dfrac{1+\sqrt{3}}{2}$

49. $\sqrt{a}(\sqrt{a}-3) = a - 3\sqrt{a}$

51. $4\sqrt{a}(a+\sqrt{a}) = 4a\sqrt{a} + 4a$

53. $(2\sqrt{3m})^2 = 4 \cdot 3m = 12m$

55. $\left(-2\sqrt{xy^2z}\right)^2 = (-2)^2 xy^2z = 4xy^2z$

57. $\sqrt[3]{m}\left(\sqrt[3]{m^2} - \sqrt[3]{m^5}\right) = \sqrt[3]{m^3} - \sqrt[3]{m^6}$

$\qquad\qquad\qquad\qquad = m - m^2$

59. $\sqrt[3]{8x^4} + \sqrt[3]{27x^4} = \sqrt[3]{8x^3} \cdot \sqrt[3]{x} + \sqrt[3]{27x^3} \cdot \sqrt[3]{x}$

$\qquad = 2x \cdot \sqrt[3]{x} + 3x \cdot \sqrt[3]{x} = 5x \cdot \sqrt[3]{x}$

61. $\left(2m \cdot \sqrt[4]{2m^2}\right)^3 = 8m^3 \cdot \sqrt[4]{8m^6}$

$\qquad = 8m^3 \cdot \sqrt[4]{m^4} \cdot \sqrt[4]{8m^2}$

$\qquad = 8m^3 \cdot m \cdot \sqrt[4]{8m^2} = 8m^4 \cdot \sqrt[4]{8m^2}$

63. $\dfrac{4}{2+\sqrt{8}} = \dfrac{4(2-\sqrt{8})}{(2+\sqrt{8})(2-\sqrt{8})} = \dfrac{8-8\sqrt{2}}{-4}$

$\qquad = 2\sqrt{2} - 2$

65. $\dfrac{5(\sqrt{2}+1)}{(\sqrt{2}-1)(\sqrt{2}+1)} + \dfrac{3(\sqrt{2}-1)}{(\sqrt{2}+1)(\sqrt{2}-1)}$

$\qquad = \dfrac{5\sqrt{2}+5}{2-1} + \dfrac{3\sqrt{2}-3}{2-1} = \ 2 + 8\sqrt{2}$

67. $\dfrac{1}{\sqrt{2}} + \dfrac{1}{\sqrt{3}} = \dfrac{\sqrt{2}}{2} + \dfrac{\sqrt{3}}{3}$

$\qquad\qquad = \dfrac{3\sqrt{2}}{3 \cdot 2} + \dfrac{2\sqrt{3}}{2 \cdot 3} = \ \dfrac{3\sqrt{2}+2\sqrt{3}}{6}$

69. $\dfrac{3}{\sqrt{2}-1} + \dfrac{4}{\sqrt{2}+1}$

$= \dfrac{3(\sqrt{2}+1)}{(\sqrt{2}-1)(\sqrt{2}+1)} + \dfrac{4(\sqrt{2}-1)}{(\sqrt{2}+1)(\sqrt{2}-1)}$

$= \dfrac{3\sqrt{2}+3}{1} + \dfrac{4\sqrt{2}-4}{1} = \ 7\sqrt{2} - 1$

71. $\dfrac{\sqrt{x}(\sqrt{x}-2)}{(\sqrt{x}+2)(\sqrt{x}-2)} + \dfrac{3\sqrt{x}(\sqrt{x}+2)}{(\sqrt{x}-2)(\sqrt{x}+2)}$

$= \dfrac{x-2\sqrt{x}}{x-4} + \dfrac{3x+6\sqrt{x}}{x-4} = \ \dfrac{4x+4\sqrt{x}}{x-4}$

73.

$$\frac{1(1-\sqrt{x})}{\sqrt{x}(1-\sqrt{x})} + \frac{1\sqrt{x}}{(1-\sqrt{x})\sqrt{x}}$$

$$= \frac{1-\sqrt{x}}{\sqrt{x}-x} + \frac{\sqrt{x}}{\sqrt{x}-x} = \frac{1}{\sqrt{x}-x}$$

$$= \frac{1(\sqrt{x}+x)}{(\sqrt{x}-x)(\sqrt{x}+x)} = \frac{x+\sqrt{x}}{x-x^2}$$

75. $\dfrac{\sqrt{2}}{\sqrt{3}} = \dfrac{\sqrt{2}\sqrt{3}}{\sqrt{3}\sqrt{3}} = \dfrac{\sqrt{6}}{3}$

77. $\dfrac{1}{\sqrt{2}-1} = \dfrac{1(\sqrt{2}+1)}{(\sqrt{2}-1)(\sqrt{2}+1)}$

$$= \frac{\sqrt{2}+1}{1}$$

79. $\dfrac{1}{\sqrt{x}-1} = \dfrac{1(\sqrt{x}+1)}{(\sqrt{x}-1)(\sqrt{x}+1)} = \dfrac{\sqrt{x}+1}{x-1}$

81. $\dfrac{3}{\sqrt{2}+x} = \dfrac{3(\sqrt{2}-x)}{(\sqrt{2}+x)(\sqrt{2}-x)}$

$$= \frac{3\sqrt{2}-3x}{2-x^2}$$

83. $\sqrt{3}+\sqrt{5} = 3.968$ **85.** $2\sqrt{3}+5\sqrt{3} = 12.124$

87. $(2\sqrt{3})(3\sqrt{2}) = 14.697$

89. $\sqrt{5}(\sqrt{5}+\sqrt{3}) = 8.873$

91. $\dfrac{-1+\sqrt{6}}{2} = 0.725$ **93.** $\dfrac{4-\sqrt{10}}{-2} = -0.419$

5.5 WARM-UPS

1. False, because $x^2 = 4$ is equivalent to the compound equation $x = 2$ or $x = -2$.

2. True, because the square of any real number is nonnegative.

3. False, 0 is a solution.

4. False, because -2 is not a solution to $x^3 = 8$.

5. True, because if x is positive the left side of the equation is negative and the right side is positive.

6. False, we should square each side.

7. False, extraneous roots are found but they do not satisfy the equation.

8. True, because we get $x = 49$, and 49 does not satisfy $\sqrt{x} = -7$.

9. True, because both square roots of 6 satisfy $x^2 - 6 = 0$.

10. True, we get extraneous roots only by raising each side to an even power.

5.5 EXERCISES

1.
$$x^3 = -1000$$
$$x = \sqrt[3]{-1000} = -10$$

Solution set: $\{-10\}$

3.
$$32m^5 - 1 = 0$$
$$32m^5 = 1$$
$$m^5 = \frac{1}{32}$$
$$m = \sqrt[5]{\frac{1}{32}} = \frac{1}{2}$$

Solution set: $\left\{\dfrac{1}{2}\right\}$

5.
$$(y-3)^3 = -8$$
$$y - 3 = \sqrt[3]{-8}$$
$$y - 3 = -2$$
$$y = 1$$

Solution set: $\{1\}$

7.
$$\frac{1}{2}x^3 + 4 = 0$$
$$\frac{1}{2}x^3 = -4$$
$$x^3 = -8$$
$$x = \sqrt[3]{-8} = -2$$

Solution set: $\{-2\}$

9.
$$x^2 = 25$$
$$x = \pm 5$$

Solution set: $\{-5, 5\}$

11.
$$x^2 - 20 = 0$$
$$x^2 = 20$$
$$x = \pm\sqrt{20} = \pm 2\sqrt{5}$$

Solution set: $\{-2\sqrt{5}, 2\sqrt{5}\}$

13.
$$x^2 = -9$$

Since an even root of a negative number is not a real number, the solution set is $\emptyset$.

15.
$$(x-3)^2 = 16$$
$$x - 3 = \pm 4$$
$$x = 3 \pm 4$$
$$x = 3 + 4 \quad \text{or} \quad x = 3 - 4$$
$$x = 7 \quad\quad \text{or} \quad\quad x = -1$$

Solution set: $\{-1, 7\}$

17.
$$(x+1)^2 = 8$$
$$x + 1 = \pm\sqrt{8}$$
$$x = -1 \pm 2\sqrt{2}$$

Solution set: $\{-1 - 2\sqrt{2}, -1 + 2\sqrt{2}\}$

19.
$$\frac{1}{2}x^2 = 5$$
$$x^2 = 10$$
$$x = \pm\sqrt{10}$$
Solution set: $\{-\sqrt{10}, \sqrt{10}\}$

21. $4(a+1)^2 = 1$
$$(a+1)^2 = \frac{1}{4}$$
$$a+1 = \pm\sqrt{\frac{1}{4}}$$
$$a = -1 \pm \frac{1}{2}$$
$$a = -1 + \frac{1}{2} \quad \text{or} \quad a = -1 - \frac{1}{2}$$
$$a = -\frac{1}{2} \quad \text{or} \quad a = -\frac{3}{2}$$
Solution set: $\left\{-\frac{3}{2}, -\frac{1}{2}\right\}$

23.
$$(y-3)^4 = 0$$
$$y - 3 = \sqrt[4]{0} = 0$$
$$y = 3$$
Solution set: $\{3\}$

25.
$$2x^6 = 128$$
$$x^6 = 64$$
$$x = \pm\sqrt[6]{64}$$
$$x = \pm 2$$
Solution set: $\{-2, 2\}$

27. $81a^4 - 16 = 0$
$$81a^4 = 16$$
$$a^4 = \frac{16}{81}$$
$$a = \pm\sqrt[4]{\frac{16}{81}} = \pm\frac{2}{3}$$
Solution set: $\left\{-\frac{2}{3}, \frac{2}{3}\right\}$

29.
$$\sqrt{x-3} = 7$$
$$(\sqrt{x-3})^2 = 7^2$$
$$x - 3 = 49$$
$$x = 52$$
Check 52 in the original equation.
Solution set: $\{52\}$

31.
$$2\sqrt{w+4} = 5$$
$$(2\sqrt{w+4})^2 = 5^2$$
$$4(w+4) = 25$$
$$4w + 16 = 25$$
$$4w = 9$$
$$w = \frac{9}{4}$$
Check $\frac{9}{4}$ in the original equation.
Solution set: $\left\{\frac{9}{4}\right\}$

33.
$$\sqrt[3]{2x+3} = \sqrt[3]{x+12}$$
$$\left(\sqrt[3]{2x+3}\right)^3 = \left(\sqrt[3]{x+12}\right)^3$$
$$2x + 3 = x + 12$$
$$x = 9$$
Check 9 in the original equation.
Solution set: $\{9\}$

35.
$$\sqrt{2t+4} = \sqrt{t-1}$$
$$(\sqrt{2t+4})^2 = (\sqrt{t-1})^2$$
$$2t + 4 = t - 1$$
$$t = -5$$
Check: $\sqrt{2(-5)+4} = \sqrt{-5-1}$

Since each side of the equation is the square root of a negative number, the solution set is $\emptyset$.

37.
$$\sqrt{4x^2 + x - 3} = 2x$$
$$\left(\sqrt{4x^2+x-3}\right)^2 = (2x)^2$$
$$4x^2 + x - 3 = 4x^2$$
$$x - 3 = 0$$
$$x = 3$$
Check 3 in the original equation.
Solution set: $\{3\}$

39.
$$\sqrt{x^2 + 2x - 6} = 3$$
$$\left(\sqrt{x^2+2x-6}\right)^2 = 3^2$$
$$x^2 + 2x - 6 = 9$$
$$x^2 + 2x - 15 = 0$$
$$(x+5)(x-3) = 0$$
$$x+5 = 0 \quad \text{or} \quad x-3 = 0$$
$$x = -5 \quad \text{or} \quad x = 3$$
Check 3 and −5 in the original equation.
Solution set: $\{-5, 3\}$

41.
$$\sqrt{2x^2 - 1} = x$$
$$\left(\sqrt{2x^2-1}\right)^2 = x^2$$
$$2x^2 - 1 = x^2$$
$$x^2 = 1$$
$$x = \pm 1$$
Checking in the original we find that if $x = -1$ we get $\sqrt{1} = -1$, which is incorrect. So the solution set is $\{1\}$.

43.
$$\sqrt{2x^2 + 5x + 6} = x$$
$$\left(\sqrt{2x^2+5x+6}\right)^2 = x^2$$
$$2x^2 + 5x + 6 = x^2$$
$$x^2 + 5x + 6 = 0$$
$$(x+2)(x+3) = 0$$

$$x = -2 \quad \text{or} \quad x = -3$$

If we use $x = -2$ in the original equation we get $\sqrt{4} = -2$, and if $x = -3$ in the original we get $\sqrt{9} = -3$. Since both of these equations are incorrect, the solution set is $\emptyset$.

45.
$$\left(\sqrt{2x^2 + 6x + 4}\right)^2 = (x+1)^2$$
$$2x^2 + 6x + 4 = x^2 + 2x + 1$$
$$x^2 + 4x + 3 = 0$$
$$(x+3)(x+1) = 0$$
$$x = -3 \quad \text{or} \quad x = -1$$

If we use $x = -3$ in the original equation we get $\sqrt{4} = -2$, which is incorrect. If $x = -1$ we get $\sqrt{0} = 0$, which is correct. The solution set is $\{-1\}$.

47.
$$\sqrt{x+3} - \sqrt{x-2} = 1$$
$$\sqrt{x+3} = 1 + \sqrt{x-2}$$
$$\left(\sqrt{x+3}\right)^2 = \left(1 + \sqrt{x-2}\right)^2$$
$$x + 3 = 1 + 2\sqrt{x-2} + x - 2$$
$$4 = 2\sqrt{x-2}$$
$$2 = \sqrt{x-2}$$
$$2^2 = \left(\sqrt{x-2}\right)^2$$
$$4 = x - 2$$
$$6 = x$$

Check 6 in the original equation. The solution set is $\{6\}$.

49.
$$\sqrt{2x+2} - \sqrt{x-3} = 2$$
$$\sqrt{2x+2} = 2 + \sqrt{x-3}$$
$$\left(\sqrt{2x+2}\right)^2 = \left(2 + \sqrt{x-3}\right)^2$$
$$2x + 2 = 4 + 4\sqrt{x-3} + x - 3$$
$$x + 1 = 4\sqrt{x-3}$$
$$(x+1)^2 = \left(4\sqrt{x-3}\right)^2$$
$$x^2 + 2x + 1 = 16(x-3)$$
$$x^2 - 14x + 49 = 0$$
$$(x-7)^2 = 0$$
$$x = 7$$

Check 7 in the original equation.
Solution set: $\{7\}$

51.
$$\sqrt{4-x} - \sqrt{x+6} = 2$$
$$\sqrt{4-x} = 2 + \sqrt{x+6}$$
$$\left(\sqrt{4-x}\right)^2 = \left(2 + \sqrt{x+6}\right)^2$$
$$4 - x = 4 + 4\sqrt{x+6} + x + 6$$
$$-6 - 2x = 4\sqrt{x+6}$$
$$-3 - x = 2\sqrt{x+6}$$

$$(-3-x)^2 = \left(2\sqrt{x+6}\right)^2$$
$$9 + 6x + x^2 = 4(x+6)$$
$$x^2 + 2x - 15 = 0$$
$$(x+5)(x-3) = 0$$
$$x = -5 \quad \text{or} \quad x = 3$$

If $x = 3$ in the original equation we get $1 - 3 = 2$, which is incorrect. If $x = -5$ we get $3 - 1 = 2$. So the solution set is $\{-5\}$.

53.
$$\left(x^{2/3}\right)^3 = 3^3$$
$$x^2 = 27$$
$$x = \pm\sqrt{27} = \pm 3\sqrt{3}$$

Solution set: $\{-3\sqrt{3},\, 3\sqrt{3}\}$

55.
$$\left(y^{-2/3}\right)^{-3} = (9)^{-3}$$
$$y^2 = \frac{1}{729}$$
$$y = \pm\sqrt{\frac{1}{729}} = \pm\frac{1}{27}$$

Solution set: $\left\{-\frac{1}{27},\, \frac{1}{27}\right\}$

57.
$$\left(w^{1/3}\right)^3 = 8^3$$
$$w = 512$$

Solution set: $\{512\}$

59.
$$\left(t^{-1/2}\right)^{-2} = (9)^{-2}$$
$$t = \frac{1}{81}$$

Solution set: $\left\{\frac{1}{81}\right\}$

61.
$$\left((3a-1)^{-2/5}\right)^{-5} = 1^{-5}$$
$$(3a-1)^2 = 1$$
$$3a - 1 = \pm 1$$
$$3a - 1 = 1 \quad \text{or} \quad 3a - 1 = -1$$
$$3a = 2 \quad \text{or} \quad 3a = 0$$
$$a = \frac{2}{3} \quad \text{or} \quad a = 0$$

Solution set: $\left\{0,\, \frac{2}{3}\right\}$

63.
$$\left((t-1)^{-2/3}\right)^{-3} = 2^{-3}$$
$$(t-1)^2 = \frac{1}{8}$$
$$t - 1 = \pm\sqrt{\frac{1}{8}} = \pm\frac{\sqrt{2}}{4}$$
$$t = 1 \pm \frac{\sqrt{2}}{4} = \frac{4}{4} \pm \frac{\sqrt{2}}{4} = \frac{4 \pm \sqrt{2}}{4}$$

Solution set: $\left\{\frac{4 - \sqrt{2}}{4},\, \frac{4 + \sqrt{2}}{4}\right\}$

65.
$$(x-3)^{2/3} = -4$$
$$\left((x-3)^{2/3}\right)^3 = (-4)^3$$
$$(x-3)^2 = -64$$
Because the square of any real number is nonnegative, the solution set is $\emptyset$.

67.
$$2x^2 + 3 = 7$$
$$2x^2 = 4$$
$$x^2 = 2$$
$$x = \pm\sqrt{2}$$
Solution set: $\{-\sqrt{2}, \sqrt{2}\}$

69.
$$\sqrt[3]{2w+3} = \sqrt[3]{w-2}$$
$$\left(\sqrt[3]{2w+3}\right)^3 = \left(\sqrt[3]{w-2}\right)^3$$
$$2w+3 = w-2$$
$$w = -5$$
Solution set: $\{-5\}$

71.
$$9x^2 - 1 = 0$$
$$x^2 = \frac{1}{9}$$
$$x = \pm\sqrt{\frac{1}{9}} = \pm\frac{1}{3}$$
Solution set: $\left\{-\frac{1}{3}, \frac{1}{3}\right\}$

73.
$$\left((w+1)^{2/3}\right)^3 = (-3)^3$$
$$(w+1)^2 = -27$$
This equation has no solution by the even root property. The solution set is $\emptyset$.

75.
$$\left((a+1)^{1/3}\right)^3 = (-2)^3$$
$$a+1 = -8$$
$$a = -9$$
Solution set: $\{-9\}$

77.
$$(4y-5)^7 = 0$$
$$4y-5 = \sqrt[7]{0} = 0$$
$$4y = 5$$
$$y = \frac{5}{4}$$
Solution set: $\left\{\frac{5}{4}\right\}$

79.
$$\sqrt{x^2+5x} = 6$$
$$\left(\sqrt{x^2+5x}\right)^2 = 6^2$$
$$x^2 + 5x = 36$$
$$x^2 + 5x - 36 = 0$$
$$(x+9)(x-4) = 0$$
$$x = -9 \quad \text{or} \quad x = 4$$
Solution set: $\{-9, 4\}$

81.
$$\sqrt{4x^2} = x+2$$
$$\left(\sqrt{4x^2}\right)^2 = (x+2)^2$$
$$4x^2 = x^2 + 4x + 4$$
$$3x^2 - 4x - 4 = 0$$
$$(3x+2)(x-2) = 0$$
$$3x+2 = 0 \quad \text{or} \quad x-2 = 0$$
$$x = -\frac{2}{3} \quad \text{or} \quad x = 2$$
Solution set: $\left\{-\frac{2}{3}, 2\right\}$

83.
$$(t+2)^4 = 32$$
$$t+2 = \pm\sqrt[4]{32} = \pm 2\cdot\sqrt[4]{2}$$
$$t = -2 \pm 2\cdot\sqrt[4]{2}$$
Solution set: $\{-2 - 2\cdot\sqrt[4]{2}, \ -2 + 2\cdot\sqrt[4]{2}\}$

85.
$$\sqrt{x^2 - 3x} = x$$
$$\left(\sqrt{x^2-3x}\right)^2 = x^2$$
$$x^2 - 3x = x^2$$
$$-3x = 0$$
$$x = 0$$
Solution set: $\{0\}$

87.
$$x^{-3} = 8$$
$$\left(x^{-3}\right)_1 = 8^{-1}$$
$$x^3 = \frac{1}{8}$$
$$x = \sqrt[3]{\frac{1}{8}} = \frac{1}{2}$$
Solution set: $\left\{\frac{1}{2}\right\}$

89.
$$a^{-2} = 3$$
$$\left(a^{-2}\right)^{-1} = 3^{-1}$$
$$a^2 = \frac{1}{3}$$
$$a = \pm\sqrt{\frac{1}{3}} = \pm\frac{\sqrt{3}}{3}$$
Solution set: $\left\{-\frac{\sqrt{3}}{3}, \frac{\sqrt{3}}{3}\right\}$

91.
$$\sqrt{x+1} - \sqrt{2x+9} = -2$$
$$\sqrt{x+1} = \sqrt{2x+9} - 2$$
$$\left(\sqrt{x+1}\right)^2 = \left(\sqrt{2x+9} - 2\right)^2$$
$$x+1 = 2x+9 - 4\sqrt{2x+9} + 4$$
$$-x - 12 = -4\sqrt{2x+9}$$
$$x + 12 = 4\sqrt{2x+9}$$

$$(x+12)^2 = (4\sqrt{2x+9})^2$$
$$x^2 + 24x + 144 = 16(2x+9)$$
$$x^2 - 8x = 0$$
$$x(x-8) = 0$$
$$x = 0 \quad \text{or} \quad x = 8$$

Solution set: $\{0, 8\}$

93. Let $x =$ the length of a side. Two sides and the diagonal of a square form a right triangle. By the Pythagorean theorem we can write the equation
$$x^2 + x^2 = 8^2$$
$$2x^2 = 64$$
$$x^2 = 32$$
$$x = \pm\sqrt{32} = \pm 4\sqrt{2}$$

The length of the side is not a negative number. So the side is $4\sqrt{2}$ feet in length.

95. Let $s =$ the length of the side of the square. Since $A = s^2$ for a square, we can write the equation
$$s^2 = 50$$
$$s = \pm\sqrt{50} = \pm 5\sqrt{2}$$

Since the side of a square is not negative, the length of the side is $5\sqrt{2}$ feet.

97. Let $d =$ the length of a diagonal of the rectangle whose sides are 30 and 40 feet. By the Pythagorean theorem we can write the equation
$$d^2 = 30^2 + 40^2$$
$$d^2 = 2500$$
$$d = \pm\sqrt{2500} = \pm 50$$

Since the diagonal is not negative, the length of the diagonal is 50 feet.

99. If the volume of the cube is 2, each side of the cube has length $\sqrt[3]{2}$, because $V = s^3$. Let $d =$ the length of the diagonal of a side. The diagonal of a side is the diagonal of a square with sides of length $\sqrt[3]{2}$. By the Pythagorean theorem we can write the equation

$$d^2 = \left(\sqrt[3]{2}\right)^2 + \left(\sqrt[3]{2}\right)^2$$

$$d^2 = \sqrt[3]{4} + \sqrt[3]{4}$$

$$d^2 = 2 \cdot \sqrt[3]{4} = \sqrt[3]{8} \cdot \sqrt[3]{4} = \sqrt[3]{32}$$

$$d = \left(\sqrt[3]{32}\right)^{1/2} = \left(32^{1/3}\right)^{1/2} = 32^{1/6}$$

The length of the diagonal is $\sqrt[6]{32}$ meters.

101. Let $x =$ the third side to the triangle whose given sides are 3 and 5. By the Pythagorean theorem we can find x:
$$x^2 + 3^2 = 5^2$$
$$x^2 = 16$$
$$x = 4$$

Since $x = 4$, the base of length 12 is divided into 2 parts, one of length 4 and the other of length 8. The side marked a is the hypotenuse of a right triangle with legs 3 and 8:
$$a^2 = 3^2 + 8^2$$
$$a^2 = 73$$
$$a = \sqrt{73} \text{ km}$$

103.
$$r = \left(\frac{S}{P}\right)^{1/n} - 1$$
$$1 + r = \left(\frac{S}{P}\right)^{1/n}$$
$$\frac{S}{P} = (1+r)^n$$
$$S = P(1+r)^n$$

Solve for P:
$$P = \frac{S}{(1+r)^n}$$
$$P = S(1+r)^{-n}$$

105.
$$\frac{11.86^2}{5.2^3} = \frac{29.46^2}{R^3}$$
$$11.86^2 R^3 = 5.2^3 \cdot 29.46^2$$
$$R^3 = \frac{5.2^3 \cdot 29.46^2}{11.86^2}$$
$$R = \sqrt[3]{\frac{5.2^3 \cdot 29.46^2}{11.86^2}} \approx 9.5 \text{ AU}$$

107.
$$x^2 = 3.24$$
$$x = \pm\sqrt{3.24} = \pm 1.8$$

Solution set: $\{-1.8, 1.8\}$

109.
$$\sqrt{x-2} = 1.73$$
$$x - 2 = (1.73)^2$$
$$x = 2 + (1.73)^2 = 4.9929$$

Solution set: $\{4.993\}$

111.
$$x^{2/3} = 8.86$$
$$\left(x^{2/3}\right)^3 = (8.86)^3$$
$$x^2 = 695.506$$
$$x = \pm\sqrt{695.506} = \pm 26.372$$

Solution set: $\{-26.372, 26.372\}$

5.6 WARM-UPS

1. True, because every real number is a complex number.   **2.** False, because $2 - \sqrt{-6} = 2 - i\sqrt{6}$.
3. False, $\sqrt{-9} = 3i$.   **4.** True, $(\pm 3i)^2 = -9$.
5. True, because we subtract complex numbers just like we subtract binomials.
6. True, because $i^4 = i^2 \cdot i^2 = (-1)(-1) = 1$.
7. True, because $(2-i)(2+i) = 4 - i^2$
$$= 4 - (-1) = 5.$$
8. False, $i^3 = i^2 \cdot i = -1 \cdot i = -i$.
9. True, $i^{48} = (i^4)^{12} = 1^{12} = 1$.
10. False, $x^2 = 0$ has only one solution.

5.6 EXERCISES

1. $(2 + 3i) + (-4 + 5i) = -2 + 8i$
3. $(2 - 3i) - (6 - 7i) = 2 - 3i - 6 + 7i$
$$= -4 + 4i$$
5. $(-1 + i) + (-1 - i) = -2$
7. $(-2 - 3i) - (6 - i) = -2 - 3i - 6 + i$
$$= -8 - 2i$$
9. $3(2 + 5i) = 3 \cdot 2 + 3 \cdot 5i = 6 + 15i$
11. $2i(i - 5) = 2i^2 - 10i = 2(-1) - 10i$
$$= -2 - 10i$$
13. $-4i(3 - i) = -12i + 4i^2$
$$= -12i + 4(-1) = -4 - 12i$$
15. $(2 + 3i)(4 + 6i) = 8 + 24i + 18i^2$
$$= 8 + 24i + 18(-1) = -10 + 24i$$
17. $(-1 + i)(2 - i) = -2 + 3i - i^2$
$$= -2 + 3i - (-1) = -1 + 3i$$
19. $(-1 - 2i)(2 + i) = -2 - 5i - 2i^2$
$$= -2 - 5i - 2(-1) = -5i$$
21. $(5 - 2i)(5 + 2i) = 25 - 4i^2$
$$= 25 - 4(-1) = 29$$
23. $(1 - i)(1 + i) = 1 - i^2 = 1 - (-1) = 2$

25. $(4 + 2i)(4 - 2i) = 16 - 4i^2$
$$= 16 - 4(-1) = 20$$
27. $(3i)^2 = 9i^2 = 9(-1) = -9$
29. $(-5i)^2 = (-5)^2 i^2 = 25(-1) = -25$
31. $(2i)^4 = 2^4 i^4 = 16(1) = 16$
33. $i^9 = (i^4)^2 \cdot i = 1^2 \cdot i = i$
35. $(3 + 5i)(3 - 5i) = 9 - 25i^2 = 9 + 25 = 34$
37. $(1 - 2i)(1 + 2i) = 1 - 4i^2 = 1 - 4(-1)$
$$= 5$$
39. $(-2 + i)(-2 - i) = 4 - i^2 = 4 - (-1)$
$$= 5$$
41. $(2 - i\sqrt{3})(2 + i\sqrt{3}) = 4 - 3i^2$
$$= 4 - 3(-1) = 7$$
43. $\dfrac{3}{4+i} = \dfrac{3(4-i)}{(4+i)(4-i)} = \dfrac{12 - 3i}{16 - i^2}$
$$= \dfrac{12 - 3i}{17} = \dfrac{12}{17} - \dfrac{3}{17}i$$
45. $\dfrac{2+i}{3-2i} = \dfrac{(2+i)(3+2i)}{(3-2i)(3+2i)} = \dfrac{6 + 7i + 2i^2}{9 - 4i^2}$
$$= \dfrac{4 + 7i}{13} = \dfrac{4}{13} + \dfrac{7}{13}i$$
47. $\dfrac{4+3i}{i} = \dfrac{(4+3i)(-i)}{(i)(-i)} = \dfrac{-4i - 3i^2}{-i^2} =$
$$= \dfrac{-4i + 3}{1} = 3 - 4i$$
49. $\dfrac{2+6i}{2} = \dfrac{2}{2} + \dfrac{6i}{2} = 1 + 3i$
51. $2 + \sqrt{-4} = 2 + i\sqrt{4} = 2 + 2i$
53. $2\sqrt{-9} + 5 = 2i\sqrt{9} + 5 = 2i \cdot 3 + 5$
$$= 5 + 6i$$
55. $7 - \sqrt{-6} = 7 - i\sqrt{6}$
57. $\sqrt{-8} + \sqrt{-18} = i\sqrt{4}\sqrt{2} + i\sqrt{9}\sqrt{2}$
$$= 2i\sqrt{2} + 3i\sqrt{2} = 5i\sqrt{2}$$
59. $\dfrac{2 + \sqrt{-12}}{2} = \dfrac{2 + i\sqrt{4}\sqrt{3}}{2} = 1 + i\sqrt{3}$
61. $\dfrac{-4 - \sqrt{-24}}{4} = \dfrac{-4 - i\sqrt{4}\sqrt{6}}{4}$
$$= \dfrac{-4}{4} - \dfrac{2i\sqrt{6}}{4} = -1 - \dfrac{1}{2}i\sqrt{6}$$

63. $x^2 = -36$

$\quad x = \pm\sqrt{-36} = \pm 6i$

Solution set: $\{\pm 6i\}$

65. $x^2 = -12$

$\quad x = \pm\sqrt{-12} = \pm i\sqrt{4}\sqrt{3} = \pm 2i\sqrt{3}$

Solution set: $\{\pm 2i\sqrt{3}\}$

67. $2x^2 + 5 = 0$

$\quad x^2 = -\dfrac{5}{2}$

$\quad x = \pm\sqrt{-\dfrac{5}{2}} = \pm\dfrac{i\sqrt{5}}{\sqrt{2}} = \pm\dfrac{i\sqrt{10}}{2}$

Solution set: $\left\{\pm\dfrac{i\sqrt{10}}{2}\right\}$

69. $3x^2 + 6 = 0$

$\quad x^2 = -2$

$\quad x = \pm\sqrt{-2} = \pm i\sqrt{2}$

Solution set: $\{\pm i\sqrt{2}\}$

71. $(2 - 3i)(3 + 4i) = 6 - i - 12i^2$
$\quad = 6 - i - 12(-1) = 18 - i$

73. $(2 - 3i) + (3 + 4i) = 5 + i$

75. $\dfrac{2 - 3i}{3 + 4i} = \dfrac{(2 - 3i)(3 - 4i)}{(3 + 4i)(3 - 4i)}$

$\quad = \dfrac{6 - 17i + 12i^2}{9 - 16i^2} = \dfrac{6 - 17i + 12(-1)}{9 - 16(-1)}$

$\quad = \dfrac{-6 - 17i}{25} = -\dfrac{6}{25} - \dfrac{17}{25}i$

77. $i(2 - 3i) = 2i - 3i^2 = 2i + 3 = 3 + 2i$

79. $(-3i)^2 = 9i^2 = -9$

81. $\sqrt{-12} + \sqrt{-3} = 2i\sqrt{3} + i\sqrt{3} = 3i\sqrt{3}$

83. $(2 - 3i)^2 = 4 - 12i + 9i^2 = -5 - 12i$

85. $\dfrac{-4 + \sqrt{-32}}{2} = \dfrac{-4 + 4i\sqrt{2}}{2} = -2 + 2i\sqrt{2}$

CHAPTER 5 REVIEW

1. $(-27)^{-2/3} = \dfrac{1}{(-3)^2} = \dfrac{1}{9}$

3. $(2^6)^{1/3} = 2^{6/3} = 2^2 = 4$

5. $100^{-3/2} = \dfrac{1}{10^3} = \dfrac{1}{1000}$

7. $\dfrac{3x^{-1/2}}{3^{-2}x^{-1}} = 3^3 x^{-1/2 + 1} = 27x^{1/2}$

$\quad -9x^{-1/2}$

9. $a^{3/2}b^3 a^2 b^{2/4} = a^{7/2}b^{7/2}$

11. $x^{1/2 + 1/4}y^{1/4 + 1} = x^{3/4}y^{5/4}$

13. $\sqrt{72x^5} = \sqrt{36x^4}\sqrt{2x} = 6x^2 \cdot \sqrt{2x}$

15. $\sqrt[3]{72x^5} = \sqrt[3]{8x^3} \cdot \sqrt[3]{9x^2} = 2x \cdot \sqrt[3]{9x^2}$

17. $\sqrt{2^6} = 2^3 = 8$

19. $\sqrt{\dfrac{2}{5}} = \dfrac{\sqrt{2}\sqrt{5}}{\sqrt{5}\sqrt{5}} = \dfrac{\sqrt{10}}{5}$

21. $\sqrt[3]{\dfrac{2}{3}} = \dfrac{\sqrt[3]{2} \cdot \sqrt[3]{9}}{\sqrt[3]{3} \cdot \sqrt[3]{9}} = \dfrac{\sqrt[3]{18}}{\sqrt[3]{27}} = \dfrac{\sqrt[3]{18}}{3}$

23. $\dfrac{2}{\sqrt{3x}} = \dfrac{2\sqrt{3x}}{\sqrt{3x}\sqrt{3x}} = \dfrac{2\sqrt{3x}}{3x}$

25. $\dfrac{\sqrt{10y^3}}{\sqrt{6}} = \dfrac{\sqrt{2}\sqrt{5y}\sqrt{y^2}}{\sqrt{2}\sqrt{3}} = \dfrac{y\sqrt{5y}}{\sqrt{3}} = \dfrac{y\sqrt{5y}\sqrt{3}}{\sqrt{3}\sqrt{3}}$

$\quad = \dfrac{y\sqrt{15y}}{3}$

27. $\dfrac{3}{\sqrt[3]{2a}} = \dfrac{3\sqrt[3]{4a^2}}{\sqrt[3]{2a}\sqrt[3]{4a^2}} = \dfrac{3\sqrt[3]{4a^2}}{2a}$

29. $\dfrac{5}{\sqrt[4]{3x^2}} = \dfrac{5 \cdot \sqrt[4]{27x^2}}{\sqrt[4]{3x^2} \cdot \sqrt[4]{27x^2}}$

$\quad = \dfrac{5 \cdot \sqrt[4]{27x^2}}{\sqrt[4]{81x^4}} = \dfrac{5 \cdot \sqrt[4]{27x^2}}{3x}$

31. $\sqrt[4]{48x^5y^{12}} = \sqrt[4]{16x^4y^{12}} \cdot \sqrt[4]{3x} = 2xy^3 \cdot \sqrt[4]{3x}$

33. $\sqrt{13}\sqrt{13} = 13$

35. $\sqrt{27} + \sqrt{45} - \sqrt{75} = 3\sqrt{3} + 3\sqrt{5} - 5\sqrt{3}$
$$= 3\sqrt{5} - 2\sqrt{3}$$

37. $\sqrt{\dfrac{1}{3}} + \sqrt{27} = \dfrac{\sqrt{3}}{3} + 3\sqrt{3} = \dfrac{\sqrt{3}}{3} + \dfrac{9\sqrt{3}}{3} = \dfrac{10\sqrt{3}}{3}$

39. $3\sqrt{2}(5\sqrt{2} - 7\sqrt{3}) = 15 \cdot 2 - 21\sqrt{6} = 30 - 21\sqrt{6}$

41. $(2 - \sqrt{3})(3 + \sqrt{2}) = 6 - 3\sqrt{3} + 2\sqrt{2} - \sqrt{6}$

43. $\sqrt[3]{40} - \sqrt[3]{5} = \sqrt[3]{8} \cdot \sqrt[3]{5} - \sqrt[3]{5}$
$$= 2 \cdot \sqrt[3]{5} - \sqrt[3]{5} = \sqrt[3]{5}$$

45. $5 \div \sqrt{2} = \dfrac{5}{\sqrt{2}} = \dfrac{5\sqrt{2}}{\sqrt{2}\sqrt{2}} = \dfrac{5\sqrt{2}}{2}$

47. $(\sqrt{3})^4 = 3^{4/2} = 3^2 = 9$

49. $\dfrac{2 - 2\sqrt{2}}{2} = \dfrac{2(1 - \sqrt{2})}{2} = 1 - \sqrt{2}$

51. $\dfrac{\sqrt{6}(1 + \sqrt{3})}{(1 - \sqrt{3})(1 + \sqrt{3})} = \dfrac{\sqrt{6} + \sqrt{18}}{1 - 3}$
$$= \dfrac{\sqrt{6} + 3\sqrt{2}}{-2} = \dfrac{-\sqrt{6} - 3\sqrt{2}}{2}$$

53. $\dfrac{2\sqrt{3}}{3\sqrt{6} - 2\sqrt{3}} = \dfrac{2\sqrt{3}(3\sqrt{6} + 2\sqrt{3})}{(3\sqrt{6} - 2\sqrt{3})(3\sqrt{6} + 2\sqrt{3})}$

$= \dfrac{6\sqrt{18} + 12}{54 - 12} = \dfrac{6 \cdot 3\sqrt{2} + 12}{42} = \dfrac{6 \cdot 3\sqrt{2} + 6 \cdot 2}{6 \cdot 7}$
$$= \dfrac{3\sqrt{2} + 2}{7}$$

55. $\left(2w \sqrt[3]{2w^2}\right)^6 = 2^6 w^6 \sqrt[3]{2^6 w^{12}}$
$$= 2^6 2^2 w^6 w^4 = 256 w^{10}$$

57. $x^2 = 16$
$$x = \pm 4$$
Solution set: $\{-4, 4\}$

59. $(a - 5)^2 = 4$
$$a - 5 = \pm 2$$
$$a = 5 \pm 2$$
$$a = 5 + 2 \quad \text{or} \quad a = 5 - 2$$
$$a = 7 \quad \text{or} \quad a = 3$$
Solution set: $\{3, 7\}$

61. $(a + 1)^2 = 5$
$$a + 1 = \pm\sqrt{5}$$
$$a = -1 \pm \sqrt{5}$$
Solution set: $\{-1 - \sqrt{5},\ -1 + \sqrt{5}\}$

63. $(m + 1)^2 = -8$
Since the square root of -8 is not a real number, the solution set is $\emptyset$.

65. $\sqrt{m - 1} = 3$
$$(\sqrt{m - 1})^2 = 3^2$$
$$m - 1 = 9$$
$$m = 10$$
Solution set: $\{10\}$

67. $\sqrt[3]{2x + 9} = 3$
$$(\sqrt[3]{2x + 9})^3 = 3^3$$
$$2x + 9 = 27$$
$$2x = 18$$
$$x = 9$$
Solution set: $\{9\}$

69. $w^{2/3} = 4$
$$(w^{2/3})^3 = 4^3$$
$$w^2 = 64$$
$$w = \pm 8$$
Solution set: $\{-8, 8\}$

71. $(m + 1)^{1/3} = 5$
$$((m + 1)^{1/3})^3 = 5^3$$
$$m + 1 = 125$$
$$m = 124$$
Solution set: $\{124\}$

73. $\sqrt{x - 3} = \sqrt{x + 2} - 1$
$$(\sqrt{x - 3})^2 = (\sqrt{x + 2} - 1)^2$$
$$x - 3 = x + 2 - 2\sqrt{x + 2} + 1$$
$$-6 = -2\sqrt{x + 2}$$
$$3 = \sqrt{x + 2}$$
$$3^2 = (\sqrt{x + 2})^2$$
$$9 = x + 2$$
$$7 = x$$
Solution set: $\{7\}$

75.
$$\sqrt{5x - x^2} = \sqrt{6}$$

$$\left(\sqrt{5x - x^2}\right)^2 = \left(\sqrt{6}\right)^2$$

$$5x - x^2 = 6$$
$$-x^2 + 5x - 6 = 0$$
$$x^2 - 5x + 6 = 0$$
$$(x - 2)(x - 3) = 0$$
$$x = 2 \quad \text{or} \quad x = 3$$
Solution set: {2, 3}

77.
$$\sqrt{x + 7} - 2\sqrt{x} = -2$$

$$\sqrt{x + 7} = 2\sqrt{x} - 2$$

$$\left(\sqrt{x + 7}\right)^2 = \left(2\sqrt{x} - 2\right)^2$$

$$x + 7 = 4x - 8\sqrt{x} + 4$$

$$8\sqrt{x} = 3x - 3$$

$$\left(8\sqrt{x}\right)^2 = (3x - 3)^2$$

$$64x = 9x^2 - 18x + 9$$

$$0 = 9x^2 - 82x + 9$$
$$0 = (9x - 1)(x - 9)$$

$$x = \frac{1}{9} \quad \text{or} \quad x = 9$$

Since $\sqrt{\frac{1}{9} + 7} - 2\sqrt{\frac{1}{9}} = \frac{8}{3} - \frac{2}{3} = 2$, 1/9 is an

extraneous root. The solution set is {9}.

79.
$$2\sqrt{x} - \sqrt{x - 3} = 3$$

$$2\sqrt{x} = \sqrt{x - 3} + 3$$

$$\left(2\sqrt{x}\right)^2 = \left(\sqrt{x - 3} + 3\right)^2$$

$$4x = x - 3 + 6\sqrt{x - 3} + 9$$

$$3x - 6 = 6\sqrt{x - 3}$$

$$x - 2 = 2\sqrt{x - 3}$$

$$(x - 2)^2 = (2\sqrt{x - 3})^2$$

$$x^2 - 4x + 4 = 4(x - 3)$$

$$x^2 - 8x + 16 = 0$$

$$(x - 4)^2 = 0$$

$$x = 4$$

Solution set: {4}

81. $(2 - 3i)(-5 + 5i) = -10 + 25i - 15i^2$
$$= -10 + 25i + 15 = 5 + 25i$$

83. $(2 + i) + (5 - 4i) = 7 - 3i$

85. $(1 - i) - (2 - 3i) = 1 - i - 2 + 3i$
$$= -1 + 2i$$

87. $\dfrac{6 + 3i}{3} = \dfrac{6}{3} + \dfrac{3i}{3} = 2 + i$

89. $\dfrac{4 - \sqrt{-12}}{2} = \dfrac{4 - 2i\sqrt{3}}{2} = 2 - i\sqrt{3}$

91. $\dfrac{(2 - 3i)(4 - i)}{(4 + i)(4 - i)} = \dfrac{5 - 14i}{17} = \dfrac{5}{17} - \dfrac{14}{17}i$

93. $x^2 + 100 = 0$

$$x^2 = -100$$
$$x = \pm\sqrt{-100} = \pm 10i$$
The solution set is {±10i}.

95. $2b^2 + 9 = 0$

$$2b^2 = -9$$

$$b^2 = -\frac{9}{2}$$

$$b = \sqrt{-\frac{9}{2}} = \pm\frac{3i}{\sqrt{2}} = \pm\frac{3i\sqrt{2}}{2}$$

The solution set is $\left\{ \pm\dfrac{3i\sqrt{2}}{2} \right\}$.

97. False, because $2^3 \cdot 3^2 = 8 \cdot 9 = 72$.

99. True, because $(\sqrt{2})^3 = \sqrt{8} = 2\sqrt{2}$.

101. True, because $8^{200}8^{200} = (8 \cdot 8)^{200} = 64^{200}$.

103. False, because $4^{1/2} = \sqrt{4} = 2$.

105. False, because $5^2 \cdot 5^2 = 5^4 = 625$.

107. False, $\sqrt{w^{10}} = |w^5|$.

109. False, $\sqrt{x^6} = |x^3|$.

111. True, $\sqrt{x^8} = x^{8/2} = x^4$.

113. False, $\sqrt{16} = 4$.

115. True, $2^{600} = (2^2)^{300} = 4^{300}$.

117. False, $\dfrac{2 + \sqrt{6}}{2} = 1 + \dfrac{\sqrt{6}}{2}$.

119. False, $\sqrt{\dfrac{4}{6}} = \sqrt{\dfrac{2}{3}} = \dfrac{\sqrt{2}\sqrt{3}}{\sqrt{3}\sqrt{3}} = \dfrac{\sqrt{6}}{3}$.

121. True, $81^{2/4} = 3^2 = 9 = \sqrt{81}$.

123. True, because $(a^4b^2)^{1/2}$ is nonnegative and a^2b could be negative, absolute value symbols are necessary.

125. To find the time for which $s = 12000$, solve the equation
$$16t^2 = 12000$$
$$t^2 = 750$$
$$t = \sqrt{750} = 5\sqrt{30}$$
The time is $5\sqrt{30}$ seconds.

127. The guy wire of length 40 is the hypotenuse of a right triangle where one leg is length 30 and the other is length x. By the Pythagorean theorem we can write
$$x^2 + 30^2 = 40^2$$
$$x^2 = 700$$
$$x = \sqrt{700} = 10\sqrt{7}$$
The wire is attached to the ground $10\sqrt{7}$ feet from the base of the antenna.

129. Let x = the length of the guy wire. The height of the antenna is 200 feet and the distance from the base of the antenna to the point on the ground where the guy wire is attached is 200 feet. The guy wire is the hypotenuse of a right triangle whose legs each have length 200. By the Pythagorean theorem we can write
$$x^2 = 200^2 + 200^2$$
$$x^2 = 80000$$
$$x = \sqrt{80000} = \sqrt{40000}\sqrt{2} = 200\sqrt{2}$$
The length of the guy wires should be $200\sqrt{2}$ feet.

131. If the volume is 40 ft^3, then each side is $\sqrt[3]{40}$ ft in length. The surface area of the six square sides is $6(\sqrt[3]{40})^2$ ft^2. Multiply by 1.1 to get $1.1 \cdot 6(\sqrt[3]{40})^2$ ft^2 as the amount of cardboard needed to make the box. Simplify:
$$6.6\left(2\sqrt[3]{5}\right)^2 = 26.4\sqrt[3]{25} \text{ ft}^2$$

133. $666.2 = 74.4(1+r)^{20}$
$$(1+r)^{20} = \frac{666.2}{74.4}$$
$$1 + r = \left(\frac{666.2}{74.4}\right)^{1/20}$$
$$r = \left(\frac{666.2}{74.4}\right)^{1/20} - 1$$
$$r \approx 0.1158 = 11.58\%$$

135. $V = \sqrt{\dfrac{841L}{CS}} = \dfrac{\sqrt{841} \cdot \sqrt{L} \cdot \sqrt{CS}}{\sqrt{CS} \cdot \sqrt{CS}}$
$$V = \frac{29\sqrt{LCS}}{CS}$$

CHAPTER 5   TEST

1. $8^{2/3} = 2^2 = 4$

2. $4^{-3/2} = \dfrac{1}{2^3} = \dfrac{1}{8}$

3. $\dfrac{\sqrt{21}}{\sqrt{7}} = \sqrt{\dfrac{21}{7}} = \sqrt{3}$

4. $2\sqrt{5} \cdot 3\sqrt{5} = 6 \cdot 5 = 30$

5. $\sqrt{20} + \sqrt{5} = 2\sqrt{5} + \sqrt{5} = 3\sqrt{5}$

6. $\sqrt{5} + \dfrac{1\sqrt{5}}{\sqrt{5}\sqrt{5}} = \sqrt{5} + \dfrac{\sqrt{5}}{5} = \dfrac{5\sqrt{5}}{5} + \dfrac{\sqrt{5}}{5}$
$$= \frac{6\sqrt{5}}{5}$$

7. $2^{1/2} \cdot 2^{1/2} = 2^1 = 2$

8. $\sqrt{72} = \sqrt{36}\sqrt{2} = 6\sqrt{2}$

9. $\sqrt{\dfrac{5}{12}} = \dfrac{\sqrt{5}\sqrt{3}}{\sqrt{12}\sqrt{3}} = \dfrac{\sqrt{15}}{\sqrt{36}} = \dfrac{\sqrt{15}}{6}$

10. $\dfrac{6 + 3\sqrt{2}}{6} = \dfrac{3(2 + \sqrt{2})}{3 \cdot 2} = \dfrac{2 + \sqrt{2}}{2}$

11. $(2\sqrt{3} + 1)(\sqrt{3} - 2)$
$$= 6 + \sqrt{3} - 4\sqrt{3} - 2 = 4 - 3\sqrt{3}$$

12. $\sqrt[4]{32a^5y^8} = \sqrt[4]{16a^4y^8} \cdot \sqrt[4]{2a} = 2ay^2 \cdot \sqrt[4]{2a}$

13. $\dfrac{1}{\sqrt[3]{2x^2}} = \dfrac{1 \cdot \sqrt[3]{4x}}{\sqrt[3]{2x^2} \cdot \sqrt[3]{4x}} = \dfrac{\sqrt[3]{4x}}{\sqrt[3]{8x^3}} = \dfrac{\sqrt[3]{4x}}{2x}$

14. $\sqrt{\dfrac{8a^9}{b^3}} = \dfrac{2a^4\sqrt{2a}}{b\sqrt{b}} = \dfrac{2a^4\sqrt{2a}\sqrt{b}}{b\sqrt{b}\sqrt{b}} = \dfrac{2a^4\sqrt{2ab}}{b^2}$

15. $\sqrt[3]{-27x^9} = -3x^{9/3} = -3x^3$

16. $\sqrt{20m^3} = \sqrt{4m^2}\sqrt{5m} = 2m\sqrt{5m}$

17. $x^{1/2}x^{1/4} = x^{1/2 + 1/4} = x^{3/4}$

18. $(9y^4x^{1/2})^{1/2} = 3y^2x^{1/4}$

19. $\sqrt[3]{40x^7} = \sqrt[3]{8x^6} \cdot \sqrt[3]{5x} = 2x^2 \cdot \sqrt[3]{5x}$

20. $(4 + \sqrt{3})^2 = 16 + 8\sqrt{3} + 3 = 19 + 8\sqrt{3}$

21. $\dfrac{2}{5 - \sqrt{3}} = \dfrac{2(5 + \sqrt{3})}{(5 - \sqrt{3})(5 + \sqrt{3})}$

$= \dfrac{2(5 + \sqrt{3})}{25 - 3} = \dfrac{2(5 + \sqrt{3})}{22} = \dfrac{5 + \sqrt{3}}{11}$

22. $\dfrac{\sqrt{6}(4\sqrt{3} - \sqrt{2})}{(4\sqrt{3} + \sqrt{2})(4\sqrt{3} - \sqrt{2})} = \dfrac{4\sqrt{18} - \sqrt{12}}{48 - 2}$

$= \dfrac{12\sqrt{2} - 2\sqrt{3}}{46} = \dfrac{6\sqrt{2} - \sqrt{3}}{23}$

23. $(3 - 2i)(4 + 5i) = 12 + 7i - 10i^2$

$= 22 + 7i$

24. $i^4 - i^5 = 1 - i^4 \cdot i = 1 - i$

25. $\dfrac{(3 - i)(1 - 2i)}{(1 + 2i)(1 - 2i)} = \dfrac{3 - 7i + 2i^2}{1 - 4i^2} = \dfrac{1 - 7i}{5}$

$= \dfrac{1}{5} - \dfrac{7}{5}i$

26. $\dfrac{-6 + \sqrt{-12}}{8} = \dfrac{-6 + 2i\sqrt{3}}{8} = \dfrac{-3 + i\sqrt{3}}{4}$

$= -\dfrac{3}{4} + \dfrac{1}{4}i\sqrt{3}$

27. $\qquad (x - 2)^2 = 49$

$\qquad\qquad x - 2 = \pm 7$

$\qquad\qquad\qquad x = 2 \pm 7$

$\qquad x = 2 + 7 \qquad \text{or} \qquad x = 2 - 7$

$\qquad x = 9 \qquad\qquad \text{or} \qquad x = -5$

Solution set: $\{-5, 9\}$

28. $\qquad 2\sqrt{x + 4} = 3$

$\qquad \left(2\sqrt{x + 4}\right)^2 = (3)^2$

$\qquad\qquad 4(x + 4) = 9$

$\qquad\qquad\qquad 4x = -7$

$\qquad\qquad\qquad x = -\dfrac{7}{2}$

Solution set: $\left\{-\dfrac{7}{4}\right\}$

29. $\qquad\qquad w^{2/3} = 4$

$\qquad\qquad (w^{2/3})^3 = 4^3$

$\qquad\qquad\qquad w^2 = 64$

$\qquad\qquad\qquad w = \pm 8$

Solution set: $\{-8, 8\}$

30. $\qquad 9y^2 + 16 = 0$

$\qquad\qquad 9y^2 = -16$

$\qquad\qquad y^2 = -\dfrac{16}{9}$

$\qquad\qquad y = \pm\sqrt{-\dfrac{16}{9}} = \pm\dfrac{4}{3}$

The solution set is $\left\{\pm\dfrac{4}{3}i\right\}$.

31. $\qquad \sqrt{2x^2 + x - 12} = x$

$\qquad \left(\sqrt{2x^2 + x - 12}\right)^2 = x^2$

$\qquad\qquad 2x^2 + x - 12 = x^2$

$\qquad\qquad x^2 + x - 12 = 0$

$\qquad\qquad (x + 4)(x - 3) = 0$

$\qquad x + 4 = 0 \qquad \text{or} \qquad x - 3 = 0$

$\qquad\qquad x = -4 \quad \text{or} \qquad\qquad x = 3$

Since -4 does not check in the original equation, the solution set is $\{3\}$.

32. $\qquad \sqrt{x - 1} + \sqrt{x + 4} = 5$

$\qquad\qquad \sqrt{x - 1} = 5 - \sqrt{x + 4}$

$\qquad\qquad \left(\sqrt{x - 1}\right)^2 = \left(5 - \sqrt{x + 4}\right)^2$

$\qquad\qquad x - 1 = 25 - 10\sqrt{x + 4} + x + 4$

$\qquad 10\sqrt{x + 4} = 30$

$\qquad \left(\sqrt{x + 4}\right)^2 = (3)^2$

$\qquad\qquad x + 4 = 9$

$\qquad\qquad x = 5$

Solution set: $\{5\}$

33. Let $x =$ the length of the side. Since the diagonal is the hypotenuse of a right triangle, we can write the equation

$$x^2 + x^2 = 3^2$$
$$2x^2 = 9$$
$$x^2 = \dfrac{9}{2}$$
$$x = \sqrt{\dfrac{9}{2}} = \dfrac{3\sqrt{2}}{\sqrt{2}\sqrt{2}} = \dfrac{3\sqrt{2}}{2}$$

The length of each side of the square is $\dfrac{3\sqrt{2}}{2}$ feet.

34. Let x = one number and $x + 11$ = the other. Since their square roots differ by 1, we can write

$$\sqrt{x + 11} - \sqrt{x} = 1$$
$$\sqrt{x + 11} = \sqrt{x} + 1$$
$$\left(\sqrt{x + 11}\right)^2 = \left(\sqrt{x} + 1\right)^2$$
$$x + 11 = x + 2\sqrt{x} + 1$$
$$10 = 2\sqrt{x}$$
$$10^2 = (2\sqrt{x})^2$$
$$4x = 100$$
$$x = 25$$
$$x + 11 = 36$$

The numbers are 25 and 36.

35. If the perimeter is 20, the sum of the length and width is 10. Let x = the length and $10 - x$ = the width. Use the Pythagorean theorem to write the equation

$$x^2 + (10 - x)^2 = \left(2\sqrt{13}\right)^2$$
$$x^2 + 100 - 20x + x^2 = 52$$
$$2x^2 - 20x + 48 = 0$$
$$x^2 - 10x + 24 = 0$$
$$(x - 4)(x - 6) = 0$$
$$x = 4 \quad \text{or} \quad x = 6$$
$$10 - x = 6 \quad \text{or} \quad 10 - x = 4$$

The length and width are 4 feet and 6 feet.

36. $R = (248.530)^{2/3} \approx 39.53$ AU

$$30.08 = T^{2/3}$$
$$T = 30.08^{3/2} = 164.97 \text{ years}$$

Tying It All Together Chapters 1-5

1.
$$3x - 6 + 5 = 7 - 4x - 12$$
$$3x - 1 = -4x - 5$$
$$7x = -4$$
$$x = -\frac{4}{7}$$
Solution set: $\left\{-\frac{4}{7}\right\}$

2.
$$\left(\sqrt{6x + 7}\right)^2 = (4)^2$$
$$6x + 7 = 16$$
$$6x = 9$$
$$x = \frac{3}{2}$$
Solution set: $\left\{\frac{3}{2}\right\}$

3.
$$|2x + 5| > 1$$
$$2x + 5 > 1 \quad \text{or} \quad 2x + 5 < -1$$
$$2x > -4 \quad \text{or} \quad 2x < -6$$
$$x > -2 \quad \text{or} \quad x < -3$$
$$(-\infty, -3) \cup (-2, \infty)$$

4.
$$8x^3 - 27 = 0$$
$$x^3 = \frac{27}{8}$$
$$x = \sqrt[3]{\frac{27}{8}} = \frac{3}{2}$$
Solution set: $\left\{\frac{3}{2}\right\}$

5.
$$2x - 3 > 3x - 4$$
$$1 > x$$
$$(-\infty, 1)$$

6.
$$\sqrt{2x - 3} - \sqrt{3x + 4} = 0$$
$$\left(\sqrt{2x - 3}\right)^2 = \left(\sqrt{3x + 4}\right)^2$$
$$2x - 3 = 3x + 4$$
$$-7 = x$$
Checking -7 gives us a square root of a negative number. So the solution set is $\emptyset$.

7.
$$6\left(\frac{w}{3} + \frac{w - 4}{2}\right) = 6\left(\frac{11}{2}\right)$$
$$2w + 3w - 12 = 33$$
$$5w = 45$$
$$w = 9$$
Solution set: $\{9\}$

8.
$$2x + 14 - 4 = x - 10 + x$$
$$2x + 10 = 2x - 10$$
$$10 = -10$$
Solution set: $\emptyset$

9.
$$(x + 7)^2 = 25$$
$$x + 7 = \pm 5$$
$$x = -7 \pm 5$$
Solution set: $\{-12, -2\}$

10.
$$a^{-1/2} = 4$$
$$\left(a^{-1/2}\right)^{-2} = (4)^{-2}$$
$$a = \frac{1}{16}$$
Solution set: $\left\{\frac{1}{16}\right\}$

11. $x - 3 > 2$ or $x < 2x + 6$

$x > 5$ or $-x < 6$

$x > 5$ or $x > -6$

$(-6, \infty)$

12. $a^{-2/3} = 16$

$\left(a^{-2/3}\right)^{-3} = (16)^{-3}$

$a^2 = 2^{-12}$

$a = \pm\sqrt{2^{-12}} = \pm 2^{-6} = \pm\dfrac{1}{64}$

Solution set: $\left\{-\dfrac{1}{64}, \dfrac{1}{64}\right\}$

13. $3x^2 - 1 = 0$

$x^2 = \dfrac{1}{3}$

$x = \pm\sqrt{\dfrac{1}{3}} = \pm\dfrac{\sqrt{3}}{3}$

Solution set: $\left\{-\dfrac{\sqrt{3}}{3}, \dfrac{\sqrt{3}}{3}\right\}$

14. $5 - 2x + 4 = 3x - 5x + 10 - 1$

$-2x + 9 = -2x + 9$

Solution set is all real numbers R.

15. $|3x - 4| < 5$

$-5 < 3x - 4 < 5$

$-1 < 3x < 9$

$-\dfrac{1}{3} < x < 3$

$\left(-\dfrac{1}{3},\ 3\right)$

16. $3x - 1 = 0$

$x = \dfrac{1}{3}$

Solution set: $\left\{\dfrac{1}{3}\right\}$

17. $\left(\sqrt{y - 1}\right)^2 = 9^2$

$y - 1 = 81$

$y = 82$

Solution set: $\{82\}$

18. $|5x - 10 + 1| = 3$

$|5x - 9| = 3$

$5x - 9 = 3$ or $5x - 9 = -3$

$5x = 12$ $\qquad$ $5x = 6$

$x = \dfrac{12}{5}$ or $x = \dfrac{6}{5}$

Solution set: $\left\{\dfrac{6}{5}, \dfrac{12}{5}\right\}$

19. $0.06x - 0.04x + 0.8 = 2.8$

$0.02x = 2.0$

$x = 100$

Solution set: $\{100\}$

20. $|3x - 1| > -2$

Since absolute value of any quantity is greater than or equal to zero, any real number satisfies this inequality. Solution set: R

21. $\dfrac{3\sqrt{2}}{x} = \dfrac{\sqrt{3}}{4\sqrt{5}}$

$\sqrt{3}x = 12\sqrt{10}$

$x = \dfrac{12\sqrt{10}}{\sqrt{3}} = \dfrac{12\sqrt{10}\sqrt{3}}{\sqrt{3}\sqrt{3}} = 4\sqrt{30}$

Solution set: $\{4\sqrt{30}\}$

22. $\dfrac{\sqrt{x} - 4}{x} = \dfrac{1}{\sqrt{x} + 5}$

$x = (\sqrt{x} - 4)(\sqrt{x} + 5)$

$x = x + \sqrt{x} - 20$

$20 = \sqrt{x}$

$400 = x$

Solution set: $\{400\}$

23. $\dfrac{3\sqrt{2} + 4}{\sqrt{2}} = \dfrac{x\sqrt{18}}{3\sqrt{2} + 2}$

$6x = (3\sqrt{2} + 4)(3\sqrt{2} + 2)$

$6x = 18 + 18\sqrt{2} + 8$

$x = \dfrac{26 + 18\sqrt{2}}{6} = \dfrac{13 + 9\sqrt{2}}{3}$

Solution set: $\left\{\dfrac{13 + 9\sqrt{2}}{3}\right\}$

24.
$$\frac{x}{2\sqrt{5}-\sqrt{2}}=\frac{2\sqrt{5}+\sqrt{2}}{x}$$
$$x^2=20-2$$
$$x=\pm\sqrt{18}=\pm3\sqrt{2}$$
Solution set: $\{-3\sqrt{2},\ 3\sqrt{2}\ \}$

25.
$$\frac{\sqrt{2x}-5}{x}=\frac{-3}{\sqrt{2x}+5}$$
$$-3x=2x-25$$
$$-5x=-25$$
$$x=5$$
Solution set: $\{5\}$

26.
$$\frac{\sqrt{6}+2}{x}=\frac{2}{\sqrt{6}+4}$$
$$2x=(\sqrt{6}+2)(\sqrt{6}+4)$$
$$2x=14+6\sqrt{6}$$
$$x=7+3\sqrt{6}$$
Solution set: $\{7+3\sqrt{6}\ \}$

27.
$$\frac{x-1}{\sqrt{6}}=\frac{\sqrt{6}}{x}$$
$$x^2-x=6$$
$$x^2-x-6=0$$
$$(x-3)(x+2)=0$$
$$x=3\quad\text{or}\quad x=-2$$
Solution set: $\{-2,\ 3\}$

28.
$$\frac{x+3}{\sqrt{10}}=\frac{\sqrt{10}}{x}$$
$$x^2+3x=10$$
$$x^2+3x-10=0$$
$$(x+5)(x-2)=0$$
$$x=-5\quad\text{or}\quad x=2$$
Solution set: $\{-5,\ 2\}$

29. $6x(x-1)(\frac{1}{x}-\frac{1}{x-1})=6x(x-1)(-\frac{1}{6})$
$$6x-6-6x=-x^2+x$$
$$x^2-x+6=0$$
$$(x-3)(x+2)=0$$
$$x-3=0\quad\text{or}\quad x+2=0$$
$$x=3\quad\text{or}\quad x=-2$$
Solution set: $\{-2,\ 3\}$

30. $3x(x-2)(\frac{1}{x^2-2x}+\frac{1}{x})=3x(x-2)\frac{2}{3}$
$$3+3x-6=2x^2-4x$$
$$-2x^2+7x-3=0$$
$$2x^2-7x+3=0$$
$$(2x-1)(x-3)=0$$
$$2x-1=0\quad\text{or}\quad x-3=0$$
$$x=\frac{1}{2}\quad\text{or}\quad x=3$$
Solution set: $\left\{\frac{1}{2},\ 3\right\}$

31. $\dfrac{-2+\sqrt{2^2-4(1)(-15)}}{2(1)}=\dfrac{-2+\sqrt{64}}{2}=3$

32. $\dfrac{-8+\sqrt{8^2-4(1)(12)}}{2(1)}=\dfrac{-8+\sqrt{16}}{2}=-2$

33. $\dfrac{-5+\sqrt{5^2-4(2)(-3)}}{2(2)}=\dfrac{-5+\sqrt{49}}{4}=\dfrac{1}{2}$

34. $\dfrac{-7+\sqrt{7^2-4(6)(-3)}}{2(6)}=\dfrac{-7+\sqrt{121}}{12}=\dfrac{1}{3}$

35. $v=-94.8+21.4x-0.761x^2$
$$v=-94.8+21.4(11)-0.761(11)^2=48.5$$

36. Moisture content of 14% will produce the maximum volume of popped corn.
Maximum volume is about 55.6 cm^3.

6.1 WARM-UPS

1. False. **2.** False, it is equivalent to $x - 3 = \pm 2\sqrt{3}$. **3.** False, because some quadratic polynomials cannot be factored. **4.** False, because one-half of 4/3 is 2/3 and 2/3 squared is 4/9. **5.** True. **6.** False, we must first divide each side of the equation by 2. **7.** False, $x = 3/2$ or $x = -5/3$. **8.** True, one-half of 3 is 3/2 and 3/2 squared is 9/4. **9.** False, $x = \pm 2i\sqrt{2}$. **10.** False, $(x - 3)^2 = 0$ is a quadratic equation with only one solution.

6.1 EXERCISES

1.
$$x^2 - x - 6 = 0$$
$$(x - 3)(x + 2) = 0$$
$$x - 3 = 0 \quad \text{or} \quad x + 2 = 0$$
$$x = 3 \quad \text{or} \quad x = -2$$
Solution set: $\{-2, 3\}$

3.
$$a^2 + 2a - 15 = 0$$
$$(a + 5)(a - 3) = 0$$
$$a + 5 = 0 \quad \text{or} \quad a - 3 = 0$$
$$a = -5 \quad \text{or} \quad a = 3$$
Solution set: $\{-5, 3\}$

5.
$$2x^2 - x - 3 = 0$$
$$(2x - 3)(x + 1) = 0$$
$$2x - 3 = 0 \quad \text{or} \quad x + 1 = 0$$
$$x = \frac{3}{2} \quad \text{or} \quad x = -1$$
Solution set: $\left\{-1, \frac{3}{2}\right\}$

7.
$$x^2 - x - 12 = 0$$
$$(x - 4)(x + 3) = 0$$
$$x - 4 = 0 \quad \text{or} \quad x + 3 = 0$$
$$x = 4 \quad \text{or} \quad x = -3$$
Solution set: $\{-3, 4\}$

9.
$$y^2 + 14y + 49 = 0$$
$$(y + 7)^2 = 0$$
$$y + 7 = 0$$
$$y = -7$$
Solution set: $\{-7\}$

11.
$$a^2 - 16 = 0$$
$$(a - 4)(a + 4) = 0$$
$$a - 4 = 0 \quad \text{or} \quad a + 4 = 0$$
$$a = 4 \quad \text{or} \quad a = -4$$
Solution set: $\{-4, 4\}$

13.
$$x^2 = 81$$
$$x = \pm\sqrt{81} = \pm 9$$
Solution set: $\{-9, 9\}$

15.
$$x^2 = \frac{16}{9}$$
$$x = \pm\sqrt{\frac{16}{9}} = \pm\frac{4}{3}$$
Solution set: $\left\{-\frac{4}{3}, \frac{4}{3}\right\}$

17.
$$(x - 3)^2 = 16$$
$$x - 3 = \pm\sqrt{16}$$
$$x = 3 \pm 4$$
$$x = 3 + 4 \quad \text{or} \quad x = 3 - 4$$
Solution set: $\{-1, 7\}$

19.
$$(z + 1)^2 = 5$$
$$z + 1 = \pm\sqrt{5}$$
$$z = -1 \pm\sqrt{5}$$
Solution set: $\{-1 - \sqrt{5}, -1 + \sqrt{5}\}$

21.
$$\left(w - \frac{3}{2}\right)^2 = \frac{7}{4}$$
$$w - \frac{3}{2} = \pm\sqrt{\frac{7}{4}}$$
$$w = \frac{3}{2} \pm \frac{\sqrt{7}}{2} = \frac{3 \pm \sqrt{7}}{2}$$
Solution set: $\left\{\frac{3 - \sqrt{7}}{2}, \frac{3 + \sqrt{7}}{2}\right\}$

23.
$$\left(y - \frac{1}{2}\right)^2 = \frac{9}{2}$$
$$y - \frac{1}{2} = \pm\sqrt{\frac{9}{2}}$$
$$y = \frac{1}{2} \pm \frac{3\sqrt{2}}{2} = \frac{1 \pm 3\sqrt{2}}{2}$$
Solution set: $\left\{\frac{1 + 3\sqrt{2}}{2}, \frac{1 - 3\sqrt{2}}{2}\right\}$

25. One-half of 2 is 1, and 1 squared is 1:
$$x^2 + 2x + 1$$

27. One-half of -3 is $-3/2$, and $-3/2$ squared is 9/4: $x^2 - 3x + \frac{9}{4}$

29. One-half of 1/4 is 1/8, and 1/8 squared is 1/64: $y^2 + \frac{1}{4}y + \frac{1}{64}$

31. One-half of 2/3 is 1/3, and 1/3 squared is 1/9: $x^2 + \frac{2}{3}x + \frac{1}{9}$

33. $x^2 + 8x + 16 = (x + 4)^2$

35. $y^2 - 5y + \frac{25}{4} = \left(y - \frac{5}{2}\right)^2$

37. $z^2 - \frac{4}{7}z + \frac{4}{49} = (z - \frac{2}{7})^2$

39. $t^2 + \frac{3}{5}t + \frac{9}{100} = (t + \frac{3}{10})^2$

41.
$$x^2 - 2x = 15$$
$$x^2 - 2x + 1 = 15 + 1$$
$$(x - 1)^2 = 16$$
$$x - 1 = \pm 4$$
$$x = 1 \pm 4$$
$$x = 5 \quad \text{or} \quad x = -3$$
Solution set: $\{-3, 5\}$

43.
$$x^2 + 8x = 20$$
$$x^2 + 8x + 16 = 20 + 16$$
$$(x + 4)^2 = 36$$
$$x + 4 = \pm 6$$
$$x = -4 \pm 6$$
$$x = 2 \quad \text{or} \quad x = -10$$
Solution set: $\{-10, 2\}$

45.
$$2x^2 - 4x = 70$$
$$x^2 - 2x = 35$$
$$x^2 - 2x + 1 = 35 + 1$$
$$(x - 1)^2 = 36$$
$$x - 1 = \pm 6$$
$$x = 1 \pm 6$$
$$x = 7 \quad \text{or} \quad x = -5$$
Solution set: $\{-5, 7\}$

47.
$$w^2 - w = 20$$
$$w^2 - w + \frac{1}{4} = 20 + \frac{1}{4}$$
$$(w - \frac{1}{2})^2 = \frac{81}{4}$$
$$w - \frac{1}{2} = \pm \frac{9}{2}$$
$$w = \frac{1}{2} \pm \frac{9}{2}$$
$$x = \frac{10}{2} = 5 \quad \text{or} \quad x = \frac{-8}{2} = -4$$
Solution set: $\{-4, 5\}$

49.
$$q^2 + 5q = 14$$
$$q^2 + 5q + \frac{25}{4} = 14 + \frac{25}{4}$$
$$(q + \frac{5}{2})^2 = \frac{81}{4}$$
$$q + \frac{5}{2} = \pm \frac{9}{2}$$
$$q = -\frac{5}{2} \pm \frac{9}{2}$$
$$q = \frac{4}{2} = 2 \quad \text{or} \quad q = \frac{-14}{2} = -7$$
Solution set: $\{-7, 2\}$

51.
$$2h^2 - h = 3$$
$$h^2 - \frac{1}{2}h = \frac{3}{2}$$
$$h^2 - \frac{1}{2}h + \frac{1}{16} = \frac{3}{2} + \frac{1}{16}$$
$$(h - \frac{1}{4})^2 = \frac{25}{16}$$
$$h - \frac{1}{4} = \pm \frac{5}{4}$$
$$h = \frac{1}{4} \pm \frac{5}{4}$$
Solution set: $\left\{-1, \frac{3}{2}\right\}$

53.
$$x^2 + 4x = 6$$
$$x^2 + 4x + 4 = 6 + 4$$
$$(x + 2)^2 = 10$$
$$x + 2 = \pm \sqrt{10}$$
$$x = -2 \pm \sqrt{10}$$
Solution set: $\{-2 - \sqrt{10}, -2 + \sqrt{10}\}$

55.
$$x^2 + 8x = 4$$
$$x^2 + 8x + 16 = 4 + 16$$
$$(x + 4)^2 = 20$$
$$x + 4 = \pm \sqrt{20}$$
$$x = -4 \pm 2\sqrt{5}$$
Solution set: $\{-4 - 2\sqrt{5}, -4 + 2\sqrt{5}\}$

57.
$$2x^2 + 3x = 4$$
$$x^2 + \frac{3}{2}x = 2$$
$$x^2 + \frac{3}{2}x + \frac{9}{16} = 2 + \frac{9}{16}$$
$$(x + \frac{3}{4})^2 = \frac{41}{16}$$
$$x + \frac{3}{4} = \pm \frac{\sqrt{41}}{4}$$
$$x = -\frac{3}{4} \pm \frac{\sqrt{41}}{4}$$
Solution set: $\left\{\frac{-3 - \sqrt{41}}{4}, \frac{-3 + \sqrt{41}}{4}\right\}$

59.
$$\sqrt{2x + 1} = x - 1$$
$$(\sqrt{2x + 1})^2 = (x - 1)^2$$
$$2x + 1 = x^2 - 2x + 1$$
$$0 = x^2 - 4x$$
$$x(x - 4) = 0$$
$$x = 0 \quad \text{or} \quad x = 4$$
Since 0 is an extraneous root, the solution set is $\{4\}$.

96

61.
$$2w = \sqrt{w+1}$$
$$(2w)^2 = (\sqrt{w+1})^2$$
$$4w^2 = w+1$$
$$4w^2 - w = 1$$
$$w^2 - \frac{1}{4}w = \frac{1}{4}$$
$$w^2 - \frac{1}{4}w + \frac{1}{64} = \frac{1}{4} + \frac{1}{64}$$
$$\left(w - \frac{1}{8}\right)^2 = \frac{17}{64}$$
$$w - \frac{1}{8} = \pm\frac{\sqrt{17}}{8}$$
$$w = \frac{1}{8} \pm \frac{\sqrt{17}}{8}$$

The number $\dfrac{1 - \sqrt{17}}{8}$ is a negative number. No negative number can be a solution to the original equation because the left side would be negative and the right side is a principal square root.

Solution set: $\left\{\dfrac{1 + \sqrt{17}}{8}\right\}$

63.
$$\frac{t}{t-2} = \frac{2t-3}{t}$$
$$t^2 = (t-2)(2t-3)$$
$$t^2 = 2t^2 - 7t + 6$$
$$-t^2 + 7t - 6 = 0$$
$$t^2 - 7t + 6 = 0$$
$$(t-6)(t-1) = 0$$
$$t = 6 \quad \text{or} \quad t = 1$$

Solution set: $\{1, 6\}$

65.
$$x^2\left(\frac{2}{x^2} + \frac{4}{x} + 1\right) = x^2(0)$$
$$2 + 4x + x^2 = 0$$
$$x^2 + 4x = -2$$
$$x^2 + 4x + 4 = -2 + 4$$
$$(x+2)^2 = 2$$
$$x + 2 = \pm\sqrt{2}$$
$$x = -2 \pm \sqrt{2}$$

Solution set: $\{-2 - \sqrt{2},\ -2 + \sqrt{2}\}$

67.
$$x^2 + 2x + 5 = 0$$
$$x^2 + 2x + 1 = -5 + 1$$
$$(x+1)^2 = -4$$
$$x + 1 = \pm\sqrt{-4}$$
$$x = -1 \pm 2i$$

The solution set is $\{-1 - 2i,\ -1 + 2i\}$.

69.
$$x^2 + 12 = 0$$
$$x^2 = -12$$
$$x = \pm\sqrt{-12}$$
$$x = \pm 2i\sqrt{3}$$

The solution set is $\{-2i\sqrt{3},\ 2i\sqrt{3}\}$

71.
$$5z^2 - 4z + 1 = 0$$
$$z^2 - \frac{4}{5}z + \frac{4}{25} = -\frac{1}{5} + \frac{4}{25}$$
$$\left(z - \frac{2}{5}\right)^2 = \frac{-1}{25}$$
$$z - \frac{2}{5} = \pm\sqrt{\frac{-1}{25}}$$
$$z = \frac{2}{5} \pm \sqrt{\frac{-1}{25}}$$
$$z = \frac{2}{5} \pm \frac{i}{5} = \frac{2 \pm i}{5}$$

The solution set is $\left\{\dfrac{2 \pm i}{5}\right\}$.

73.
$$4x^2 = -25$$
$$x^2 = -\frac{25}{4}$$
$$x = \pm\sqrt{-\frac{25}{4}} = \pm\frac{5}{2}i$$

Solution set: $\left\{-\dfrac{5}{2}i,\ \dfrac{5}{2}i\right\}$

75.
$$\left(p + \frac{1}{2}\right)^2 = \frac{9}{4}$$
$$p + \frac{1}{2} = \pm\frac{3}{2}$$
$$p = -\frac{1}{2} \pm \frac{3}{2}$$

The solution set is $\left\{-2,\ 1\right\}$

77.
$$5t^2 + 4t = 3$$
$$t^2 + \frac{4}{5}t = \frac{3}{5}$$
$$t^2 + \frac{4}{5}t + \frac{4}{25} = \frac{3}{5} + \frac{4}{25}$$
$$\left(t + \frac{2}{5}\right)^2 = \frac{19}{25}$$
$$t + \frac{2}{5} = \pm\frac{\sqrt{19}}{5}$$
$$t = -\frac{2}{5} \pm \frac{\sqrt{19}}{5}$$

Solution set: $\left\{ \dfrac{-2 - \sqrt{19}}{5},\ \dfrac{-2 + \sqrt{19}}{5} \right\}$

79.
$$m^2 + 2m - 24 = 0$$
$$m^2 + 2m = 24$$
$$m^2 + 2m + 1 = 24 + 1$$
$$(m+1)^2 = 25$$
$$m + 1 = \pm 5$$
$$m = -1 \pm 5$$
$$m = -1 - 5 = -6 \quad \text{or} \quad m = -1 + 5 = 4$$
The solution set is $\{-6,\ 4\}$.

81.
$$\left(a + \frac{2}{3}\right)^2 = -\frac{32}{9}$$
$$a + \frac{2}{3} = \pm i\sqrt{\frac{32}{9}}$$
$$a = -\frac{2}{3} \pm \frac{4i\sqrt{2}}{3} = \frac{-2 \pm 4i\sqrt{2}}{3}$$

Solution set: $\left\{ \dfrac{-2 - 4i\sqrt{2}}{3},\ \dfrac{-2 + 4i\sqrt{2}}{3} \right\}$

83.
$$-x^2 + x = -6$$
$$x^2 - x = 6$$
$$x^2 - x + \frac{1}{4} = 6 + \frac{1}{4}$$
$$\left(x - \frac{1}{2}\right)^2 = \frac{25}{4}$$
$$x - \frac{1}{2} = \pm\frac{5}{2}$$
$$x = \frac{1}{2} \pm \frac{5}{2}$$

Solution set: $\{-2,\ 3\}$

85.
$$x^2 - 6x + 10 = 0$$
$$x^2 - 6x + 9 = -10 + 9$$
$$(x - 3)^2 = -1$$
$$x - 3 = \pm\sqrt{-1} = \pm i$$
$$x = 3 \pm i$$
The solution set is $\{3 - i,\ 3 + i\}$.

87.
$$(2x - 5)^2 = (\sqrt{7x + 7})^2$$
$$4x^2 - 20x + 25 = 7x + 7$$
$$4x^2 - 27x + 18 = 0$$
$$(4x - 3)(x - 6) = 0$$
$$4x - 3 = 0 \quad \text{or} \quad x - 6 = 0$$
$$x = \frac{3}{4} \quad \text{or} \quad x = 6$$

Since 3/4 does not check, the solution set is $\{6\}$.

89.
$$4x(x-1)\left(\frac{1}{x} + \frac{1}{x-1}\right) = 4x(x-1)\left(\frac{1}{4}\right)$$
$$4x - 4 + 4x = x^2 - x$$
$$-x^2 + 9x - 4 = 0$$
$$x^2 - 9x + 4 = 0$$
$$x^2 - 9x + \frac{81}{4} = -4 + \frac{81}{4}$$
$$\left(x - \frac{9}{2}\right)^2 = \frac{65}{4}$$
$$x - \frac{9}{2} = \pm\frac{\sqrt{65}}{2}$$
$$x = \frac{9}{2} \pm \frac{\sqrt{65}}{2}$$

Solution set: $\left\{ \dfrac{9 - \sqrt{65}}{2},\ \dfrac{9 + \sqrt{65}}{2} \right\}$

91.
$$x^2 + 6x = -k$$
$$x^2 + 6x + 9 = 9 - k$$
$$(x + 3)^2 = 9 - k$$
$$x + 3 = \pm\sqrt{9 - k}$$
$$x = -3 \pm \sqrt{9 - k}$$

Solution set: $\{-3 - \sqrt{9-k},\ -3 + \sqrt{9-k}\}$

93.
$$x^2 + 2kx + w = 0$$
$$x^2 + 2kx + k^2 = -w + k^2$$
$$(x + k)^2 = k^2 - w$$
$$x + k = \pm\sqrt{k^2 - w}$$
$$x = -k \pm \sqrt{k^2 - w}$$

95. $\left(x + \frac{m}{2}\right)^2 = \frac{m^2 - 4}{4}$

$$x + \frac{m}{2} = \pm \frac{\sqrt{m^2 - 4}}{2}$$

$$x = -\frac{m}{2} \pm \frac{\sqrt{m^2 - 4}}{2}$$

$$x = \frac{-m \pm \sqrt{m^2 - 4}}{2}$$

97. $(2 + \sqrt{3})^2 - 4(2 + \sqrt{3}) + 1$
$= 4 + 4\sqrt{3} + 3 - 8 - 4\sqrt{3} + 1 = 0$
$(2 - \sqrt{3})^2 - 4(2 - \sqrt{3}) + 1$
$= 4 - 4\sqrt{3} + 3 - 8 + 4\sqrt{3} + 1 = 0$

99. $(i + 1)^2 - 2(i + 1) + 2$
$= -1 + 2i + 1 - 2i - 2 + 2 = 0$
$(1 - i)^2 - 2(1 - i) + 2$
$= 1 - 2i - 1 - 2 + 2i + 2 = 0$

101. $1211.1 \cdot 8700 = 2.81A^2 \cdot 200$

$$A^2 = \frac{1211.1 \cdot 8700}{2.81 \cdot 200}$$

$$A = \sqrt{\frac{1211.1 \cdot 8700}{2.81 \cdot 200}} \approx 136.9 \text{ ft/sec}$$

103. $17{,}568 = 1500x - 3x^2$
$3x^2 - 1500x = -17{,}568$
$x^2 - 500x = -5856$
$x^2 - 500x + 62500 = -5856 + 62500$
$(x - 250)^2 = 56644$
$x - 250 = \pm 238$
$x = 250 \pm 238$
$x = 12 \quad \text{or} \quad x = 488$
Since x is less than 25, the answer is 12.

6.2 WARM-UPS

1. True. **2.** False, before identifying a, b, and c, we must write the equation as $3x^2 - 4x + 7 = 0$. **3.** True, this is just the quadratic formula with $a = d$, $b = e$, and $c = f$.
4. False, the quadratic formula works to solve any quadratic equation. **5.** True, because $(-3)^2 - 4(2)(-4) = 9 + 32 = 41$. **6.** True, if the discriminant is 0, then the quadratic equation has one real solution. **7.** True.
8. True, because we can write the equation in the form $-1x^2 + 2x + 0 = 0$.
9. False, x and $6 - x$ have a sum of 6.
10. False, there can be one real solution, 2 real solutions, or 2 imaginary solutions.

6.2 EXERCISES

1. $a = 1$, $b = 5$, $c = 6$

$$x = \frac{-5 \pm \sqrt{5^2 - 4(1)(6)}}{2(1)} = \frac{-5 \pm \sqrt{1}}{2}$$

$$= \frac{-5 \pm 1}{2} = \frac{-4}{2}, \frac{-6}{2}$$

Solution set: $\{-3, -2\}$

3. $y^2 + y - 6 = 0$

$a = 1$, $b = 1$, $c = -6$

$$y = \frac{-1 \pm \sqrt{1^2 - 4(1)(-6)}}{2(1)} = \frac{-1 \pm \sqrt{25}}{2}$$

$$= \frac{-1 \pm 5}{2} = \frac{4}{2}, \frac{-6}{2}$$

Solution set: $\{-3, 2\}$

5. $a = 6$, $b = -7$, $c = -3$

$$z = \frac{7 \pm \sqrt{(-7)^2 - 4(6)(-3)}}{2(6)} = \frac{7 \pm \sqrt{121}}{12}$$

$$= \frac{7 \pm 11}{12} = \frac{18}{12}, \frac{-4}{12}$$

Solution set: $\left\{-\frac{1}{3}, \frac{3}{2}\right\}$

7. $a = 4$, $b = -4$, $c = 1$

$$x = \frac{4 \pm \sqrt{16 - 4(4)(1)}}{2(4)} = \frac{4 \pm 0}{8} = \frac{1}{2}$$

Solution set: $\left\{\frac{1}{2}\right\}$

9. $a = 9$, $b = -6$, $c = 1$

$$x = \frac{6 \pm \sqrt{36 - 4(9)(1)}}{2(9)} = \frac{6 \pm \sqrt{0}}{18} = \frac{1}{3}$$

Solution set: $\left\{\frac{1}{3}\right\}$

11. $16x^2 + 24x + 9 = 0$

$a = 16$, $b = 24$, $c = 9$

$$x = \frac{-24 \pm \sqrt{576 - 4(16)(9)}}{2(16)} = \frac{-24 \pm \sqrt{0}}{32} = -\frac{3}{4}$$

Solution set: $\left\{-\frac{3}{4}\right\}$

13. $a = 1, b = 8, c = 6$

$$v = \frac{-8 \pm \sqrt{8^2 - 4(1)(6)}}{2(1)} = \frac{-8 \pm \sqrt{40}}{2}$$

$$= \frac{-8 \pm 2\sqrt{10}}{2} = -4 \pm \sqrt{10}$$

Solution set: $\{-4 \pm \sqrt{10}\,\}$

15. $x^2 + 5x - 1 = 0$: $a = 1, b = 5, c = -1$

$$x = \frac{-5 \pm \sqrt{5^2 - 4(1)(-1)}}{2(1)} = \frac{-5 \pm \sqrt{29}}{2}$$

Solution set: $\left\{\dfrac{-5 \pm \sqrt{29}}{2}\right\}$

17. $a = 2, b = -6, c = 1$

$$t = \frac{6 \pm \sqrt{36 - 4(2)(1)}}{2(2)} = \frac{6 \pm \sqrt{28}}{4} = \frac{6 \pm 2\sqrt{7}}{4}$$

$$= \frac{2(3 \pm \sqrt{7})}{2(2)} = \frac{3 \pm \sqrt{7}}{2}$$

Solution set: $\left\{\dfrac{3 \pm \sqrt{7}}{2}\right\}$

19. $2t^2 - 6t + 5 = 0$

$$t = \frac{6 \pm \sqrt{36 - 4(2)(5)}}{2(2)} = \frac{6 \pm \sqrt{-4}}{4} = \frac{6 \pm 2i}{4}$$

$$= \frac{3 \pm i}{2} \qquad \text{Solution set: } \left\{\dfrac{3 \pm i}{2}\right\}$$

21. $-2x^2 + 3x - 6 = 0$: $a = -2, b = 3, c = -6$

$$x = \frac{-3 \pm \sqrt{(3)^2 - 4(-2)(-6)}}{2(-2)} = \frac{-3 \pm \sqrt{-39}}{-4}$$

$$= \frac{-3 \pm i\sqrt{39}}{-4} = \frac{3 \pm i\sqrt{39}}{4}$$

Solution set: $\left\{\dfrac{3 \pm i\sqrt{39}}{4}\right\}$

23. $\frac{1}{2}x^2 + 13 = 5x,\quad x^2 - 10x + 26 = 0$

$$x = \frac{10 \pm \sqrt{100 - 4(1)(26)}}{2(1)} = \frac{10 \pm \sqrt{-4}}{2}$$

$$= \frac{10 \pm 2i}{2} = 5 \pm i \quad \text{Solution set: } \{5 \pm i\}$$

25. $x^2 - 6x + 2 = 0$: $a = 1, b = -6, c = 2$

$b^2 - 4ac = 36 - 4(1)(2) = 28$

So there are two real solutions to the equation.

27. $2x^2 - 5x + 6 = 0$: $a = 2, b = -5, c = 6$

$b^2 - 4ac = 25 - 4(2)(6) = -23$

So there are no real solutions to the equation.

29. $4m^2 - 20m + 25 = 0$: $a = 4, b = -20, c = 25$

$b^2 - 4ac = 400 - 4(4)(25) = 0$

So there is one real solution to the equation.

31. $a = 1, b = -\frac{1}{2}, c = \frac{1}{4}$

$b^2 - 4ac = \frac{1}{4} - 4(1)(\frac{1}{4}) = -\frac{3}{4}$

So there are no real solutions to the equation.

33. $a = -3, \ b = 5, \ c = 6$

$b^2 - 4ac = 25 - 4(-3)(6) = 97$

So there are two real solutions to the equation.

35. $16z^2 - 24z + 9 = 0$: $a = 16, b = -24, c = 9$

$b^2 - 4ac = (-24)^2 - 4(16)(9) = 0$

So there is one real solution to the equation.

37. $5x^2 - 7 = 0$: $a = 5, b = 0, c = -7$

$b^2 - 4ac = 0 - 4(5)(-7) = 140$

So there are two real solutions to the equation.

39. $x^2 - x = 0$: $a = 1, b = -1, c = 0$

$b^2 - 4ac = 1 - 4(1)(0) = 1$

So there are two real solutions to the equation.

41.
$$6\left(\frac{1}{3}x^2 + \frac{1}{2}x\right) = 6\left(\frac{1}{3}\right)$$
$$2x^2 + 3x = 2$$
$$2x^2 + 3x - 2 = 0$$
$$a = 2, b = 3, c = -2$$

$$x = \frac{-3 \pm \sqrt{9 - 4(2)(-2)}}{2(2)} = \frac{-3 \pm 5}{4} = \frac{2}{4}, \frac{-8}{4}$$

Solution set: $\left\{-2, \frac{1}{2}\right\}$

43.
$$\frac{w}{w-2} = \frac{w}{w-3}$$
$$w^2 - 3w = w^2 - 2w$$
$$0 = w$$

Solution set: $\{0\}$

45.
$$\frac{9(3x-5)^2}{4} = 1$$
$$(3x-5)^2 = \frac{4}{9}$$
$$3x - 5 = \pm\frac{2}{3}$$

$$3x = 5 + \frac{2}{3} \quad \text{or} \quad 3x = 5 - \frac{2}{3}$$

$$3x = \frac{17}{3} \quad \text{or} \quad 3x = \frac{13}{3}$$

$$x = \frac{17}{9} \quad \text{or} \quad x = \frac{13}{9}$$

Solution set: $\left\{\dfrac{13}{9}, \dfrac{17}{9}\right\}$

47.
$$1 + \frac{20}{x^2} = \frac{8}{x}$$
$$x^2\left(1 + \frac{20}{x^2}\right) = x^2 \cdot \frac{8}{x}$$
$$x^2 - 8x + 20 = 0$$
$$x^2 - 8x + 16 = -20 + 16$$
$$(x-4)^2 = -4$$
$$x - 4 = \pm 2i$$
$$x = 4 \pm 2i$$
Solution set: $\{4 \pm 2i\}$

49.
$$(x-8)(x+4) = -42$$
$$x^2 - 4x + 10 = 0$$
$$x^2 - 4x + 4 = -10 + 4$$
$$(x-2)^2 = -6$$
$$x - 2 = \pm i\sqrt{6}$$
$$x = 2 \pm i\sqrt{6}$$
Solution set: $\{2 \pm i\sqrt{6}\}$

51.
$$y = \frac{3(2y+5)}{8(y-1)}$$
$$8y^2 - 8y = 6y + 15$$
$$8y^2 - 14y - 15 = 0$$
$$(4y+3)(2y-5) = 0$$
$$4y + 3 = 0 \quad \text{or} \quad 2y - 5 = 0$$
$$y = -\frac{3}{4} \quad \text{or} \quad y = \frac{5}{2}$$
Solution set: $\left\{-\frac{3}{4}, \frac{5}{2}\right\}$

53.
$$x = \frac{-3.2 \pm \sqrt{(3.2)^2 - 4(1)(-5.7)}}{2(1)}$$
$$= \frac{-3.2 \pm \sqrt{33.04}}{2} \qquad \{-4.474,\ 1.274\}$$

55.
$$x = \frac{7.4 \pm \sqrt{(-7.4)^2 - 4(1)(13.69)}}{2(1)}$$
$$= \frac{7.4 \pm \sqrt{0}}{2(1)} = 3.7 \qquad \{3.7\}$$

57.
$$x = \frac{-6.72 \pm \sqrt{(6.72)^2 - 4(1.85)(3.6)}}{2(1.85)}$$
$$= \frac{-6.72 \pm \sqrt{18.5184}}{3.7} \qquad \{-2.979,\ -0.653\}$$

59.
$$x = \frac{-14379 \pm \sqrt{14379^2 - 4(3)(243)}}{2(3)}$$
$$= \frac{-14379 \pm 14378.8986}{6}$$
Solution set: $\{-4792.983,\ -0.017\}$

61.
$$x = \frac{-0.00075 \pm \sqrt{0.00075^2 - 4 \cdot 1(-0.0062)}}{2(1)}$$
$$= \frac{-0.00075 \pm 0.1574819}{2} \qquad \{-0.079,\ 0.078\}$$

63. Let x = one number and $x + 1$ = the other. Since their product is 16, we can write
$$x(x+1) = 16$$
$$x^2 + x = 16$$
$$x^2 + x - 16 = 0$$
$$x = \frac{-1 \pm \sqrt{1^2 - 4(1)(-16)}}{2(1)} = \frac{-1 \pm \sqrt{65}}{2}$$
Since the numbers are positive, $x = \frac{-1 + \sqrt{65}}{2}$

and $x + 1 = \frac{-1 + \sqrt{65}}{2} + \frac{2}{2} = \frac{1 + \sqrt{65}}{2}$.

The numbers are $\frac{1 + \sqrt{65}}{2}$ and $\frac{-1 + \sqrt{65}}{2}$.

65. Let x = one of the numbers. If the numbers are to have a sum of 6, then $6 - x$ = the other number. Since their product is 4, we can write the equation
$$x(6-x) = 4$$
$$-x^2 + 6x - 4 = 0$$
$$x^2 - 6x + 4 = 0$$
$$x = \frac{6 \pm \sqrt{36 - 4(1)(4)}}{2(1)} = \frac{6 \pm 2\sqrt{5}}{2} = 3 \pm \sqrt{5}$$

If $x = 3 + \sqrt{5}$, then
$6 - x = 6 - (3 + \sqrt{5}) = 3 - \sqrt{5}$.
If $x = 3 - \sqrt{5}$, then
$6 - x = 6 - (3 - \sqrt{5}) = 3 + \sqrt{5}$.
So the numbers are $3 + \sqrt{5}$ and $3 - \sqrt{5}$.

67. Let x = the width and $x + 1$ = the length. Since the diagonal is the hypotenuse of a right triangle, we can write
$$x^2 + (x+1)^2 = (\sqrt{3})^2$$
$$x^2 + x^2 + 2x + 1 = 3$$
$$2x^2 + 2x - 2 = 0$$
$$x^2 + x - 1 = 0$$

$$x = \frac{-1 \pm \sqrt{1 - 4(1)(-1)}}{2(1)} = \frac{-1 \pm \sqrt{5}}{2}$$

Since the width of a rectangle is a positive number, we have width $= \frac{-1 + \sqrt{5}}{2}$ feet and length $= \frac{-1 + \sqrt{5}}{2} + \frac{2}{2} = \frac{1 + \sqrt{5}}{2}$ feet.

69. Let $x =$ the width, and $x + 4 =$ the length. Since the area is 10 square feet, we can write

$$x(x + 4) = 10$$
$$x^2 + 4x - 10 = 0$$

$$x = \frac{-4 \pm \sqrt{16 - 4(1)(-10)}}{2(1)} = \frac{-4 \pm 2\sqrt{14}}{2}$$
$$= -2 \pm \sqrt{14}$$

Since the width must be positive, the width is $-2 + \sqrt{14}$ feet, and the length is $-2 + \sqrt{14} + 4 = 2 + \sqrt{14}$ feet.

71. The time it takes to reach the ground is found by solving the equation

$$-16t^2 + 16t + 96 = 0$$
$$t^2 - t - 6 = 0$$
$$(t - 3)(t + 2) = 0$$
$$t = 3 \quad \text{or} \quad t = -2$$

The pine cone reaches the earth 3 seconds after it is tossed.

73. The time it takes to reach the river is found by solving the equation

$$-16t^2 - 30t + 1000 = 0$$
$$8t^2 + 15t - 500 = 0$$

$$t = \frac{-15 \pm \sqrt{15^2 - 4(8)(-500)}}{2(8)}$$

$$= \frac{-15 \pm \sqrt{16225}}{16} = \frac{-15 \pm 127.377}{16}$$

Since the time is positive, the time is $\frac{-15 + 127.377}{16} = 7.02$ seconds.

75. Let $x =$ the number originally purchased. She paid $\frac{200}{x}$ dollars for each melon. She sold them for $\frac{200}{x} + 1.50$ dollars each. When she sold $x - 30$ of them her revenue was \$200:

$$(x - 30)\left(\frac{200}{x} + 1.50\right) = 200$$
$$200 - \frac{6000}{x} + 1.50x - 45 = 200$$
$$1.50x - 45 - \frac{6000}{x} = 0$$
$$1.50x^2 - 45x - 6000 = 0$$
$$x^2 - 30x - 4000 = 0$$

$$x = \frac{30 \pm \sqrt{900 - 4 \cdot 1(-4000)}}{2} = \frac{30 \pm 130}{2}$$
$$= 80 \text{ or } -50$$

She originally purchased 80 watermelons.

77. If x is the width of the border, then
$$(30 - 2x)(40 - 2x) = 704$$
$$1200 - 140x + 4x^2 = 704$$
$$4x^2 - 140x + 496 = 0$$
$$x^2 - 35x + 124 = 0$$
$$x = \frac{35 \pm \sqrt{35^2 - 4 \cdot 1 \cdot 124}}{2} = \frac{35 \pm \sqrt{729}}{2}$$
$$= \frac{35 \pm 27}{2} = 31 \text{ or } 4$$

The width of the border is 4 inches.

6.3 WARM-UPS

1. True, because if $w = x^2$ the equation becomes $w^2 - 5w + 6 = 0$. **2.** False, the equation cannot be solved by substitution. **3.** False, we can use factoring.

4. True, because $(x^{1/6})^2 = x^{2/6} = x^{1/3}$.

5. False, we should let $w = \sqrt{x}$.

6. False, because $(2^{1/2})^2 = 2^{2/2} = 2^1 = 2$.

7. False, his rate is $1/x$ of the fence per hour.

8. True, because $R = D/T$. **9.** False, against a 5 mph current it will go 5 mph. **10.** False, the dimensions of the bottom will be $11 - 2x$ by $14 - 2x$.

6.3 EXERCISES

1. $b^2 - 4ac = (-1)^2 - 4(2)4 = -31$
Polynomial is prime.
3. $b^2 - 4ac = 6^2 - 4(2)(-5) = 76$
Polynomial is prime.
5. $b^2 - 4ac = 19^2 - 4(6)(-36) = 1225$
Since $\sqrt{1225} = 35$, the polynomial is not prime.
$6x^2 + 19x - 36 = (3x - 4)(2x + 9)$
7. $b^2 - 4ac = 25 - 4(4)(-12) = 217$
Polynomial is prime.
9. $b^2 - 4ac = (-18)^2 - 4 \cdot 8(-45) = 1764$
Since $\sqrt{1764} = 42$, the polynomial is not prime.
$8x^2 - 18x - 45 = (4x - 15)(2x + 3)$
11. $(x - 3)(x + 7) = 0$
$\qquad x^2 + 4x - 21 = 0$
13. $(x - 4)(x - 1) = 0$
$\qquad x^2 - 5x + 4 = 0$
15. $(x - \sqrt{5})(x + \sqrt{5}) = 0$
$\qquad\qquad x^2 - 5 = 0$
17. $(x - 4i)(x + 4i) = 0$
$\qquad\qquad x^2 + 16 = 0$
19. $(x - i\sqrt{2})(x + i\sqrt{2}) = 0$
$\qquad\qquad\qquad x^2 + 2 = 0$
21. $(2x - 1)(3x - 1) = 0$
$\qquad\qquad 6x^2 - 5x + 1 = 0$
23. Let $w = 2a - 1$.
$$w^2 + 2w - 8 = 0$$
$$(w + 4)(w - 2) = 0$$
$w + 4 = 0 \quad$ or $\quad w - 2 = 0$
$\qquad w = -4 \quad$ or $\qquad w = 2$
$2a - 1 = -4 \quad$ or $\quad 2a - 1 = 2$
$\quad 2a = -3 \quad$ or $\qquad 2a = 3$
$\quad a = -\frac{3}{2} \quad$ or $\qquad a = \frac{3}{2}$

Solution set: $\left\{ -\frac{3}{2}, \frac{3}{2} \right\}$

25. Let $y = w - 1$.
$$y^2 + 5y + 5 = 0$$

$$y = \frac{-5 \pm \sqrt{25 - 4(1)(5)}}{2(1)} = \frac{-5 \pm \sqrt{5}}{2}$$

$$w - 1 = \frac{-5 \pm \sqrt{5}}{2}$$

$$w = \frac{-5 \pm \sqrt{5}}{2} + 1 = \frac{-5 \pm \sqrt{5}}{2} + \frac{2}{2} = \frac{-3 \pm \sqrt{5}}{2}$$

Solution set: $\left\{ \frac{-3 \pm \sqrt{5}}{2} \right\}$

27. Let $w = x^2$ and $w^2 = x^4$.
$$w^2 - 14w + 45 = 0$$
$$(w - 5)(w - 9) = 0$$
$w - 5 = 0 \quad$ or $\quad w - 9 = 0$
$\quad w = 5 \quad$ or $\qquad w = 9$

$\quad x^2 = 5 \quad$ or $\qquad x^2 = 9$
$\quad x = \pm\sqrt{5} \quad$ or $\quad x = \pm 3$
Solution set: $\{ \pm\sqrt{5}, \pm 3 \}$
29. Let $w = x^3$, and $w^2 = x^6$.
$$w^2 + 7w = 8$$
$$w^2 + 7w - 8 = 0$$
$$(w - 1)(w + 8) = 0$$
$w - 1 = 0 \quad$ or $\qquad w + 8 = 0$
$\quad w = 1 \quad$ or $\qquad w = -8$
$\quad x^3 = 1 \quad$ or $\qquad x^3 = -8$
$\quad x = 1 \quad$ or $\qquad x = -2$
Solution set: $\{-2, 1\}$

31. Let $w = x^2 + 2x$.
$$w^2 - 7w + 12 = 0$$
$$(w - 3)(w - 4) = 0$$
$w - 3 = 0 \qquad$ or $\qquad w - 4 = 0$
$\quad w = 3 \qquad$ or $\qquad w = 4$
$x^2 + 2x = 3 \qquad$ or $\qquad x^2 + 2x = 4$
$x^2 + 2x - 3 = 0 \qquad$ or $\quad x^2 + 2x - 4 = 0$

$(x + 3)(x - 1) = 0 \quad$ or $\quad x = \dfrac{-2 \pm \sqrt{4 - 4(1)(-4)}}{2}$

$x = -3$ or $x = 1 \quad$ or $\quad x = \dfrac{-2 \pm 2\sqrt{5}}{2}$
Solution set: $\{-1 \pm \sqrt{5}, -3, 1\}$
33. Let $w = y^2 + y$.
$$(y^2 + y)^2 - 8(y^2 + y) + 12 = 0$$
$$w^2 - 8w + 12 = 0$$
$$(w - 6)(w - 2) = 0$$
$\quad w = 6 \qquad$ or $\qquad w = 2$
$y^2 + y = 6 \qquad$ or $\qquad y^2 + y = 2$
$y^2 + y - 6 = 0 \quad$ or $\qquad y^2 + y - 2 = 0$
$(y + 3)(y - 2) = 0$ or $(y + 2)(y - 1) = 0$
$y = -3$ or $y = 2 \quad$ or $\quad y = -2$ or $y = 1$
Solution set: $\{-3, -2, 1, 2\}$

35. Let $w = x^{1/4}$ and $w^2 = x^{1/2}$.
$$w^2 - 5w + 6 = 0$$
$$(w - 3)(w - 2) = 0$$
$w - 3 = 0 \quad$ or $\qquad w - 2 = 0$
$\quad w = 3 \quad$ or $\qquad w = 2$
$x^{1/4} = 3 \quad$ or $\qquad x^{1/4} = 2$
$(x^{1/4})^4 = 3^4 \quad$ or $\quad (x^{1/4})^4 = 2^4$
$\quad x = 81 \quad$ or $\qquad x = 16$
Solution set: $\{16, 81\}$

37. Let $w = x^{1/2}$ and $w^2 = x$.
$$2w^2 - 5w - 3 = 0$$
$$(2w + 1)(w - 3) = 0$$
$$2w + 1 = 0 \quad \text{or} \quad w - 3 = 0$$
$$w = -\frac{1}{2} \quad \text{or} \quad w = 3$$
$$x^{1/2} = -\frac{1}{2} \quad \text{or} \quad x^{1/2} = 3$$
$$x = \frac{1}{4} \quad \text{or} \quad x = 9$$

The solution 1/4 does not check in the original equation. So the solution set is $\{9\}$.

39. Let $t = x^{-1}$:
$$x^{-2} + x^{-1} - 6 = 0$$
$$t^2 + t - 6 = 0$$
$$(t + 3)(t - 2) = 0$$
$$t = -3 \quad \text{or} \quad t = 2$$
$$x^{-1} = -3 \quad \text{or} \quad x^{-1} = 2$$
$$x = -\frac{1}{3} \quad \text{or} \quad x = \frac{1}{2}$$
Solution set: $\left\{-\frac{1}{3}, \frac{1}{2}\right\}$

41. Let $w = x^{1/6}$ and $w^2 = x^{1/3}$.
$$w - w^2 + 2 = 0$$
$$w^2 - w - 2 = 0$$
$$(w - 2)(w + 1) = 0$$
$$w = 2 \quad \text{or} \quad w = -1$$
$$x^{1/6} = 2 \quad \text{or} \quad x^{1/6} = -1$$
$$(x^{1/6})^6 = 2^6 \quad \text{or} \quad (x^{1/6})^6 = (-1)^6$$
$$x = 64 \quad \text{or} \quad x = 1$$

The original equation is not satisfied for $x = 1$. So the solution set is $\{64\}$.

43. Let $w = \frac{1}{y - 1}$.
$$w^2 + w = 6$$
$$w^2 + w - 6 = 0$$
$$(w + 3)(w - 2) = 0$$
$$w = -3 \quad \text{or} \quad w = 2$$
$$\frac{1}{y - 1} = -3 \quad \text{or} \quad \frac{1}{y - 1} = 2$$
$$-3y + 3 = 1 \quad \text{or} \quad 2y - 2 = 1$$
$$-3y = -2 \quad \text{or} \quad 2y = 3$$
$$y = \frac{2}{3} \quad \text{or} \quad y = \frac{3}{2}$$
Solution set: $\left\{\frac{2}{3}, \frac{3}{2}\right\}$

45. Let $w = \sqrt{2x^2 - 3}$ and $w^2 = 2x^2 - 3$.
$$w^2 - 6w + 8 = 0$$
$$(w - 4)(w - 2) = 0$$
$$w = 4 \quad \text{or} \quad w = 2$$
$$\sqrt{2x^2 - 3} = 4 \quad \text{or} \quad \sqrt{2x^2 - 3} = 2$$
Square each side:
$$2x^2 - 3 = 16 \quad \text{or} \quad 2x^2 - 3 = 4$$
$$2x^2 = 19 \quad \text{or} \quad 2x^2 = 7$$
$$x^2 = \frac{19}{2} \quad \text{or} \quad x^2 = \frac{7}{2}$$
$$x = \pm\sqrt{\frac{19}{2}} \quad \text{or} \quad x = \pm\sqrt{\frac{7}{2}}$$
$$x = \pm\frac{\sqrt{38}}{2} \quad \text{or} \quad x = \pm\frac{\sqrt{14}}{2}$$
Solution set: $\left\{\pm\frac{\sqrt{14}}{2}, \pm\frac{\sqrt{38}}{2}\right\}$

47. Let $t = x^{-1}$ in $x^{-2} - 2x^{-1} - 1 = 0$
$$t^2 - 2t - 1 = 0$$
$$t = \frac{2 \pm \sqrt{4 - 4(1)(-1)}}{2} = \frac{2 \pm 2\sqrt{2}}{2} = 1 \pm \sqrt{2}$$
$$x = \frac{1}{1 + \sqrt{2}} \quad \text{or} \quad x = \frac{1}{1 - \sqrt{2}}$$
$$x = \frac{1(1 - \sqrt{2})}{(1 + \sqrt{2})(1 - \sqrt{2})} \quad \text{or} \quad x = \frac{1(1 + \sqrt{2})}{(1 - \sqrt{2})(1 + \sqrt{2})}$$
$$x = \frac{1 - \sqrt{2}}{-1} \quad \text{or} \quad x = \frac{1 + \sqrt{2}}{-1}$$
$$x = -1 + \sqrt{2} \quad \text{or} \quad x = -1 - \sqrt{2}$$
Solution set: $\{-1 + \sqrt{2}, -1 - \sqrt{2}\}$

49. $x^2 + ix + 2 = 0$
$$x = \frac{-i \pm \sqrt{i^2 - 4(1)(2)}}{2} = \frac{-i \pm \sqrt{-9}}{2}$$
$$= \frac{-i \pm 3i}{2} = i \text{ or } -2i$$
Solution set: $\{-2i, i\}$

51. $ix^2 + 2x - 3i = 0$
$$x = \frac{-2 \pm \sqrt{2^2 - 4(i)(-3i)}}{2i} = \frac{-2 \pm \sqrt{-8}}{2i}$$
$$= \frac{-2 \pm 2i\sqrt{2}}{2i} = \frac{-1}{i} \pm \sqrt{2} = i \pm \sqrt{2}$$
Solution set: $\{i \pm \sqrt{2}\}$

53. $x^2 + \sqrt{13}x - 3 = 0$

$$x = \frac{-\sqrt{13} \pm \sqrt{(\sqrt{13})^2 - 4(1)(-3)}}{2} = \frac{-\sqrt{13} \pm \sqrt{25}}{2}$$

$$= \frac{-\sqrt{13} \pm 5}{2}$$

Solution set: $\left\{ \dfrac{-\sqrt{13} \pm 5}{2} \right\}$

55. Let $x =$ Gary's travel time and $x + 1 =$ Harry's travel time. Since $R = D/T$, Gary's speed is $300/x$ and Harry's speed is $300/(x+1)$. Since Gary travels 10 mph faster, we can write the equation

$$\frac{300}{x} = \frac{300}{x+1} + 10$$

$$x(x+1)\left(\frac{300}{x}\right) = x(x+1)\left(\frac{300}{x+1} + 10\right)$$

$$300x + 300 = 300x + 10x^2 + 10x$$
$$0 = 10x^2 + 10x - 300$$

$$0 = x^2 + x - 30$$
$$0 = (x+6)(x-5)$$
$$x + 6 = 0 \quad \text{or} \quad x - 5 = 0$$
$$x = -6 \quad \text{or} \quad x = 5$$

Gary travels 5 hours and they arrive at 2 P.M.

57. Let $x =$ her speed before lunch and $x - 4 =$ her speed after lunch. Since $T = D/R$, her time before lunch was $60/x$ and her time after lunch was $46/(x-4)$. Since she put in one hour more after lunch, we can write the equation

$$\frac{46}{x-4} - 1 = \frac{60}{x}$$

$$x(x-4)\left(\frac{46}{x-4} - 1\right) = x(x-4)\left(\frac{60}{x}\right)$$

$$46x - x^2 + 4x = 60x - 240$$
$$-x^2 - 10x + 240 = 0$$
$$x^2 + 10x - 240 = 0$$

$$x = \frac{-10 \pm \sqrt{100 - 4(1)(-240)}}{2(1)} = \frac{-10 \pm \sqrt{1060}}{2}$$

$$= \frac{-10 \pm 2\sqrt{265}}{2} = -5 \pm \sqrt{265}$$

Since $-5 - \sqrt{265}$ is a negative number, we disregard that answer. Her speed before lunch was $-5 + \sqrt{265} \approx 11.3$ mph, and her speed after lunch was 4 mph slower:
$-9 + \sqrt{265} \approx 7.3$ mph.

59. Let $x =$ Andrew's time and $x + 3 =$ John's time. Andrew's rate is $1/x$ job/hr and John's rate is $1/(x+3)$ job/hr. In 8 hours Andrew does $8/x$ job and John does $8/(x+3)$ job.

$$\frac{8}{x} + \frac{8}{x+3} = 1$$

$$x(x+3)\left(\frac{8}{x} + \frac{8}{x+3}\right) = x(x+3)1$$

$$8x + 24 + 8x = x^2 + 3x$$
$$0 = x^2 - 13x - 24$$

$$x = \frac{13 \pm \sqrt{169 - 4(1)(-24)}}{2(1)} = \frac{13 \pm \sqrt{265}}{2}$$

Since $\dfrac{13 - \sqrt{265}}{2}$ is a negative number, we disregard that solution. Andrew's time is $\dfrac{13 + \sqrt{265}}{2} \approx 14.6$ hours and John's time is 3 hours more :
$$3 + \frac{13 + \sqrt{265}}{2} = \frac{19 + \sqrt{265}}{2} \approx 17.6 \text{ hours}$$

61. Let $x =$ the amount of increase. The new length and width will be $30 + x$ and $20 + x$. Since the new area is to be 1000, we can write the equation

$$(20 + x)(30 + x) = 1000$$
$$x^2 + 50x - 400 = 0$$

$$x = \frac{-50 \pm \sqrt{2500 - 4(1)(-400)}}{2(1)}$$

$$= \frac{-50 \pm \sqrt{4100}}{2} = \frac{-50 \pm 10\sqrt{41}}{2}$$
$$= -25 \pm 5\sqrt{41}$$

Disregard the negative solution.
Use $x = -25 + 5\sqrt{41}$ to get
$30 + x = 5 + 5\sqrt{41} \approx 37.02$ feet, and
$20 + x = -5 + 5\sqrt{41} \approx 27.02$ feet.

63. Let $x =$ the number of hours for A to empty the pool and $x + 2 =$ the number of hours for B to empty the pool. A's rate is $\frac{1}{x}$ pool/hr and B's rate is $\frac{1}{x+2}$ pool/hr. A works for 9 hrs and does $\frac{9}{x}$ of the pool while B works for 6 hours and does $\frac{6}{x+2}$ of the pool. Since the pool is half full

$$\frac{9}{x}+\frac{6}{x+2}=\frac{1}{2}$$

$$2x(x+2)\left(\frac{9}{x}+\frac{6}{x+2}\right)=2x(x+2)\frac{1}{2}$$

$$18x+36+12x=x^2+2x$$

$$-x^2+28x+36=0$$

$$x^2-28x-36=0$$

$$x=\frac{28\pm\sqrt{28^2-4(1)(-36)}}{2}=\frac{28\pm\sqrt{928}}{2}$$

$$=\frac{28\pm4\sqrt{58}}{2}=14\pm2\sqrt{58}=-1.2 \text{ or } 29.2$$

A would take 29.2 hours working alone.

65. $$\frac{10}{w}=\frac{w}{10-w}$$

$$100-10w=w^2$$
$$w^2+10w-100=0$$

$$w=\frac{-10\pm\sqrt{10^2-4(1)(-100)}}{2}=\frac{-10\pm10\sqrt{5}}{2}$$

$$w=-5\pm5\sqrt{5}\approx-16.2 \text{ or } 6.2$$

So the width is 6.2 meters.

6.4 WARM-UPS

1. False, solution set is $(-\infty,-2)\cup(2,\infty)$.
2. False, we do not multiply each side by a variable.   **3.** False, that is not how we solve a quadratic inequality.   **4.** False, we can solve any quadratic inequality.   **5.** True, that is why we make the sign graph.   **6.** True, because inequalities change direction when multiplied by a negative number and we do not know if the variable is positive or negative.
7. True, because the solution is based on rules for multiplying or dividing signed numbers.
8. True, multiply each side by 2.
9. True, subtract 1 from each side.
10. False, because 4 causes the denominator to be 0 and so it cannot be in the solution set.

6.4 EXERCISES

1. $(x+3)(x-2)<0$

$$
\begin{array}{ll}
x+3 & \texttt{-----0++++++++++++}\\
x-2 & \texttt{-------------0++++}\\
\end{array}
$$
$$\langle \underset{-3}{}\quad\underset{2}{}\rangle$$

The product is negative only if the factors have opposite signs. That happens when x is chosen between -3 and 2.
$(-3, 2)$

The product is negative only if the factors have opposite signs.

3. $y^2-4>0$
$(y-2)(y+2)>0$

$$
\begin{array}{ll}
y+2 & \texttt{-----0+++++++++++}\\
y-2 & \texttt{-------------0++++}\\
\end{array}
$$
$$\langle \underset{-2}{}\quad\underset{2}{}\rangle$$

The product is positive if both factors are positive or both are negative. $(-\infty,-2)\cup(2,\infty)$

5. $2u^2+5u-12\geq0$
$(2u-3)(u+4)\geq0$

$$
\begin{array}{ll}
2u-3 & \texttt{-----0+++++++++++}\\
u+4 & \texttt{-------------0++++}\\
\end{array}
$$
$$\langle \underset{-4}{}\quad\underset{3/2}{}\rangle$$

The product is positive only if the factors have the same sign. That happens if u is chosen to the left of -4 or to the right of 3/2. The product is 0 if u is either -4 or 3/2.
$(-\infty,-4]\cup[\frac{3}{2},\infty)$

7. $4x^2-8x\geq0$
$4x(x-2)\geq0$

$$
\begin{array}{ll}
4x & \texttt{-----0+++++++++++}\\
x-2 & \texttt{-------------0++++}\\
\end{array}
$$
$$\langle \underset{0}{}\quad\underset{2}{}\rangle$$

The product is positive only if the factors have the same sign.
$(-\infty,0]\cup[2,\infty)$

9. $5x(1-2x) < 0$

```
5x    -----0+++++++++++
1-2x  +++++++++++0-----
<------------------------->
      0          1/2
```

The product is negative only if the factors have opposite signs.
$(-\infty, 0) \cup (\frac{1}{2}, \infty)$

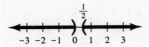

11. $x^2 + 6x + 9 \geq 0$
$(x+3)(x+3) \geq 0$

```
x+3  ------------0++++
x+3  ------------0++++
<------------------------->
            -3
```

The product is positive only if the factors have the same sign. That happens for every value of x except -3, in which case the product is 0. The solution set is the set of all real numbers, $(-\infty, \infty)$.

13. $\frac{x}{x-3} > 0$

```
x    -----0+++++++++++
x-3  ------------0++++
<------------------------->
     0          3
```

The quotient is positive only if the numerator and denominator have the same sign.
$(-\infty, 0) \cup (3, \infty)$

15. $\frac{x+2}{x} \leq 0$

```
x+2  -----0+++++++++++
x    ------------0++++
<------------------------->
     -2         0
```

The quotient is negative only if the numerator and denominator have opposite signs. Since the denominator must not be 0, we do not include 0 in the solution set.
$[-2, 0)$

17. $\frac{t-3}{t+6} > 0$

```
t+6  -----0+++++++++++
t-3  ------------0++++
<------------------------->
     -6         3
```

The quotient is positive only if the numerator and denominator have the same sign.
$(-\infty, -6) \cup (3, \infty)$

19. $\frac{x}{x+2} + 1 > 0$

$\frac{x}{x+2} + \frac{x+2}{x+2} > 0$

$\frac{2x+2}{x+2} > 0$

```
x+2   -----0+++++++++++
2x+2  ------------0++++
<------------------------->
      -2         -1
```

The quotient is positive only if the numerator and denominator have the same sign.
$(-\infty, -2) \cup (-1, \infty)$

21. $\frac{2}{x-5} > \frac{1}{x+4}$

$\frac{2}{x-5} - \frac{1}{x+4} > 0$

$\frac{2(x+4)}{(x-5)(x+4)} - \frac{1(x-5)}{(x+4)(x-5)} > 0$

$\frac{x+13}{(x-5)(x+4)} > 0$

```
x+4   -------0+++++++++
x+13  --0+++++++++++++++
x-5   ------------0++++
<------------------------->
  -13      -4       5
```

This quotient will be positive only if an even number of the factors have negative values.

$(-13, -4) \cup (5, \infty)$

23. $\frac{m(m-1)}{(m-5)(m-1)} + \frac{3(m-5)}{(m-1)(m-5)} > 0$

$\frac{m^2 + 2m - 15}{(m-1)(m-5)} > 0$

$\frac{(m+5)(m-3)}{(m-1)(m-5)} > 0$

```
m−3  ----------0++++++
m−1  -----0+++++++++
m+5  --0+++++++++++++
m−5  --------------0+++
```
```
<────┬────┬────┬────┬──────>
    −5    1    3    5
```

This quotient is positive only if there is an even number of factors with negative signs.
$(-\infty, -5) \cup (1, 3) \cup (5, \infty)$

```
◄─┼─┼)┼┼┼┼┼┼(─┼)┼(─┼─►
 −7 −5 −3 −2  1  3  5  7
```

25.
$$\frac{x}{x-3} + \frac{8}{x-6} \le 0$$

$$\frac{x(x-6)}{(x-3)(x-6)} + \frac{8(x-3)}{(x-6)(x-3)} \le 0$$

$$\frac{x^2 + 2x - 24}{(x-3)(x-6)} \le 0$$

$$\frac{(x+6)(x-4)}{(x-3)(x-6)} \le 0$$

```
x−4  ----------0++++++
x−3  ------0+++++++++++
x+6  --0+++++++++++++++
x−6  --------------0++
```
```
<────┬────┬────┬────┬──────>
    −6    3    4    6
```

The quotient is negative only if an odd number of factors have negative signs. Note that 3 and 6 cause the denominator to have a value of 0, and so they are excluded from the solution set.
$[-6, 3) \cup [4, 6)$

```
◄─[┼┼┼┼┼┼┼┼┼)┼[┼)─►
 −6 −4 −2  0  2  4  6
```

27. To solve $x^2 - 2x - 4 > 0$, first solve
$x^2 - 2x - 4 = 0$:

$$x = \frac{2 \pm \sqrt{4 - 4(1)(-4)}}{2(1)} = \frac{2 \pm 2\sqrt{5}}{2}$$

$$= 1 \pm \sqrt{5}$$

The two solutions to the equation divide the number line into three regions. Choose one number in each region, say −10, 0, and 10. Now evaluate $x^2 - 2x - 4$ for the numbers −10, 0, and 10.

$$(-10)^2 - 2(-10) - 4 = 116$$

$$0^2 - 2(0) - 4 = -4$$

$$10^2 - 2(10) - 4 = 76$$

The signs of these answers indicate the solution to the inequality.
$(-\infty, 1 - \sqrt{5}) \cup (1 + \sqrt{5}, \infty)$

```
◄────────)───(────────►
       1−√5    1+√5
```

29. To solve $2x^2 - 6x + 3 \ge 0$, we first solve $2x^2 - 6x + 3 = 0$.

$$x = \frac{6 \pm \sqrt{36 - 4(2)(3)}}{2(2)} = \frac{3 \pm \sqrt{3}}{2}$$

The two solutions to the equation divide the number line into three regions. Choose a test point in each region, say −10, 2, and 10. Evaluate the polynomial $2x^2 - 6x + 3$ at each of these points.

$$2(-10)^2 - 6(-10) + 3 = 263$$

$$2(2)^2 - 6(2) + 3 = -1$$

$$2(10)^2 - 6(10) + 3 = 143$$

The signs of these answers indicate the regions that satisfy the inequality.

$(-\infty, \frac{3-\sqrt{3}}{2}] \cup [\frac{3+\sqrt{3}}{2}, \infty)$

```
◄──────┤────────├──────►
    3−√3       3+√3
     ⎯⎯         ⎯⎯
      2          2
```

31. To solve $y^2 - 3y - 9 \le 0$ we first solve $y^2 - 3y - 9 = 0$:

$$y = \frac{3 \pm \sqrt{9 - 4(1)(-9)}}{2} = \frac{3 \pm 3\sqrt{5}}{2}$$

Pick test points −10, 4, and 10, and evaluate $y^2 - 3y - 9$ for these numbers.

$$(-10)^2 - 3(-10) - 9 = 121$$

$$4^2 - 3(4) - 9 = -5$$

$$10^2 - 3(10) - 9 = 61$$

The signs of these answers indicate which regions satisfy the original inequality.

$$\left[\frac{3 - 3\sqrt{5}}{2}, \frac{3 + 3\sqrt{5}}{2}\right]$$

```
◄──────┤────────├──────►
    3−3√5      3+3√5
     ⎯⎯⎯        ⎯⎯⎯
      2          2
```

33.
$$x^2 - 9 \le 0$$
$$(x-3)(x+3) \le 0$$

```
x+3  -----0++++++++++++
x−3  -------------0++++
```
```
<────┬──────────────┬──────>
    −3              3
```

The product is negative only if the factors have opposite signs. That happens if x is chosen between -3 and 3. At ± 3 the product is 0. The solution set and its graph follow: $[-3, 3]$

35.
$$16 - x^2 > 0$$
$$-1(16 - x^2) > -1(0)$$
$$x^2 - 16 < 0$$
$$(x - 4)(x + 4) < 0$$

```
x+4  -----0+++++++++++
 x-4 -------------0++++
<_____>
     -4            4
```

The product is negative only if the value of x is between -4 and 4. The solution set and its graph follow: $(-4, 4)$

37.
$$x^2 - 4x \geq 0$$
$$x(x - 4) \geq 0$$

```
x    -----0+++++++++++
 x-4 -----------0++++
<_____>
     0            4
```

The product is positive only if the factors have the same sign. $(-\infty, 0] \cup [4, \infty)$

39.
$$6w^2 - 15 < w$$
$$6w^2 - w - 15 < 0$$
$$(2w + 3)(3w - 5) < 0$$

```
2w+3  ----0+++++++++++
 3w-5 ------------0++++
<_____>
     -3/2         5/3
```

The product is negative only if the factors have opposite signs. $\left(-\frac{3}{2}, \frac{5}{3}\right)$

41.
$$z^2 \geq 4z + 12$$
$$z^2 - 4z - 12 \geq 0$$
$$(z - 6)(z + 2) \geq 0$$

```
z+2  -----0+++++++++++
 z-6 -------------0++++
<_____>
     -2           6
```

The product is positive only if the factors have the same sign. $(-\infty, -2] \cup [6, \infty)$

43.
$$q^2 + 8q + 16 > 10q + 31$$
$$q^2 - 2q - 15 > 0$$
$$(q + 3)(q - 5) > 0$$

```
q+3  -----0+++++++++++
 q-5 -------------0++++
<_____>
     -3           5
```

The product is positive only if the factors have the same sign. $(-\infty, -3) \cup (5, \infty)$

45.
$$\tfrac{1}{2}x^2 \geq 4 - x$$
$$x^2 \geq 8 - 2x$$
$$x^2 + 2x - 8 \geq 0$$
$$(x + 4)(x - 2) \geq 0$$

```
x+4  -----0+++++++++++
 x-2 -------------0++++
<_____>
     -4           2
```

The product is positive only if the factors have the same sign. $(-\infty, -4] \cup [2, \infty)$

47. $\dfrac{x - 4}{x + 3} \leq 0$

```
x+3  ----0+++++++++++
 x-4 -------------0++++
<_____>
     -3           4
```

The quotient is negative if the factors have opposite signs. $(-3, 4]$

49. $(x - 2)(x + 1)(x - 5) \geq 0$

```
x-2  ------0+++++++++
x+1  --0+++++++++++++
 x-5 -----------0++++
<_____>
    -1    2      5
```

The product of these three factors is positive only if an even number of the factors have negative signs. This happens between −1 and 2, and also above 5 where no factors have negative signs.   [−1, 2] ∪ [5, ∞)

51.
$$x^2(x+3) - 1(x+3) < 0$$
$$(x^2 - 1)(x+3) < 0$$
$$(x-1)(x+1)(x+3) < 0$$

```
x+1 ------0++++++++++
x+3 --0++++++++++++++
x-1 ------------0++++
   -3      -1        1
```

The product of these three factors is negative only if an odd number of them have negative signs. This happens to the left of −3 and also between −1 and 1. $(-\infty, -3) \cup (-1, 1)$

53. To solve $0.23x^2 + 6.5x + 4.3 < 0$, we first solve $0.23x^2 + 6.5x + 4.3 = 0$.

$$x = \frac{-6.5 \pm \sqrt{(6.5)^2 - 4(0.23)(4.3)}}{2(0.23)}$$

$x = -27.58$ or $x = -0.68$

Test the numbers −30, −1, and 0.

$0.23(-30)^2 + 6.5(-30) + 4.3 = 16.3$

$0.23(-1)^2 + 6.5(-1) + 4.3 = -1.97$

$0.23(0)^2 + 6.5(0) + 4.3 = 4.3$

According to the signs of the values of the polynomial at the test points, the value of the polynomial is negative between the two solutions to the equation. $(-27.58, -0.68)$

55. $\dfrac{x}{x-2} + \dfrac{1}{x+3} > 0$

$$\frac{x(x+3)}{(x-2)(x+3)} + \frac{1(x-2)}{(x+3)(x-2)} > 0$$

$$\frac{x^2 + 4x - 2}{(x+3)(x-2)} > 0$$

Solve $x^2 + 4x - 2 = 0$:

$$x = \frac{-4 \pm \sqrt{16 - 4(1)(-2)}}{2} = \frac{-4 \pm 2\sqrt{6}}{2}$$
$$= -2 \pm \sqrt{6} = -4.4, \ 0.4$$

The numbers −4.4, −3, 0.4, and 2 divide the number line into 5 regions. Pick a number in each region and test it in the original inequality. We get the solution set
$(-\infty, -4.4) \cup (-3, 0.4) \cup (2, \infty)$.

57. To solve $x^2 + 5x - 50 > 0$, we first solve $x^2 + 5x - 50 = 0$:
$$(x + 10)(x - 5) = 0$$
$$x = -10 \text{ or } x = 5$$
Test a point in each region of the number line, to find that the profit is positive if $x < -10$ or if $x > 5$. He cannot sell negative mobile homes so he must sell more than 5 to have a positive profit. He should sell 6, 7, 8, etc.

59. We must solve the inequality

$$-16t^2 + 96t + 6 > 86$$

$$-16t^2 + 96t - 80 > 0$$

$$t^2 - 6t + 5 < 0$$

Solve the equation $t^2 - 6t + 5 = 0$.
$$(t-5)(t-1) = 0$$
$$t = 5 \text{ or } t = 1$$
Using a test point we find that the inequality is satisfied for t between 1 and 5 seconds. So the arrow is more than 86 feet high for 4 seconds.

61. $S = -16t^2 + \dfrac{240\sqrt{2}}{\sqrt{2}}t + 0$

Solve

$$-16t^2 + \frac{240\sqrt{2}}{\sqrt{2}}t > 864$$

$$-16t^2 + 240t - 864 > 0$$

$$t^2 - 15t + 54 < 0$$

Solve the equation

$$(t - 6)(t - 9) = 0$$

$$t = 6 \text{ or } t = 9$$

Evaluate $t^2 - 15t + 54$ at the test points 0, 7, and 10.

$0^2 - 15(0) + 54 = 54$
$7^2 - 15(7) + 54 = -2$
$10^2 - 15(10) + 54 = 4$

So the inequality is satisfied for t in the interval (6, 9). The projectile was above 864 ft for 3 sec.

CHAPTER 6 REVIEW

1. $x^2 - 2x - 15 = 0$
$(x-5)(x+3) = 0$
$x - 5 = 0$ or $x + 3 = 0$
$x = 5$ or $x = -3$
Solution set: $\{-3, 5\}$

3. $2x^2 + x - 15 = 0$
$(2x-5)(x+3) = 0$
$2x - 5 = 0$ or $x + 3 = 0$
$x = \frac{5}{2}$ or $x = -3$
Solution set: $\{-3, \frac{5}{2}\}$

5. $w^2 - 25 = 0$
$(w-5)(w+5) = 0$
$w - 5 = 0$ or $w + 5 = 0$
$w = 5$ or $w = -5$
Solution set: $\{-5, 5\}$

7. $4x^2 - 12x + 9 = 0$
$(2x-3)^2 = 0$
$2x - 3 = 0$
$x = \frac{3}{2}$
Solution set: $\{\frac{3}{2}\}$

9. $x^2 = 12$
$x = \pm\sqrt{12} = \pm 2\sqrt{3}$
Solution set: $\{\pm 2\sqrt{3}\}$

11. $(x-1)^2 = 9$
$x - 1 = \pm 3$
$x = 1 \pm 3$
Solution set: $\{-2, 4\}$

13. $(x-2)^2 = \frac{3}{4}$
$x - 2 = \pm\sqrt{\frac{3}{4}}$
$x = 2 \pm \frac{\sqrt{3}}{2} = \frac{4}{2} \pm \frac{\sqrt{3}}{2}$
Solution set: $\left\{\frac{4 \pm \sqrt{3}}{2}\right\}$

15. $4x^2 = 9$
$x^2 = \frac{9}{4}$
$x = \pm\sqrt{\frac{9}{4}} = \pm\frac{3}{2}$
Solution set: $\{\pm 3/2\}$

17. $x^2 - 6x = -8$
$x^2 - 6x + 9 = -8 + 9$
$(x-3)^2 = 1$
$x - 3 = \pm 1$
$x = 3 \pm 1$
Solution set: $\{2, 4\}$

19. $x^2 - 5x = -6$
$x^2 - 5x + \frac{25}{4} = -\frac{24}{4} + \frac{25}{4}$
$(x - \frac{5}{2})^2 = \frac{1}{4}$
$x - \frac{5}{2} = \pm\frac{1}{2}$
$x = \frac{5}{2} \pm \frac{1}{2}$ $\left(\frac{6}{2} \text{ or } \frac{4}{2}\right)$
Solution set: $\{2, 3\}$

21. $2x^2 - 7x = -3$
$x^2 - \frac{7}{2}x = -\frac{3}{2}$
$x^2 - \frac{7}{2}x + \frac{49}{16} = -\frac{24}{16} + \frac{49}{16}$
$(x - \frac{7}{4})^2 = \frac{25}{16}$
$x - \frac{7}{4} = \pm\frac{5}{4}$
$x = \frac{7}{4} \pm \frac{5}{4}$ $\left(\frac{12}{4} \text{ or } \frac{2}{4}\right)$
Solution set: $\{\frac{1}{2}, 3\}$

23. $x^2 + 4x = -1$
$x^2 + 4x + 4 = -1 + 4$
$(x+2)^2 = 3$
$x + 2 = \pm\sqrt{3}$
$x = -2 \pm \sqrt{3}$
Solution set: $\{-2 \pm \sqrt{3}\}$

25. $x^2 - 3x - 10 = 0$

$x = \dfrac{3 \pm \sqrt{9 - 4(1)(-10)}}{2(1)} = \dfrac{3 \pm \sqrt{49}}{2} = \dfrac{3 \pm 7}{2}$

Solution set: $\{-2, 5\}$

27. $6x^2 - 7x - 3 = 0$

$x = \dfrac{7 \pm \sqrt{49 - 4(6)(-3)}}{2(6)} = \dfrac{7 \pm \sqrt{121}}{12} = \dfrac{7 \pm 11}{12}$

Solution set: $\left\{-\frac{1}{3}, \frac{3}{2}\right\}$

29. $x^2 + 4x + 2 = 0$

$x = \dfrac{-4 \pm \sqrt{16 - 4(1)(2)}}{2(1)} = \dfrac{-4 \pm 2\sqrt{2}}{2} = -2 \pm \sqrt{2}$

Solution set: $\{-2 \pm \sqrt{2}\}$

31. $3x^2 - 5x + 1 = 0$

$$x = \frac{5 \pm \sqrt{25 - 4(3)(1)}}{2(3)} = \frac{5 \pm \sqrt{13}}{6}$$

Solution set: $\left\{ \frac{5 \pm \sqrt{13}}{6} \right\}$

33. $b^2 - 4ac = (-20)^2 - 4(25)(4) = 0$

One real solution.

35. $b^2 - 4ac = (-3)^2 - 4(1)(7) = -19$

No real solutions

37. $2x^2 - 5x + 1 = 0$

$b^2 - 4ac = (-5)^2 - 4(2)(1) = 17$

Two real solution

39. $2x^2 - 4x + 3 = 0$

$$x = \frac{4 \pm \sqrt{16 - 4(2)(3)}}{2(2)} = \frac{4 \pm \sqrt{-8}}{4}$$

$$= \frac{4 \pm 2i\sqrt{2}}{4} = \frac{3 \pm i\sqrt{2}}{2}$$

Solution set: $\left\{ \frac{2 \pm i\sqrt{2}}{2} \right\}$

41. $2x^2 - 3x + 3 = 0$

$$x = \frac{3 \pm \sqrt{9 - 4(2)(3)}}{2(2)} = \frac{3 \pm \sqrt{-15}}{4} = \frac{3 \pm i\sqrt{15}}{4}$$

Solution set: $\left\{ \frac{3 \pm i\sqrt{15}}{4} \right\}$

43. $3x^2 + 2x + 2 = 0$

$$x = \frac{-2 \pm \sqrt{4 - 4(3)(2)}}{2(3)} = \frac{-2 \pm \sqrt{-20}}{6}$$

$$= \frac{-2 \pm 2i\sqrt{5}}{6} = \frac{-1 \pm i\sqrt{5}}{3}$$

Solution set: $\left\{ \frac{-1 \pm i\sqrt{5}}{3} \right\}$

45. $x^2 + 6x + 16 = 0$

$$x = \frac{-6 \pm \sqrt{36 - 4(1)(16)}}{2(1)} = \frac{-6 \pm \sqrt{-28}}{2}$$

$$= \frac{-6 \pm 2i\sqrt{7}}{2} = -3 \pm i\sqrt{7}$$

Solution set: $\{-3 \pm i\sqrt{7}\}$

47. $b^2 - 4ac = (-10)^2 - 4(8)(-3) = 196$
Since 196 is a perfect square, the polynomial is not prime.
$8x^2 - 10x - 3 = (4x + 1)(2x - 3)$

49. $b^2 - 4ac = (-5)^2 - 4(4)(2) = -7$
Since -7 is a not a perfect square, the polynomial is prime.

51. $b^2 - 4ac = (10)^2 - 4(8)(-25) = 900$
Since 900 is a perfect square, the polynomial is not prime.
$8y^2 + 10y - 25 = (4y - 5)(2y + 5)$

53. $(x + 3)(x + 6) = 0$
$x^2 + 9x + 18 = 0$

55. $(x + 5\sqrt{2})(x - 5\sqrt{2}) = 0$
$x^2 - 50 = 0$

57.
$$x^6 + 7x^3 - 8 = 0$$
$$(x^3 + 8)(x^3 - 1) = 0$$
$$x^3 + 8 = 0 \quad \text{or} \quad x^3 - 1 = 0$$
$$x^3 = -8 \quad \text{or} \quad x^3 = 1$$
$$x = -2 \quad \text{or} \quad x = 1$$
Solution set: $\{-2, 1\}$

59.
$$x^4 - 13x^2 + 36 = 0$$
$$(x^2 - 4)(x^2 - 9) = 0$$
$$x^2 - 4 = 0 \quad \text{or} \quad x^2 - 9 = 0$$
$$x^2 = 4 \quad \text{or} \quad x^2 = 9$$
$$x = \pm 2 \quad \text{or} \quad x = \pm 3$$
Solution set: $\{\pm 2, \pm 3\}$

61. Let $w = x^2 + 3x$.
$$w^2 - 28w + 180 = 0$$
$$(w - 10)(w - 18) = 0$$
$$w = 10 \quad \text{or} \quad w = 18$$
$$x^2 + 3x = 10 \quad \text{or} \quad x^2 + 3x = 18$$
$$x^2 + 3x - 10 = 0 \quad \text{or} \quad x^2 + 3x - 18 = 0$$
$$(x + 5)(x - 2) = 0 \quad \text{or} \quad (x + 6)(x - 3) = 0$$
$$x = -5 \quad \text{or} \quad x = 2 \quad \text{or} \quad x = -6 \quad \text{or} \quad x = 3$$
Solution set: $\{-6, -5, 2, 3\}$

63. Let $w = \sqrt{x^2 - 6x}$ and $w^2 = x^2 - 6x$.
$$w^2 + 6w - 40 = 0$$
$$(w + 10)(w - 4) = 0$$
$$w = -10 \quad \text{or} \quad w = 4$$
$$\sqrt{x^2 - 6x} = -10 \quad \text{or} \quad \sqrt{x^2 - 6x} = 4$$
No real solution here. $\qquad x^2 - 6x = 16$

$$x^2 - 6x - 16 = 0$$
$$(x - 8)(x + 2) = 0$$
$$x = 8 \quad \text{or} \quad x = -2$$

Solution set: $\{-2, 8\}$

65. Let $w = t^{-1}$ and $w^2 = t^{-2}$.
$$w^2 + 5w - 36 = 0$$
$$(w + 9)(w - 4) = 0$$
$$w = -9 \quad \text{or} \quad w = 4$$
$$t^{-1} = -9 \quad \text{or} \quad t^{-1} = 4$$
$$t = -\frac{1}{9} \quad \text{or} \quad t = \frac{1}{4}$$
Solution set: $\{-\frac{1}{9}, \frac{1}{4}\}$

67. Let $y = \sqrt{w}$ and $y^2 = w$.
$$y^2 - 13y + 36 = 0$$
$$(y - 9)(y - 4) = 0$$
$$y = 9 \quad \text{or} \quad y = 4$$
$$\sqrt{w} = 9 \quad \text{or} \quad \sqrt{w} = 4$$
$$w = 81 \quad \text{or} \quad w = 16$$
Solution set: $\{16, 81\}$

69.
$$a^2 + a > 6$$
$$a^2 + a - 6 > 0$$
$$(a + 3)(a - 2) > 0$$

```
a+ 3  -----0+++++++++++
 a-2  -------------0++++
<_____>
     -3            2
```

The inequality is satisfied if both factors are the same sign. $(-\infty, -3) \cup (2, \infty)$

```
<+|+)+|+|+|+|(+|+>
-4-3-2-1 0 1 2 3
```

71.
$$x^2 - x - 20 \leq 0$$
$$(x - 5)(x + 4) \leq 0$$

```
x+ 4  -----0+++++++++++
 x-5  -------------0++++
<_____>
     -4            5
```

The inequality is satisfied if the factors have opposite signs, or if one of the factors is zero. $[-4, 5]$

```
<[+|+|+|+|+|+|+|+]+>
 -4-3-2-1 0 1 2 3 4 5
```

73.
$$w^2 - w < 0$$
$$w(w - 1) < 0$$

```
w     -----0+++++++++++
 w-1  -------------0++++
<_____>
      0            1
```

The inequality is satisfied if the factors have opposite signs. $(0, 1)$

```
<+|+|(+)+|+|+>
 -2 -1 0 1 2 3
```

75.
$$\frac{x - 4}{x + 2} \geq 0$$

```
x+ 2  ----0+++++++++++++
 x-4  -------------0++++
<_____>
     -2            4
```

The inequality is satisfied if the factors have the same sign, or if the numerator is equal to zero.
$(-\infty, -2) \cup [4, \infty)$

```
<+|+)+|+|+|+|[+|+>
-4-3-2-1 0 1 2 3 4 5 6
```

77.
$$\frac{x - 2}{x + 3} - 1 < 0$$
$$\frac{x - 2}{x + 3} - \frac{x + 3}{x + 3} < 0$$
$$\frac{-5}{x + 3} < 0$$

Since the numerator is definitely negative, the inequality is satisfied only if the denominator is positive: $x + 3 > 0$, or $x > -3$.
$(-3, \infty)$

```
<+|+|(+|+|+|+|+>
 -5 -4 -3 -2 -1 0 1
```

79.
$$\frac{3}{x + 2} - \frac{1}{x + 1} > 0$$
$$\frac{3(x + 1)}{(x + 2)(x + 1)} - \frac{1(x + 2)}{(x + 1)(x + 2)} > 0$$
$$\frac{2x + 1}{(x + 2)(x + 1)} > 0$$

```
x+1  ------0+++++++++
x+2  --0+++++++++++++++
2x+1 ------------0++++
<_____>
   -2    -1    -1/2
```

The inequality is satisfied when an even number of the factors have negative signs.

$(-2, -1) \cup (-\frac{1}{2}, \infty)$

```
        -1/2
<+|+|(+)+(+|+|+>
 -3 -2 -1 0 1 2
```

81.
$$144x^2 - 120x + 25 = 0$$
$$(12x - 5)^2 = 0$$
$$12x - 5 = 0$$
$$x = \frac{5}{12}$$
Solution set: $\left\{\frac{5}{12}\right\}$

83. $(2x+3)^2 + 7 = 12$
$$(2x+3)^2 = 5$$
$$2x+3 = \pm\sqrt{5}$$
$$x = \frac{-3 \pm \sqrt{5}}{2}$$

Solution set: $\left\{\frac{-3 \pm \sqrt{5}}{2}\right\}$

85. $9x^2\left(1 + \frac{20}{9x^2}\right) = 9x^2 \cdot \frac{8}{3x}$
$$9x^2 + 20 = 24x$$
$$9x^2 - 24x + 20 = 0$$
$$x = \frac{24 \pm \sqrt{(-24)^2 - 4(9)(20)}}{2(9)} = \frac{24 \pm 12i}{18}$$
$$= \frac{4 \pm 2i}{3}$$

Solution set: $\left\{\frac{4 \pm 2i}{3}\right\}$

87. $\sqrt{3x^2 + 7x - 30} = x$
$$3x^2 + 7x - 30 = x^2$$
$$2x^2 + 7x - 30 = 0$$
$$(2x-5)(x+6) = 0$$
$$2x - 5 = 0 \quad \text{or} \quad x + 6 = 0$$
$$x = \frac{5}{2} \quad \text{or} \qquad x = -6$$

Since -6 does not check, solution set is $\left\{\frac{5}{2}\right\}$.

89. $2(2x+1)^2 + 5(2x+1) = 3$
$$2y^2 + 5y - 3 = 0$$
$$(2y-1)(y+3) = 0$$
$$2y - 1 = 0 \quad \text{or} \quad y + 3 = 0$$
$$y = \frac{1}{2} \quad \text{or} \qquad y = -3$$
$$2x + 1 = \frac{1}{2} \quad \text{or} \quad 2x + 1 = -3$$
$$x = -\frac{1}{4} \quad \text{or} \qquad x = -2$$

Solution set: $\left\{-2, -\frac{1}{4}\right\}$

91. $x^{1/2} - 15x^{1/4} + 50 = 0 \qquad \text{Let } y = x^{1/4}$
$$y^2 - 15y + 50 = 0$$
$$(y-5)(y-10) = 0$$
$$y = 5 \quad \text{or} \qquad y = 10$$
$$x^{1/4} = 5 \quad \text{or} \quad x^{1/4} = 10$$
$$x = 5^4 \quad \text{or} \qquad x = 10^4$$
$$x = 625 \quad \text{or} \quad x = 10,000$$

Solution set: $\{625, 10,000\}$

93. If $x =$ one of the numbers, then $x + 4 =$ the other number. Since their product is 4, we can write the equation
$$x(x+4) = 4$$
$$x^2 + 4x = 4$$
$$x^2 + 4x + 4 = 4 + 4$$
$$(x+2)^2 = 8$$
$$x + 2 = \pm 2\sqrt{2}$$
$$x = -2 \pm 2\sqrt{2}$$

Disregard $-2 - 2\sqrt{2}$ since it is not positive. If $x = -2 + 2\sqrt{2}$ then $x + 4 = 2 + 2\sqrt{2}$.

The two numbers are $-2 + 2\sqrt{2} \approx 0.83$ and $2 + 2\sqrt{2} \approx 4.83$.

95. Let $x =$ the height and $x + 4 =$ the width. The diagonal 19 is the hypotenuse of a right triangle with legs x and $x + 4$. By the Pythagorean theorem we can write
$$x^2 + (x+4)^2 = 19^2$$
$$x^2 + x^2 + 8x + 16 = 361$$
$$2x^2 + 8x - 345 = 0$$
$$x = \frac{-8 \pm \sqrt{64 - 4(2)(-345)}}{2(2)} = \frac{-8 \pm \sqrt{2824}}{4}$$
$$= \frac{-8 \pm 2\sqrt{706}}{4} = \frac{-4 \pm \sqrt{706}}{2}$$

Since the height must be positive, we have
$$x = \frac{-4 + \sqrt{706}}{2} \text{ and}$$
$$x + 4 = \frac{8}{2} + \frac{-4 + \sqrt{706}}{2} = \frac{4 + \sqrt{706}}{2}.$$

So the width is $\frac{4 + \sqrt{706}}{2} \approx 15.3$ inches and the height is $\frac{-4 + \sqrt{706}}{2} \approx 11.3$ inches.

97. Let $x =$ the width of the border. The dimensions of the printed area will be $8 - 2x$ by $10 - 2x$. Since the printed area is to be 24 square inches, we can write the equation
$$(8 - 2x)(10 - 2x) = 24$$
$$80 - 36x + 4x^2 = 24$$
$$4x^2 - 36x + 56 = 0$$

114

$$x^2 - 9x + 14 = 0$$
$$(x-7)(x-2) = 0$$
$$x = 7 \qquad \text{or} \qquad x = 2$$

Since 7 inches is too wide for a border on an 8 by 10 piece of paper, the border must be 2 inches wide.

99. Let x = width and $x + 4$ = the length.
$$x(x+4) = 45$$
$$x^2 + 4x - 45 = 0$$
$$(x+9)(x-5) = 0$$
$$x + 9 = 0 \quad \text{or} \quad x - 5 = 0$$
$$x = -9 \quad \text{or} \qquad x = 5$$
$$x + 4 = 9$$

The table is 5 ft wide and 9 feet long.

101.
$$12 = -16t^2 + 32t$$
$$16t^2 - 32t + 12 = 0$$
$$4t^2 - 8t + 3 = 0$$

$$t = \frac{8 \pm \sqrt{64 - 4(4)(3)}}{2(4)} = \frac{8 \pm \sqrt{16}}{8} = 1 \pm \frac{1}{2}$$
$$= 1.5 \text{ or } 0.5$$

His height was 12 ft for $t = 0.5$ sec and $t = 1.5$ sec.

CHAPTER 6   TEST

1. $(-3)^2 - 4(2)(2) = 9 - 16 = -7$

The equation has no real solutions.

2. $(5)^2 - 4(-3)(-1) = 25 - 12 = 13$

The equation has 2 real solutions.

3. $(-4)^2 - 4(4)(1) = 16 - 16 = 0$

The equation has 1 real solution.

4. $2x^2 + 5x - 3 = 0$

$$x = \frac{-5 \pm \sqrt{25 - 4(2)(-3)}}{2(2)} = \frac{-5 \pm \sqrt{49}}{4}$$

$$= \frac{-5 \pm 7}{4} \qquad \left(\frac{2}{4} \text{ or } \frac{-12}{4}\right)$$

Solution set: $\left\{-3, \frac{1}{2}\right\}$

5. $\qquad x^2 + 6x + 6 = 0$

$$x = \frac{-6 \pm \sqrt{36 - 4(1)(6)}}{2} = \frac{-6 \pm \sqrt{12}}{2}$$

$$= \frac{-6 \pm 2\sqrt{3}}{2} = -3 \pm \sqrt{3}$$

Solution set: $\{-3 \pm \sqrt{3}\}$

6. $\qquad x^2 + 10x + 25 = 0$
$$(x+5)^2 = 0$$
$$x + 5 = 0$$
$$x = -5$$

Solution set: $\{-5\}$

7. $\qquad 2x^2 + x = 6$

$$x^2 + \frac{1}{2}x = 3$$

$$x^2 + \frac{1}{2}x + \frac{1}{16} = 3 + \frac{1}{16}$$

$$\left(x + \frac{1}{4}\right)^2 = \frac{49}{16}$$

$$x + \frac{1}{4} = \pm\frac{7}{4}$$

$$x = -\frac{1}{4} \pm \frac{7}{4} \qquad \left(\frac{6}{4} \text{ or } \frac{-8}{4}\right)$$

Solution set: $\left\{-2, \frac{3}{2}\right\}$

8. $\qquad x(x+1) = 12$
$$x^2 + x - 12 = 0$$
$$(x+4)(x-3) = 0$$
$$x = -4 \qquad \text{or} \qquad x = 3$$

Solution set: $\{-4, 3\}$

9. $\qquad a^4 - 5a^2 + 4 = 0$
$$(a^2 - 4)(a^2 - 1) = 0$$
$$a^2 - 4 = 0 \quad \text{or} \quad a^2 - 1 = 0$$
$$a^2 = 4 \quad \text{or} \qquad a^2 = 1$$
$$a = \pm 2 \quad \text{or} \qquad a = \pm 1$$

Solution set: $\{\pm 1, \pm 2\}$

10. Let $w = \sqrt{x-2}$ and $w^2 = x - 2$.
$$w^2 - 8w + 15 = 0$$
$$(w-3)(w-5) = 0$$
$$w = 3 \qquad \text{or} \qquad w = 5$$
$$\sqrt{x-2} = 3 \quad \text{or} \quad \sqrt{x-2} = 5$$
$$x - 2 = 9 \quad \text{or} \quad x - 2 = 25$$
$$x = 11 \quad \text{or} \qquad x = 27$$

Solution set: $\{11, 27\}$

11.
$$x^2 + 36 = 0$$
$$x^2 = -36$$
$$x = \pm\sqrt{-36} = \pm 6i$$

Solution set: $\{\pm 6i\}$

12. $x^2 + 6x + 10 = 0$

$$x = \frac{-6 \pm \sqrt{36 - 4(1)(10)}}{2(1)} = \frac{-6 \pm \sqrt{-4}}{2}$$

$$= \frac{-6 \pm 2i}{2} = -3 \pm i$$

Solution set: $\{-3 \pm i\}$

13. $3x^2 - x + 1 = 0$

$$x = \frac{1 \pm \sqrt{1 - 4(3)(1)}}{2(3)} = \frac{1 \pm \sqrt{-11}}{6}$$

Solution set: $\left\{ \frac{1 \pm i\sqrt{11}}{6} \right\}$

14. $(x + 4)(x - 6) = 0$
$$x^2 - 2x - 24 = 0$$

15. $(x - 5i)(x + 5i) = 0$
$$x^2 + 25 = 0$$

16.
$$w^2 + 3w - 18 < 0$$
$$(w + 6)(w - 3) < 0$$

```
w+6   ----0++++++++++++
w-3   ------------0++++
    <------------------------>
      -6           3
```

The inequality is satisfied when the factors have opposite signs. $(-6, 3)$

```
<---(+++++++++)--->
  -6-5-4-3-2-1 0 1 2 3
```

17.
$$\frac{2}{x-2} - \frac{3}{x+1} < 0$$
$$\frac{2(x+1)}{(x-2)(x+1)} - \frac{3(x-2)}{(x+1)(x-2)} < 0$$
$$\frac{-x+8}{(x-2)(x+1)} < 0$$

```
x-2   -------0++++++++++
x+1   --0++++++++++++++++
-x+8  ++++++++++++0----
    <------------------------>
      -1    2         8
```

This quotient will be negative only if an odd number of factors are negative.
$(-1, 2) \cup (8, \infty)$

```
<---(+++)+++++++(+++--->
  -1 0 1 2 3 4 5 6 7 8 9 10
```

18. Let $x =$ the width and $x + 2 =$ the length. Since the area is 16 square feet, we can write the equation
$$x(x + 2) = 16$$
$$x^2 + 2x - 16 = 0$$
$$x = \frac{-2 \pm \sqrt{4 - 4(1)(-16)}}{2} = \frac{-2 \pm \sqrt{68}}{2}$$
$$= \frac{-2 \pm 2\sqrt{17}}{2} = -1 \pm \sqrt{17}$$

Since the width must be positive,   we   have
$x = -1 + \sqrt{17}$ and $x + 2 = 1 + \sqrt{17}$. The width is
$-1 + \sqrt{17}$ feet   and   the   length   is $1 + \sqrt{17}$
feet.

19. Let $x =$ time for the new computer and $x + 1 =$ time for the old computer. New computer's rate is $\frac{1}{x}$ payroll/hr and old computer's rate is $\frac{1}{x+1}$ payroll/hr. In 3 hrs new computer does $\frac{3}{x}$ payroll and old computer does $\frac{3}{x+1}$ payroll.

$$\frac{3}{x} + \frac{3}{x+1} = 1$$
$$x(x+1)\left(\frac{3}{x} + \frac{3}{x+1}\right) = x(x+1)1$$
$$3x + 3 + 3x = x^2 + x$$
$$0 = x^2 - 5x - 3$$

$$x = \frac{5 \pm \sqrt{5^2 - 4(1)(-3)}}{2} = \frac{5 \pm \sqrt{37}}{2} \approx 5.5 \text{ or } -0.5$$

It takes the new computer 5.5 hrs to do the payroll by itself.

Tying It All Together Chapters 1 - 6

1.
$$2x - 15 = 0$$
$$2x = 15$$
$$x = \frac{15}{2}$$

Solution set: $\left\{ \frac{15}{2} \right\}$

2.
$$2x^2 - 15 = 0$$
$$2x^2 = 15$$
$$x^2 = \frac{15}{2}$$
$$x = \pm\sqrt{\frac{15}{2}} = \pm\frac{\sqrt{15}\sqrt{2}}{\sqrt{2}\sqrt{2}} = \pm\frac{\sqrt{30}}{2}$$

Solution set: $\left\{ \pm\frac{\sqrt{30}}{2} \right\}$

3.
$$2x^2 + x - 15 = 0$$
$$(2x - 5)(x + 3) = 0$$
$$2x - 5 = 0 \quad \text{or} \quad x + 3 = 0$$
$$x = \frac{5}{2} \quad \text{or} \quad x = -3$$

Solution set: $\left\{ -3, \frac{5}{2} \right\}$

4.
$$2x^2 + 4x - 15 = 0$$
$$x = \frac{-4 \pm \sqrt{16 - 4(2)(-15)}}{2(2)}$$
$$= \frac{-4 \pm \sqrt{136}}{4} = \frac{-4 \pm 2\sqrt{34}}{4} = \frac{-2 \pm \sqrt{34}}{2}$$

Solution set: $\left\{ \frac{-2 \pm \sqrt{34}}{2} \right\}$

5.
$$|4x + 11| = 3$$
$$4x + 11 = 3 \quad \text{or} \quad 4x + 11 = -3$$
$$4x = -8 \quad \text{or} \quad 4x = -14$$
$$x = -2 \quad \text{or} \quad x = -\frac{7}{2}$$

Solution set: $\left\{ -\frac{7}{2}, -2 \right\}$

6.
$$|4x^2 + 11x| = 3$$
$$4x^2 + 11x = 3 \quad \text{or} \quad 4x^2 + 11x = -3$$
$$4x^2 + 11x - 3 = 0 \quad \text{or} \quad 4x^2 + 11x + 3 = 0$$
$$(4x - 1)(x + 3) = 0 \quad \text{or} \quad x = \frac{-11 \pm \sqrt{73}}{8}$$
$$4x - 1 = 0 \quad \text{or} \quad x + 3 = 0$$
$$x = \frac{1}{4} \quad \text{or} \quad x = -3$$

Solution set: $\left\{ -3, \frac{1}{4}, \frac{-11 \pm \sqrt{73}}{8} \right\}$

7.
$$\sqrt{x} = x - 6$$
$$(\sqrt{x})^2 = (x - 6)^2$$
$$x = x^2 - 12x + 36$$
$$0 = x^2 - 13x + 36$$
$$0 = (x - 4)(x - 9)$$
$$x - 4 = 0 \quad \text{or} \quad x - 9 = 0$$
$$x = 4 \quad \text{or} \quad x = 9$$

Since $\sqrt{4} = 4 - 6$ is incorrect and $\sqrt{9} = 9 - 6$ is correct, the solution set is $\{9\}$.

8.
$$\left((2x - 5)^{2/3} \right)^3 = 4^3$$
$$(2x - 5)^2 = 64$$
$$2x - 5 = \pm 8$$
$$2x = 5 \pm 8$$
$$x = \frac{5 \pm 8}{2}$$

Solution set: $\left\{ -\frac{3}{2}, \frac{13}{2} \right\}$

9.
$$1 - 2x < 5 - x$$
$$-x < 4$$
$$x > -4$$

$(-4, \infty)$

10. $(1 - 2x)(5 - x) \leq 0$

```
1-2x  ++++0-----------
5-x   +++++++++0-----
<------|---------|------>
       1/2       5
```

The inequality is satisfied when the factors have opposite signs.

$[\frac{1}{2}, 5]$

11. $\frac{1-2x}{5-x} \leq 0$ Same as last exercise, but 5 because of the denominator $5 - x$.

$[\frac{1}{2}, 5)$

12.
$$|5 - x| < 3$$
$$-3 < 5 - x < 3$$
$$-8 < -x < -2$$
$$8 > x > 2$$

$(2, 8)$

13. $3x - 1 < 5$ and $-3 \leq x$
$$x < 2 \quad \text{and} \quad x \geq -3$$

$[-3, 2)$

14. $x - 3 < 1$ or $2x \geq 8$

$\qquad x < 4$ or $x \geq 4$

$(-\infty, \infty)$

15. $\qquad 2x - 3y = 9$

$\qquad -3y = -2x + 9$

$\qquad y = \dfrac{-2x + 9}{-3}$

$\qquad y = \dfrac{2}{3}x - 3$

16. $\qquad \dfrac{y - 3}{x + 2} = -\dfrac{1}{2}$

$\qquad y - 3 = -\dfrac{1}{2}(x + 2)$

$\qquad y - 3 = -\dfrac{1}{2}x - 1$

$\qquad y = -\dfrac{1}{2}x + 2$

17. $\qquad 3y^2 + cy + d = 0$

$\qquad y = \dfrac{-c \pm \sqrt{c^2 - 4(3)(d)}}{2(3)}$

$\qquad y = \dfrac{-c \pm \sqrt{c^2 - 12d}}{6}$

18. $\qquad my^2 - ny - w = 0$

$\qquad y = \dfrac{-(-n) \pm \sqrt{(-n)^2 - 4(m)(-w)}}{2m}$

$\qquad y = \dfrac{n \pm \sqrt{n^2 + 4mw}}{2m}$

19. $\qquad 30\left(\dfrac{1}{3}x - \dfrac{2}{5}y\right) = 30\left(\dfrac{5}{6}\right)$

$\qquad 10x - 12y = 25$

$\qquad -12y = -10x + 25$

$\qquad y = \dfrac{-10}{-12}x + \dfrac{25}{-12}$

$\qquad y = \dfrac{5}{6}x - \dfrac{25}{12}$

20. $\qquad y - 3 = -\dfrac{2}{3}(x - 4)$

$\qquad y - 3 = -\dfrac{2}{3}x + \dfrac{8}{3}$

$\qquad y = -\dfrac{2}{3}x + \dfrac{8}{3} + \dfrac{9}{3}$

$\qquad y = -\dfrac{2}{3}x + \dfrac{17}{3}$

21. $m = \dfrac{3 - 7}{2 - 5} = \dfrac{-4}{-3} = \dfrac{4}{3}$

22. $m = \dfrac{5 - (-6)}{-3 - 4} = -\dfrac{11}{7}$

23. $m = \dfrac{0.8 - 0.4}{0.3 - 0.5} = \dfrac{0.4}{-0.2} = -2$

24. $m = \dfrac{\dfrac{3}{5} - \left(-\dfrac{4}{3}\right)}{\dfrac{1}{2} - \dfrac{1}{3}} = \dfrac{\dfrac{29}{15}}{\dfrac{1}{6}} = \dfrac{29}{15} \cdot \dfrac{6}{1} = \dfrac{58}{5}$

25. At \$20 per ticket,

$\qquad n = 48{,}000 - 400(20) = 40{,}000.$

At \$25 per ticket

$\qquad n = 48{,}000 - 400(25) = 38{,}000$

If 35,000 tickets are sold, then the price is \$32.50 per ticket.

26. If $p = \$20$, then $R = 20(48{,}000 - 400 \cdot 20)$

$\quad = \$800{,}000$

If $p = \$25$, then $R = 25(48{,}000 - 400 \cdot 25)$

$\quad = \$950{,}000$

$\qquad 1{,}280{,}000 = p(48{,}000 - 400p)$

$\qquad 1{,}280{,}000 = 48{,}000p - 400p^2$

$\quad 400p^2 - 48{,}000p + 1{,}280{,}000 = 0$

$\qquad p^2 - 120p + 3{,}200 = 0$

$\qquad (p - 80)(p - 40) = 0$

$\qquad p = 80 \qquad$ or $\qquad p = 40$

A revenue of \$1.28 million occurs at a price of \$40 and at a price of \$80.

The price that determines the maximum revenue is \$60 per ticket.

7.1 WARM-UPS

1. False, because $3(5) - 2(2) = -4$ is not correct.
2. False, the vertical axis is called the y-axis.
3. False, because the point $(0, 0)$ is not in any quadrant.   **4.** True, because its first coordinate is 0.   **5.** True, because the graph of $x = k$ for any real number k is a vertical line.
6. True, because the graph of $y = k$ for any real number k is a horizontal line.
7. False, because the correct formula is $\sqrt{(x_2 - x_1)^2 + (y_2 - y_1)^2}$.
8. False, because the distance is 7.   **9.** True, because $\sqrt{(-2-3)^2 + (6-6)^2} = \sqrt{25} = 5$.
10. True, if $x = 0$, we get $y = 2(0) - 3 = -3$.

7.1 EXERCISES

1. To plot $(2, 5)$, start at the origin and go 2 units to the right and up 5. The point is in quadrant I.
3. To plot $(-3, -1/2)$, start at the origin and go 3 units to the left and down 1/2 unit. The point is in quadrant III.
5. To plot $(0, 4)$, start at the origin and go 4 units upward. The point is on the y-axis.
7. To plot $(\pi, -1)$, start at the origin and go approximately 3.14 units to the right and 1 unit downward. The point is in quadrant IV.
9. To plot $(-4, 3)$, start at the origin and go 4 units to the left and 3 units upward. The point is in quadrant II.
11. To plot $(\pi/2, 0)$, start at the origin and go approximately 1.57 units to the right. The point is on the x-axis.
13. To plot $(0, -1)$, start at the origin and go 1 unit downward. The point is on the y-axis.
15. To plot $(0, 0)$ place a dot at the intersection of the x-axis and y-axis.
17. $y = x + 1$

If $x = 0$, $y = 0 + 1 = 1$.

If $x = 1$, $y = 1 + 1 = 2$.

If $x = 2$, $y = 2 + 1 = 3$.

If $x = -1$, $y = -1 + 1 = 0$.

Plot $(0, 1)$, $(1, 2)$, $(2, 3)$, $(-1, 0)$, and draw a line through the points.

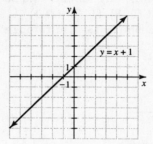

19. $y = -2x + 3$
If $x = 0$, $y = -2(0) + 3 = 3$.
If $x = 1$, $y = -2(1) + 3 = 1$.
If $x = 2$, $y = -2(2) + 3 = -1$.
If $x = -1$, $y = -2(-1) + 3 = 5$.
Plot $(0, 3)$, $(1, 1)$, $(2, -1)$, and $(-1, 5)$ and draw a line through the points.

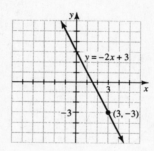

21. $y = x$ Since the x and y-coordinates are equal, plot $(0, 0)$, $(1, 1)$, $(2, 2)$, $(-1, -1)$ and draw a line through the points.

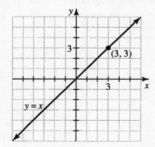

23. $y = 3$
The x-coordinate can be any number and the y-coordinate is 3. Plot (0, 3), (1, 3), (2, 3), (−1, 3), and draw a horizontal line through the points.

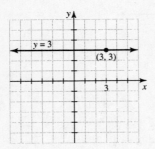

25. For the equation $y = 1 - x$, plot the points (0, 1), (1, 0), (2, −1), and (−2, 3). Draw a line through the points.

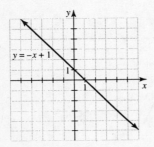

27. For $x = 2$, the y-coordinate can be any number, but the x-coordinate is 2. Plot (2, 0), (2, 1), (2, 3), (2, −1), and draw a vertical line through the points.

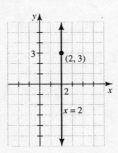

29. For the equation $y = \frac{1}{2}x - 1$, plot the points (0, −1), (2, 0), (4, 1), and (−4, −3). Draw a line through the points.

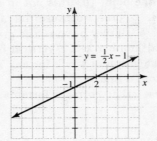

31. $x - 4 = 0$
$\qquad x = 4$
Plot (4, 5), (4, 2), (4, −1), (4, 1), and draw a line through the points.

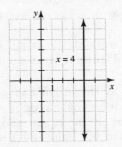

33. $3x + y = 5$
$\qquad y = -3x + 5$
Plot (0, 5), (1, 2), (2, −1), (−1, 8), and draw a line through the points.

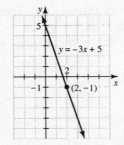

35. $2x + y = 1$

$\qquad y = -2x + 1$

Plot the points $(0, 1)$, $(1, -1)$, $(2, -3)$, $(-1, 3)$.
Draw a line through the points.

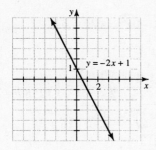

37. $x - y = -3$

$\qquad y = x + 3$

Plot the points $(0, 3)$, $(1, 4)$, $(2, 5)$, $(-1, 2)$.
Draw a line through the points.

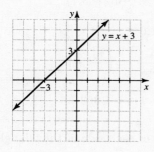

39. $y = 1.35x - 4.27$

If $x = -1$, $y = 1.35(-1) - 4.27 = -5.62$

If $x = 0$, $y = -4.27$

If $x = 2$, $y = 1.35(2) - 4.27 = -1.57$

Plot $(-1, -5.62)$, $(0, -4.27)$, $(2, -1.57)$, and
draw a line through the points.

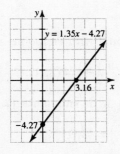

41. If $x = 0$, $\quad 4(0) - 3y = 12$

$\qquad\qquad\qquad -3y = 12$

$\qquad\qquad\qquad\qquad y = -4$

If $y = 0$, $\quad 4x - 3(0) = 12$

$\qquad\qquad\qquad 4x = 12$

$\qquad\qquad\qquad\quad x = 3$

Plot $(0, -4)$ and $(3, 0)$.

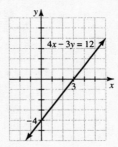

43. If $x = 0$, $\quad 0 - y + 5 = 0$

$\qquad\qquad\qquad\qquad 5 = y$

If $y = 0$, $\qquad x - 0 + 5 = 0$

$\qquad\qquad\qquad\qquad x = -5$

Plot $(0, 5)$ and $(-5, 0)$, and draw a line through
the points.

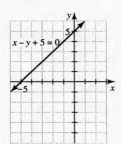

45. If $x = 0$, $2(0) + 3y = 5$

$\qquad\qquad\qquad\qquad y = 5/3$

If $y = 0$, $\quad 2x + 3(0) = 5$

$\qquad\qquad\qquad\quad x = 5/2$

Draw a line through $(0, 5/3)$ and $(5/2, 0)$.

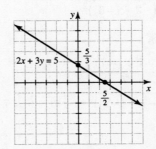

47. If $x = 0$, $y = \frac{3}{5}(0) + \frac{2}{3}$, or $y = \frac{2}{3}$. If $y = 0$,

$$0 = \frac{3}{5}x + \frac{2}{3}$$
$$-\frac{3}{5}x = \frac{2}{3}$$
$$x = -\frac{10}{9}$$

Plot $(0, 2/3)$ and $(-10/9, 0)$, and draw a line through the points.

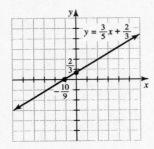

49. If $x = 2$, then $y = -3(2) + 6 = 0$.

If $y = -3$, then $-3 = -3x + 6$, or $x = 3$.

The points are $(2, 0)$ and $(3, -3)$.

51. If $x = -4$, then $\frac{1}{2}(-4) - \frac{1}{3}y = 9$, or

$y = -33$. If $y = 6$, then $\frac{1}{2}x - \frac{1}{3} \cdot 6 = 9$, or $x = 22$.

The points are $(-4, -33)$ and $(22, 6)$.

53. $M = \left(\frac{6+4}{2}, \frac{5+2}{2}\right) = \left(5, \frac{7}{2}\right)$

$L = \sqrt{(6-4)^2 + (5-2)^2} = \sqrt{4+9} = \sqrt{13}$

55. $M = \left(\frac{3+1}{2}, \frac{5+(-3)}{2}\right) = (2, 1)$

$L = \sqrt{(3-1)^2 + (5-(-3))^2} = \sqrt{4+64}$

$$= \sqrt{68} = \sqrt{4}\sqrt{17} = 2\sqrt{17}$$

57. $M = \left(\frac{4+(-3)}{2}, \frac{-2+(-6)}{2}\right) = \left(\frac{1}{2}, -4\right)$

$L = \sqrt{((4-(-3))^2 + (-2-(-6))^2}$

$$= \sqrt{49+16} = \sqrt{65}$$

59. $M = \left(\frac{2+(-5)}{2}, \frac{-7+(-7)}{2}\right) = \left(-\frac{3}{2}, -7\right)$

$L = \sqrt{(2-(-5))^2 + (-7-(-7))^2} = \sqrt{49+0} = 7$

61. $M = \left(\frac{0+4}{2}, \frac{-3+0}{2}\right) = \left(2, -\frac{3}{2}\right)$

$L = \sqrt{(0-4)^2 + (-3-0)^2} = \sqrt{16+9} = \sqrt{25} = 5$

63. $M = \left(\frac{0+(-6)}{2}, \frac{0+8}{2}\right) = (-3, 4)$

$L = \sqrt{(-6-0)^2 + (8-0)^2} = \sqrt{36+64}$

$$= \sqrt{100} = 10$$

65. $M = \left(\frac{1/2+1/3}{2}, \frac{1/4+1/2}{2}\right) = \left(\frac{5}{12}, \frac{3}{8}\right)$

$L = \sqrt{\left(\frac{1}{2}-\frac{1}{3}\right)^2 + \left(\frac{1}{4}-\frac{1}{2}\right)^2} = \sqrt{\frac{1}{36}+\frac{1}{16}}$

$$= \sqrt{\frac{13}{144}} = \frac{\sqrt{13}}{12}$$

67. $M = \left(\frac{3.5+(-1.4)}{2}, \frac{-6.2+3.7}{2}\right)$

$$= (1.05, -1.25)$$

$L = \sqrt{(3.5-(-1.4))^2 + (-6.2-3.7)^2}$

$$= \sqrt{122.02} = 11.046$$

69. $M = \left(\frac{\frac{\pi}{2}+\pi}{2}, \frac{0+1}{2}\right) = \left(\frac{3\pi}{4}, \frac{1}{2}\right)$

$L = \sqrt{(\frac{\pi}{2}-\pi)^2 + (0-1)^2} = 1.862$

71. Find the length of each side of the triangle:

$$\sqrt{(-1-0)^2 + (-1-6)^2} = \sqrt{50}$$

$$\sqrt{(-1-5)^2 + (-1-5)^2} = \sqrt{72}$$

$$\sqrt{(0-5)^2 + (6-5)^2} = \sqrt{26}$$

Since all three sides have different lengths, the triangle is not isosceles.

73. Midpoint of the diagonal with endpoints

$(0, 0)$ and $(6, 2)$ is $\left(\frac{0+6}{2}, \frac{0+2}{2}\right) = (3, 1)$.

Midpoint of the diagonal with endpoints $(2, 2)$

and $(4, 0)$ is $\left(\frac{2+4}{2}, \frac{2+0}{2}\right) = (3, 1)$.

Since the midpoints are the same, the diagonals bisect each other.

75. Find the lengths of the sides.

$$\sqrt{(4-0)^2 + (4-8)^2} = \sqrt{32}$$

$$\sqrt{(4-(-2))^2 + (4-(-2))^2} = \sqrt{72}$$

$$\sqrt{(-2-0)^2 + (-2-8)^2} = \sqrt{104}$$

Check the Pythagorean theorem:

$$(\sqrt{32})^2 + (\sqrt{72})^2 = 32 + 72 = 104$$

Since the square of the longest side is 104, the triangle is a right triangle.

77. Find the length of each segment:

$$\sqrt{(-2-1)^2 + (-13-(-4))^2} = \sqrt{90} = 3\sqrt{10}$$

$$\sqrt{(1-3)^2 + (-4-2)^2} = \sqrt{40} = 2\sqrt{10}$$

$$\sqrt{(-2-3)^2 + (-13-2)^2} = \sqrt{250} = 5\sqrt{10}$$

Since $3\sqrt{10} + 2\sqrt{10} = 5\sqrt{10}$, the three points are on a straight line.

79. Plot the 4 points and find the length of each side of the quadrilateral:

$$\sqrt{(1-3)^2 + (5-9)^2} = \sqrt{20}$$

$$\sqrt{(3-4)^2 + (9-7)^2} = \sqrt{5}$$

$$\sqrt{(4-2)^2 + (7-3)^2} = \sqrt{20}$$

$$\sqrt{(1-2)^2 + (5-3)^2} = \sqrt{5}$$

Since the opposite sides are equal in length, the quadrilateral is a parallelogram.

81. Plot the 4 points and find the length of each side of the quadrilateral.

$$\sqrt{(-2-(-1))^2 + (4-6)^2} = \sqrt{5}$$

$$\sqrt{(-1-5)^2 + (6-3)^2} = \sqrt{45}$$

$$\sqrt{(5-4)^2 + (3-1)^2} = \sqrt{5}$$

$$\sqrt{(-2-4)^2 + (4-1)^2} = \sqrt{45}$$

Since the opposite sides are equal in length it is a parallelogram.
Find the length of the diagonal:

$$\sqrt{(-2-5)^2 + (4-3)^2} = \sqrt{50}$$

Since $(\sqrt{5})^2 + (\sqrt{45})^2 = (\sqrt{50})^2$ is correct, we can conclude that there is a right angle in the parallelogram and the parallelogram is a rectangle.

83. $P = 12{,}463 + 720n$
If $n = 0$, $P = 12{,}463 + 720(0) = 12{,}463$
If $n = 5$, $P = 12{,}463 + 720(5) = 16{,}063$
If $n = 15$, $P = 12{,}463 + 720(15) = 23{,}263$
Z28 lists for \$12,463 in 1985, \$16,063 in 1990, and \$23,263 in 2000.

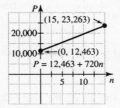

85. $C = 0.26m + 42$
If $m = 400$, $C = 0.26(400) + 42 = 146$.
The charge for a car driven 400 miles is \$146.
If $m = 0$, $C = 0.26(0) + 42 = 42$.
If $m = 1000$, $C = 0.26(1000) + 42 = 302$.
Plot $(0, 42)$ and $(1000, 302)$, and draw a line segment with those endpoints.

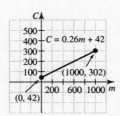

7.2 WARM-UPS

1. True.   **2.** False, slope is rise divided by run.   **3.** False, because it is horizontal line with 0 slope.   **4.** True, because it is a vertical line.   **5.** False, slope of a line can be any real number.   **6.** False, the slope is 2/5.   **7.** False, it has slope 1.   **8.** True.   **9.** False, lines with slope 2/3 and −3/2 are perpendicular.   **10.** False, because two vertical parallel lines do not have equal slopes.

7.2 EXERCISES

1. In going from $(-3, 0)$ to $(0, 2)$ we rise 2 and run 3. The slope is 2/3.

3. The slope for this line is undefined because it is a vertical line.

5. In going from $(0, 0)$ to $(1, -1)$, we rise −1 and run 1. The slope is −1.

7. In going from $(0, -3)$ to $(2, 0)$ we rise 3 and run 2. The slope is 3/2.

9. In going from $(0, 2)$ to $(2, 0)$ we rise −2 and run 2. The slope is −1.

11. $m = \frac{6-1}{2-5} = -\frac{5}{3}$

13. $m = \frac{-1-3}{-3-4} = \frac{-4}{-7} = \frac{4}{7}$

15. $m = \frac{2-7}{-2-(-1)} = \frac{-5}{-1} = 5$

17. $m = \frac{-5-0}{3-0} = \frac{-5}{3} = -\frac{5}{3}$

19. $m = \frac{3-0}{0-5} = -\frac{3}{5}$

21. $m = \frac{-1-(-\frac{1}{2})}{\frac{3}{4}-(-\frac{1}{2})} = \frac{-\frac{1}{2}}{\frac{5}{4}} = -\frac{1}{2} \cdot \frac{4}{5} = -\frac{2}{5}$

23. $m = \frac{212-209}{6-7} = \frac{3}{-1} = -3$

25. $m = \frac{7-7}{4-(-12)} = \frac{0}{16} = 0$

27. $\frac{6-(-6)}{2-2} = \frac{12}{0}$ The slope is undefined because 0 is in the denominator.

29. $m = \frac{11.9 - 8.4}{24.3 - 3.57} = \frac{3.5}{20.73} = 0.169$

31. $m = \frac{3-5}{\sqrt{2} - \frac{\sqrt{2}}{2}} = \frac{-2}{0.7071} = -2.828$

33. $m = \frac{1-0}{\frac{\pi}{4} - \frac{\pi}{2}} = \frac{1}{-0.7854} = -1.273$

35. The line through $(-5, 1)$ and $(3, -2)$ has slope
$$m = \frac{1-(-2)}{-5-3} = \frac{3}{-8} = -\frac{3}{8}.$$
Any line perpendicular to this line has slope 8/3 because 8/3 is the opposite of the reciprocal of −3/8. Line l has slope 8/3. Draw line l through $(3, 4)$ using a slope of 8/3.

37. The slope of the line through $(-3, -2)$ and $(4, 1)$ is
$$m = \frac{-2-1}{-3-4} = \frac{3}{7}.$$
Any line parallel to this line also has slope 3/7. So line l has slope 3/7.

39. The opposite of the reciprocal of 4/5 is −5/4. Line l has slope −5/4. Draw a line with slope 4/5 through the origin and a line with slope −5/4 through the origin to see that they look perpendicular.

41. Draw a quadrilateral with the four given points as vertices. Find the slope of each side.

$m_1 = \frac{3-1}{0-(-6)} = \frac{1}{3}$ $m_2 = \frac{3-1}{0-4} = -\frac{1}{2}$

$m_3 = \frac{1-(-1)}{4-(-2)} = \frac{1}{3}$ $m_4 = \frac{-1-1}{-2-(-6)} = -\frac{1}{2}$

Since opposite sides of the quadrilateral have equal slopes, the quadrilateral is a parallelogram.

43. Draw a quadrilateral with the four given points as vertices. Find the slope of each side.

$$m_1 = \frac{6-2}{3-(-3)} = \frac{2}{3} \qquad m_2 = \frac{6-4}{3-6} = -\frac{2}{3}$$

$$m_3 = \frac{4-(-1)}{6-(-1)} = \frac{5}{7} \qquad m_4 = \frac{-1-2}{-1-(-3)} = -\frac{3}{2}$$

Since there are no parallel sides, this quadrilateral is not a trapezoid.

45. Draw a quadrilateral with the four given points as vertices. Find the slope of each side.

$$m_1 = \frac{6-4}{0-(-4)} = \frac{1}{2} \qquad m_2 = \frac{6-0}{0-3} = -2$$

$$m_3 = \frac{0-(-2)}{3-(-1)} = \frac{1}{2} \qquad m_4 = \frac{-2-4}{-1-(-4)} = -2$$

From the slopes, we see that opposite sides are parallel and adjacent sides are perpendicular. The quadrilateral is a rectangle.

47. Draw a triangle with the three given points as vertices. Find the slope of each side.

$$m_1 = \frac{6-3}{-1-(-3)} = \frac{3}{2} \qquad m_2 = \frac{6-0}{-1-0} = -6$$

$$m_3 = \frac{3-0}{-3-0} = -1$$

From the slopes we see that none of the line segments are perpendicular. The triangle is not a right triangle.

49.
$$m = \frac{20,115 - 12,674}{1993 - 1985} = 930.125$$

The slope is the average yearly increase in price. In the year 2000 the car will cost
$20,115 + 7(930.125) \approx \$26,626$

51. The slope of the line perpendicular to the line with slope 0.247 is the opposite of the reciprocal of 0.247:
$$-\frac{1}{0.247} = -4.049$$

53. Find the slope:
$$\frac{k-(-5)}{2-(-3)} = \frac{1}{2}$$

$$\frac{k+5}{5} = \frac{1}{2}$$
$$2k + 10 = 5$$
$$k = -\frac{5}{2}$$

59. Draw a quadrilateral with the four given vertices. The slope of the diagonal joining $(-3, -1)$ and $(5, 3)$ is $m_1 = \frac{3-(-1)}{5-(-3)} = \frac{1}{2}$.
The slope of the diagonal joining $(0, 3)$ and $(2, -1)$ is $m_2 = \frac{3-(-1)}{0-2} = -2$.
Since the opposite of the reciprocal of 1/2 is -2, the diagonals are perpendicular.

7.3 WARM-UPS

1. True. **2.** False, the line through (a, b) with slope m has equation $y - b = m(x - a)$. **3.** False, $y = mx + b$ goes through $(0, b)$ and has slope m. **4.** True, because the y-intercept has x-coordinate 0. **5.** True, because the x-intercept has y-coordinate 0. **6.** False, because vertical lines do not have equations in slope intercept form. **7.** True, because if we solve the equation for y, we get $y = (-3/2)x + (7/2)$. **8.** False, because a line perpendicular to a line with slope 3 has slope $-1/3$. **9.** False, because if $x = 0$ in this equation, we get $y = 5/2$. **10.** True, because even the vertical lines can be expressed in standard form.

7.3 EXERCISES

1. Use the given point and the given slope in the point-slope formula:
$$y - (-3) = 2(x - 2)$$
$$y + 3 = 2x - 4$$
$$y = 2x - 7$$

3. Use the point $(-2, 3)$ and the slope $-1/2$ in the point-slope formula:
$$y - 3 = -\frac{1}{2}(x - (-2))$$
$$y - 3 = -\frac{1}{2}x - 1$$
$$y = -\frac{1}{2}x + 2$$

5. The slope of the line through (2, 3) and (−5, 6) is
$$m = \frac{3-6}{2-(-5)} = -\frac{3}{7}.$$

Use (2, 3) and slope −3/7 in the point-slope formula:
$$y - 3 = -\frac{3}{7}(x-2)$$
$$y - 3 = -\frac{3}{7}x + \frac{6}{7}$$
$$y = -\frac{3}{7}x + \frac{6}{7} + 3$$
$$y = -\frac{3}{7}x + \frac{27}{7}$$

7. The slope of the line through (−3, 1) and (5, −1) is
$$m = \frac{1-(-1)}{-3-5} = -\frac{1}{4}.$$

So the line we want has slope 4 and goes through (3, 4):
$$y - 4 = 4(x-3)$$
$$y - 4 = 4x - 12$$
$$y = 4x - 8$$

9. The line through (9, −3) and (−3, 6) has slope
$$m = \frac{6-(-3)}{-3-9} = -\frac{3}{4}$$

Use the same slope −3/4 and (0, 0) in the point-slope formula:
$$y - 0 = -\frac{3}{4}(x-0)$$
$$y = -\frac{3}{4}x$$

11. The line goes through (0, 2) and (2, 3). The slope is $\frac{1}{2}$ and the y-intercept is (0, 2). Using slope-intercept form, we can write $y = \frac{1}{2}x + 2$.

13. This is a vertical line with an x-intercept of (1, 0) and so its equation is $x = 1$.

15. This line has a y-intercept of (0, 0) and also goes through (3, −3). Its slope is −1. Its equation in slope-intercept form is $y = -x$.

17. This line goes through (2, 0) and (0, −3). Its slope is 3/2 and its equation is $y = \frac{3}{2}x - 3$.

19. This line goes through (0, 2) and (2, 0). Since its slope is −1, its equation is $y = -x + 2$.

21.
$$3(y) = 3(\tfrac{1}{3}x - 2)$$
$$3y = x - 6$$
$$-x + 3y = -6$$
$$x - 3y = 6$$

23.
$$2(y-5) = 2 \cdot \tfrac{1}{2}(x+3)$$
$$2y - 10 = x + 3$$
$$-x + 2y = 13$$
$$x - 2y = -13$$

25.
$$6\left(y + \tfrac{1}{2}\right) = 6 \cdot \tfrac{1}{3}(x-4)$$
$$6y + 3 = 2(x-4)$$
$$-2x + 6y = -11$$
$$2x - 6y = 11$$

27.
$$100(0.05x + 0.06y - 8.9) = 100(0)$$
$$5x + 6y - 890 = 0$$
$$5x + 6y = 890$$

29.
$$y + 55 = -0.02x + 12$$
$$0.02x + y = -43$$
$$50(0.02x + y) = 50(-43)$$
$$x + 50y = -2150$$

31.
$$2x + 5y = 1$$
$$5y = -2x + 1$$
$$y = -\frac{2}{5}x + \frac{1}{5}$$
The slope is $-\frac{2}{5}$ and the y-intercept is $(0, \frac{1}{5})$.

33.
$$3x - y - 2 = 0$$
$$3x - 2 = y$$
The slope is 3 and the y-intercept is (0, −2).

35.
$$y + 3 = 5$$
$$y = 2$$
The slope is 0 and the y-intercept is (0, 2).

37.
$$y - 2 = 3(x-1)$$
$$y - 2 = 3x - 3$$
$$y = 3x - 1$$
The slope is 3 and the y-intercept is (0, −1).

39.
$$\frac{y-5}{x+4} = \frac{3}{2}$$
$$2(y-5) = 3(x+4)$$
$$2y - 10 = 3x + 12$$
$$2y = 3x + 22$$
$$y = \frac{3}{2}x + 11$$
The slope is $\frac{3}{2}$ and the y-intercept is (0, 11).

41.
$$y - \frac{1}{2} = \frac{1}{3}\left(x + \frac{1}{4}\right)$$
$$y - \frac{1}{2} = \frac{1}{3}x + \frac{1}{12}$$
$$y = \frac{1}{3}x + \frac{1}{12} + \frac{1}{2}$$
$$y = \frac{1}{3}x + \frac{7}{12}$$

The slope is $\frac{1}{3}$ and the y-intercept is $(0, \frac{7}{12})$.

43.
$$y - 6000 = 0.01(x + 5700)$$
$$y - 6000 = 0.01x + 57$$
$$y = 0.01x + 6057$$

The slope is 0.01 and the y-intercept is (0, 6057).

45.
$$10(0.3x - 0.2y) = 10(10)$$
$$3x - 2y = 100$$
$$-2y = -3x + 100$$
$$y = \frac{3}{2}x - 50$$

The slope is $\frac{3}{2}$ and the y-intercept is $(0, -50)$.

47. Since the y-intercept is (0, 5) and the slope is 1/2, we can use slope-intercept form to write the equation.
$$y = \frac{1}{2}x + 5$$
To get integral coefficients multiply by 2.
$$2y = x + 10$$
$$-x + 2y = 10$$
$$x - 2y = -10 \quad \text{(Standard form)}$$

49. The slope of the line through (2, 0) and (0, 4) is
$$m = \frac{4 - 0}{0 - 2} = -2.$$
Use slope-intercept form to write the equation of the line with slope -2 and y-intercept (0, 4):
$$y = -2x + 4$$
$$2x + y = 4 \quad \text{(Standard form)}$$

51. Any line parallel to $y = 2x + 6$ has slope 2. Use point-slope formula to find the equation of a line with slope 2 and going through $(-2, 1)$:
$$y - 1 = 2(x - (-2))$$
$$y - 1 = 2x + 4$$
$$-2x + y = 5$$
$$2x - y = -5 \quad \text{(Standard form)}$$

53. Write $2x + 4y = 1$ in slope-intercept form to identify the slope:
$$4y = -2x + 1$$
$$y = -\frac{1}{2}x + \frac{1}{4}$$

Any line parallel to $2x + 4y = 1$ has slope $-1/2$. Use the point-slope formula to find the equation

of the line with slope $-1/2$ through $(-3, 5)$:
$$y - 5 = -\frac{1}{2}(x - (-3))$$
$$y - 5 = -\frac{1}{2}x - \frac{3}{2}$$
$$2y - 10 = -x - 3$$
$$x + 2y = 7 \quad \text{(Standard form)}$$

55. A line perpendicular to $y = (1/2)x - 3$ has slope -2. Use the point-slope formula to find the equation of a line with slope -2 going through (1, 1):
$$y - 1 = -2(x - 1)$$
$$y - 1 = -2x + 2$$
$$2x + y = 3 \quad \text{(Standard form)}$$

57. Write $x + 3y = 4$ in slope-intercept form to determine its slope:
$$3y = -x + 4$$
$$y = -\frac{1}{3}x + \frac{4}{3}$$

The slope of $x + 3y = 4$ is $-1/3$. Any line perpendicular to $x + 3y = 4$ has slope 3. Use the point-slope formula to find the equation of the line with slope 3 through $(-2, 3)$.
$$y - 3 = 3(x - (-2))$$
$$y - 3 = 3x + 6$$
$$-3x + y = 9$$
$$3x - y = -9 \quad \text{(Standard form)}$$

59. Any line parallel to the x-axis has slope 0. If it goes through (2, 5), it has y-intercept (0, 5). Using slope-intercept form we get
$$y = 0 \cdot x + 5$$
$$y = 5$$

61. The line $x = 3$ is a vertical line. Its x-intercept is (3, 0). A vertical line through $(-3, -2)$ has an x-intercept of $(-3, 0)$ and equation $x = -3$.

63. The slope of $y = (1/2)x$ is 1/2 and the y-intercept is (0, 0). To graph the line start at (0, 0) and go up 1 unit and 2 units to the right to find another point on the line, (2, 1). Draw the line through (0, 0) and (2, 1).

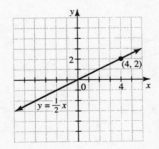

65. The slope of $y = 2x - 3$ is 2 and the y-intercept is $(0, -3)$. The slope of 2 is obtained from 2/1. To graph the line start at $(0, -3)$ and rise 2 units and go 1 unit to the right to get to the point $(1, -1)$. Draw a line through the points $(0, -3)$ and $(1, -1)$.

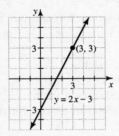

67. The slope of $y = (-2/3)x + 2$ is $-2/3$ and the y-intercept is $(0, 2)$. To graph the line start at $(0, 2)$ and go down 2 units and then 3 units to the right to locate the point $(3, 0)$. Draw a line through $(0, 2)$ and $(3, 0)$.

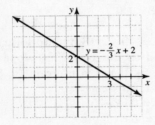

69. The equation $3y + x = 0$ can be written as $y = (-1/3)x$. Its slope is $-1/3$ and its y-intercept is $(0, 0)$. To graph the line start at $(0, 0)$ and go 1 unit down and 3 units to the right to locate the point $(3, -1)$. Draw a line through $(0, 0)$ and $(3, -1)$.

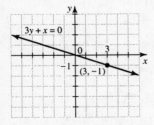

71. Write the equation as $y = x - 3$ to see that the slope is 1 and the y-intercept is $(0, -3)$. Since $1 = 1/1$, we start at $(0, -3)$ and go up 1 unit and 1 unit to the right to locate the point $(1, -2)$. Draw a line through $(0, -3)$ and $(1, -2)$.

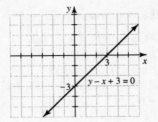

73. Solve the equation for y:
$$3x - 2y = 6$$
$$-2y = -3x + 6$$
$$y = \frac{3}{2}x - 3$$

The slope is 3/2 and the y-intercept is $(0, -3)$. Start at $(0, -3)$ and go 3 units upward and then 2 units to the right to locate the point $(2, 0)$. Draw a line through $(0, -3)$ and $(2, 0)$.

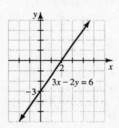

75. Solve the equation for y: $y + 2 = 0$, or $y = -2$. The slope is 0 and the y-intercept is $(0, -2)$. Draw a horizontal line through $(0, -2)$.

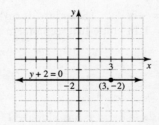

128

77. Solve the equation for x:
$$x + 3 = 0$$
$$x = -3$$
There is no slope-intercept form for this line because it is a vertical line. The x-intercept is $(-3, 0)$. Draw a vertical line through $(-3, 0)$.

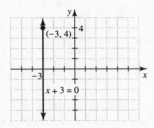

79. Solve $x + 3y = 7$ for y to get $y = -\frac{1}{3}x + \frac{7}{3}$. The lines are perpendicular because there slopes are 3 and $-1/3$.

81. Solve $2x - 4y = 9$ for y to get $y = \frac{1}{2}x - \frac{9}{4}$. Solve $\frac{1}{3}x = \frac{2}{3}y - 8$ for y to get $y = \frac{1}{2}x + 12$. The lines are parallel because there slopes are equal.

83. Solve $x - 6 = 9$ for x to get $x = 15$. Solve $y - 4 = 12$ for y to get $y = 16$. The lines are perpendicular, because one is horizontal and the other is vertical.

85. The slope of the line through $(0, 60)$ and $(120, 200)$ is
$$m = \frac{200 - 60}{120 - 0} = \frac{140}{120} = \frac{7}{6}.$$
Using slope-intercept form with a t-intercept of $(0, 60)$ we get $t = \frac{7}{6}s + 60$.

After 30 seconds the temperature will be
$$t = \frac{7}{6}(30) + 60 = 35 + 60 = 95°F.$$

87. The slope of the line through $(8.86, 1204)$ and $(7.07, 659)$ is
$$m = \frac{1204 - 659}{8.86 - 7.07} \approx 304.47$$
Use the point-slope formula for the equation of a line with slope 304.47 and the point $(8.86, 1204)$.
$$w - 1204 = 304.47(d - 8.86)$$
$$w - 1204 = 304.47d - 2697.60$$
$$w = 304.47d - 1493.6$$
Use $d = 8.24$ in the formula to find w.
$$w = 304.47(8.24) - 1493.6 \approx 1015.23$$
When depth is 8.24 ft, the flow is 1015 ft^3/sec.

7.4 WARM-UPS

1. True, because $-3 > -3(2) + 2$.

2. False, because $3x - y > 2$ is equivalent to $y < 3x - 2$, which is below the line $3x - y = 2$.

3. True, because $3x + y < 5$ is equivalent to $y < -3x + 5$.

4. False, the region $x < -3$ is to the left of the vertical line $x = -3$.

5. True, because the word "and" is used.

6. True, because the word "or" is used.

7. False, because $(2, -5)$ does not satisfy the inequality $y > -3x + 5$. Note that $-5 > -3(2) + 5$ is incorrect.

8. True, because $(-3, 2)$ satisfies $y \leq x + 5$.

9. False, it is equivalent to the compound inequality $2x - y \leq 4$ and $2x - y \geq -4$.

10. True, because in general $|x| > k$ (for a positive k) is equivalent to $x > k$ or $x < -k$.

7.4 EXERCISES

1. Graph the equation $y = x + 2$. Use its slope of 1 and y-intercept of $(0, 2)$. The **graph of** $y < x + 2$ is the region below the line $y = x + 2$. The line is dashed because of the inequality symbol.

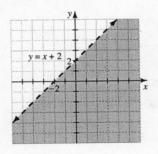

3. First graph the line $y = -2x + 1$. Start at its y-intercept $(0, 1)$ and use a slope of -2. The graph of the inequality $y \leq -2x + 1$ is the region below the line and including the line. For this reason the line is drawn solid.

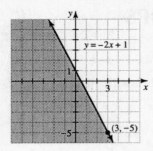

line with y-intercept $(0, -2)$ and slope $3/4$. Since the inequality symbol is $\geq$, shade the region above the line.

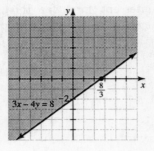

5. The inequality $x + y > 3$ is equivalent to $y > -x + 3$. First use slope of -1 and a y-intercept of $(0, 3)$ to graph $y = -x + 3$. The graph of $y > -x + 3$ is the region above the line. The line is drawn dashed because of the inequality symbol.

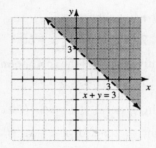

11. If we solve $x - y > 0$ for y we get $y < x$. To graph $y = x$, use a y-intercept of $(0, 0)$ and a slope of 1. Draw a dashed line and shade below the line for the graph of $y < x$.

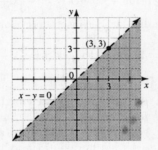

7. The inequality $2x + 3y < 9$ is equivalent to $y < -\frac{2}{3}x + 3$. Use a slope of $-2/3$ and a y-intercept of 3 to graph the line $y = -\frac{2}{3}x + 3$. Shade the region below the line.

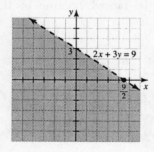

13. The graph of $x \geq 1$ consists of the vertical line $x = 1$ together with the region to the right of the line. Draw a solid vertical line through $(1, 0)$ and shade the region to the right.

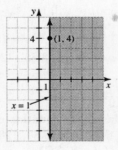

9. Solve the inequality for y:

$$3x - 4y \leq 8$$

$$-4y \leq -3x + 8$$

$$y \geq \frac{3}{4}x - 2$$

To graph the equation $y = \frac{3}{4}x - 2$, draw a solid

15. The graph of $y < 3$ is the region below the horizontal line $y = 3$. Draw a dashed horizontal line through $(0, 3)$ and shade the region below.

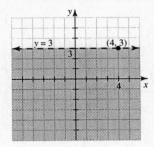

17. First find the x and y-intercepts to determine the graph of the line $2x - 3y = 5$.
If $x = 0$, $2(0) - 3y = 5$ or $y = -5/3$.
If $y = 0$, $2x - 3(0) = 5$ or $x = 5/2$.
Draw a dashed line through $(0, -5/3)$ and $(5/2, 0)$. Test $(0, 0)$: $2(0) - 3(0) < 5$.
Since $(0, 0)$ satisfies the inequality, we shade the side of the line that includes $(0, 0)$.

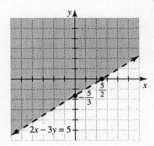

19. Graph $x + y + 3 = 0$ by using the x and y-intercepts, $(0, -3)$ and $(-3, 0)$. Draw a solid line through these two points. Test $(0, 0)$ in the inequality: $0 + 0 + 3 \geq 0$
Since $(0, 0)$ satisfies the inequality, we shade the side of the line that includes $(0, 0)$.

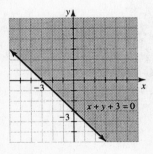

21. Find x and y-intercepts for $\frac{1}{2}x + \frac{1}{3}y = 1$:

If $x = 0$, we get $\frac{1}{3}y = 1$ or $y = 3$.
If $y = 0$, we get $\frac{1}{2}x = 1$ or $x = 2$.

Draw a dashed line through the intercepts $(0, 3)$ and $(2, 0)$. Test the point $(0, 0)$ in the inequality:
$$\frac{1}{2}(0) + \frac{1}{3}(0) < 1$$
Since $(0, 0)$ satisfies the inequality, we shade the region that includes $(0, 0)$.

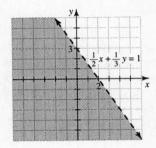

23. To graph $y > x$ and $y > -2x + 3$, we first draw dashed lines for $y = x$ and for $y = -2x + 3$. Test one point in each of the four regions to see if it satisfies the compound inequality. Test $(5, 0)$, $(0, 5)$, $(-5, 0)$ and $(0, -5)$. Only $(0, 5)$ satisfies both inequalities. So shade the region containing $(0, 5)$.

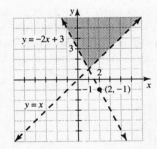

25. First graph the equations $y = x + 3$ and $y = -x + 2$. Test the points $(0, 5)$, $(5, 0)$, $(0, -5)$, and $(-5, 0)$ in the compound inequality. Only $(-5, 0)$ fails to satisfy the compound inequality. So shade all regions except the one containing $(-5, 0)$.

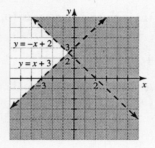

27. First graph the equations $x + y = 5$ and $x - y = 3$. Test one point in each of the 4 regions. Only points in the region containing $(0, 0)$ satisfy both inequalities. Shade that region including the boundary lines.

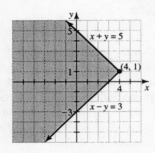

29. Graph the equations $x - 2y = 4$ and $2x - 3y = 6$. Only the region containing the point $(0, -5)$ fails to satisfy the compound inequality. Shade that region including the boundary lines.

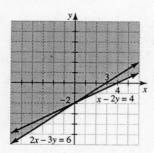

31. Graph the horizontal line $y = 2$ and the vertical line $x = 3$. Only points in the region containing $(0, 5)$ satisfy both inequalities. Shade that region with dashed boundary lines.

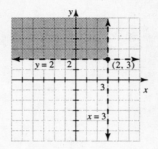

33. Graph $y = x$ and $x = 2$. Only points in the region containing $(0, 5)$ satisfy both inequalities. Shade that region and include the boundary lines.

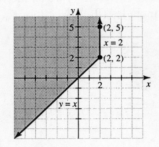

35. Graph $2x = y + 3$ and $y = 2 - x$. Only points in the region containing $(0, -5)$ fail to satisfy the compound inequality. Shade all regions except the one containing $(0, -5)$. Use dashed lines for the boundaries because of the inequality symbols.

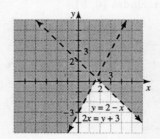

132

37. Graph the lines $y = x - 1$ and $y = x + 3$. Only points in the region containing $(0, 0)$ satisfy the compound inequality. Shade that region and use dashed boundary lines.

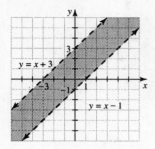

39. Graph the lines $y = 0$, $y = x$, and $x = 1$. Only points inside the triangular region bounded by the lines satisfy the compound inequality. Shade that region and use solid boundary lines.

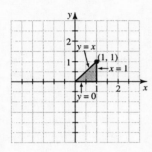

41. Graph $x = 1$, $x = 3$, $y = 2$, and $y = 5$. Only points inside the rectangular region satisfy the compound inequality. Shade that region and include the boundary lines.

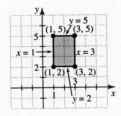

43. Graph the equations $x + y = 2$ and $x + y = -2$. Only points between these two parallel lines satisfy the absolute value inequality.

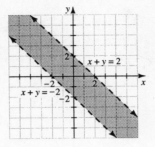

45. Graph the parallel lines $2x + y = 1$ and $2x + y = -1$. Points between the lines do not satisfy the absolute value inequality. So shade the other two regions.

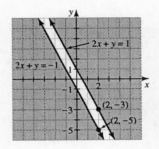

47. Rewrite the inequality:

$$x - y - 3 > 5 \qquad \text{or} \qquad x - y - 3 < -5$$
$$-y > -x + 8 \quad \text{or} \qquad -y < -x - 2$$
$$y < x - 8 \qquad \text{or} \qquad y > x + 2$$

Graph the lines $y = x - 8$ and $y = x + 2$. Points above $y = x + 2$ together with points below $y = x - 8$ satisfy the absolute value inequality.

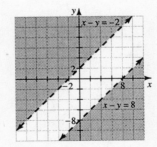

49. Graph the parallel lines $x - 2y = 4$ and $x - 2y = -4$. Points in the region between the lines satisfy the inequality.

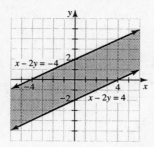

51. Graph the vertical lines $x = 2$ and $x = -2$. Points in the region between the lines do not satisfy the inequality but points in the other two regions do.

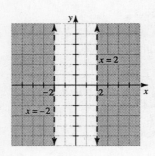

53. Graph the horizontal lines $y = 1$ and $y = -1$. Only points in the region between the lines satisfy $|y| < 1$.

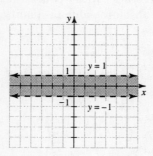

55. Graph the equation $y = |x|$. Only points in the region below that graph satisfy $y < |x|$.

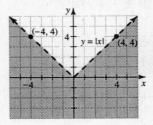

57. Graph the lines $x = 2$, $x = -2$, $y = 3$, and $y = -3$. Only points inside the rectangular region bounded by the lines satisfy the compound inequality $|x| < 2$ and $|y| < 3$.

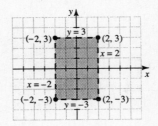

59. The inequality $|x - 3| < 1$ is equivalent to $2 < x < 4$. The inequality $|y - 2| < 1$ is equivalent to $1 < y < 3$. Graph the lines $x = 2$, $x = 4$, $y = 1$, and $y = 3$. Only points inside the square region bounded by these lines satisfy the inequalities $|x - 3| < 1$ and $|y - 2| < 1$.

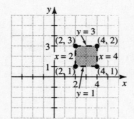

134

61. Let x = the number of compact cars and y = the number of full-size cars. We have
$$15,000x + 20,000y \leq 120,000$$
$$3x + 4y \leq 24$$
We also have $x \geq 0$ and $y \geq 0$ because they cannot purchase a negative number of cars. Graph $3x + 4y \leq 24$, $x \geq 0$, and $y \geq 0$.

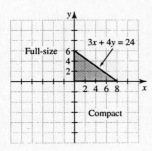

63. Graph $3x + 4y \leq 24$, $x \geq 0$, $y \geq 0$, and $y > x$:

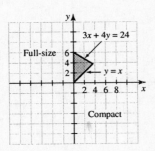

65. Graph $h \leq 187 - 0.85a$ and $h \geq 154 - 0.70a$ for $20 \leq a \leq 75$.

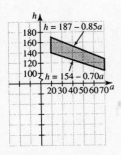

67. Let d = the number of days of newspaper advertising and t = the number times an ad is aired on TV.

$$300d + 1000t \leq 9000$$
$$3d + 10t \leq 90$$

Graph $3d + 10t \leq 90$, $d \geq 0$, and $t \geq 0$:

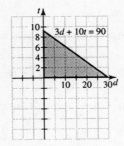

Chapter 7 Review

1. The point $(-3, -2)$ lies in quadrant III.

3. The point $(\pi, 0)$ lies on the x-axis. The quadrants do not include points on any axis.

5. The point $(0, -1)$ is on the y-axis.

7. The point $(\sqrt{2}, -3)$ is in quadrant IV.

9. $M = \left(\dfrac{2+1}{2}, \dfrac{5+3}{2}\right) = \left(\dfrac{3}{2}, 4\right)$

$L = \sqrt{(2-1)^2 + (5-3)^2} = \sqrt{1+4} = \sqrt{5}$

11. $M = \left(\dfrac{-6+0}{2}, \dfrac{0+(-8)}{2}\right) = (-3, -4)$

$L = \sqrt{(-6-0)^2 + (0-(-8))^2} = \sqrt{36+64} = 10$

13. If $x = 0$, $y = -3(0) + 2 = 2$.
The ordered pair is $(0, 2)$.
If $y = 0$, then $0 = -3x + 2$.
$$3x = 2$$
$$x = \frac{2}{3}$$
The ordered pair is $(\frac{2}{3}, 0)$.

If $x = 4$, $y = -3(4) + 2 = -10$.
The ordered pair is $(4, -10)$.
If $y = -3$, then $-3 = -3x + 2$.
$$3x = 5$$
$$x = \frac{5}{3}$$
The ordered pair is $(\frac{5}{3}, -3)$.

135

15. $m = \dfrac{9-6}{-2-(-5)} = \dfrac{3}{3} = 1$

17. $m = \dfrac{-2-1}{-3-4} = \dfrac{-3}{-7} = \dfrac{3}{7}$

19. $m = \dfrac{-1-(-4)}{5-(-3)} = \dfrac{3}{8}$

The slope of any line parallel to one with slope 3/8 is also 3/8.

21. $m = \dfrac{-6-5}{4-(-3)} = \dfrac{-11}{7} = -\dfrac{11}{7}$

The slope of any line perpendicular to one with slope $-11/7$ is 7/11.

23. The slope for $y = -3x + 4$ is -3, and the y-intercept is $(0, 4)$.

25. Write the equation in slope-intercept form.

$$y - 3 = \tfrac{2}{3}(x - 1)$$

$$y - 3 = \tfrac{2}{3}x - \tfrac{2}{3}$$

$$y = \tfrac{2}{3}x + \tfrac{7}{3}$$

The slope is $\tfrac{2}{3}$ and the y-intercept is $\left(0, \tfrac{7}{3}\right)$.

27.
$$y = 2/3 \, x - 4$$
$$3(y) = 3(\tfrac{2}{3}x - 4) \qquad 4(3) = 3(\tfrac{2}{3}x - 4)$$
$$3y =$$
$$3y = 2x - 12$$
$$-2x + 3y = -12$$
$$2x - 3y = 12 \quad \text{(Standard form)}$$

29.
$$2(y - 1) = 2 \cdot \tfrac{1}{2}(x + 3)$$
$$2y - 2 = x + 3$$
$$-x + 2y = 5$$
$$x - 2y = -5 \quad \text{(Standard form)}$$

31. Start with point-slope form:
$$y - (-3) = \tfrac{1}{2}(x - 1)$$
$$y + 3 = \tfrac{1}{2}x - \tfrac{1}{2}$$
$$2y + 6 = x - 1$$
$$-x + 2y = -7$$
$$x - 2y = 7$$

33. Start with point-slope form:
$$y - 6 = -\tfrac{3}{4}(x - (-2))$$
$$y - 6 = -\tfrac{3}{4}x - \tfrac{3}{2}$$
$$4(y - 6) = 4(-\tfrac{3}{4}x - \tfrac{3}{2})$$
$$4y - 24 = -3x - 6$$
$$3x + 4y = 18$$

35. Any line with slope 0 is a horizontal line. Since it contains $(3, 5)$ its equation is $y = 5$.

37. To graph $y = 2x - 3$ note that the y-intercept is $(0, -3)$ the slope is $2 = 2/1$. Start at $(0, -3)$ and go up 2 units and 1 unit to the right to locate the second point $(1, -1)$. Draw a line through the two points.

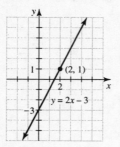

39. If $x = 0$, $3(0) - 2y = -6$, or $y = 3$
If $y = 0$, then $3x - 2(0) = -6$, $x = -2$.
Draw a line through the intercepts $(0, 3)$ and $(-2, 0)$.

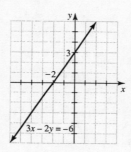

41. The equation $y - 3 = 10$ is equivalent to $y = 13$. Its graph is a horizontal line with a y-intercept of $(0, 13)$.

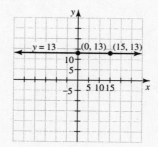

43. If $x = 0$, then $5(0) - 3y = 7$, or $y = -7/3$. If $y = 0$, then $5x - 3(0) = 7$, or $x = 7/5$. Draw a line through the intercepts $(0, -7/3)$ and $(7/5, 0)$.

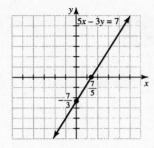

45. For $5x + 4y = 100$, the intercepts are $(0, 25)$ and $(20, 0)$.

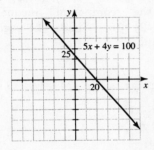

47. For $x - 80y = 400$ the intercepts are $(0, -5)$ and $(400, 0)$.

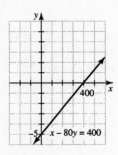

49. Use the slope $3 = 3/1$ and y-intercept $(0, -2)$ to graph the equation $y = 3x - 2$. Start at $(0, -2)$ and go up 3 units and 1 to the right. Draw a dashed line through the two points $(0, -2)$ and $(1, 1)$. Shade the region above the line to indicate the graph of $y > 3x - 2$.

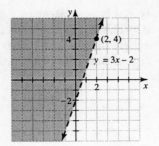

51. First graph $x - y = 5$ using the intercepts $(0, -5)$ and $(5, 0)$. The line should be solid because of the inequality symbol $\leq$. Test the point $(0, 0)$ in the inequality $x - y \leq 5$:
$$0 - 0 \leq 5$$
Since $(0, 0)$ satisfies the inequality, shade the region containing $(0, 0)$.

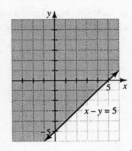

53. The inequality $3x > 2$ is equivalent to $x > \frac{2}{3}$. First graph the vertical line $x = \frac{2}{3}$ as a dashed line through $(\frac{2}{3}, 0)$. The graph of $x > \frac{2}{3}$ is the region to the right of this line.

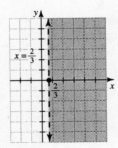

55. The inequality $4y \leq 0$ is equivalent to $y \leq 0$. Graph the line $y = 0$. The line $y = 0$ coincides with the x-axis. Draw the line solid. Points on the line and in the region below the line satisfy the inequality $y \leq 0$.

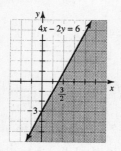

57. Solve $4x - 2y \geq 6$ for y:
$$-2y \geq -4x + 6$$
$$y \leq 2x - 3$$
All points on the line $y = 2x - 3$ together with the points below the line satisfy the inequality. Use y-intercept $(0, -3)$ and slope 2 to graph the solid line.

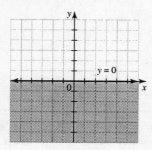

59. The x and y-intercepts for the line $5x - 2y = 9$ are $(0, -\frac{9}{2})$ and $(\frac{9}{5}, 0)$. Draw a dashed line through the intercepts. Check $(0, 0)$ in the inequality $5x - 2y < 9$:
$$5(0) - 2(0) < 9 \text{ is correct.}$$
Shade the region containing $(0, 0)$.

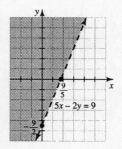

61. First graph the lines $y = 3$ and $y - x = 5$. Test the points $(0, 0)$, $(0, 4)$, $(0, 6)$, and $(-6, 0)$. Only $(0, 4)$ satisfies $y > 3$ and $y - x < 5$. Shade the region containing $(0, 4)$.

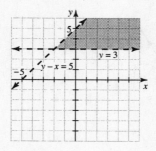

63. First graph the lines $3x + 2y = 8$ and $3x - 2y = 6$. Test the points $(0, 0)$, $(0, 5)$, $(0, -6)$, and $(5, 0)$. The points $(0, 0)$, $(0, 5)$, and $(5, 0)$ satisfy the compound inequality $3x + 2y \geq 8$ or $3x - 2y \leq 6$. Shade the regions containing those points.

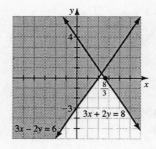

65. First graph the equations $x + 2y = 10$ and $x + 2y = -10$. Test the points $(0, 0)$, $(0, 8)$, and $(0, -8)$ in the inequality $|x + 2y| < 10$. Only $(0, 0)$ satisfies the inequality. So shade the region containing $(0, 0)$.

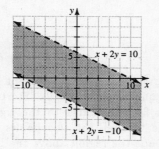

138

67. First graph the two vertical lines $x = 5$ and $x = -5$. Test the points $(0, 0)$, $(0, 8)$, $(-8, 0)$ in the inequality $|x| \leq 5$. Only $(0, 0)$ satisfies the inequality. So shade the region containing $(0, 0)$, the region between the parallel lines.

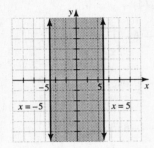

69. First graph the lines $y - x = 2$ and $y - x = -2$. Test the points $(0, 0)$, $(0, 4)$, and $(0, -4)$ in the inequality $|y - x| > 2$. The points $(0, 4)$ and $(0, -4)$ satisfy the inequality. Shade the regions containing those points.

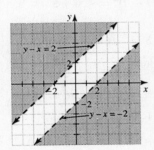

71. This line contains the points $(2, 0)$ and $(0, -6)$. The line has slope

$$m = \frac{-6 - 0}{0 - 2} = 3.$$

In slope-intercept form its equation is
$$y = 3x - 6,$$
and in standard form $3x - y = 6$.

73. Use the point-slope formula.

$$y - 4 = -\frac{1}{2}(x - (-1))$$

$$y - 4 = -\frac{1}{2}x - \frac{1}{2}$$

$$2y - 8 = -x - 1$$
$$x + 2y = 7$$

75. The line through $(2, -6)$ and $(2, 5)$ is a vertical line. All vertical lines are of the form $x = k$. So the equation is $x = 2$.

77. Since $x = 5$ is a vertical line, a line perpendicular to it is horizontal. The equation of a horizontal line through $(0, 0)$ is $y = 0$.

79. Any line parallel to $y = 2x + 1$ has slope 2. Use the point-slope formula with the point $(-1, 4)$ and slope 2:

$$y - 4 = 2(x - (-1))$$

$$y - 4 = 2x + 2$$

$$-2x + y = 6$$

$$2x - y = -6$$

81. The line with y-intercept $(0, 6)$ and slope 3 has equation $y = 3x + 6$. Written in standard form it is $3x - y = -6$.

83. A horizontal line has slope 0. If it goes through $(2, 5)$, its y-intercept is $(0, 5)$. Its equation is $y = 0 \cdot x + 5$ or simply $y = 5$.

85. Draw a quadrilateral with the 4 points as vertices. Find the slope of each side:

$$m_1 = \frac{-5 - (-1)}{-5 - (-3)} = 2 \qquad m_2 = \frac{-1 - 2}{-3 - 6} = \frac{-3}{-9} = \frac{1}{3}$$

$$m_3 = \frac{2 - (-2)}{6 - 4} = 2 \qquad m_4 = \frac{-2 - (-5)}{4 - (-5)} = \frac{3}{9} = \frac{1}{3}$$

Since the slopes of the opposite sides are equal, it is a parallelogram.

87. Use the distance formula to find the length of each side of the triangle.

$$d_1 = \sqrt{(-4 - 4)^2 + (-3 - (-1))^2}$$
$$= \sqrt{64 + 4} = \sqrt{68} = 2\sqrt{17}$$

$$d_2 = \sqrt{(4 - 2)^2 + (-1 - 7)^2}$$
$$= \sqrt{4 + 64} = \sqrt{68} = 2\sqrt{17}$$

$$d_3 = \sqrt{(2 - (-4))^2 + (7 - (-3))^2}$$
$$= \sqrt{36 + 100} = \sqrt{136} = 2\sqrt{34}$$

Since two sides have length $2\sqrt{17}$, it is an isosceles triangle.

89. Find the length of each side of the triangle:

$$d_1 = \sqrt{(-1 - 3)^2 + (0 - 0)^2} = \sqrt{16} = 4$$

$$d_2 = \sqrt{(3-1)^2 + (0-2\sqrt{3})^2} = \sqrt{4+12} = 4$$
$$d_3 = \sqrt{(-1-1)^2 + (0-2\sqrt{3})^2} = \sqrt{4+12} = 4$$

Since all sides have length 4, it is an equilateral triangle.

91. To find the perimeter, find the length of each side and add the results:

$$L_1 = \sqrt{(-4-4)^2 + (-3-(-1))^2} = \sqrt{64+4}$$
$$= \sqrt{68} = 2\sqrt{17}$$

$$L_2 = \sqrt{(4-2)^2 + (-1-7)^2} = \sqrt{4+64} = 2\sqrt{17}$$

$$L_3 = \sqrt{(2-(-4))^2 + (7-(-3))^2} = \sqrt{36+100}$$
$$= \sqrt{136} = 4\sqrt{34}$$

$$P = 2\sqrt{17} + 2\sqrt{17} + 4\sqrt{34} = 4\sqrt{17} + 2\sqrt{34}$$

93. Find the slope of the line through (20, 200) and (70, 150).

$$m = \frac{200-150}{20-70} = \frac{50}{-50} = -1$$

Use the point-slope formula with $m = -1$ and the point (20, 200):

$$h - h_1 = m(a - a_1)$$
$$h - 200 = -1(a - 20)$$
$$h - 200 = -a + 20$$
$$h = 220 - a$$

If $a = 40$, then $h = 220 - 40 = 180$.

95. To determine the number of days it would take for the rental charge to equal $1080, we solve the equation $1080 = 26 + 17d$.

$$1054 = 17d$$
$$62 = d$$

It will take 62 days for the rental charge to equal the cost of the air hammer. If $d = 1$, then $C = 26 + 17(1) = 43$. If $d = 30$, then $C = 26 + 17(30) = 536$. The graph of this function for d ranging from 1 to 30 is a straight line segment joining (1, 43) and (30, 536).

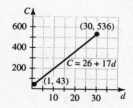

CHAPTER 7   TEST

1. If $x = 0$, then $2(0) + y = 5$.
$$y = 5$$
If $y = 0$, then $2x + 0 = 5$.
$$x = 5/2$$
If $x = 4$, then $2(4) + y = 5$.
$$y = -3$$
If $y = -8$, then $2x + (-8) = 5$.
$$2x = 13$$
$$x = 13/2$$
The pairs are $(0, 5)$, $(\frac{5}{2}, 0)$, $(4, -3)$, and $(\frac{13}{2}, -8)$.

2. $M = \left(\frac{3+5}{2}, \frac{-2+(-6)}{2}\right) = (4, -4)$

$$L = \sqrt{(3-5)^2 + (-2-(-6))^2} = \sqrt{4+16}$$
$$= \sqrt{20} = 2\sqrt{5}$$

3. $M = \left(\frac{0+(-4)}{2}, \frac{3+0}{2}\right) = \left(-2, \frac{3}{2}\right)$

$$L = \sqrt{(0-(-4))^2 + (3-0)^2} = \sqrt{16+9} = 5$$

4. $m = \frac{7-1}{-3-2} = \frac{6}{-5} = -\frac{6}{5}$

5. $m = \frac{-6-2}{0-4} = \frac{-8}{-4} = 2$

6. For the line $y = -\frac{1}{2}x - 2$, the slope is $-\frac{1}{2}$, and the y-intercept is $(0, -2)$.

7. Solve the equation $8x - 5y = -10$ for y:

$$-5y = -8x - 10$$
$$y = \frac{8}{5}x + 2$$

The slope is $\frac{8}{5}$ and the y-intercept is $(0, 2)$.

8. The equation of the line with y-intercept $(0, 3)$ and slope $-1/2$ is

$$y = -\frac{1}{2}x + 3$$
$$2y = -x + 6$$
$$x + 2y = 6$$

9. Use the point-slope form to write the equation of the line through $(-3, 5)$ with slope -4:

$$y - 5 = -4(x - (-3))$$
$$y - 5 = -4(x + 3)$$
$$y - 5 = -4x - 12$$
$$4x + y = -7$$

10. First find the slope of the line through (4, 1) and (−2, −5):

$$m = \frac{1 - (-5)}{4 - (-2)} = \frac{6}{6} = 1$$

Use (4, 1) and slope 1 in the point-slope form:

$$y - 1 = 1(x - 4)$$
$$y - 1 = x - 4$$
$$-x + y = -3$$
$$x - y = 3$$

11. Solve $3x - 5y = 7$ for y:

$$-5y = -3x + 7$$
$$y = \frac{3}{5}x - \frac{7}{5}$$

Since this line has slope 3/5, any line perpendicular to it has slope −5/3. The line through (2, 3) with slope −5/3 has equation

$$y - 3 = -\frac{5}{3}(x - 2)$$
$$y - 3 = -\frac{5}{3}x + \frac{10}{3}$$
$$3y - 9 = -5x + 10$$
$$5x + 3y = 19$$

12. The line $y = -1$ has slope 0, and any line parallel to it has slope 0. The equation of the line through (−4, 6) with slope 0 is $y = 6$.

13. The line shown goes through the points (0, 2) and (−4, 0). We can see from the graph that its slope is 2/4 or 1/2 and its y-intercept is (0, 2). In slope-intercept form its equation is

$$y = \frac{1}{2}x + 2$$
$$2y = x + 4$$
$$-x + 2y = 4$$
$$x - 2y = -4$$

14. Plot the three points and then find the slope of each side of the triangle.

$$m_1 = \frac{-2 - (-1)}{-1 - 2} = \frac{-1}{-3} = \frac{1}{3}$$

$$m_2 = \frac{-1 - 2}{2 - 1} = \frac{-3}{1} = -3$$

The two sides with slopes −3 and $\frac{1}{3}$ are perpendicular and so it is a right triangle.

15. Plot the four points and find the slopes of the sides of the quadrilateral.

$$m_1 = \frac{2 - 0}{6 - 0} = \frac{1}{3} \qquad m_2 = \frac{2 - 0}{6 - 5} = 2$$

$$m_3 = \frac{-2 - 0}{-1 - 5} = \frac{1}{3} \qquad m_4 = \frac{-2 - 0}{-1 - 0} = 2$$

Since the opposite sides have equal slopes, the quadrilateral is a parallelogram.

16. The line $y = \frac{1}{3}x - 2$ has a y-intercept of (0, −2) and a slope of 1/3. Start at (0, −2) and rise 1 unit and go 3 units to the right to locate a second point on the line.

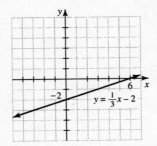

17. To graph $5x - 2y = 7$ find the x-intercept and y-intercept. If $x = 0$, we get $-2y = 7$ or $y = -7/2$. If $y = 0$, we get $5x = 7$ or $x = 7/5$. Draw a line going through the intercepts (0, −7/2) and (7/5, 0).

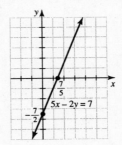

18. First graph the equation $y = (-1/2)x + 3$, using y-intercept of (0, 3), a slope of −1/2, and a dashed line. The graph of $y > (-1/2)x + 3$ is the region above this line.

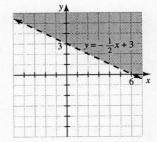

19. First graph the line $3x - 2y - 6 = 0$, using x-intercept $(2, 0)$, y-intercept $(0, -3)$, and a solid line. Test the point $(0, 0)$ in $3x - 2y - 6 \leq 0$: $3(0) - 2(0) - 6 \leq 0$ is correct. Shade the region containing $(0, 0)$.

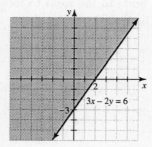

20. Graph the vertical line $x = 2$ using a dashed line. Graph $x + y = 0$ using a dashed line. Test the points $(-2, 0)$, $(1, 0)$, $(5, 0)$, and $(3, -5)$. Only $(5, 0)$ satisfies $x > 2$ and $x + y > 0$. So shade the region containing $(5, 0)$.

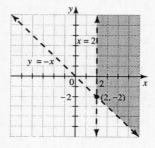

21. First graph the parallel lines $2x + y = 3$ and $2x + y = -3$, using solid lines. Test the points $(-5, 0)$, $(0, 0)$ and $(5,0)$ in $|2x + y| \geq 3$. Both $(-5, 0)$ and $(5, 0)$ satisfy the inequality, so shade the regions containing those points.

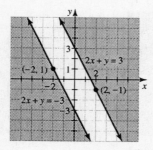

22. The first coordinate is a and the second coordinate is V. We want the equation of the line containing the two points $(0, 22,000)$ and $(3, 16,000)$. The slope of the line is

$$m = \frac{22,000 - 16,000}{0 - 3} = -2,000.$$

Using the slope and y-intercept, we can write the equation $V = -2000a + 22,000$.

Tying It All Together Chapters 1 - 7

1. $-5^4 = -(5 \cdot 5 \cdot 5 \cdot 5) = -625$

2. $125^{2/3} = (125^{1/3})^2 = 5^2 = 25$

3. $\dfrac{4 + \sqrt{8}}{4} = \dfrac{4 + 2\sqrt{2}}{4} = \dfrac{2(2 + \sqrt{2})}{2 \cdot 2} = \dfrac{2 + \sqrt{2}}{2}$

4. $\dfrac{8x - 2}{4} = \dfrac{2(4x - 1)}{2 \cdot 2} = \dfrac{4x - 1}{2}$

5. $(x + h)^2 = x^2 + 2xh + h^2$

6. $(a + 1)^2 - (a + 1) = a^2 + 2a + 1 - a - 1$

$$= a^2 + a$$

7. $\dfrac{(x + h)^2 - x^2}{h} = \dfrac{x^2 + 2xh + h^2 - x^2}{h}$

$$= \dfrac{2xh + h^2}{h} = 2x + h$$

8. $\dfrac{3(x + h)^2 - 2(x + h) - (3x^2 - 2x)}{h}$

$$= \dfrac{3x^2 + 6xh + 3h^2 - 2x - 2h - 3x^2 + 2x}{h}$$

$$= \dfrac{6xh + 3h^2 - 2h}{h} = 6x + 3h - 2$$

9. The line $y = (1/3)x$ has y-intercept $(0, 0)$ and slope $1/3$. Start at the origin and go up 1 unit and 3 to the right to get a second point on the line.

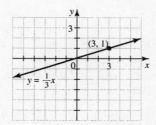

142

10. The line $y = 3x$ has slope $3 = 3/1$ and y-intercept $(0, 0)$. Start at the origin and go up 3 and 1 to the right to get a second point on the line.

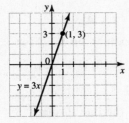

11. The line $y = 2x + 5$ has slope $2 = 2/1$ and y-intercept $(0, 5)$. Start at $(0, 5)$ and go up 2 and 1 to the right to get a second point on the line.

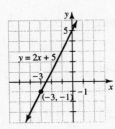

12. Note that $y = \dfrac{x - 5}{2}$ can be written as $y = \dfrac{1}{2}x - \dfrac{5}{2}$. Use a slope of $1/2$ and a y-intercept of $(0, -5/2)$. Start at $(0, -5/2)$ and go up 1 and 2 to the right to locate a second point on the line.

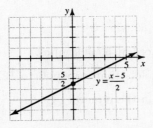

13.
$$2x - 5 = 9$$
$$2x = 14$$
$$x = 7$$

The solution set is $\{7\}$.

14.
$$3(y - 2) + 4 = -5$$
$$3y - 6 + 4 = -5$$
$$3y = -3$$
$$y = -1$$

The solution set is $\{-1\}$.

15.
$$|2x - 3| = 0$$
$$2x - 3 = 0$$
$$2x = 3$$
$$x = \frac{3}{2}$$

The solution set is $\left\{\dfrac{3}{2}\right\}$.

16.
$$\sqrt{x - 4} = 0$$
$$x - 4 = 0$$
$$x = 4$$

The solution set is $\{4\}$.

17.
$$x^2 + 9 = 0$$
$$x^2 = -9$$

Since the square of every real number is nonnegative, the solution set is $\emptyset$.

18.
$$x^2 - 8x + 12 = 0$$
$$(x - 6)(x - 2) = 0$$
$$x - 6 = 0 \quad \text{or} \quad x - 2 = 0$$
$$x = 6 \quad \text{or} \quad x = 2$$

The solution set is $\{2, 6\}$.

19.
$$2x - 1 > 7$$
$$2x > 8$$
$$x > 4$$

143

20.
$$5 - 3x \leq -1$$
$$-3x \leq -6$$
$$x \geq 2$$

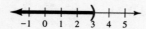

21. $x - 5 \leq 4$ and $3x - 1 < 8$
$$x \leq 9 \text{ and } \quad 3x < 9$$
$$x \leq 9 \text{ and } \quad\quad x < 3$$
The last compound inequality is equivalent to the simple inequality $x < 3$.

22. $2x \leq -6$ or $5 - 2x < -7$
$$x \leq -3 \text{ or } \quad -2x < -12$$
$$x \leq -3 \text{ or } \quad\quad x > 6$$

23.
$$|x - 3| < 2$$
$$-2 < x - 3 < 2$$
$$1 < x < 5$$

24.
$$|1 - 2x| \geq 7$$
$$1 - 2x \geq 7 \quad \text{or} \quad 1 - 2x \leq -7$$
$$-2x \geq 6 \text{ or } \quad -2x \leq -8$$
$$x \leq -3 \text{ or } \quad\quad x \geq 4$$

25. If x is negative, $\sqrt{x}$ is not a real number. So the domain is $[0, \infty)$.

26. If $\sqrt{6 - 2x}$ is to be a real number, then we must have $6 - 2x \geq 0$, or $-2x \geq -6$, or $x \leq 3$. The domain is $(-\infty, 3]$.

27. The equation $x^2 + 1 = 0$ has no real solutions. So the denominator is nonzero for any real number x. The domain is the set of all real numbers $(-\infty, \infty)$.

28. To find the domain we must solve
$$x^2 - 10x + 9 = 0.$$
$$(x - 1)(x - 9) = 0$$
$$x - 1 = 0 \quad \text{or} \quad x - 9 = 0$$
$$x = 1 \quad \text{or} \quad\quad x = 9$$
The numbers 1 and 9 cannot be used for x, so the domain is $(-\infty, 1) \cup (1, 9) \cup (9, \infty)$.

29. a) $C = 3000 + 0.12x$

b) Find the equation of the line through $(0, 0.15)$ and $(100{,}000, 0.25)$
$$m = \frac{0.25 - 0.15}{100{,}000 - 0} = 0.000001$$
P-intercept is $(0, 0.15)$

$$P = 0.000001x + 0.15$$

c) $T = \frac{C}{x} + P$
$$T = \frac{3000 + 0.12x}{x} + 0.000001x + 0.15$$
$$T = \frac{3000}{x} + 0.000001x + 0.27$$
If $x = 20{,}000$, then $T = \$0.44$
If $x = 30{,}000$, then $T = \$0.40$
If $x = 90{,}000$, then $T = \$0.39$

d) $\frac{3000}{x} + 0.000001x + 0.27 = 0.38$
$$\frac{3000}{x} + 0.000001x = 0.11$$
$$3000 + 0.000001x^2 = 0.11x$$
$$0.000001x^2 - 0.11x + 3000 = 0$$

$$x = \frac{0.11 \pm \sqrt{0.11^2 - 4(0.000001)(3000)}}{2(0.000001)}$$

$$= 50{,}000 \text{ or } 60{,}000$$

The car will be replaced at 60,000 miles.

e) The total cost is less than or equal to $0.38 per mile for mileage in the interval $[50{,}000, 60{,}000]$

8.1 Warm-ups

1. False, $\{(1, 2), (1, 3)\}$ is not a function.
2. True, because $C = \pi D$.   **3.** True, because no two ordered pairs have the same first coordinate and different second coordinates.
4. False, because $\{(1, 2), (1, 3)\}$ is a relation but not a function.   **5.** False, because $(1, 5)$ and $(1, 7)$ have the same first coordinate and different second coordinates.   **6.** False, because 0 is also in the domain of the function.
7. True, because the absolute value of any number is greater than or equal to zero.
8. True, because $y = (1/4)x$ guarantees that no two ordered pairs can have the same first coordinate and different second coordinates.
9. False, because both $(16, 2)$ and $(16, -2)$ are in the function.   **10.** True, because $h(-2) = (-2)^2 - 3 = 4 - 3 = 1$.

8.1 EXERCISES

1. A set of ordered pairs is a function unless there are two pairs with the same first coordinate and different second coordinates. So this set is a function.
3. This set of ordered pairs is not a function because $(-2, 4)$ and $(-2, 6)$ have the same x-coordinate and different y-coordinates.
5. This set of ordered pairs is not a function because $(\pi, -1)$ and $(\pi, 1)$ have the same x-coordinate and different y-coordinates.
7. This set is a function because no two ordered pairs have the same first coordinate and different second coordinates.
9. This set is a function because for each value of x, $y = (x - 1)^2$ determines only one value of y.

11. This set is not a function because $(2, 2)$ and $(2, -2)$ both belong to this set.

13. The set is a function because each value of s determines only one value of t by the formula $t = s$.
15. If we solve $x = 5y + 2$ for y, we get $y = \dfrac{x - 2}{5}$. So the relation is a function because each value of x determines only one y value.

17. Ordered pairs $(2, 1)$ and $(2, -1)$ both satisfy $x = 2y^2$. So $x = 2y^2$ does not define y as a function of x.

19. The equation $y = 3x - 4$ defines y as a function of x because each value of x determines only one value of y.
21. This equation does not define a function because ordered pairs such as $(0, 1)$ and $(0, -1)$ satisfy $x^2 + y^2 = 1$.
23. The equation $y = \sqrt{x}$ is a function because to each value of x there corresponds only one value of y.
25. This relation is not a function because both $(2, -1)$ and $(2, 1)$ satisfy $x = 2|y|$.
27. The equation $y = x^{1/3}$ is a function because to each value of x there corresponds only one value of y, namely the cube root of x.
29. The domain is the set of first coordinates, $\{2\}$, and the range is the set of second coordinates, $\{3, 5, 7\}$.
31. Since we can find the absolute value of any number, the domain is R. Since the absolute values of real numbers are always greater than or equal to 0, the range is $[0, \infty)$.
33. Since x is equal to the positive square root of y, the values of x must be greater than or equal to 0. The domain is $[0, \infty)$. Since square root is only defined on nonnegative numbers, the values of y must be nonnegative. The range is $[0, \infty)$.
35. Since s is greater than or equal to 8, the domain is $[8, \infty)$. Since t is the cube root of a number greater than or equal to 8, t is greater than or equal to 2. The range is $[2, \infty)$.
37. $f(4) = 3(4) - 2 = 12 - 2 = 10$
39. $g(-2) = (-2)^2 - 3(-2) + 2$
$\qquad = 4 + 6 + 2 = 12$
41. $h(-3) = |-3 + 2| = |-1| = 1$
43. Since $f(x) = 3x - 2$ and $f(x) = 5$, we have
$$5 = 3x - 2.$$
$$7 = 3x$$
$$\frac{7}{3} = x$$
45. If $g(x) = x^2 - 3x + 2$ and $g(x) = 0$, then
$$x^2 - 3x + 2 = 0.$$
$$(x - 2)(x - 1) = 0$$
$$x - 2 = 0 \quad \text{or} \quad x - 1 = 0$$
$$x = 2 \quad \text{or} \qquad x = 1$$

47. If $h(x) = |x + 2|$ and $h(x) = 3$, then
$$|x + 2| = 3$$
$$x + 2 = 3 \quad \text{or} \quad x + 2 = -3$$
$$x = 1 \quad \text{or} \qquad x = -5$$
49. Since $f(x) = 4x - 1$, $f(a) = 4a - 1$.

51. Since $f(x) = 4x - 1$,
$f(x + 2) = 4(x + 2) - 1 = 4x + 7$.

53. Since $g(x) = x^2 - 2x + 4$,
$$g(x + 3) = (x + 3)^2 - 2(x + 3) + 4$$
$$= x^2 + 6x + 9 - 2x - 6 + 4$$
$$= x^2 + 4x + 7$$

55. $g(x + h) = (x + h)^2 - 2(x + h) + 4$
$$= x^2 + 2xh + h^2 - 2x - 2h + 4$$

57. $\dfrac{f(3) - f(1)}{3 - 1} = \dfrac{3^2 - 1^2}{2} = \dfrac{8}{2} = 4$

59. $\dfrac{h(8) - h(4)}{8 - 4} = \dfrac{104 - 20}{4} = \dfrac{84}{4} = 21$

61. $f(a + 1) = (a + 1)^2 - (a + 1) + 2$
$$= a^2 + a + 2$$
$f(a) = a^2 - a + 2$
$$\dfrac{f(a + 1) - f(a)}{a + 1 - a} = \dfrac{a^2 + a + 2 - (a^2 - a + 2)}{1}$$
$$= \dfrac{2a}{1} = 2a$$

63. $g(x + h) = 3(x + h) - 5 = 3x + 3h - 5$
$g(x) = 3x - 5$
$$\dfrac{g(x + h) - g(x)}{x + h - x} = \dfrac{3x + 3h - 5 - (3x - 5)}{h}$$
$$= \dfrac{3h}{h} = 3$$

65. $f(3.46) = \sqrt{3.46 + 2} = 2.337$

67. $g(-3.5) = 3(-3.5)^2 - 8(-3.5) + 2$
$$= 66.75$$

69. $\sqrt{x + 2} = 5.6$
$x + 2 = 5.6^2$
$x = 5.6^2 - 2 = 29.36$

71. $3x^2 - 8x + 2 = 0$
$$x = \dfrac{8 \pm \sqrt{8^2 - 4(3)(2)}}{2(3)} = 2.387 \text{ or } 0.279$$

73. $f(a - 5) - f(a) = 3(a - 5) - 2 - (3a - 2)$
$$= 3a - 17 - 3a + 2 = -15$$

75. $g(a + 2) - g(a) = (a + 2)^2 - (a + 2) - (a^2 - a)$
$$= a^2 + 4a + 4 - a - 2 - a^2 + a = 4a + 2$$

77. $\dfrac{f(x + h) - f(x)}{h} = \dfrac{3h}{h} = 3$

79. $\dfrac{g(x + h) - g(x)}{(x + h) - x} = \dfrac{2xh + h^2 - h}{h} = 2x + h - 1$

81. The area of a square is found by squaring the length of a side: $A = s^2$.

83. The cost of the purchase, C, is the price per yard times the number of yards purchased: $C = 3.98y$.

85. The total cost, C, is found by multiplying the number of toppings, n, by the cost for each topping, $0.50, and adding on the $14.95 base charge: $C = 0.50n + 14.95$

87. We must find the linear equation through the points (2, 78) and (4, 86). First find the slope:
$$m = \dfrac{86 - 78}{4 - 2} = 4.$$
Now use point slope form:
$h - 78 = 4(t - 2)$
$h - 78 = 4t - 8$
$h = 4t + 70 \quad \text{for } 0 \leq h \leq 8$

89. $\dfrac{5675 - 11,640}{1994 - 1990} = -\$1491.25 \text{ per year}$

91. $\dfrac{60 - 0}{5.4 - 0} \approx 11.11 \text{ mph per second}$

8.2 WARM-UPS

1. True, because the graph of a function consists of all ordered pairs that are in the function.

2. True, that is why they are called linear functions.

3. True, because absolute value functions are generally v-shaped.

4. False, because 0 is not in the domain of the function $f(x) = 1/x$.

5. False, because the x-coordinate of the vertex is $x = -b/(2a)$. **6.** False, the domain of a quadratic function is $(-\infty, \infty)$ or R.

7. True, because f(x) is just another name for the dependent variable y.

8. True, because if a vertical line crosses a graph more than once, the graph is not the graph of function.

9. False, because 1 is also in the domain of the function.

10. True, because any real number can be used in a quadratic function.

8.2 EXERCISES

1. The graph of $h(x) = -2$ is the same as the graph of the horizontal line $y = -2$.

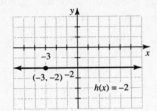

The domain of the function is R. The only y-coordinate used is -2, and so the range is $\{-2\}$.

3. The graph of $f(x) = 2x - 1$ is the same as the graph of the linear equation $y = 2x - 1$. To draw the graph, start at the y-intercept $(0, -1)$, and use the slope of $2 = 2/1$. Rise 2 and run 1 to locate a second point on the line.

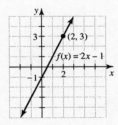

From the graph, we can see that the domain is $(-\infty, \infty)$ and the range is also $(-\infty, \infty)$.

5. Graph the line $y = (1/2)x + 2$ by locating the y-intercept $(0, 2)$ and using a slope of $1/2$. The domain is $(-\infty, \infty)$ and the range is $(-\infty, \infty)$.

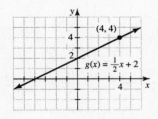

7. The graph of $y = -\frac{2}{3}x + 3$ is a straight line with y-intercept $(0, 3)$ and slope $-2/3$.

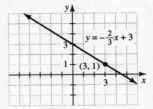

The domain is $(-\infty, \infty)$ and **range is** $(-\infty, \infty)$.

9. The graph of $y = -0.3x + 6.5$ is a straight line with y-intercept $(0, 6.5)$ and slope $-3/10$.

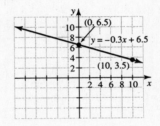

The domain is $(-\infty, \infty)$ and range is $(-\infty, \infty)$.

11. The graph of $y = |x| + 1$ contains the points $(0, 1)$, $(2, 3)$, and $(-2, 3)$. Plot these points and draw the v-shaped graph.

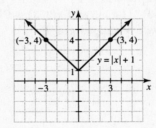

The domain is $(-\infty, \infty)$, and from the graph we can see that the range is $[1, \infty)$.

13. The graph of $h(x) = |x + 1|$ includes the points $(0, 1)$, $(1, 2)$, $(-1, 0)$, and $(-2, 1)$.

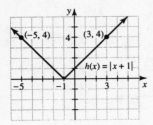

From the graph we can see that the domain is $(-\infty, \infty)$. In the vertical direction the graph is on or above the x-axis. Since all y-coordinates on the graph are greater than or equal to 0, the range is $[0, \infty)$.

15. The graph of $g(x) = |3x|$ includes the points $(0, 0)$, $(1, 3)$, and $(-1, 3)$. Plot these points and draw the graph.

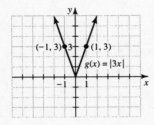

The graph indicates that the domain is $(-\infty, \infty)$. Since all of the y-coordinates on the graph are nonnegative, the range is $[0, \infty)$.

17. The graph of $f(x) = |2x - 1|$ includes the points $(0, 1)$, $(1/2, 0)$, $(1, 1)$, and $(2, 3)$. Plot these points and draw a v-shaped graph.

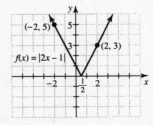

From the graph we can see that the domain is $(-\infty, \infty)$, and the range is $[0, \infty)$.

19. The graph of $f(x) = |x - 2| + 1$ includes the points $(2, 1)$, $(3, 2)$, $(4, 3)$, $(1, 2)$, and $(0, 3)$. Plot these points and draw a v-shaped graph. Form the graph we can see that the domain is $(-\infty, \infty)$. Since all y-coordinates are greater than or equal to 1, the range is $[1, \infty)$.

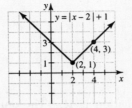

21. The graph of $g(x) = x^2 + 2$ includes the points $(-2, 6)$, $(-1, 3)$, $(0, 2)$, $(1, 3)$, and $(2, 6)$. Plot these points and draw the parabola through them.

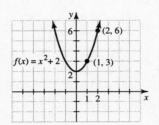

The domain is $(-\infty, \infty)$. Since no y-coordinate on the graph is below 2, the range is $[2, \infty)$.

23. The graph of $f(x) = 2x^2$ includes the points $(0, 0)$, $(1, 2)$, $(-1, 2)$, $(2, 8)$, and $(-2, 8)$. Plot these points and draw a parabola through them.

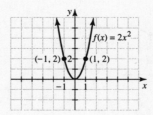

The domain is $(-\infty, \infty)$ and the range is $[0, \infty)$.

148

25. The graph of $y = 6 - x^2$ includes the points $(-3, -3)$, $(0, 6)$, $(3, -3)$, $(-2, 2)$, and $(2, 2)$. Draw a parabola through these points.

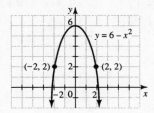

The domain is $(-\infty, \infty)$ and range is $(-\infty, 6]$.

27. Find vertex for $g(x) = x^2 - x - 2$:

$x = \dfrac{-b}{2a} = \dfrac{1}{2(1)} = \dfrac{1}{2}$, $y = (\tfrac{1}{2})^2 - \tfrac{1}{2} - 2 = -\dfrac{9}{4}$

The graph includes the points $(-1, 0)$, $(0, -2)$, $(1/2, -9/4)$, $(2, 0)$, and $(3, 4)$. Plot these points and draw a parabola through them.

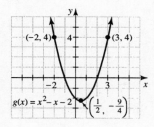

Vertex is $(1/2, -9/4)$. The domain is $(-\infty, \infty)$. The range is $[-9/4, \infty)$.

29. Find vertex for $h(x) = 2x^2 + 9x + 4$:

$x = \dfrac{-9}{2(2)} = -\dfrac{9}{4}$, $y = -6.125$

Plot $(-1, -3)$, $(0, 4)$, $(1, 15)$, $(-2, -6)$.

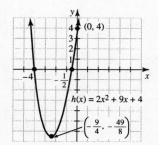

Vertex: $(-2.25, -6.125)$. Domain $(-\infty, \infty)$. Range: $[-6.125, \infty)$

31. Find vertex for $y = -x^2 + 2x + 8$

$x = \dfrac{-2}{2(-1)} = 1$, $y = -1^2 + 2 \cdot 1 + 8 = 9$

Plot $(-2, 0)$, $(0, 8)$, $(1, 9)$, $(2, 8)$, and $(4, 0)$.

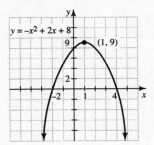

Vertex $(1, 9)$, domain $(-\infty, \infty)$, range $(-\infty, 9]$.

33. Find vertex for $y = -2x^2 - 3x + 2$

$x = \dfrac{3}{2(-2)} = -\dfrac{3}{4}$, $y = \dfrac{25}{8}$

Plot $(-2, 0)$, $(0, 2)$, $(1, -3)$, $(1/2, 0)$, and $(-3/4, 25/8)$.

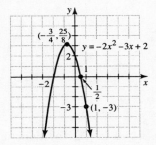

Vertex $(-3/4, 25/8)$, domain $(-\infty, \infty)$, range $(-\infty, 25/8]$.

35. The graph of $g(x) = 2\sqrt{x}$ includes the points $(0, 0)$, $(1, 2)$, and $(4, 4)$. Plot these points and draw a curve through them.

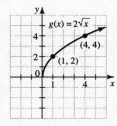

Since we must have $x \geq 0$ in $\sqrt{x}$, the domain is $[0, \infty)$. The range is $[0, \infty)$

149

37. The graph of $f(x) = \sqrt{x-1}$ includes $(1, 0)$, $(2, 1)$, and $(5, 2)$. Plot these points and draw a smooth curve through them. The curve is half of a parabola, positioned on its side.

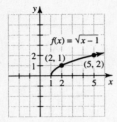

Since we must have $x - 1 \geq 0$, the domain is $[1, \infty)$. The range is $[0, \infty)$.

39. The graph of $h(x) = -\sqrt{x}$ includes the points $(0, 0)$, $(1, -1)$, and $(4, -2)$. Plot these points and draw a curve through them.

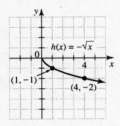

Since we must have $x \geq 0$ in $\sqrt{x}$, the domain is $[0, \infty)$. The range is $(-\infty, 0]$.

41. The graph of $y = \sqrt{x} + 2$ includes the points $(0, 2)$, $(1, 3)$, and $(4, 4)$. Graph these points and draw a curve through them.

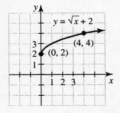

Because $\sqrt{x}$ is a real number only for nonnegative values of x, the domain is $[0, \infty)$. From the graph we see that y-coordinates go no lower than 2, so the range is $[2, \infty)$.

43. The graph of $x = |y|$ includes the points $(0, 0)$, $(1, -1)$, $(1, 1)$, $(2, -2)$, and $(2, 2)$. Note that in this case it is easier to pick the y-coordinate and then find the appropriate x-coordinate using $x = |y|$. Draw the v-shaped graph through these points.

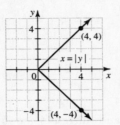

Since x is nonnegative, the domain is $[0, \infty)$. The range is $(-\infty, \infty)$.

45. To find pairs that satisfy $x = -y^2$, pick the y-coordinate first. The points $(0, 0)$, $(-1, 1)$, $(1, 1)$, $(-4, 2)$, and $(-4, -2)$ are on the graph.

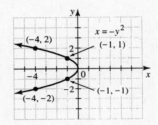

The graph indicats that x is nonpositive, so the domain is $(-\infty, 0]$. The range is $(-\infty, \infty)$.

47. The equation $x = 5$ is the equation of a vertical line with an x-intercept of $(5, 0)$.

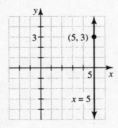

All points on this line have x-coordinate 5, so the domain is $\{5\}$. Every real number occurs as a y-coordinate, so the range is $(-\infty, \infty)$.

150

49. Rewrite $x + 9 = y^2$ as $x = y^2 - 9$. Select some y-coordinates first. The points $(-9, 0)$, $(0, 3)$, and $(0, -3)$ are on the parabola.

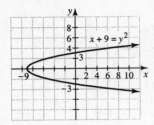

Since no x-coordinate is below -9, the domain is $[-9, \infty)$. The range is $(-\infty, \infty)$.

51. For the equation $x = \sqrt{y}$, we must select nonnegative values for y and find the corresponding value for x. The points $(0, 0)$, $(1, 1)$, and $(2, 4)$ are on the graph.

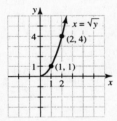

Since the x-coordinates are also nonnegative, the domain is $[0, \infty)$. Since we could only use nonnegative values for y, the range is $[0, \infty)$.

53. The graph of $x = (y - 1)^2$ includes the points $(0, 1)$, $(1, 0)$, $(1, 2)$, $(4, 3)$, and $(4, -1)$. Plot the points and draw a curve through them. From the graph we see that the domain is $[0, \infty)$ and the range is $(-\infty, \infty)$.

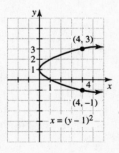

55. It is easy to find a vertical line that crosses this graph more than once, so this graph is not the graph of a function.

57. No vertical line crosses this graph more than once, so this graph is the graph of a function.

59. It is easy to find a vertical line that crosses this graph more than once, so this graph is not the graph of a function.

61. To graph $f(x) = 1 - |x|$, we arbitrarily select a value for x and find the corresponding value for y. The points $(-2, -1)$, $(-1, 0)$, $(0, 1)$, $(1, 0)$, and $(2, -1)$ are on the graph.

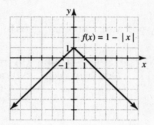

Since any real number could be used for x, the domain is $(-\infty, \infty)$. From the graph we see that the y-coordinates are not higher than 1, so the range is $(-\infty, 1]$.

63. The graph of $y = (x - 3)^2 - 1$ includes the points $(1, 3)$, $(2, 0)$, $(3, -1)$, $(4, 0)$, and $(5, 3)$.

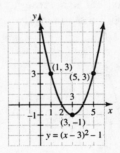

The domain is $(-\infty, \infty)$. Since the y-coordinates are greater than or equal to -1, the range is $[-1, \infty)$.

65. Graph of $y = |x + 3| + 1$ goes through $(-4, 2)$, $(-3, 1)$, $(-2, 2)$, $(-1, 3)$, $(0, 4)$.

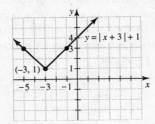

The domain is $(-\infty, \infty)$ and range is $[1, \infty)$.

67. Graph of $y = \sqrt{x} - 3$ goes through $(0, -3)$, $(1, -2)$, and $(4, -1)$. Domain is $[0, \infty)$ and range is $[-3, \infty)$.

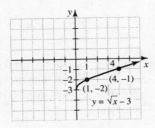

69. The graph of $y = 3x - 5$ is a straight line with y-intercept $(0, -5)$, and a slope of 3.

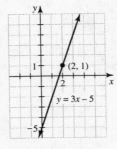

The domain is $(-\infty, \infty)$. From the graph we see that any real number could occur as a y-coordinate, so the range is $(-\infty, \infty)$.

71. Find vertex for $y = -x^2 + 5x - 4$:

$x = \dfrac{-5}{2(-1)} = 2.5$, $y = 2.25$

Graph goes through $(2.5, 2.25)$, $(2, 2)$, $(3, 2)$.

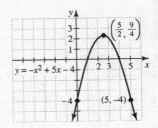

Domain is $(-\infty, \infty)$ and range is $(-\infty, 2.25]$.

73. The points $(0, 0)$, $(1, 48)$, $(2, 64)$, $(3, 48)$, and $(4, 0)$ satisfy $h(t) = -16t^2 + 64t$.

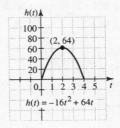

The x-coordinate of the vertex is $x = -b/(2a) = -64/(-32) = 2$. The maximum height reached by the object is the y-coordinate of the vertex, 64 feet.

75. The graph of $C(n) = 0.015n^2 - 4.5n + 400$ includes $(0, 400)$, $(150, 62.5)$, and $(200, 100)$. The x-coordinate of the vertex is $x = -b/(2a) = -(-4.5)/(0.03) = 150$. The minimum cost is the y-coordinate of the vertex, $62.50.

77. For 1998, $t = 8$ and
$h(8) = 103(8) + 93 = 917$.
If $h(t) = 402$, then
$$103t + 93 = 402$$
$$103t = 309$$
$$t = 3$$
So 402 permits were issued in 1993.

81. Graph of $f(x) = \sqrt{x^2}$ is same as graph of $y = |x|$.

152

8.3 WARM-UPS

1. True, because $(f - g)(x) = f(x) - g(x)$

$= x - 2 - (x + 3) = -5.$ **2.** True, because

$(f/g)(2) = f(2)/g(2) = (2 + 4)/(3 \cdot 2) = 6/6 = 1.$

3. False, because if $f(x) = 3x$ and $g(x) = x + 1$,

then $f(g(x)) = 3(x + 1) = 3x + 3$ and

$g(f(x)) = 3x + 1.$

4. False, because $(f \circ g)(x) = f(x + 2)$

$= (x + 2)^2 = x^2 + 4x + 4.$

5. False, because if $f(x) = 3x$ and $g(x) = x + 1$,

then $(f \circ g)(x) = f(x + 1) = 3x + 3$

and $(f \cdot g)(x) = 3x(x + 1) = 3x^2 + 3x.$

6. False, because $g[f(x)] = \sqrt{x} - 9$ and

$f[g(x)] = \sqrt{x - 9}$ which are not equal for all

values of x.

7. True, because $(f \circ g)(x) = f(\frac{x}{3}) = 3 \cdot \frac{x}{3} = x.$

8. True, because b determines a and a

determines c so b determines the value of c.

9. True, $F(x) = \sqrt{x - 5}$ is a composition of

$g(x) = x - 5$ and $h(x) = \sqrt{x}.$

10. True, because

$(g \circ h)(x) = g[h(x)] = g[x - 1] = (x - 1)^2 = F(x).$

8.3 EXERCISES

1. $(f + g)(x) = f(x) + g(x)$
$$= 4x - 3 + x^2 - 2x$$
$$= x^2 + 2x - 3$$

3. $(f \cdot g)(x) = f(x) \cdot g(x)$
$$= (4x - 3)(x^2 - 2x)$$
$$= 4x^3 - 11x^2 + 6x$$

5. Use $x = 3$ in the formula of Exercise 1:
$$(f + g)(3) = 3^2 + 2(3) - 3 = 12$$

7. $(f - g)(-3) = f(-3) - g(-3)$
$$= 4(-3) - 3 - [(-3)^2 - 2(-3)]$$
$$= -12 - 3 - [9 + 6]$$
$$= -30$$

9. Use $x = -1$ in the formula of Exercise 3:
$(f \cdot g)(-1) = 4(-1)^3 - 11(-1)^2 + 6(-1)$
$$= -4 - 11 - 6$$
$$= -21$$

11. $(f/g)(4) = \dfrac{f(4)}{g(4)} = \dfrac{4(4) - 3}{4^2 - 2(4)} = \dfrac{13}{8}$

13. Since $a = 2x - 6$, and $y = 3a - 2$ we have
$$y = 3(2x - 6) - 2$$
$$y = 6x - 18 - 2$$
$$y = 6x - 20$$

15. Let $d = (x + 1)/2$ in $y = 2d + 1$.
$$y = 2\left(\frac{x + 1}{2}\right) + 1$$
$$y = x + 1 + 1$$
$$y = x + 2$$

17. Let $m = x + 1$ in $y = m^2 - 1$.
$$y = (x + 1)^2 - 1$$
$$y = x^2 + 2x + 1 - 1$$
$$y = x^2 + 2x$$

19. $y = \dfrac{\frac{2x + 3}{1 - x} - 3}{\frac{2x + 3}{1 - x} + 2}$

$y = \dfrac{\left(\frac{2x + 3}{1 - x} - 3\right)(1 - x)}{\left(\frac{2x + 3}{1 - x} + 2\right)(1 - x)}$

$y = \dfrac{2x + 3 - 3(1 - x)}{2x + 3 + 2(1 - x)}$

$y = \dfrac{2x + 3 - 3 + 3x}{2x + 3 + 2 - 2x} = \dfrac{5x}{5}$

$y = x$

21. $(g \circ f)(1) = g[f(1)] = g(-1)$
$$= (-1)^2 + 3(-1) = -2$$

23. $(f \circ g)(1) = f[g(1)] = f[4] = 2(4) - 3 = 5$

25. $(f \circ f)(4) = f[f(4)] = f[5] = 2(5) - 3 = 7$

27. $(h \circ f)(5) = h[f(5)] = h[7] = \dfrac{7 + 3}{2} = 5$

29. $(f \circ h)(5) = f[h(5)] = f[4] = 2(4) - 3 = 5$

31. $(g \circ h)(-1) = g[h(-1)] = g[1] = 4$

33. $(f \circ g)(2.36) = f[g(2.36)] = f[12.6496]$

$= 22.2992$

35. $(g \circ f)(x) = g[f(x)] = g[2x - 3]$

$\quad = (2x - 3)^2 + 3(2x - 3)$

$\quad = 4x^2 - 12x + 9 + 6x - 9 = 4x^2 - 6x$

37. $f[g(x)] = f[x^2 + 3x] = 2(x^2 + 3x) - 3$

$\quad = 2x^2 + 6x - 3$

39. $(h \circ f)(x) = h[f(x)] = h[2x - 3]$

$\quad\quad\quad = \dfrac{2x - 3 + 3}{2} = x$

41. $(f \circ f)(x) = f[f(x)] = f[2x - 3]$

$\quad\quad\quad = 2(2x - 3) - 3 = 4x - 9$

43. $(h \circ h)(x) = h[h(x)] = h\left[\dfrac{x + 3}{2}\right]$

$\quad = \dfrac{\dfrac{x + 3}{2} + 3}{2} = \dfrac{\left(\dfrac{x + 3}{2} + 3\right)2}{2 \cdot 2}$

$\quad = \dfrac{x + 3 + 6}{4} = \dfrac{x + 9}{4}$

45. $F(x) = \sqrt{x - 3} = \sqrt{h(x)} = f[h(x)]$

Therefore, $F = f \circ h$.

47. $G(x) = x^2 - 6x + 9 = (x - 3)^2$

$\quad\quad\quad = [h(x)]^2 = g[h(x)]$

Therefore $G = g \circ h$.

49. $H(x) = x^2 - 3 = g(x) - 3 = h[g(x)]$

Therefore $H = h \circ g$.

51. $J(x) = x - 6 = x - 3 - 3$

$\quad\quad\quad = h(x) - 3 = h[h(x)]$

Therefore, $J = h \circ h$.

53. $K(x) = x^4 = (x^2)^2 = [g(x)]^2 = g[g(x)]$

Therefore, $K = g \circ g$.

55. $f[g(x)] = f\left(\dfrac{x-5}{3}\right) = 3\left(\dfrac{x-5}{3}\right) + 5 = x$

$g[f(x)] = g(3x + 5) = \dfrac{3x + 5 - 5}{3} = x$

57. $f[g(x)] = f\left(\sqrt[3]{x+9}\right) = \left(\sqrt[3]{x+9}\right)^3 - 9$

$\quad\quad\quad = x + 9 - 9 = x$

$g[f(x)] = g[x^3 - 9] = \sqrt[3]{x^3 - 9 + 9}$

$\quad\quad\quad = \sqrt[3]{x^3} = x$

59. $f[g(x)] = f\left(\dfrac{x+1}{1-x}\right) = \dfrac{\dfrac{x+1}{1-x} - 1}{\dfrac{x+1}{1-x} + 1}$

$= \dfrac{\left(\dfrac{x+1}{1-x} - 1\right)(1-x)}{\left(\dfrac{x+1}{1-x} + 1\right)(1-x)} = \dfrac{x+1-(1-x)}{x+1+1-x} = \dfrac{2x}{2}$

$\quad\quad\quad\quad\quad\quad\quad\quad\quad\quad\quad\quad = x$

$g[f(x)] = g\left(\dfrac{x-1}{x+1}\right) = \dfrac{\dfrac{x-1}{x+1} + 1}{1 - \dfrac{x-1}{x+1}}$

$= \dfrac{\left(\dfrac{x-1}{x+1} + 1\right)(x+1)}{\left(1 - \dfrac{x-1}{x+1}\right)(x+1)} = \dfrac{x-1+x+1}{x+1-(x-1)}$

$\quad\quad\quad\quad\quad\quad\quad\quad\quad = \dfrac{2x}{2} = x$

61. $f[g(x)] = f\left(\dfrac{1}{x}\right) = \dfrac{1}{\dfrac{1}{x}} = x$

$g[f(x)] = g\left(\dfrac{1}{x}\right) = \dfrac{1}{\dfrac{1}{x}} = x$

63. Let $s =$ the length of a side of the square.

$\quad 15^2 = s^2 + s^2$

$\quad 2s^2 = 225$

$\quad s^2 = \dfrac{225}{2}$

$\quad s = \sqrt{\dfrac{225}{2}}$

Since $A = s^2$, we have $A = \dfrac{225}{2} = 112.5$ in.^{2}

In general if d is the diagonal of a square and s is the length of a side, we have $d^2 = s^2 + s^2$ or $s^2 = \dfrac{d^2}{2}$. Since $A = s^2$, we have $A = \dfrac{d^2}{2}$.

65. $P(x) = R(x) - C(x)$

$\quad\quad = x^2 - 10x + 30 - (2x^2 - 30x + 200)$

$\quad\quad = -x^2 + 20x - 170$

67. Substitute $F = 0.25I$ into $J = 0.10F$.

$\quad J = 0.10(0.25I)$

$\quad J = 0.025I$

154

8.4 WARM-UPS

1. False, because the inverse function is obtained by interchanging the coordinates in every ordered pair. **2.** False, because the points $(0, 3)$ and $(1, 3)$ both satisfy $f(x) = 3$.
3. False, because the inverse of multiplication by 2 is division by 2. If $g(x) = 2x$, then $g^{-1}(x) = x/2$. **4.** True, because if we interchange the coordinates in a function that is not one-to-one, we do not obtain a function.
5. True, because in an inverse function the domain and range of the function are reversed.
6. False, because it is not one-to-one. Both $(2, 16)$ and $(-2, 16)$ satisfy $f(x) = x^4$. **7.** True, because taking the opposite of a number twice gives back the original number. **8.** True, because the inverse function interchanges the coordinates in all of the ordered pairs of the function. **9.** True, the inverse of $k(x) = 3x - 6$

is $k^{-1}(x) = (x + 6)/3 = (1/3)x + 2$.

10. False, because $f^{-1}(x) = (x + 4)/3$.

8.4 EXERCISES

1. The function is not one-to-one because of the pairs $(-2, 2)$ and $(2, 2)$. Therefore, the function is not invertible.
3. The function is one-to-one. The inverse is $\{(4, 16), (3, 9), (0, 0)\}$.
It is obtained by interchanging the coordinates in each ordered pair.
5. The function is not one-to-one because of the pairs $(5, 0)$ and $(6, 0)$. Therefore, it is not invertible.
7. This function is one-to-one. Its inverse is obtained by interchanging the coordinates in each ordered pair: $\{(0, 0), (2, 2), (9, 9)\}$.
9. This function is not one-to-one because we can draw a horizontal line that crosses the graph twice.

11. This function is one-to-one because we cannot draw a horizontal line that crosses the graph more than once.

13. Since $(f \circ g)(x) = f(g(x)) = f(0.5x) = 2(0.5x)$ $= x$ and $(g \circ f)(x) = g(f(x)) = g(2x) = 0.5(2x)$ $= x$, the functions are inverses of each other.

15. $(f \circ g)(x) = f(g(x)) = f(\frac{1}{2}x + 5)$

$= 2(\frac{1}{2}x + 5) - 10 = x + 10 - 10 = x$

$(g \circ f)(x) = g(f(x)) = g(2x - 10) = \frac{1}{2}(2x - 10) + 5$

$= x - 5 + 5 = x$

Therefore, the functions are inverses of each other.

17. $(f \circ g)(x) = f(g(x)) = f(-x) = -(-x) = x$
$(g \circ f)(x) = g(f(x)) = g(-x) = -(-x) = x$
Therefore, the functions are inverses of each other.

19. $(g \circ f)(x) = g(f(x)) = g(x^4) = (x^4)^{1/4} = |x|$

For example, $g(f(-2)) = g(16) = 2$. So the

functions are not inverses of each other.

21. $y = 5x$, $x = 5y$, $y = \frac{1}{5}x$

So $f^{-1}(x) = \frac{x}{5}$.

23. $y = x - 9$, $x = y - 9$, $y = x + 9$

So $g^{-1}(x) = x + 9$.

25. $y = 5x - 9$, $x = 5y - 9$, $5y = x + 9$, $y = \frac{x + 9}{5}$

So $k^{-1}(x) = \frac{x + 9}{5}$.

27. $y = \frac{2}{x}$, $x = \frac{2}{y}$, $xy = 2$, $y = \frac{2}{x}$

So $m^{-1}(x) = \frac{2}{x}$.

29. $y = \sqrt[3]{x - 4}$, $x = \sqrt[3]{y - 4}$, $x^3 = y - 4$,

$y = x^3 + 4$ So $f^{-1}(x) = x^3 + 4$.

31. $y = \frac{3}{x - 4}$, $x = \frac{3}{y - 4}$, $x(y - 4) = 3$, $y - 4 = \frac{3}{x}$

$y = \frac{3}{x} + 4$ So $f^{-1}(x) = \frac{3}{x} + 4$.

33. $f(x) = \sqrt[3]{3x + 7}$

$y = \sqrt[3]{3x + 7}$

$x = \sqrt[3]{3y + 7}$

$x^3 = 3y + 7$

$x^3 - 7 = 3y$

$\frac{x^3 - 7}{3} = y$

$f^{-1}(x) = \frac{x^3 - 7}{3}$

35.
$$f(x) = \frac{x+1}{x-2}$$
$$y = \frac{x+1}{x-2}$$
$$x = \frac{y+1}{y-2}$$
$$x(y-2) = y+1$$
$$xy - 2x = y+1$$
$$xy - y = 2x+1$$
$$y(x-1) = 2x+1$$
$$y = \frac{2x+1}{x-1}$$

$$f^{-1}(x) = \frac{2x+1}{x-1}$$

37.
$$f(x) = \frac{x+1}{3x-4}$$
$$y = \frac{x+1}{3x-4}$$
$$x = \frac{y+1}{3y-4}$$
$$x(3y-4) = y+1$$
$$3xy - 4x = y+1$$
$$3xy - y = 1+4x$$
$$y(3x-1) = 1+4x$$
$$y = \frac{1+4x}{3x-1}$$

$$f^{-1}(x) = \frac{1+4x}{3x-1}$$

39. The function $p(x) = \sqrt[4]{x}$ finds the fourth root. The inverse must be the fourth power. Since the domain and range of p are both the set of nonnegative real numbers, $p^{-1}(x) = x^4$ for $x \geq 0$.

41. $y = (x-2)^2$, $x = (y-2)^2$, $y - 2 = \pm\sqrt{x}$
$y = 2 \pm \sqrt{x}$. Since $x \geq 2$ in the function, we must have $y \geq 2$ in the inverse function. So $y = 2 - \sqrt{x}$ is not the inverse and $f^{-1}(x) = 2 + \sqrt{x}$.

43. $y = x^2 + 3$, $x = y^2 + 3$, $y^2 = x - 3$,
$y = \pm\sqrt{x-3}$. Since $x \geq 0$ in the function, $y \geq 0$ in the inverse function. So $f^{-1}(x) = \sqrt{x-3}$.

45.
$$f(x) = \sqrt{x+2}$$
$$y = \sqrt{x+2}$$
$$x = \sqrt{y+2}$$
$$x^2 = y+2$$
$$x^2 - 2 = y$$

Since $y \geq 0$ in the function, $x \geq 0$ in the inverse. So $f^{-1}(x) = x^2 - 2$ for $x \geq 0$.

47. The inverse of $f(x) = 2x + 3$ is a function that subtracts 3 and then divides by 2, $f^{-1}(x) = \frac{x-3}{2} = \frac{1}{2}x - \frac{3}{2}$. Use y-intercepts and slopes to graph each straight line.

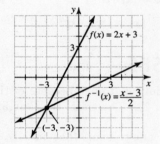

49. The graph of $f(x) = x^2 - 1$ for $x \geq 0$ contains the points $(0, -1)$, $(1, 0)$, and $(2, 3)$. For the inverse function we add 1 and then take the square root, $f^{-1}(x) = \sqrt{x+1}$. The function f^{-1} contains $(-1, 0)$, $(0, 1)$, and $(3, 2)$.

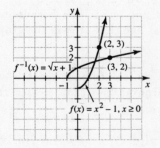

51. The graph of f is a line through (0, 0) with slope 5, and the graph of $f^{-1}(x) = x/5$ is a straight line through (0, 0) with slope 1/5.

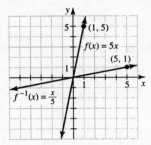

53. The inverse of cubing is the cube root. So $f^{-1}(x) = \sqrt[3]{x}$. The graph of f contains the points (0, 0), (1, 1), (−1, −1), (2, 8), and (−2, −8). The graph of f^{-1} contains the points (0, 0), (1, 1), (−1, −1), (8, 2), and (−8, −2).

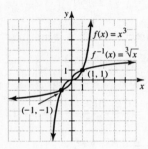

55. The inverse of subtracting 2 and then taking a square root is squaring and then adding 2. So $f^{-1}(x) = x^2 + 2$ for $x \geq 0$. The graph of f contains (2, 0), (3, 1), and (6, 2). The graph of f^{-1} contains (0, 2), (1, 3), and (2, 6).

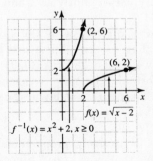

57. $(f^{-1} \circ f)(x) = f^{-1}(f(x)) = f^{-1}(x^3 - 1)$
$$= \sqrt[3]{x^3 - 1 + 1} = x$$

59. $(f^{-1} \circ f)(x) = f^{-1}(f(x)) = f^{-1}(\frac{1}{2}x - 3)$
$$= 2(\frac{1}{2}x - 3) + 6 = x - 6 + 6 = x$$

61. $(f^{-1} \circ f)(x) = f^{-1}(f(x)) = f^{-1}(\frac{1}{x} + 2)$
$$= \frac{1}{\frac{1}{x} + 2 - 2} = \frac{1}{\frac{1}{x}} = x$$

63. $(f^{-1} \circ f)(x) = f^{-1}(f(x)) = f^{-1}(\frac{x + 1}{x - 2})$

$$= \frac{2(\frac{x + 1}{x - 2}) + 1}{\frac{x + 1}{x - 2} - 1} = \frac{\left(\frac{2x + 2}{x - 2} + 1\right)(x - 2)}{\left(\frac{x + 1}{x - 2} - 1\right)(x - 2)} = \frac{2x + 2 + x - 2}{x + 1 - x + 2}$$

$$= \frac{3x}{3} = x$$

65. $S = \sqrt{22.5(150)} \approx 58$ mph
$S^2 = 30D$ or $D = \frac{S^2}{30}$

67. $T(x) = x + 0.09x + 125$ or

$T(x) = 1.09x + 125$, $\quad T^{-1}(x) = \frac{x - 125}{1.09}$

69. $f(x) = x^n$ is invertible if n is an odd positive integer.

71. The functions f and g are not inverses of each other because the function $y_2 = \sqrt{y_1} = \sqrt{x^2}$ is the absolute value function and not the identity function.

8.5 WARM-UPS

1. True, because direct variation means that a is a constant multiple of b. **2.** False, because a inversely proportional to b means $a = k/b$. **3.** False, because jointly proportional means $a = kbc$ for some constant k. **4.** True, a is a constant multiple of the square root of c. **5.** False, b is directly proportional to a means that $b = ka$. **6.** True, because a is a multiple of b and b is divided by c. **7.** False, a is jointly proportional to c and the square of b means that $a = kcb^2$. **8.** False, $a = kc/\sqrt{b}$ is correct. **9.** False, b is directly proportional to a and inversely proportional to the square of c means that $b = ka/c^2$. **10.** True, b is equal to a constant divided by the square of c.

8.5 EXERCISES

1. If a varies directly as m, then a is equal to a constant multiple of m, $a = km$.

3. If d varies inversely as e, then d is equal to a constant divided by e, $d = k/e$.

5. If I varies jointly as r and t, then I is a constant multiple of the product of r and t, $I = krt$.

7. If m is directly proportional to the square of p, then m is a constant multiple of the square of p, $m = kp^2$.

9. If B is directly proportional to the cube root of w, then B is a constant multiple of the cube root of w, $B = k \cdot \sqrt[3]{w}$.

11. If t is inversely proportional to the square of x, then t is equal to a constant divided by the square of x, $t = \dfrac{k}{x^2}$.

13. If v varies directly as m and inversely as n, then v is equal to a constant multiple of m divided by n, $v = \dfrac{km}{n}$.

15. If y varies directly as x, then $y = kx$. If $y = 6$ when $x = 4$, then $6 = k \cdot 4$, or $k = \dfrac{6}{4} = \dfrac{3}{2}$. Therefore, $y = kx$ can be written as $y = \dfrac{3}{2}x$.

17. If A varies inversely as B, then $A = \dfrac{k}{B}$. If $A = 10$ when $B = 3$, then $10 = \dfrac{k}{3}$ or $k = 30$. Therefore, $A = \dfrac{k}{B}$ can be written as $A = \dfrac{30}{B}$.

19. If m varies inversely as the square root of p, then $m = \dfrac{k}{\sqrt{p}}$. If $m = 12$ when $p = 9$, then $12 = \dfrac{k}{\sqrt{9}}$ or $k = 12\sqrt{9} = 36$. Therefore, $m = \dfrac{k}{\sqrt{p}}$ can be written as $m = \dfrac{36}{\sqrt{p}}$.

21. If A varies jointly as t and u, then $A = ktu$. If $A = 6$ when $t = 5$ and $u = 3$, then $6 = k(5)(3)$ or $k = \dfrac{2}{5}$. Therefore, $A = \dfrac{2}{5}tu$.

23. If y varies directly as x and inversely as z, then $y = \dfrac{kx}{z}$. If $y = 2.37$ when $x = \pi$ and $z = \sqrt{2}$, then $2.37 = \dfrac{k\pi}{\sqrt{2}}$ or $k = \dfrac{2.37\sqrt{2}}{\pi} = 1.067$. So, $y = \dfrac{1.067x}{z}$.

25. If y varies directly as x, then $y = kx$. Since $y = 7$ when $x = 5$, we have $7 = k(5)$ or $k = \dfrac{7}{5}$. So, $y = kx$ can be written as $y = \dfrac{7}{5}x$. To find y when $x = -3$, we get $y = \dfrac{7}{5}(-3) = -\dfrac{21}{5}$.

27. If w varies inversely as z, then $w = \dfrac{k}{z}$. Since $w = 6$ when $z = 2$, we have $6 = \dfrac{k}{2}$ or $k = 12$. Therefore, $w = \dfrac{12}{z}$. Now when $z = -8$, we get $w = \dfrac{12}{-8} = -\dfrac{3}{2}$.

29. If A varies jointly as F and T, then $A = kFT$. Since $A = 6$ when $F = 3\sqrt{2}$ and $T = 4$, we have $6 = k3\sqrt{2}(4)$ or $k = \dfrac{\sqrt{2}}{4}$. Therefore, $A = \dfrac{\sqrt{2}}{4}FT$. When $F = 2\sqrt{2}$ and $T = \dfrac{1}{2}$, we get $A = \dfrac{\sqrt{2}}{4}(2\sqrt{2})(\dfrac{1}{2}) = \dfrac{1}{2}$.

31. If D varies directly with t and inversely with the square of s, then $D = \dfrac{kt}{s^2}$. Since $D = 12.35$ when $t = 2.8$ and $s = 2.48$, we get

$$12.35 = \dfrac{k2.8}{(2.48)^2} \text{ or } k = \dfrac{12.35(2.48)^2}{2.8} = 27.128.$$

When $t = 5.63$ and $s = 6.81$, we get

$$D = \dfrac{(27.128)(5.63)}{(6.81)^2} = 3.293.$$

33. Inverse variation, y varies inversely as x.

35. Direct variation, y varies directly as x.

37. Combined variation, y varies directly as x and inversely as w.

39. Joint variation, y varies jointly as x and z.

41. The cost c varies directly with the size of the lawn s means that $c = ks$. Since $c = \$280$ when $s = 4000$, we get $280 = k(4000)$ or $k = 0.07$. The formula is now written as $c = 0.07s$. For a 6000 square foot lawn, the cost is $c = 0.07(6000) = \$420$.

43. The volume v is inversely proportional to the weight w means that $v = k/w$. Since the volume is 6 when the weight is 30, we have $6 = k/30$ or $k = 180$. Therefore, $v = 180/w$. If the weight is 20 kg, then $v = 180/20 = 9 \text{ cm}^3$.

45. The price is jointly proportional to the radius and length means that $P = kRL$. Since a 12 foot culvert with a 6 inch radius costs \$324, we have $324 = k(6)(12)$ or $k = 4.5$. Therefore, the formula is $P = 4.5RL$. The price of a 10 foot culvert with an 8 inch radius is $P = 4.5(8)(10) = \$360$.

47. The price varies jointly as the length and the square of the diameter means that $P = kLD^2$. Since an 18 foot rod with a diameter of 2 inches has a price of \$12.60, we get $12.60 = k(18)(2)^2$ or $k = 0.175$. Now the formula is written $P = 0.175LD^2$. The cost of a 12 foot rod with a 3 inch diameter is $P = 0.175(12)(3)^2 = \$18.90$.

49. The distance varies directly as the square of the time means that $d = kt^2$. If an object falls 0.16 feet in 0.1 seconds, then $0.16 = k(0.1)^2$ or $k = 16$. So the formula is $d = 16t^2$. In 0.5 seconds the object falls $d = 16(0.5)^2 = 4$ feet.

51. The force is inversely proportional to the length means that $F = k/L$. If a force of 2000 pounds is needed at 2 feet, then $2000 = k/2$ or $k = 4000$. The formula is now written as $F = 4000/L$. At 10 feet the force would be $F = 4000/10 = 400$ pounds.

53. Since $t = \frac{kc}{n}$ and $t = 8$ when $n = 3$ and $c = 6$, we have $8 = \frac{k6}{3}$, or $k = 4$. So $t = \frac{4c}{n}$. Now if $c = 9$ and $n = 5$, $t = \frac{4 \cdot 9}{5} = 7.2$ days.

55. y_1 increasing, y_2 decreasing

8.6 WARM-UPS

1. True, when dividing by $x - 5$ we use 5 for synthetic division. **2.** True, when dividing by $x + 7$, or $x - (-7)$, -7 is used in synthetic division. **3.** False, because the remainder of synthetic division is 2 and not 0.

4. True, because 2 is a solution to $x^3 - 8 = 0$ implies that $x - 2$ is a factor of $x^3 - 8$ and so the remainder is 0.

5. True, because of the factor theorem.

6. True, because of the factor theorem. If c is a solution to $P(x) = 0$, then $x - c$ is a factor of $P(x)$.

7. True, because $1^{35} - 3(1)^{24} + 2(1)^{18} = 0$ and so by the factor theorem $x - 1$ is a factor of the polynomial.

8. True, because $(-1)^3 - 3(-1)^2 + (-1) + 5 = 0$ and so by the factor theorem $x + 1$ is a factor of the polynomial.

9. True, because $1^3 - 5(1) + 4 = 0$ implies that $x - 1$ is a factor and so the remainder is 0.

10. True, because of the factor theorem.

8.6 EXERCISES

1.

$$
\begin{array}{r|rrrr}
1 & 1 & -1 & 1 & -1 \\
  &   & 1 & 0 & 1 \\
\hline
  & 1 & 0 & 1 & 0 \\
\end{array}
$$

Since the remainder is 0, 1 is a zero of the polynomial function.

3.

$$
\begin{array}{r|rrrrr}
-3 & -1 & -3 & -2 & 0 & 18 \\
   &   & 3 & 0 & 6 & -18 \\
\hline
   & -1 & 0 & -2 & 6 & 0 \\
\end{array}
$$

Since the remainder is 0, -3 is a zero of the function.

5.

$$
\begin{array}{r|rrrr}
2 & 2 & -4 & -5 & 9 \\
  &   & 4 & 0 & 10 \\
\hline
  & 2 & 0 & -5 & 19 \\
\end{array}
$$

Since the remainder is 19 and not 0, 2 is not a zero of the function.

7.

$$
\begin{array}{r|rrrr}
-3 & 1 & 5 & 2 & -12 \\
   &   & -3 & -6 & 12 \\
\hline
   & 1 & 2 & -4 & 0 \\
\end{array}
$$

Since the remainder is 0, -3 is a solution to the equation.

9.

$$
\begin{array}{r|rrrrr}
-2 & 1 & 3 & -5 & -10 & 5 \\
   &   & -2 & -2 & 14 & -8 \\
\hline
   & 1 & 1 & -7 & 4 & -3 \\
\end{array}
$$

Since the remainder is -3 and not 0, -2 is not a solution to the equation.

11.

$$
\begin{array}{r|rrrrr}
4 & -2 & 0 & 30 & 5 & 12 \\
  &   & -8 & -32 & -8 & -12 \\
\hline
  & -2 & -8 & -2 & -3 & 0 \\
\end{array}
$$

Since the remainder is 0, 4 is a solution to the equation.

13.

$$
\begin{array}{r|rrr}
3 & 0.8 & -0.3 & -6.3 \\
  &   & 2.4 & 6.3 \\
\hline
  & 0.8 & 2.1 & 0 \\
\end{array}
$$

Since the remainder is not 0, 3 is not a solution to the equation.

15.

$$3 \overline{\smash{\big)}\begin{array}{rrrr} 1 & 0 & -6 & -9 \\ & 3 & 9 & 9 \\ \hline 1 & 3 & 3 & 0 \end{array}}$$

Since the remainder is 0, $x-3$ is a factor. The other factor is the quotient.

$$x^3 - 6x - 9 = (x-3)(x^2 + 3x + 3)$$

17.

$$-5 \overline{\smash{\big)}\begin{array}{rrrr} 1 & 9 & 23 & 15 \\ & -5 & -20 & -15 \\ \hline 1 & 4 & 3 & 0 \end{array}}$$

Since the remainder is 0, $x+5$ is a factor of the polynomial. The quotient is the other factor.

$$\begin{aligned} x^3 + 9x^2 + 23x + 15 &= (x+5)(x^2 + 4x + 3) \\ &= (x+5)(x+3)(x+1) \end{aligned}$$

19.

$$2 \overline{\smash{\big)}\begin{array}{rrrr} 1 & -8 & 4 & -6 \\ & 2 & -12 & 16 \\ \hline 1 & -6 & 8 & 10 \end{array}}$$

Since the remainder is 10 and not 0, $x-2$ is not a factor of the polynomial.

21.

$$-1 \overline{\smash{\big)}\begin{array}{rrrrr} 1 & 1 & 0 & -8 & -8 \\ & -1 & 0 & 0 & 8 \\ \hline 1 & 0 & 0 & -8 & 0 \end{array}}$$

Since the remainder is 0, $x+1$ is a factor. The other factor is the quotient.

$$\begin{aligned} x^4 + x^3 - 8x - 8 &= (x+1)(x^3 - 8) \\ &= (x+1)(x-2)(x^2 + 2x + 4) \end{aligned}$$

23.

$$0.5 \overline{\smash{\big)}\begin{array}{rrrr} 2 & -3 & -11 & 6 \\ & 1 & -1 & -6 \\ \hline 2 & -2 & -12 & 0 \end{array}}$$

Since the remainder is 0, $x-0.5$ is a factor.

$$\begin{aligned} 2x^3 - 3x^2 - 11x + 6 &= (x-0.5)(2x^2 - 2x - 12) \\ &= (x-0.5)(2)(x^2 - x - 6) \\ &= (x-0.5)(2)(x-3)(x+2) \\ &= (2x-1)(x-3)(x+2) \end{aligned}$$

25. Try integers between -5 and 5 with synthetic division until we find a solution. Try 1 first.

$$1 \overline{\smash{\big)}\begin{array}{rrrr} 1 & 0 & -13 & 12 \\ & 1 & 1 & -12 \\ \hline 1 & 1 & -12 & 0 \end{array}}$$

$$\begin{aligned} x^3 - 13x + 12 &= 0 \\ (x-1)(x^2 + x - 12) &= 0 \\ (x-1)(x+4)(x-3) &= 0 \end{aligned}$$

$$x-1 = 0 \ \text{ or } \ x+4 = 0 \ \text{ or } \ x-3 = 0$$
$$x = 1 \quad \text{ or } \quad x = -4 \quad \text{ or } \quad x = 3$$

The solution set is $\{-4, 1, 3\}$.

27. Try integers between -5 and 5 with synthetic division to find a solution. Try 2 first.

$$2 \overline{\smash{\big)}\begin{array}{rrrr} 2 & -9 & 7 & 6 \\ & 4 & -10 & -6 \\ \hline 2 & -5 & -3 & 0 \end{array}}$$

$$\begin{aligned} 2x^3 - 9x^2 + 7x + 6 &= 0 \\ (x-2)(2x^2 - 5x - 3) &= 0 \\ (x-2)(2x+1)(x-3) &= 0 \end{aligned}$$

$$x-2 = 0 \ \text{ or } \ 2x+1 = 0 \ \text{ or } \ x-3 = 0$$
$$x = 2 \ \text{ or } \qquad x = -1/2 \ \text{ or } \ x = 3$$

The solution set is $\left\{-\frac{1}{2}, 2, 3\right\}$.

29. The only integral solution between -5 and 5 is -4:

$$-4 \overline{\smash{\big)}\begin{array}{rrrr} 2 & -3 & -50 & -24 \\ & -8 & 44 & 24 \\ \hline 2 & -11 & -6 & 0 \end{array}}$$

$$\begin{aligned} 2x^3 - 3x^2 - 50x - 24 &= 0 \\ (x+4)(2x^2 - 11x - 6) &= 0 \\ (x+4)(2x+1)(x-6) &= 0 \end{aligned}$$

$$x+4 = 0 \ \text{ or } \ 2x+1 = 0 \ \text{ or } \ x-6 = 0$$
$$x = -4 \ \text{ or } \qquad x = -1/2 \ \text{ or } \ x = 6$$

The solution set is $\left\{-4, -\frac{1}{2}, 6\right\}$.

31. Check to see if 1 is a solution.

$$1 \overline{\smash{\big)}\begin{array}{rrrr} 1 & 5 & 3 & -9 \\ & 1 & 6 & 9 \\ \hline 1 & 6 & 9 & 0 \end{array}}$$

$$\begin{aligned} x^3 + 5x^2 + 3x - 9 &= 0 \\ (x-1)(x^2 + 6x + 9) &= 0 \\ (x-1)(x+3)^2 &= 0 \end{aligned}$$

$$x-1 = 0 \ \text{ or } \ x+3 = 0$$
$$x = 1 \ \text{ or } \qquad x = -3$$

The solution set is $\{-3, 1\}$.

33. Check to see if 2 is a solution.

$$2 \overline{\smash{\big)}\begin{array}{rrrrr} 1 & -4 & 3 & 4 & -4 \\ & 2 & -4 & -2 & 4 \\ \hline 1 & -2 & -1 & 2 & 0 \end{array}}$$

160

$$x^4 - 4x^3 + 3x^2 + 4x - 4 = 0$$
$$(x-2)(x^3 - 2x^2 - x + 2) = 0$$
$$(x-2)(x^2[x-2] - 1[x-2]) = 0$$
$$(x-2)(x^2 - 1)(x-2) = 0$$
$$(x-2)(x-1)(x+1)(x-2) = 0$$
$$x - 2 = 0 \text{ or } x - 1 = 0 \text{ or } x + 1 = 0$$
$$x = 2 \text{ or } \quad x = 1 \text{ or } \quad x = -1$$

The solution set is $\{-1, 1, 2\}$.

37. $\{-2, 1, 5\}, \left\{-\frac{3}{2}, \frac{3}{2}, \frac{5}{2}\right\}$

CHAPTER 8 REVIEW

1. The relation is not a function because two ordered pairs have the same first coordinate and different second coordinates.

3. The relation is a function because the condition that $y = x^2$ guarantees that no two ordered pairs will have the same x-coordinate and different y-coordinates.

5. This relation is not a function because the ordered pairs $(16, 2)$ and $(16, -2)$ are both in the relation.

7. The domain is the set of x-coordinates $\{1, 2\}$, and the range is the set of y-coordinates $\{3\}$.

9. Since any number can be used for t, the domain is the set of all real numbers, $(-\infty, \infty)$. Since any real number can occur as $f(t)$, the range is the set of all real numbers, $(-\infty, \infty)$.

11. Since any number can be used for x, the domain is $(-\infty, \infty)$. Since $y = |x|$, the values of y will be nonnegative and so the range is $[0, \infty)$.

13. $f(0) = 2(0) - 5 = -5$

15. $g(0) = 0^2 + 0 - 6 = -6$

17. $g\left(\frac{1}{2}\right) = \left(\frac{1}{2}\right)^2 + \frac{1}{2} - 6 = \frac{1}{4} + \frac{2}{4} - \frac{24}{4} = -\frac{21}{4}$

19. $f(a) = 2a - 5$

21. $g(a - 1) = (a - 1)^2 + (a - 1) - 6$
$\qquad\qquad = a^2 - 2a + 1 + a - 1 - 6$
$\qquad\qquad = a^2 - a - 6$

23. If $f(a) = 1$, then $2a - 5 = 1$.
$$2a = 6$$
$$a = 3$$

25. Average rate of change is
$$\frac{f(8) - f(2)}{8 - 2} = \frac{20 \cdot 8^2 - 20 \cdot 2^2}{6} = 200$$

27. Average rate of change is
$$\frac{g(x+h) - g(x)}{x + h - x} = \frac{(x+h)^2 + 2(x+h) - (x^2 + 2x)}{h}$$
$$= \frac{x^2 + 2xh + h^2 + 2x + 2h - x^2 - 2x}{h}$$
$$= \frac{2xh + h^2 + 2h}{h} = 2x + h + 2$$

29. The graph of $f(x) = 3x - 4$ is a straight line with y-intercept $(0, -4)$ and slope 3. Since any number can be used for x, the domain is $(-\infty, \infty)$. From the graph we see that any real number can occur as a y-coordinate, so the range is $(-\infty, \infty)$.

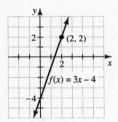

31. The graph of $h(x) = |x| - 2$ includes the points $(0, -2)$, $(1, -1)$, $(2, 0)$, $(-1, -1)$, and $(-2, 0)$. Draw a v-shaped graph through these points.

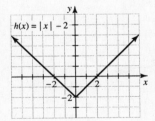

Since any real number can be used for x, the domain is $(-\infty, \infty)$. From the graph we see that all y-coordinates are greater than or equal to -2, and so the range is $[-2, \infty)$.

33. The graph of $y = x^2 - 2x + 1$ includes the points $(0, 1)$, $(1, 0)$, $(2, 1)$, $(3, 4)$, and $(-1, 4)$. Draw a parabola through these points.

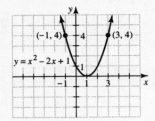

Since any number can be used for x, the domain is $(-\infty, \infty)$. Since the y-coordinates on the graph are nonnegative, the range is $[0, \infty)$.

35. The graph of $k(x) = \sqrt{x} + 2$ includes the points $(0, 2)$, $(1, 3)$, and $(4, 4)$.

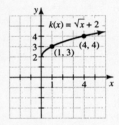

Since x must be nonnegative, the domain is $[0, \infty)$. The graph shows that no y-coordinate is less than 2, and so the range is $[2, \infty)$.

37. The graph of $y = 2/x$ includes the points $(1, 2)$, $(2, 1)$, $(4, 1/2)$, $(1/2, 4)$, $(-1, -2)$, $(-2, -1)$, $(-4, -1/2)$, and $(-1/2, -4)$.

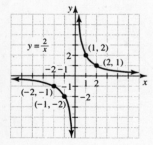

Since x must be nonzero, the domain is $(-\infty, 0) \cup (0, \infty)$. From the graph we see that y is also nonzero, and so the range is $(-\infty, 0) \cup (0, \infty)$.

39. The graph of $x = 2$ is the vertical line with x-intercept $(2, 0)$.

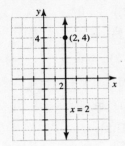

Since the only x-coordinate used on this graph is 2, the domain is $\{2\}$. Since all real numbers occur as y-coordinates, the range is $(-\infty, \infty)$.

41. The graph of $x = |y| + 1$ includes the points $(1, 0)$, $(2, 1)$, $(2, -1)$, $(3, 2)$, and $(3, -2)$. Draw a v-shaped graph through these points.

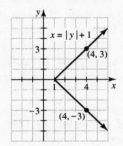

From the graph we see that the x-coordinates are not less than 1, and so the domain is $[1, \infty)$. Any number can be used for y, and so the range is $(-\infty, \infty)$.

43. $x = \dfrac{-(-4)}{2(1)} = 2$, $y = 2^2 - 4(2) + 3 = -1$

Vertex $(2, -1)$, upward

45. $x = \dfrac{-(8)}{2(-2)} = 2$, $y = -2(2)^2 + 8 \cdot 2 + 9 = 17$

Vertex $(2, 17)$, downward

47. $f(-3) = 3(-3) + 5 = -4$

49. $(h \circ f)(\sqrt{2}) = h(f(\sqrt{2})) = h(3\sqrt{2} + 5)$

$$= \frac{3\sqrt{2} + 5 - 5}{3} = \frac{3\sqrt{2}}{3} = \sqrt{2}$$

162

51. $(g \circ f)(2) = g(f(2)) = g(11) = 11^2 - 2(11)$
$$= 99$$

53. $(f+g)(3) = f(3) + g(3)$
$$= 3 \cdot 3 + 5 + 3^2 - 2 \cdot 3 = 17$$

55. $(f \cdot g)(x) = f(x) \cdot g(x) = (3x + 5)(x^2 - 2x)$
$$= 3x^3 + 5x^2 - 6x^2 - 10x = 3x^3 - x^2 - 10x$$

57. $(f \circ f)(0) = f(f(0)) = f(5) = 3 \cdot 5 + 5 = 20$

59. $F(x) = |x + 2| = |g(x)| = f(g(x))$
$$= (f \circ g)(x)$$

Therefore, F is the same function as f composite g, and we write $F = f \circ g$.

61. $H(x) = x^2 + 2 = h(x) + 2 = g(h(x))$
$$= (g \circ h)(x)$$

Therefore, H is the same function as g composite h, and we write $H = g \circ h$.

63. $I(x) = x + 4 = x + 2 + 2 = g(x) + 2 = g(g(x))$

$= (g \circ g)(x)$

Therefore, I is the same function as g composite g, and we write $I = g \circ g$.

65. This function is not invertible, because it is not one-to-one. The ordered pairs have different first coordinates and the same second coordinate.

67. The function $f(x) = 8x$ is invertible. The inverse of multiplication by 8 is division by 8 and $f^{-1}(x) = x/8$.

69. The function $g(x) = 13x - 6$ is one-to-one and so it is invertible. The inverse of multiplying by 13 and subtracting 6 is adding 6 and then dividing by 13, $g^{-1}(x) = \dfrac{x + 6}{13}$.

71.
$$j(x) = \frac{x + 1}{x - 1}$$
$$y = \frac{x + 1}{x - 1}$$
$$x = \frac{y + 1}{y - 1}$$
$$x(y - 1) = y + 1$$
$$xy - x = y + 1$$
$$xy - y = x + 1$$
$$y(x - 1) = x + 1$$
$$y = \frac{x + 1}{x - 1}$$
$$j^{-1}(x) = \frac{x + 1}{x - 1}$$

73. The function $m(x) = (x - 1)^2$ is not invertible because it is not one-to-one. The ordered pairs $(2, 1)$ and $(0, 1)$ are both in the function.

75. The function $f(x) = 3x - 1$ is inverted by adding 1 and then dividing by 3:
$$f^{-1}(x) = \frac{x + 1}{3} = \frac{1}{3}x + \frac{1}{3}$$

Use slope and y-intercept to graph both functions on the same coordinate system.

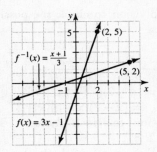

77. The function $f(x) = x^3/2$ is inverted by multiplying by 2 and then taking the cube root:
$$f^{-1}(x) = \sqrt[3]{2x}$$
The graph of f includes the points $(0, 0)$, $(1, 1/2)$, $(-1, -1/2)$, $(2, 4)$, and $(-2, 4)$. The graph of f^{-1} includes the points $(0, 0)$, $(1/2, 1)$, $(-1/2, -1)$, $(4, 2)$, and $(-4, -2)$. Plot these points and graph both functions on the same coordinate system.

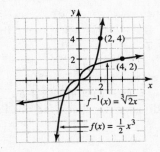

79. If y varies directly as m, then $y = km$. If $y = -3$ when $m = 1/4$, then
$$-3 = k \cdot \frac{1}{4}$$
or
$$-12 = k.$$

Since $y = -12m$, when $m = -2$ we get $y = -12(-2) = 24$.

163

81. If c varies directly as m and inversely as n, then $c = km/n$. If $c = 20$ when $m = 10$ and $n = 4$, then

$$20 = \frac{k(10)}{4}$$

or $\qquad 8 = k.$

Since $c = 8m/n$, when $m = 6$ and $n = -3$, we get

$$c = \frac{8(6)}{-3} = -16.$$

83. Use synthetic division to check the possible solutions. Try 2 first.

$$\begin{array}{r|rrrr} 2 & 1 & -4 & -11 & 30 \\ & & 2 & -4 & -30 \\ \hline & 1 & -2 & -15 & 0 \end{array}$$

$$\begin{aligned} x^3 - 4x^2 - 11x + 30 &= (x-2)(x^2 - 2x - 15) \\ &= (x-2)(x-5)(x+3) \end{aligned}$$

There is one solution for each factor. The solution set is $\{-3, 2, 5\}$.

85. Use synthetic division to check the possible solutions. Try -2 first.

$$\begin{array}{r|rrrr} -2 & 1 & -5 & -2 & 24 \\ & & -2 & 14 & -24 \\ \hline & 1 & -7 & 12 & 0 \end{array}$$

$$\begin{aligned} x^3 - 5x^2 - 2x + 24 &= (x+2)(x^2 - 7x + 12) \\ &= (x+2)(x-3)(x-4) \end{aligned}$$

There is one solution for each factor of the polynomial. The solution set is $\{-2, 3, 4\}$.

87. Use synthetic division to check for possible factors.

$$\begin{array}{r|rrrr} 3 & 1 & 4 & -11 & -30 \\ & & 3 & 21 & 30 \\ \hline & 1 & 7 & 10 & 0 \end{array}$$

$$\begin{aligned} x^3 + 4x^2 - 11x - 30 &= (x-3)(x^2 + 7x + 10) \\ &= (x-3)(x+2)(x+5) \end{aligned}$$

89. Use synthetic division to check for possible factors.

$$\begin{array}{r|rrrr} -2 & 1 & 0 & -5 & -6 \\ & & -2 & 4 & 2 \\ \hline & 1 & -2 & -1 & -4 \end{array}$$

Since the remainder is not 0, $x + 2$ is not a factor of the polynomial.

91. The cost is a linear function of the number of pounds n, so the straight line goes through the two points (400, 1800) and (600, 2300). The slope of the line is

$$m = \frac{2300 - 1800}{600 - 400} = \frac{500}{200} = \frac{5}{2}.$$

Use point-slope form to get the equation of the line.

$$C - 1800 = \frac{5}{2}(n - 400)$$

$$C - 1800 = \frac{5}{2}n - 1000$$

$$C = \frac{5}{2}n + 800$$

93. $d = kt^2$
If $d = 144$ when $t = 3$, then $144 = k3^2$, or $k = 16$. So $d = 16t^2$. In 4 seconds the ball travels $d = 16 \cdot 4^2 = 256$ ft.

95. The area of the circle is $A = \pi r^2$. Let $s =$ the length of the side of the square. Since the square is inscribed in the circle, the length of the diagonal of the square is the same as the diameter of the circle 2r. Since the diagonal of the square is the hypotenuse of a right triangle whose sides are s and s, we can write the following equation.

$$s^2 + s^2 = (2r)^2$$

$$2s^2 = 4r^2$$

$$s^2 = 2r^2$$

Since the area of the square is s^2, we have $B = s^2$, or $B = 2r^2$. Since $r^2 = \frac{A}{\pi}$, we can write $B = 2 \cdot \frac{A}{\pi}$, or $B = \frac{2A}{\pi}$.

97. Substitute $k = 5w - 6$ into the equation $a = 3k + 2$:

$$a = 3(5w - 6) + 2$$
$$a = 15w - 18 + 2$$
$$a = 15w - 16$$

99. The formula for the profit is $P(x) = -x^2 + 50x - 400$. The x-coordinate of the vertex is

$$x = -\frac{b}{2a} = -\frac{50}{2(-1)} = 25.$$

So 25 magazine subscriptions will maximize his profit. His profit for selling 25 subscriptions is $P(25) = -(25)^2 + 50(25) - 400 = \$225$.

101. Average rate of change $= \dfrac{60 - 0}{8.5 - 0}$

≈ 7.1 mph per second.

CHAPTER 8 TEST

1. The relation is not a function because $(-2, 5)$ and $(-2, 3)$ have the same first coordinate and different second coordinates.

2. The relation is a function, because not two ordered pairs have the same first coordinate and different second coordinates.

3. The relation is a function because the formula $y = -\sqrt{x}$ determines a unique second coordinate for each allowable value of x.

4. The relation is not a function, because the ordered pairs $(7, -1)$ and $(7, 1)$ both satisfy $x = |y| + 6$.

5. The domain of $f(x) = |x|$ is $(-\infty, \infty)$ because any real number can be used in place of x. Since the y-coordinates are nonnegative, the range is $[0, \infty)$.

6. In the equation $x = \sqrt{y}$, the values of y must be nonnegative. The values of x are the square roots of nonnegative numbers and they will be nonnegative. So the domain is $[0, \infty)$ and the range is $[0, \infty)$.

7. $f(-3) = -2(-3) + 5 = 6 + 5 = 11$

8. $g(a + 1) = (a + 1)^2 + 4 = a^2 + 2a + 1 + 4$
$$= a^2 + 2a + 5$$

9. $(g \circ f)(x) = g(f(x)) = g(-2x + 5)$
$$= (-2x + 5)^2 + 4 = 4x^2 - 20x + 29$$

10. Since $f(x) = -2x + 5$, the inverse function is to subtract 5 and then divide by -2:
$$f^{-1}(x) = \frac{x - 5}{-2} = -\frac{1}{2}x + \frac{5}{2}$$

11. $(g+f)(2) = g(2) + f(2) = 2^2 + 4 + (-2 \cdot 2 + 5)$
$$= 9$$

12. $(f \cdot g)(1) = f(1) \cdot g(1) = (-2 \cdot 1 + 5)(1^2 + 4)$
$$= (3)(5) = 15$$

13. Since $(f^{-1} \circ f)(x) = x$ for any value of x, we have $(f^{-1} \circ f)(-1782) = -1782$.

14. If $g(x) = 9$, then $x^2 + 4 = 9$, $x^2 = 5$, $x = \pm\sqrt{5}$.

15. The graph of $g(x) = -\frac{2}{3}x + 1$ is a straight line with y-intercept $(0, 1)$ and slope $-\frac{2}{3}$.

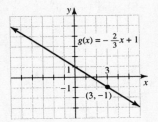

16. The graph of $y = |x| - 4$ includes the points $(0, -4)$, $(2, -2)$, $(-2, 2)$, $(4, 0)$, and $(-4, 0)$. Draw the v-shaped graph through these points.

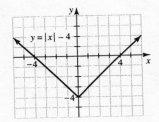

17. The graph of $y = 5 - x^2$ goes through $(\pm 2, 1)$, $(\pm 1, 4)$, and $(0, 5)$.

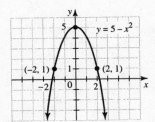

18. The graph of $f(x) = x^2 + 2x - 8$ is a parabola whose vertex has x-coordinate

$$x = -\frac{2}{2(1)} = -1.$$

The parabola goes through the points $(-1, -9)$, $(0, -8)$, $(2, 0)$, $(-2, -8)$, and $(-4, 0)$.

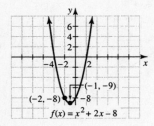

19. The graph of $h(x) = 2\sqrt{x} - 1$ includes the points $(0, -1)$, $(1, 1)$, and $(4, 3)$.

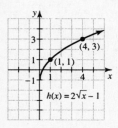

20. The graph of $x = y^2$ includes the points $(0, 0)$, $(1, 1)$, $(1, -1)$, $(4, 2)$, $(4, -2)$.

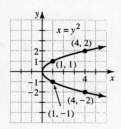

21. Average rate of change of f(x)

$$\frac{f(4) - f(1)}{4 - 1} = \frac{32 - (-1)}{3} = \frac{33}{3} = 11$$

22.
$$f(x) = \frac{2x + 1}{x - 1}$$
$$y = \frac{2x + 1}{x - 1}$$
$$x = \frac{2y + 1}{y - 1}$$
$$x(y - 1) = 2y + 1$$
$$xy - x = 2y + 1$$
$$xy - 2y = x + 1$$
$$y(x - 2) = x + 1$$
$$y = \frac{x + 1}{x - 2}$$
$$f^{-1}(x) = \frac{x + 1}{x - 2}$$

23. Since the volume varies directly as the cube of the radius, we have $V = kr^3$. If $V = 36\pi$ when $r = 3$, we have $36\pi = k3^3$ or $k = 4\pi/3$. To find the volume of a sphere with a radius of 2, use $r = 2$ and $k = 4\pi/3$ in the formula $V = kr^3$:

$$V = \frac{4\pi}{3}(2)^3 = \frac{32\pi}{3} \text{ cubic feet}$$

24. The shipping and handling fee, S, is found by multiplying the weight n by $0.50 and then adding $3.00, $S = 0.50n + 3$.

25. If y varies directly as x and inversely as the square root of z, then $y = kx/\sqrt{z}$. If $y = 12$ when $x = 7$ and $z = 9$, we have

$$12 = \frac{k(7)}{\sqrt{9}} \text{ or } k = \frac{36}{7}.$$

26. Since the function $A(t) = -16t^2 + 40t + 6$ is a quadratic function its maximum value is the second coordinate of the vertex. The value of t for which the altitude is a maximum is the first coordinate of the vertex:

$$t = -\frac{b}{2a} = -\frac{40}{2(-16)} = \frac{40}{32} = \frac{5}{4} \text{ seconds}$$

27.

$$\begin{array}{r|rrrrr}
-1 & 2 & -5 & 3 & 6 & -4 \\
   &   & -2 & 7 & -10 & 4 \\
\hline
   & 2 & -7 & 10 & -4 & 0
\end{array}$$

Because the remainder is 0, $x + 1$ is a factor of the polynomial.

28. Use synthetic division to check each integer from 1 to 10 to see if it is a solution. If the remainder is not 0, try a different number. We will try 3 first.

$$3 \begin{array}{|rrrr} 1 & -12 & 47 & -60 \\ & 3 & -27 & 60 \\ \hline 1 & -9 & 20 & 0 \end{array}$$

$$x^3 - 12x^2 + 47x - 60 = 0$$
$$(x - 3)(x^2 - 9x + 20) = 0$$
$$(x - 3)(x - 4)(x - 5) = 0$$
$$x - 3 = 0 \text{ or } x - 4 = 0 \text{ or } x - 5 = 0$$
$$x = 3 \text{ or } x = 4 \text{ or } x = 5$$

The solution set to the equation is $\{3, 4, 5\}$.

Tying It All Together Chapters 1 - 8

1. $125^{-2/3} = 5^{-2} = \frac{1}{25}$

2. $\left(\frac{8}{27}\right)^{-1/3} = \left(\frac{27}{8}\right)^{1/3} = \frac{3}{2}$

3. $\sqrt{18} - \sqrt{8} = 3\sqrt{2} - 2\sqrt{2} = \sqrt{2}$

4. $x^5 \cdot x^3 = x^{5+3} = x^8$

5. $16^{1/4} = 2$, because $2^4 = 16$.

6. $\frac{x^{12}}{x^3} = x^{12-3} = x^9$

7. $x^2 = 9$
$$x = \pm\sqrt{9} = \pm 3$$
The solution set is $\{\pm 3\}$.

8. $x^2 = 8$, $x = \pm\sqrt{8} = \pm 2\sqrt{2}$
The solution set is $\{\pm 2\sqrt{2}\}$.

9. $x^2 - x = 0$, $x(x - 1) = 0$
$$x = 0 \text{ or } x - 1 = 0$$
The solution set is $\{0, 1\}$.

10. $x^2 - 4x - 6 = 0$

$$x = \frac{4 \pm \sqrt{(-4)^2 - 4(1)(-6)}}{2(1)}$$
$$= \frac{4 \pm \sqrt{40}}{2} = \frac{4 \pm 2\sqrt{10}}{2} = 2 \pm \sqrt{10}$$

The solution set is $\{2 \pm \sqrt{10}\}$.

11. $x^{1/4} = 3$
$$(x^{1/4})^4 = 3^4$$
$$x = 81$$

Since we raised each side to an even power we must check. Since the fourth root of 81 is 3, the solution set is $\{81\}$.

12. If we raise each side to the sixth power, we will get $x = 64$. However, the sixth root of 64 is 2 and not -2. So 64 does not check. The solution set is $\emptyset$.

13. $|x| = 8$
$$x = 8 \text{ or } x = -8$$
The solution set is $\{\pm 8\}$.

14. $|5x - 4| = 21$
$$5x - 4 = 21 \text{ or } 5x - 4 = -21$$
$$5x = 25 \text{ or } 5x = -17$$
$$x = 5 \text{ or } x = -17/5$$
The solution set is $\{-17/5, 5\}$.

15. $x^3 = 8$
$$(x^3)^{1/3} = 8^{1/3}$$
$$x = 2$$
The solution set is $\{2\}$.

16. $(3x - 2)^3 = 27$
$$3x - 2 = 3$$
$$3x = 5$$
$$x = 5/3$$
The solution set is $\{5/3\}$.

17. $\sqrt{2x - 3} = 9$
$$(\sqrt{2x - 3})^2 = 9^2$$
$$2x - 3 = 81$$
$$2x = 84$$
$$x = 42$$
The solution set is $\{42\}$.

18. $\sqrt{x - 2} = x - 8$
$$x - 2 = (x - 8)^2$$
$$x - 2 = x^2 - 16x + 64$$
$$x^2 - 17x + 66 = 0$$
$$(x - 6)(x - 11) = 0$$
$$x = 6 \text{ or } x = 11$$

Only 11 satisfies the original equation. The solution set is $\{11\}$.

19. The graph of the set of points that satisfy the equation $y = 5$ is a horizontal straight line with y-intercept $(0, 5)$.

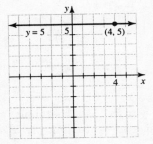

167

20. The graph of the set of points that satisfy $y = 2x - 5$ is a straight line with y-intercept $(0, -5)$ and slope 2.

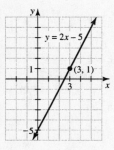

21. The graph of the set of points that satisfy $x = 5$ is a vertical line with x-intercept $(5, 0)$.

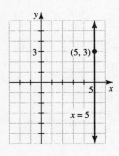

22. The graph of $3y = x$ is the same as the graph of $y = (1/3)x$, a straight line with y-intercept $(0, 0)$ and slope 1/3.

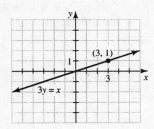

23. The graph of $y = 5x^2$ is a parabola with vertex at $(0, 0)$ and also containing $(1, 5)$ and $(-1, 5)$.

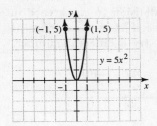

24. The graph of $y = -2x^2$ is a parabola with vertex at $(0, 0)$ and also containing the points $(1, -2)$, and $(-1, -2)$.

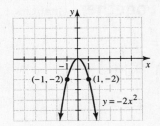

25. Because $2^2 = 4$, and $2^3 = 8$, we have $(2, 4)$ and $(3, 8)$. Because $2^1 = 2$ and $2^4 = 16$, we have $(1, 2)$ and $(4, 16)$.

26. Because $4^{1/2} = 2$ and $4^{-1} = 1/4$, we have $(1/2, 2)$ and $(-1, 1/4)$. Because $4^2 = 16$ and $4^0 = 1$, we have $(2, 16)$ and $(0, 1)$.

27. a) $B = 655 + 9.56(180/2.2) + 1.85(73 \cdot 2.54) - 4.68(30) = 1639.8$ calories

b) $B = 655 + 9.56(170/2.2) + 1.85(72 \cdot 2.54) - 4.68A$

$B = 1732 - 4.68A$

Graph goes through $(20, 1638)$ and $(70, 1404)$.

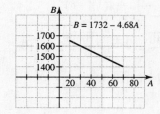

9.1 WARM-UPS

1. False, because $f(-1/2) = 4^{-1/2} = 1/2$.
2. True, because $(1/3)^{-1} = 3$. **3.** False, because the variable is not in the exponent.
4. True, because $(1/2)^x = (2^{-1})^x = 2^{-x}$.
5. True, because it is a one-to-one function.
6. False, the equation $(1/3)^x = 0$ has on solution. **7.** True, because $e^0 = 1$. **8.** False, because the domain of the exponential function $f(x) = 2^x$ is the set of all real numbers.
9. True, because $2^{-x} = 1/(2^x)$.
10. False, at the end of 3 years the investment is worth $500(1.005)^{36}$ dollars

9.1 EXERCISES

1. $f(2) = 4^2 = 16$

3. $f(1/2) = 4^{1/2} = 2$

5. $g(-2) = (1/3)^{-2+1} = (1/3)^{-1} = 3$

7. $g(0) = (1/3)^{0+1} = (1/3)^1 = \frac{1}{3}$

9. $h(0) = -2^0 = -1$

11. $h(-2) = -2^{-2} = -1/4$

13. $h(0) = 10^0 = 1$

15. $h(2) = 10^2 = 100$

17. $j(1) = e^1 = 2.718$

19. $j(-2) = e^{-2} = 0.135$

21. The graph of $f(x) = 4^x$ includes the points $(0, 1)$, $(1, 4)$, $(2, 16)$, and $(-1, 1/4)$.

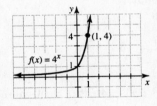

23. The graph of $h(x) = (1/3)^x$ includes $(0, 1)$, $(1, 1/3)$, $(2, 1/9)$, $(-1, 3)$, and $(-2, 9)$.

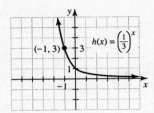

25. The graph of $y = 10^x$ includes the points $(0, 1)$, $(1, 10)$, and $(-1, 1/10)$.

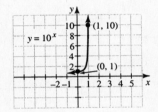

27. The graph of $y = 10^{x+2}$ includes the points $(0, 100)$, $(-1, 10)$, $(-2, 1)$, and $(-3, 1/10)$.

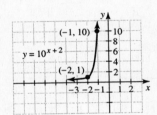

29. The graph of $f(x) = -2^x$ includes $(0, -1)$, $(1, -2)$, $(2, -4)$, $(-1, -1/2)$, and $(-2, -1/4)$.

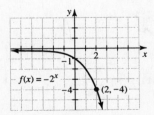

169

31. The graph of $g(x) = 2^{-x}$ includes the points $(0, 1)$, $(-1, 2)$, $(-2, 4)$, and $(1, 1/2)$.

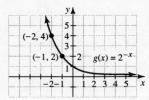

33. The graph of $f(x) = -e^x$ includes the points $(0, -1)$, $(1, -2.7)$, $(2, -7.4)$, $(-1, -0.4)$.

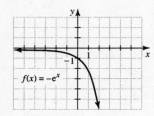

35. The graph of $H(x) = 10^{|x|}$ includes the points $(0, 1)$, $(1, 10)$, and $(-1, 10)$.

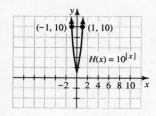

37. The graph of $P = 5000(1.05)^t$ includes the points $(0, 5000)$, $(20, 13266)$, and $(-10, 3070)$.

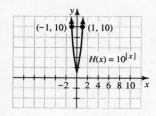

39.
$$2^x = 64$$
$$2^x = 2^6$$
$$x = 6 \qquad \text{By the one-to-one property}$$
The solution set is $\{6\}$.

41.
$$10^x = 0.001$$
$$10^x = 10^{-3}$$
$$x = -3$$
The solution set is $\{-3\}$.

43.
$$2^x = \frac{1}{4}$$
$$2^x = 2^{-2}$$
$$x = -2$$
The solution set is $\{-2\}$.

45.
$$\left(\frac{2}{3}\right)^{x-1} = \frac{9}{4}$$
$$\left(\frac{2}{3}\right)^{x-1} = \left(\frac{2}{3}\right)^{-2}$$
$$x - 1 = -2$$
$$x = -1$$
The solution set is $\{-1\}$.

47.
$$5^{-x} = 25$$

$$5^{-x} = 5^2$$
$$-x = 2 \qquad \text{By the one-to-one property}$$
$$x = -2$$
The solution set is $\{-2\}$.

49.
$$-2^{1-x} = -8$$
$$-2^{1-x} = -2^3$$
$$2^{1-x} = 2^3$$
$$1 - x = 3$$
$$x = -2$$
The solution set is $\{-2\}$.

51.
$$10^{|x|} = 1000$$
$$10^{|x|} = 10^3$$
$$|x| = 3$$
$$x = 3 \quad \text{or} \quad x = -3$$
The solution set is $\{-3, 3\}$.

53. If $f(x) = 2^x$ and $f(x) = 4 = 2^2$, then we must have $x = 2$ by the one-to-one property of exponential functions.

55. Note that $4^{2/3} = (2^2)^{2/3} = 2^{4/3}$. So if $2^x = 2^{4/3}$, then $x = \frac{4}{3}$.

57. Note that $9 = 3^2 = (1/3)^{-2}$. So if $(1/3)^x = (1/3)^{-2}$, then $x = -2$.

59. Note that $1 = (1/3)^0$. So if $(1/3)^x = (1/3)^0$, then $x = 0$.

170

61. Note that $16 = 4^2$. So if $h(x) = 4^{2x-1} = 4^2$, then

$$2x - 1 = 2.$$
$$2x = 3$$
$$x = \frac{3}{2}$$

63. Since $1 = 4^0$, $h(x) = 4^{2x-1} = 4^0$ implies that $2x - 1 = 0$ or $x = \frac{1}{2}$.

65. Compounded quarterly for 10 years means that interest will be paid 40 times at 1.25%.
$$S = 6000\left(1 + \frac{0.05}{4}\right)^{40} = 6000(1.0125)^{40}$$
$$= \$9861.72$$

67. $\$10,000(1 + 0.1624)^{20} \approx \$202,820.58$

The $10,000 investment was worth $100,000 in 1989.

69. When the book is new, $t = 0$, and the value is $V = 45 \cdot 2^{-0.9(0)} \approx \$45$. When $t = 2$, the value is $V = 45 \cdot 2^{-0.9(2)} \approx \$12.92$.

71. A deposit of $500 for 3 years at 7% compounded continuously amounts to $S = 500 \cdot e^{0.07(3)} \approx \$616.84$.

73. A deposit of $80,000 at 7.5% compounded continuously for 1 year amounts to $S = 80,000 \cdot e^{0.075(1)} \approx \$86,230.73$. The interest earned is $86,230.73 - $80,000 = $6230.73.

75. The amount at time $t = 0$ is $A = 300 \cdot e^{-0.06(0)} = 300$ grams. The amount present after 20 years, $t = 20$, is $A = 300 \cdot e^{-0.06(20)} \approx 90.4$ grams.
One-half of the substance decays in about 12 years

77. 2.66666667, 0.0516, 2.8×10^{-5}

9.2 WARM-UPS

1. True, because 3 is the exponent of a that produces 2. **2.** False, because $b = 8^a$ is equivalent to $\log_8(b) = a$. **3.** True, because the inverse of the base a exponential function is the base a logarithm function. **4.** True, because the inverse of the base e exponential function is the base e logarithm function. **5.** False, the domain is $(0, \infty)$. **6.** False, $\log_{25}(5) = 1/2$. **7.** False, $\log(-10)$ is undefined because -10 is not in the domain of the base 10 logarithm function. **8.** False, $\log(0)$ is undefined. **9.** True, because $\log_5(125) = 3$ and $5^3 = 125$. **10.** True, because $(1/2)^{-5} = 2^5 = 32$.

9.2 EXERCISES

1. $\log_2(8) = 3$ is equivalent to $2^3 = 8$.

3. $10^2 = 100$ is equivalent to $\log(100) = 2$.

5. $y = \log_5(x)$ is equivalent to $5^y = x$.

7. $2^a = b$ is equivalent to $\log_2(b) = a$.

9. $\log_3(x) = 10$ is equivalent to $3^{10} = x$.

11. $e^3 = x$ is equivalent to $\ln(x) = 3$.

13. Because $2^2 = 4$, $\log_2(4) = 2$.

15. Because $2^4 = 16$, $\log_2(16) = 4$.

17. Because $2^6 = 64$, $\log_2(64) = 6$.

19. Because $4^3 = 64$, $\log_4(64) = 3$.

21. Because $2^{-2} = \frac{1}{4}$, $\log_2(1/4) = -2$.

23. Because $10^2 = 100$, $\log(100) = 2$.

25. Because $10^{-2} = 0.01$, $\log(0.01) = -2$.

27. Because $(1/3)^1 = 1/3$, $\log_{1/3}(1/3) = 1$.

29. Because $(1/3)^{-3} = 27$, $\log_{1/3}(27) = -3$.

31. Because $e^3 = e^3$, $\ln(e^3) = 3$.

33. Because $e^2 = e^2$, $\ln(e^2) = 2$.

35. Use a calculator with a base 10 logarithm key to find $\log(5) \approx 0.6990$.

37. Use a calculator with a natural logarithm key to find $\ln(6.238) \approx 1.8307$.

39. The graph of $f(x) = \log_3(x)$ includes the points (3, 1), (1, 0), and (1/3, −1).

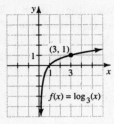

41. The graph of $y = \log_4(x)$ includes the points (4, 1), (1, 0), and (1/4, −1).

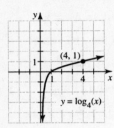

43. The graph of $y = \log_{1/4}(x)$ includes the points (1, 0), (4, −1), and (1/4, 1).

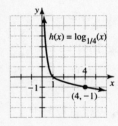

45. The graph of $h(x) = \log_{1/5}(x)$ includes the points (1, 0), (5, −1), and (1/5, 1).

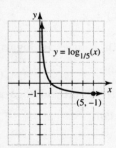

47. The inverse of $f(x) = 6^x$ is $f^{-1}(x) = \log_6(x)$.

49. The inverse of $f(x) = \ln(x)$ is $f^{-1}(x) = e^x$.

51. If $f(x) = \log_{1/2}(x)$ then $f^{-1}(x) = \left(\frac{1}{2}\right)^x$.

53. $x = (1/2)^{-2} = 2^2 = 4$

The solution set is {4}.

55.
$$5 = 25^x$$
$$x = \log_{25}(5) = 1/2$$

The solution set is $\left\{\frac{1}{2}\right\}$.

57.
$$\log(x) = -3$$
$$x = 10^{-3} = 0.001$$
The solution set is {0.001}.

59.
$$\log_x(36) = 2$$
$$x^2 = 36$$
$$x = \pm 6$$

Omit −6 because the base of any logarithm function is positive. The solution set is {6}.

61.
$$\log_x(5) = -1$$
$$x^{-1} = 5$$
$$(x^{-1})^{-1} = 5^{-1}$$
$$x = \frac{1}{5}$$

The solution set is $\left\{\frac{1}{5}\right\}$.

63.
$$\log(x^2) = \log(9)$$
$$x^2 = 9$$
$$x = \pm 3$$

Both 3 and −3 check in the original equation.

The solution set is $\{\pm 3\}$.

65.
$$3 = 10^x$$
$$x = \log(3) \approx 0.4771$$

The solution set is {0.4771}.

67.
$$10^x = \frac{1}{2}$$
$$x = \log(1/2) = \log(0.5) \approx -0.3010$$

The solution set is {−0.3010}.

69.
$$e^x = 7.2$$
$$x = \ln(7.2) \approx 1.9741$$

The solution set is {1.9741}.

71. Use the continuous compounding formula.

$$10{,}000 = 5000 \cdot e^{0.12t}$$
$$2 = e^{0.12t}$$
$$0.12t = \ln(2)$$
$$t = \frac{\ln(2)}{0.12} \approx 5.776 \text{ years}$$

73. To earn \$1000 in interest, the principal must increase from \$6000 to \$7000 in t years.

$$7000 = 6000 \cdot e^{0.08t}$$
$$7/6 = e^{0.08t}$$
$$0.08t = \ln(7/6)$$
$$t = \frac{\ln(7/6)}{0.08} \approx 1.9269 \text{ years}$$

75. $r = \frac{1}{14}\ln\left(\frac{302{,}000}{10{,}000}\right) \approx 0.243 = 24.3\%$

In 2000, $A = 10{,}000e^{20(0.243)} \approx \$1{,}290{,}242.$

77. $pH = -\log_{10}(10^{-4.1}) = -(-4.1) = 4.1$

79. $pH = -\log(9.77 \times 10^{-9}) \approx 8.01$

81. $L = 10 \cdot \log(0.001 \times 10^{12}) = 90 \text{ db}$

83. $f^{-1}(x) = 2^{x-5} + 3$

Domain $(-\infty, \infty)$, Range $(3, \infty)$

85. $y = \ln(e^x) = x$ for $-\infty < x < \infty$

$y = e^{\ln(x)} = x$ for $0 < x < \infty$

9.3 WARM-UPS

1. True, because $\log_2(x^2/8) = \log_2(x^2) - \log_2(8)$ $= \log_2(x^2) - 3.$ **2.** False, because $\log(100) = 2$ and $\log(10) = 1$, and $2 \div 1 \neq 2 - 1.$

3. True, $\ln(\sqrt{2}) = \ln(2^{1/2}) = \frac{1}{2} \cdot \ln(2) = \frac{\ln(2)}{2}.$

4. True, because $\log_3(17)$ is the exponent that we place on 3 to obtain 17.   **5.** False, because $\log_2(1/8) = -3$ and $\log_2(8) = 3.$ **6.** True, because $\ln(8) = \ln(2^3) = 3 \cdot \ln(2).$ **7.** False, because $\ln(1) = 0.$ **8.** False, because $\log(100) = 2$ and $\log(10) = 1.$ **9.** False, because $\log_2(8) = 3$, $\log_2(2) = 1$, and $\log_2(4) = 2.$ **10.** True, because
$\ln(2) + \ln(3) - \ln(7) = \ln(6) - \ln(7) = \ln(6/7).$

9.3 EXERCISES

1. $\log(3) + \log(7) = \log(3 \cdot 7) = \log(21)$

3. $\log_3(\sqrt{5}) + \log_3(\sqrt{x}) = \log_3(\sqrt{5x})$

5. $\log(x^2) + \log(x^3) = \log(x^2 \cdot x^3) = \log(x^5)$

7. $\ln(2) + \ln(3) + \ln(5) = \ln(2 \cdot 3 \cdot 5) = \ln(30)$

9. $\log(x) + \log(x + 3) = \log(x^2 + 3x)$

11. $\log_2(x - 3) + \log_2(x + 2) = \log_2(x^2 - x - 6)$

13. $\log(8) - \log(2) = \log(8/2) = \log(4)$

15. $\log_2(x^6) - \log_2(x^2) = \log_2\left(\frac{x^6}{x^2}\right) = \log_2(x^4)$

17. $\log(\sqrt{10}) - \log(\sqrt{2}) = \log\left(\frac{\sqrt{10}}{\sqrt{2}}\right) = \log(\sqrt{5})$

19. $\ln(4h - 8) - \ln(4) = \ln\left(\frac{4h - 8}{4}\right) = \ln(h - 2)$

21. $\log_2\left(\frac{w^2 - 4}{w + 2}\right) = \log_2\left(\frac{(w - 2)(w + 2)}{w + 2}\right)$

$= \log_2(w - 2)$

23. $\ln\left(\frac{x^2 + x - 6}{x + 3}\right) = \ln\left(\frac{(x + 3)(x - 2)}{x + 3}\right)$

$= \ln(x - 2)$

25. $\log(27) = \log(3^3) = 3 \cdot \log(3)$

27. $\log(\sqrt{3}) = \log(3^{1/2}) = \frac{1}{2} \cdot \log(3)$

29. $\log(3^x) = x \cdot \log(3)$

31. $\log_2(2^{10}) = 10$

33. $5^{\log_5(19)} = 19$

35. $\log(10^8) = 8$

37. $e^{\ln(4.3)} = 4.3$

39. $\log(15) = \log(3 \cdot 5) = \log(3) + \log(5)$

41. $\log(5/3) = \log(5) - \log(3)$

43. $\log(25) = \log(5^2) = 2 \cdot \log(5)$

45. $\log(75) = \log(5^2 \cdot 3) = 2 \cdot \log(5) + \log(3)$

47. $\log\left(\frac{1}{3}\right) = \log(1) - \log(3)$
$= 0 - \log(3) = -\log(3)$

49. $\log(0.2) = \log(1/5) = \log(1) - \log(5)$
$= 0 - \log(5) = -\log(5)$

51. $\log(xyz) = \log(x) + \log(y) + \log(z)$

53. $\log_2(8x) = \log_2(8) + \log_2(x) = 3 + \log_2(x)$

55. $\ln(x/y) = \ln(x) - \ln(y)$

57. $\log(10x^2) = \log(10) + \log(x^2) = 1 + 2 \cdot \log(x)$

59. $\log_5\left[\dfrac{(x-3)^2}{\sqrt{w}}\right] = \log_5\left[(x-3)^2\right] - \log_5(\sqrt{w})$

$$= 2 \cdot \log_5(x-3) - \tfrac{1}{2} \cdot \log_5(w)$$

61. $\ln\left[\dfrac{yz\sqrt{x}}{w}\right] = \ln[yz\sqrt{x}] - \ln(w)$

$$= \ln(y) + \ln(z) + \ln(\sqrt{x}) - \ln(w)$$

$$= \ln(y) + \ln(z) + \tfrac{1}{2} \cdot \ln(x) - \ln(w)$$

63. $\log(x) + \log(x-1) = \log(x^2 - x)$

65. $\ln(3x-6) - \ln(x-2) = \ln\left(\dfrac{3x-6}{x-2}\right) = \ln(3)$

67. $\ln(x) - \ln(w) + \ln(z) = \ln(xz) - \ln(w)$

$$= \ln\left(\tfrac{xz}{w}\right)$$

69. $3 \cdot \ln(y) + 2 \cdot \ln(x) - \ln(w)$

$$= \ln(y^3) + \ln(x^2) - \ln(w)$$

$$= \ln(x^2 y^3) - \ln(w) = \ln\left(\dfrac{x^2 y^3}{w}\right)$$

71. $\tfrac{1}{2} \cdot \log(x-3) - \tfrac{2}{3} \cdot \log(x+1)$

$$= \log\left((x-3)^{1/2}\right) - \log\left((x+1)^{2/3}\right)$$

$$= \log\left(\dfrac{(x-3)^{1/2}}{(x+1)^{2/3}}\right)$$

73. $\tfrac{2}{3} \cdot \log_2(x-1) - \tfrac{1}{4} \cdot \log_2(x+2)$

$$= \log_2\left((x-1)^{2/3}\right) - \log_2\left((x+2)^{1/4}\right)$$

$$= \log_2\left(\dfrac{(x-1)^{2/3}}{(x+2)^{1/4}}\right)$$

75. False, because

$$\log(56) = \log(7 \cdot 8) = \log(7) + \log(8).$$

77. True, because $\log_2(4^2) = \log_2(16) = 4$ and $(\log_2(4))^2 = (2)^2 = 4$.

79. True, because $\ln(25) = \ln(5^2) = 2 \cdot \ln(5)$.

81. False, because $\log_2(64) = 6$ and $\log_2(8) = 3$.

83. True, because $\log(1/3) = \log(1) - \log(3)$

$$= 0 - \log(3) = -\log(3).$$

85. True, $\log_2(16^5) = 5 \cdot \log_2(16) = 5 \cdot 4 = 20$.

87. True, $\log(10^3) = 3 \cdot \log(10) = 3 \cdot 1 = 3$.

89. False, because $\log(100 + 3) = \log(103)$ and

$$2 + \log(3) = \log(100) + \log(3) = \log(300).$$

91. $r = \dfrac{\ln(A) - \ln(P)}{t} = \tfrac{1}{t} \cdot \ln(A/P)$

$= \ln\left((A/P)^{1/t}\right)$ If $A = 3P$ when $t = 5$, then

$r = \ln\left((3P/P)^{1/5}\right) = \ln\left(3^{1/5}\right) \approx 0.2197 = 21.97\%$.

93. b

9.4 WARM-UPS

1. True, because $\log(x-2) + \log(x+2) = \log[(x-2)(x+2)] = \log(x^2 - 4)$. **2.** True, because of the one-to-one property of logarithms. **3.** True, because of the one-to-one property of exponential functions. **4.** False, because the bases are different and the one-to-one property does not apply. **5.** True, because $\log_2(x^2 - 3x + 5) = 3$ is equivalent to $x^2 - 3x + 5 = 2^3 = 8$. **6.** True, $a^x = y$ is equivalent to $\log_a(y) = x$. **7.** True, if $5^x = 23$, then $\ln(5^x) = \ln(23)$, or $x \cdot \ln(5) = \ln(23)$.
8. False, because $\log_3(5) = \dfrac{\ln(5)}{\ln(3)}$.
9. True, $\log_6(2) = \dfrac{\ln(2)}{\ln(6)}$ and $\log_6(2) = \dfrac{\log(2)}{\log(6)}$.
10. False, $\log(5) \approx 0.699$ and $\ln(5) \approx 1.609$.

9.4 EXERCISES

1. $\log_2(x+1) = 3$
$x + 1 = 2^3$
$x + 1 = 8$
$x = 7$
The solution set is $\{7\}$.

3. $\log(x) + \log(5) = 1$

$$\log(5x) = 1$$
$$5x = 10^1$$
$$x = 2$$

The solution set is $\{2\}$.

5. $\log_2(x-1) + \log_2(x+1) = 3$

$$\log_2[(x-1)(x+1)] = 3$$
$$\log_2(x^2 - 1) = 3$$
$$x^2 - 1 = 2^3$$
$$x^2 = 9$$
$$x = \pm 3$$

If $x = -3$ in the original equation, $\log_2(-3-1)$ is undefined. The solution set is $\{3\}$.

7. $\log_2(x-1) - \log_2(x+2) = 2$

$$\log_2\left(\frac{x-1}{x+2}\right) = 2$$
$$\frac{x-1}{x+2} = 2^2$$
$$x - 1 = 4(x+2)$$
$$x - 1 = 4x + 8$$
$$-9 = 3x$$
$$-3 = x$$

If $x = -3$ in the original equation, we get the logarithm of a negative number, which is undefined. The solution set is $\emptyset$.

9. $\log_2(x-4) + \log_2(x+2) = 4$

$$\log_2(x^2 - 2x - 8) = 4$$
$$x^2 - 2x - 8 = 2^4$$
$$x^2 - 2x - 8 = 16$$
$$x^2 - 2x - 24 = 0$$
$$(x-6)(x+4) = 0$$
$$x - 6 = 0 \quad \text{or} \quad x + 4 = 0$$
$$x = 6 \quad \text{or} \quad x = -4$$

If $x = -4$ in the original equation, we get $\log_2(-8)$, which is undefined. The solution set is $\{6\}$.

11. $\ln(x) + \ln(x+5) = \ln(x+1) + \ln(x+3)$

$$\ln(x^2 + 5x) = \ln(x^2 + 4x + 3)$$
$$x^2 + 5x = x^2 + 4x + 3$$
$$x = 3$$

The solution set is $\{3\}$.

13.

$\log(x+3) + \log(x+4) = \log(x^3 + 13x^2) - \log(x)$

$$\log(x^2 + 7x + 12) = \log(x^2 + 13x)$$
$$x^2 + 7x + 12 = x^2 + 13x$$
$$12 = 6x$$
$$2 = x$$

The solution set is $\{2\}$.

15. $2 \cdot \log(x) = \log(20 - x)$

$$\log(x^2) = \log(20 - x)$$
$$x^2 = 20 - x$$
$$x^2 + x - 20 = 0$$
$$(x-4)(x+5) = 0$$
$$x - 4 = 0 \quad \text{or} \quad x + 5 = 0$$
$$x = 4 \quad \text{or} \quad x = -5$$

If $x = -5$ in the original equation we get a logarithm of a negative number, which is undefined. The solution set is $\{4\}$.

17. $3^x = 7$

$$x = \log_3(7)$$

The solution set is $\{\log_3(7)\}$.

19.

$$2^{3x+4} = 4^{x-1}$$
$$2^{3x+4} = (2^2)^{x-1}$$
$$2^{3x+4} = (2)^{2x-2}$$
$$3x + 4 = 2x - 2$$
$$x = -6$$

The solution set is $\{-6\}$.

21.

$$(1/3)^x = 3^{1+x}$$
$$(3^{-1})^x = 3^{1+x}$$
$$3^{-x} = 3^{1+x}$$
$$-x = 1 + x$$
$$-2x = 1$$
$$x = -1/2$$

The solution set is $\left\{-\frac{1}{2}\right\}$.

23.

$$2^x = 3^{x+5}$$
$$\ln(2^x) = \ln(3^{x+5})$$
$$x \cdot \ln(2) = (x+5)\ln(3)$$
$$x \cdot \ln(2) = x \cdot \ln(3) + 5 \cdot \ln(3)$$
$$x \cdot \ln(2) - x \cdot \ln(3) = 5 \cdot \ln(3)$$
$$x(\ln(2) - \ln(3)) = 5 \cdot \ln(3)$$
$$x = \frac{5 \cdot \ln(3)}{\ln(2) - \ln(3)} \quad \text{Exact solution}$$

Use a calculator to find $\dfrac{5 \cdot \ln(3)}{\ln(2) - \ln(3)} \approx -13.548$.

25.

$$5^{x+2} = 10^{x-4}$$
$$\log(5^{x+2}) = \log(10^{x-4})$$
$$(x+2)\log(5) = x - 4$$

$$x \cdot \log(5) + 2 \cdot \log(5) = x - 4$$
$$x \cdot \log(5) - x = -4 - 2 \cdot \log(5)$$
$$x[\log(5) - 1] = -4 - 2 \cdot \log(5)$$

$$x = \frac{-4 - 2 \cdot \log(5)}{\log(5) - 1} = \frac{4 + 2 \cdot \log(5)}{1 - \log(5)} \approx 17.932$$

27.
$$8^x = 9^{x-1}$$
$$\ln(8^x) = \ln(9^{x-1})$$
$$x \cdot \ln(8) = (x-1)\ln(9)$$
$$x \cdot \ln(8) = x \cdot \ln(9) - \ln(9)$$
$$x \cdot \ln(8) - x \cdot \ln(9) = -\ln(9)$$
$$x[\ln(8) - \ln(9)] = -\ln(9)$$

$$x = \frac{-\ln(9)}{\ln(8) - \ln(9)} = \frac{\ln(9)}{\ln(9) - \ln(8)} \approx 18.655$$

29. $\log_2(3) = \dfrac{\ln(3)}{\ln(2)} \approx 1.5850$

31. $\log_3(1/2) = \dfrac{\ln(0.5)}{\ln(3)} \approx -0.6309$

33. $\log_{1/2}(4.6) = \dfrac{\ln(4.6)}{\ln(0.5)} \approx -2.2016$

35. $\log_{0.1}(0.03) = \dfrac{\ln(0.03)}{\ln(0.1)} \approx 1.5229$

37. $x \cdot \ln(2) = \ln(7)$

$$x = \frac{\ln(7)}{\ln(2)} \quad \text{This is the exact solution.}$$

Use a calculator to find the approximate value of x, $x \approx 2.8074$.

39. $3x - x \cdot \ln(2) = 1$
$$x(3 - \ln(2)) = 1$$
$$x = \frac{1}{3 - \ln(2)} \quad \text{Exact solution}$$
Use a calculator to find an approximate value for x: $\dfrac{1}{3 - \ln(2)} \approx 0.4335$.

41. $3^x = 5$
$$x = \log_3(5) = \frac{\ln(5)}{\ln(3)} \quad \text{This is the exact solution.}$$
Use a calculator to find an approximate value for x that satisfies the equation, $\dfrac{\ln(5)}{\ln(3)} \approx 1.4650$.

43.
$$2^{x-1} = 9$$
$$\ln(2^{x-1}) = \ln(9)$$
$$(x-1)\ln(2) = \ln(9)$$
$$x - 1 = \frac{\ln(9)}{\ln(2)}$$

$$x = 1 + \frac{\ln(9)}{\ln(2)} \quad \text{Exact solution}$$

Use a calculator to find an approximate value for x: $\quad 1 + \dfrac{\ln(9)}{\ln(2)} \approx 4.1699$.

45. $3^x = 20$
$$x = \log_3(20) \approx 2.7268$$

47. $\log_3(x) + \log_3(5) = 1$
$$\log_3(5x) = 1$$
$$5x = 3^1$$
$$x = \frac{3}{5}$$

49.
$$8^x = 2^{x+1}$$
$$(2^3)^x = 2^{x+1}$$
$$2^{3x} = 2^{x+1}$$
$$3x = x + 1$$
$$2x = 1$$
$$x = \frac{1}{2}$$

51. Use the formula $S = P(1+i)^n$.
$$1500 = 1000(1 + 0.01)^n$$
$$1.5 = (1.01)^n$$
$$n = \log_{1.01}(1.5) = \frac{\ln(1.5)}{\ln(1.01)} \approx 40.749$$
It takes approximately 41 months.

53. $y = 114.308e^{(0.265 \cdot 15.8)} \approx 7524 \text{ ft}^3/\text{sec}$

55.
$$40 = 28e^{0.05t}$$
$$\frac{10}{7} = e^{0.05t}$$
$$0.05t = \ln(10/7)$$
$$t = \frac{\ln(10/7)}{0.05} \approx 7.133$$

There will be 40 million people above the poverty level in approximately 7.1 years.

57.
$$28e^{0.05t} = 20e^{0.07t}$$
$$\ln(28e^{0.05t}) = \ln(20e^{0.07t})$$
$$\ln(28) + \ln(e^{0.05t}) = \ln(20) + \ln(e^{0.07t})$$
$$\ln(28) + 0.05t = \ln(20) + 0.07t$$
$$-0.02t = \ln(20) - \ln(28)$$
$$t = \frac{\ln(20) - \ln(28)}{-0.02} \approx 16.824$$

The number of people above the poverty level will equal the number below the poverty level in approximately 16.8 years.

59.
$$pH = -\log(H^+)$$
$$3.7 = -\log(H^+)$$
$$-3.7 = \log(H^+)$$
$$H^+ = 10^{-3.7} \approx 2.0 \times 10^{-4}$$

61. $d = 0.9183$

63. 2.2894

65. (2.71, 6.54)

67. (1.03, 0.04), (4.74, 2.24)

CHAPTER 9 REVIEW

1. $f(-2) = 5^{-2} = \frac{1}{5^2} = \frac{1}{25}$

3. $f(3) = 5^3 = 125$

5. $g(1) = 10^{1-1} = 10^0 = 1$

7. $g(0) = 10^{0-1} = 10^{-1} = \frac{1}{10}$

9. $h(-1) = (1/4)^{-1} = 4$

11. $h(1/2) = (1/4)^{1/2} = \sqrt{\frac{1}{4}} = \frac{1}{2}$

13. If $f(x) = 25$, then $5^x = 25$, $x = 2$.

15. If $g(x) = 1000$, then $10^{x-1} = 1000$.
$$10^{x-1} = 10^3$$
$$x - 1 = 3$$
$$x = 4$$

17. If $h(x) = 32$, then $(1/4)^x = 32$.
$$(2^{-2})^x = 2^5$$
$$2^{-2x} = 2^5$$
$$-2x = 5$$
$$x = -\frac{5}{2}$$

19. If $h(x) = 1/16$, then $(1/4)^x = 1/16$, or $(1/4)^x = (1/4)^2$, $x = 2$.

21. $f(1.34) = 5^{1.34} \approx 8.6421$

23. $g(3.25) = 10^{3.25-1} = 10^{2.25} \approx 177.828$

25. $h(2.82) = (1/4)^{2.82} = (0.25)^{2.82} \approx 0.02005$

27. $h(\sqrt{2}) = (1/4)^{\sqrt{2}} = (0.25)^{1.414} \approx 0.1408$

29. The graph of $f(x) = 5^x$ includes the points (0, 1), (1, 5), and (−1, 1/5).

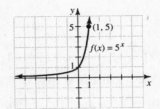

31. The graph of $y = \left(1/5\right)^x$ includes the points (1, 1/5), (0, 1) and (−1, 5).

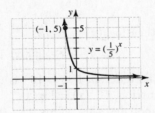

33. The graph of $y = 3^{-x}$ includes the points (0, 1), (1, 1/3), and (−1, 3).

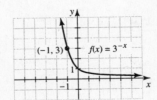

35. The graph of $y = 1 + 2^x$ includes the points $(0, 2)$, $(1, 3)$, $(2, 5)$, and $(-1, 1.5)$.

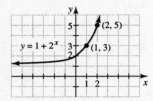

37. $\log(n) = m$

39. $k^h = t$

41. $f(1/8) = \log_2(1/8) = -3$, because $2^{-3} = 1/8$.

43. $g(0.1) = \log(0.1) = -1$, because $10^{-1} = 0.1$.

45. $g(100) = \log(100) = 2$, because $10^2 = 100$.

47. $h(1) = \log_{1/2}(1) = 0$, because $(1/2)^0 = 1$.

49. If $f(x) = 8$, then $\log_2(x) = 8$, $x = 2^8 = 256$.

51. $f(77) = \log_2(77) = \dfrac{\ln(77)}{\ln(2)} \approx 6.267$

53. $h(33.9) = \log_{1/2}(33.9) = \dfrac{\ln(33.9)}{\ln(0.5)} \approx -5.083$

55. If $f(x) = 2.475$, then $\log_2(x) \approx 2.475$.
$$x = 2^{2.475} \approx 5.560$$

57. The inverse of the function $f(x) = 10^x$ is the base 10 logarithm function, $f^{-1}(x) = \log(x)$. The graph of $f(x)$ includes the points $(1, 10)$, $(0, 1)$, and $(-1, 0.1)$. The graph of $f^{-1}(x)$ includes the points $(10, 1)$, $(1, 0)$, and $(0.1, -1)$.

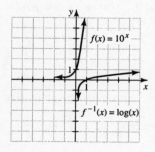

59. The inverse of the function $f(x) = e^x$ is $f^{-1}(x) = \ln(x)$. The graph of $f(x)$ includes the points $(1, e)$, $(0, 1)$, and $(-1, 1/e)$. The graph of $f^{-1}(x)$ includes the points $(e, 1)$, $(1, 0)$, and $(1/e, -1)$.

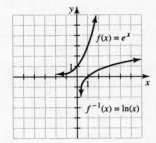

61. $\log(x^2 y) = \log(x^2) + \log(y)$
$$= 2 \cdot \log(x) + \log(y)$$

63. $\ln(16) = \ln(2^4) = 4 \cdot \ln(2)$

65. $\log_5(1/x) = \log_5(1) - \log_5(x) = -\log_5(x)$

67. $\frac{1}{2} \cdot \log(x + 2) - 2 \cdot \log(x - 1)$
$$= \log\left((x+2)^{1/2}\right) - \log\left((x-1)^2\right)$$
$$= \log\left(\frac{\sqrt{x+2}}{(x-1)^2}\right)$$

69. $\log_2(x) = 8$
$$x = 2^8 = 256$$

The solution set is $\{256\}$.

71. $\log_2(8) = x$
$$3 = x$$

The solution set is $\{3\}$.

73. $x^3 = 8$
$$x = \sqrt[3]{8} = 2$$

The solution set is $\{2\}$.

75. $\log_x(27) = 3$
$$x^3 = 27$$
$$x = \sqrt[3]{27} = 3$$

The solution set is $\{3\}$.

77.
$$x \cdot \ln(3) - x = \ln(7)$$
$$x[\ln(3) - 1] = \ln(7)$$
$$x = \frac{\ln(7)}{\ln(3) - 1}$$

The solution set is $\left\{ \dfrac{\ln(7)}{\ln(3) - 1} \right\}$.

79.
$$3^x = 5^{x-1}$$
$$\ln(3^x) = \ln(5^{x-1})$$
$$x \cdot \ln(3) = (x-1)\ln(5)$$
$$x \cdot \ln(3) = x \cdot \ln(5) - \ln(5)$$
$$x \cdot \ln(3) - x \cdot \ln(5) = -\ln(5)$$
$$x[\ln(3) - \ln(5)] = -\ln(5)$$
$$x = \frac{-\ln(5)}{\ln(3) - \ln(5)} = \frac{\ln(5)}{\ln(5) - \ln(3)}$$

The solution set is $\left\{ \dfrac{\ln(5)}{\ln(5) - \ln(3)} \right\}$.

81.
$$4^{2x} = 2^{x+1}$$
$$(2^2)^{2x} = 2^{x+1}$$
$$2^{4x} = 2^{x+1}$$
$$4x = x + 1$$
$$3x = 1$$
$$x = \frac{1}{3}$$

The solution set is $\left\{ \dfrac{1}{3} \right\}$.

83.
$$\ln(x+2) - \ln(x-10) = \ln(2)$$
$$\ln\left(\frac{x+2}{x-10}\right) = \ln(2)$$
$$\frac{x+2}{x-10} = 2$$
$$x + 2 = 2(x-10)$$
$$x + 2 = 2x - 20$$
$$22 = x$$

The solution set is $\{22\}$.

85.
$$\log(x) - \log(x-2) = 2$$
$$\log\left(\frac{x}{x-2}\right) = 2$$
$$\frac{x}{x-2} = 10^2$$
$$x = 100(x-2)$$

$$x = 100x - 200$$
$$-99x = -200$$
$$x = \frac{-200}{-99} = \frac{200}{99}$$

The solution set is $\left\{\dfrac{200}{99}\right\}$.

87.
$$6^x = 12$$
$$x = \log_6(12) = \frac{\ln(12)}{\ln(6)} \approx 1.3869$$

The solution set is $\{1.3869\}$.

89.
$$3^{x+1} = 5$$
$$x + 1 = \log_3(5)$$
$$x = -1 + \log_3(5) = -1 + \frac{\ln(5)}{\ln(3)} \approx 0.4650$$

The solution set is $\{0.4650\}$.

91. Use the formula $S = P(1+i)^n$ with $i = 11.5\% = 0.115$, $n = 15$, and $P = \$10,000$.

$$S = 10,000(1.115)^{15} \approx \$51,182.68$$

93. Use the formula $A = A_0 e^{-0.0003t}$ with $A_0 = 218$ and $t = 1000$.

$$A = 218e^{-0.0003(1000)} = 218e^{-0.3} \approx 161.5 \text{ grams}$$

95. The amount in Melissa's account is given by the formula $S = 1000(1.05)^t$ for any number of years t. The amount in Frank's account is given by the formula $S = 900e^{0.07t}$ for any number of years t. To find when they have the same amount, we set the two expressions equal and solve for t.

$$1000(1.05)^t = 900e^{0.07t}$$
$$\ln\left(1000(1.05)^t\right) = \ln\left(900e^{0.07t}\right)$$
$$\ln(1000) + t \cdot \ln(1.05) = \ln(900) + \ln(e^{0.07t})$$
$$\ln(1000) + t \cdot \ln(1.05) = \ln(900) + 0.07t$$
$$t \cdot \ln(1.05) - 0.07t = \ln(900) - \ln(1000)$$
$$t[\ln(1.05) - 0.07] = \ln(900) - \ln(1000)$$

$$t = \frac{\ln(900) - \ln(1000)}{\ln(1.05) - 0.07} \approx 4.9675$$

The amounts will be equal in approximately 5 years.

97. $114.308e^{0.265(20.6 - 6.87)} \approx 4347.5 \text{ ft}^3/\text{sec}$

CHAPTER 9 TEST

1. $f(2) = 5^2 = 25$

2. $f(-1) = 5^{-1} = \frac{1}{5}$

3. $f(0) = 5^0 = 1$

4. $g(125) = \log_5(125) = 3$, because $5^3 = 125$.

5. $g(1) = \log_5(1) = 0$, because $5^0 = 1$.

6. $g(1/5) = \log_5(1/5) = -1$, because $5^{-1} = \frac{1}{5}$.

7. The graph of $y = 2^x$ includes the points $(0, 1)$, $(1, 2)$, $(2, 4)$, and $(-1, 1/2)$.

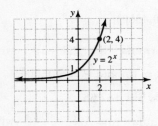

8. The graph of $f(x) = \log_2(x)$ includes the points $(1, 0)$, $(2, 1)$, $(4, 2)$, and $(1/2, -1)$.

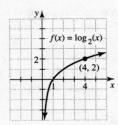

9. The graph of $y = \left(\frac{1}{3}\right)^x$ includes the points $(1, 1/3)$, $(0, 1)$, and $(-1, 3)$.

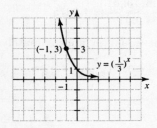

10. The graph of $g(x) = \log_{1/3}(x)$ includes the points $(1, 0)$, $(3, -1)$, and $(1/3, 1)$.

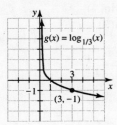

11. $\log_a(MN) = \log_a(M) + \log_a(N) = 6 + 4 = 10$

12. $\log_a(M^2/N) = 2 \cdot \log_a(M) - \log_a(N)$
$$= 2 \cdot 6 - 4 = 8$$

13. $\dfrac{\log_a(M)}{\log_a(N)} = \dfrac{6}{4} = \dfrac{3}{2}$

14. $\log_a(a^3 M^2) = 3 \cdot \log_a(a) + 2 \cdot \log_a(M)$
$$= 3 \cdot 1 + 2 \cdot 6 = 15$$

15. $\log_a(1/N) = \log_a(1) - \log_a(N)$
$$= 0 - 4 = -4$$

16. $\quad 3^x = 12$
$\quad\quad x = \log_3(12)$
The solution set is $\{\log_3(12)\}$.

17. $\quad \log_3(x) = 1/2$
$\quad\quad x = 3^{1/2} = \sqrt{3}$
The solution set is $\{\sqrt{3}\}$.

18. $\quad\quad\quad\quad 5^x = 8^{x-1}$

$\quad \ln(5^x) = \ln(8^{x-1})$

$\quad x \cdot \ln(5) = (x-1)\ln(8)$

$\quad x \cdot \ln(5) = x \cdot \ln(8) - \ln(8)$

$x \cdot \ln(5) - x \cdot \ln(8) = -\ln(8)$

$x[\ln(5) - \ln(8)] = -\ln(8)$

$x = \dfrac{-\ln(8)}{\ln(5) - \ln(8)} = \dfrac{\ln(8)}{-\ln(5) + \ln(8)}$

The solution set is $\left\{\dfrac{\ln(8)}{\ln(8) - \ln(5)}\right\}$.

19.
$$\log(x) + \log(x+15) = 2$$
$$\log(x^2 + 15x) = 2$$
$$x^2 + 15x = 10^2$$
$$x^2 + 15x - 100 = 0$$
$$(x+20)(x-5) = 0$$
$$x+20 = 0 \quad \text{or} \quad x-5 = 0$$
$$x = -20 \quad \text{or} \quad x = 5$$

If $x = -20$ in the original equation, we get a logarithm of a negative number, which is undefined. So the solution set is $\{5\}$.

20.
$$2 \cdot \ln(x) = \ln(3) + \ln(6-x)$$
$$\ln(x^2) = \ln(18 - 3x)$$
$$x^2 = 18 - 3x$$
$$x^2 + 3x - 18 = 0$$
$$(x+6)(x-3) = 0$$
$$x+6 = 0 \quad \text{or} \quad x-3 = 0$$
$$x = -6 \quad \text{or} \quad x = 3$$

If $x = -6$ in the original equation, we get a logarithm of a negative number, which is undefined. So the solution set is $\{3\}$.

21.
$$20^x = 5$$
$$x = \log_{20}(5) = \frac{\ln(5)}{\ln(20)} \approx 0.5372$$

The solution set is $\{0.5372\}$.

22.
$$\log_3(x) = 2.75$$

$$x = 3^{2.75} \approx 20.5156$$

The solution set is $\{20.5156\}$

23. To find the number present initially, let $t = 0$ in the formula:

$N = 10e^{0.4(0)} = 10e^0 = 10 \cdot 1 = 10$

To find the number present after 24 hours, let $t = 24$ in the formula:

$N = 10e^{0.4(24)} = 10e^{9.6} \approx 147,648$

24. To find how long it takes for the population to double, we must find the value of t for which $e^{0.4t} = 2$. Solve for t.

$$0.4t = \ln(2)$$
$$t = \frac{\ln(2)}{0.4} \approx 1.733$$

The bacteria population doubles in 1.733 hours.

Tying It All Together Chapters 1 - 9

1.
$$(x-3)^2 = 8$$
$$x - 3 = \pm\sqrt{8}$$
$$x = 3 \pm 2\sqrt{2}$$
The solution set is $\{3 \pm 2\sqrt{2}\}$.

2. $\log_2(x-3) = 8$
$$x - 3 = 2^8$$
$$x = 256 + 3 = 259$$
The solution set is $\{259\}$.

3. $2^{x-3} = 8$
$$2^{x-3} = 2^3$$
$$x - 3 = 3$$
$$x = 6$$
The solution set is $\{6\}$.

4. $2x - 3 = 8$
$$2x = 11$$
$$x = \frac{11}{2}$$
The solution set is $\left\{\frac{11}{2}\right\}$.

5. $|x-3| = 8$
$$x - 3 = 8 \quad \text{or} \quad x - 3 = -8$$
$$x = 11 \quad \text{or} \quad x = -5$$
The solution set is $\{-5, 11\}$

6. $\sqrt{x-3} = 8$
$$(\sqrt{x-3})^2 = 8^2$$
$$x - 3 = 64$$
$$x = 67$$
The solution set is $\{67\}$.

7. $\log_2(x-3) + \log_2(x) = \log_2(18)$

$\qquad \log_2(x^2 - 3x) = \log_2(18)$

$\qquad\qquad x^2 - 3x = 18$

$\qquad\qquad x^2 - 3x - 18 = 0$

$\qquad (x-6)(x+3) = 0$

$\qquad\quad x - 6 = 0 \text{ or } x+3 = 0$

$\qquad\qquad x = 6 \text{ or } \quad x = -3$

If $x = -3$ in the original equation, then we get a logarithm of a negative number, which is undefined. The solution set is $\{6\}$.

8. $\quad 2 \cdot \log_2(x-3) = \log_2(5-x)$

$\qquad \log_2[(x-3)^2] = \log_2(5-x)$

$\qquad\qquad (x-3)^2 = 5-x$

$\qquad\quad x^2 - 6x + 9 = 5 - x$

$\qquad\quad x^2 - 5x + 4 = 0$

$\qquad (x-4)(x-1) = 0$

$\qquad\quad x - 4 = 0 \text{ or } x - 1 = 0$

$\qquad\qquad x = 4 \text{ or } x = 1$

If $x = 1$ in the original equation, we get a logarithm of a negative number. The solution set is $\{4\}$.

9. $\quad 60\left(\frac{1}{2}x - \frac{2}{3}\right) = 60\left(\frac{3}{4}x + \frac{1}{5}\right)$

$\qquad 30x - 40 = 45x + 12$

$\qquad\quad -52 = 15x$

$\qquad -\frac{52}{15} = x$

The solution set is $\left\{-\frac{52}{15}\right\}$.

10. To solve $3x^2 - 6x + 2 = 0$ use the quadratic formula.

$$x = \frac{6 \pm \sqrt{(-6)^2 - 4(3)(2)}}{2(3)} = \frac{6 \pm \sqrt{12}}{6}$$

$$= \frac{6 \pm 2\sqrt{3}}{6} = \frac{3 \pm \sqrt{3}}{3}$$

The solution set is $\left\{\frac{3 \pm \sqrt{3}}{3}\right\}$.

11. The inverse of dividing by 3 is multiplying by 3. So $f^{-1}(x) = 3x$.

12. The inverse of the base 3 logarithm function is the base 3 exponential function,
$$g^{-1}(x) = 3^x.$$

13. The inverse of multiplying by 2 and then subtracting 4 is adding 4 and then dividing by 2,
$$f^{-1}(x) = \frac{x+4}{2}.$$

14. The inverse of the square root function is the squaring function. To keep the domain of one function equal to the range of the inverse function we must restrict the squaring function to the nonnegative numbers:
$$h^{-1}(x) = x^2 \text{ for } x \geq 0$$

15. The reciprocal function is its own inverse,
$$j^{-1}(x) = \frac{1}{x}.$$

16. The inverse of the base 5 exponential function is the base 5 logarithm function,
$$k^{-1}(x) = \log_5(x).$$

17. We will find the inverse for m by using the technique of interchanging x and y and then solving for y.

$\qquad m(x) = e^{x-1}$

$\qquad\quad y = e^{x-1}$

$\qquad\quad x = e^{y-1}$

$\qquad y - 1 = \ln(x)$

$\qquad\quad y = 1 + \ln(x)$

$\qquad m^{-1}(x) = 1 + \ln(x)$

18. The inverse of the natural logarithm function is the base e exponential function,
$$n^{-1}(x) = e^x.$$

19. The graph of $y = 2x$ is a straight line with slope 2 and y-intercept $(0, 0)$. Start at the origin and rise 2 and go 1 to the right to find a second point on the line.

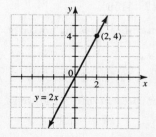

182

20. The graph of $y = 2^x$ includes the points $(0, 1)$, $(1, 2)$, $(2, 2)$, and $(-1, 1/2)$.

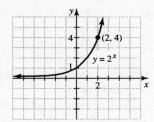

21. The graph of $y = x^2$ is a parabola through $(0, 0)$, $(1, 1)$, $(2, 4)$, $(-1, 1)$, and $(-2, 4)$.

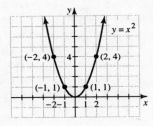

22. The graph of $y = \log_2(x)$ includes the points $(1, 0)$, $(2, 1)$, $(1/2, -1)$ and $(4, 2)$.

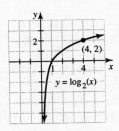

23. The graph of $y = \frac{1}{2}x - 4$ is a straight line with slope 1/2 and y-intercept $(0, -4)$. Start at $(0, -4)$, rise 1 and go 2 to the right to locate a second point on the line.

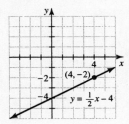

24. The graph of $y = |2 - x|$ is a v-shaped graph through $(-2, 4)$, $(-1, 3)$, $(0, 2)$, $(1, 1)$, $(2, 0)$, $(3, 1)$, $(4, 2)$, and $(5, 3)$.

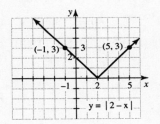

25. The graph of $y = 2 - x^2$ is a parabola with a vertex at $(0, 2)$. It also includes the points $(1, 1)$, $(-1, 1)$, $(2, -2)$, and $(-2, -2)$.

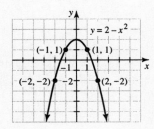

26. Note that e^2 is not a variable. The value of e^2 is approximately 7.389. The graph of $y = e^2$ is a straight line with 0 slope and y-intercept (0, 7.389).

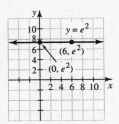

27. The graph of $n = 2.13t + 88.38$ is a straight line through (0, 88.38) and (10, 109.68). The graph of $n = 88.43e^{0.023t}$ is an exponential curve through (0, 88.43), (5, 99.32), (10, 111.3), and (15, 124.9).

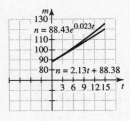

b) If $t = 7$, $n = 2.13(7) + 88.38 = 103.3$ million
If $t = 7$, $n = 88.43e^{0.023(7)} = 103.9$ million

28. $d_1 = 0.135v$, $d_2 = 0.326v$, $d_1 = 200.2$ m.

10.1 WARM-UPS

1. True, because if $x = 1$ and $y = 2$, then $2(1) + 2 = 4$ is correct. **2.** False, because if $x = 1$ and $y = 2$, then $3(1) - 2 = 6$ is incorrect. **3.** False, because (2, 3) does not satisfy $4x - y = -5$. **4.** True. **5.** True, because when we substitute, one of the variables is eliminated. **6.** True, because each of these lines has slope 3 and they are parallel. **7.** True, because the lines are not parallel and they have different y-intercepts. **8.** True, because the lines are parallel and have different y-intercepts. **9.** True, because dependent equations have the same solution sets. **10.** True, because a system of independent equations has one point in its solution set.

10.1 EXERCISES

1. The graph of $y = 2x$ is a straight line with y-intercept (0, 0) and slope 2. The graph of $y = -x + 3$ is a straight line with y-intercept (0, 3) and slope -1. The graphs appear to intersect at (1, 2). Check that (1, 2) satisfies both equations. The solution set is $\{(1, 2)\}$.
3. The graph of $y = 2x - 1$ is a straight line with y-intercept (0, -1) and slope 2. The graph of $y = \frac{1}{2}x - 1$ is a straight line with y-intercept (0, -1) and slope $\frac{1}{2}$. The graphs intersect at (0, -1). The solution set to the system of equations is $\{(0, -1)\}$.
5. The graph of $y = x - 3$ is a straight line with y-intercept (0, -3) and slope 1. The graph of $y = \frac{1}{2}x - 2$ is a straight line with y-intercept (0, -2) and slope $\frac{1}{2}$. The lines appear to intersect at (2, -1). To be sure, we check that (2, -1) satisfies both of the original equations. The solution set is $\{(2, -1)\}$.
7. If we rewrite these equations in slope-intercept form, we get $y = x + 1$ and $y = x + 3$. The graphs are parallel lines with slopes of 1 and y-intercepts of (0, 1) and (0, 3). Since the lines do not intersect, the solution set is the empty set, $\emptyset$.
9. Solving $x + 2y = 8$ for y yields $y = -\frac{1}{2}x + 4$, which is the same equation as the first. So the equations have the same graph. The solution set is $\{(x, y) \mid y = -\frac{1}{2}x + 4\}$.

11. The lines in graph (c) intersect at $(3, -2)$ and $(3, -2)$ satisfies both equations.

13. The lines in graph (b) intersect at $(-3, -2)$ and $(-3, -2)$ satisfies both equations.

15. Substitute $y = x - 5$ into $2x - 5y = 1$.

$$2x - 5(x - 5) = 1$$
$$2x - 5x + 25 = 1$$
$$-3x = -24$$
$$x = 8$$

Use $x = 8$ in $y = x - 5$ to find y.

$$y = 8 - 5$$
$$y = 3$$

The solution set is $\{(8, 3)\}$. The equations are independent.

17. Substitute $x = 2y - 7$ into $3x + 2y = -5$.

$$3(2y - 7) + 2y = -5$$
$$6y - 21 + 2y = -5$$
$$8y = 16$$
$$y = 2$$

Use $y = 2$ in $x = 2y - 7$ to find x.

$$x = 2(2) - 7 = -3$$

The solution set is $\{(-3, 2)\}$. The equations are independent.

19. Write $x - y = 5$ as $x = y + 5$ and substitute into $2x = 2y + 14$.

$$2(y + 5) = 2y + 14$$
$$2y + 10 = 2y + 14$$
$$10 = 14$$

Since this last equation is incorrect no matter what values are used for x and y, the equations are inconsistent and the solution set is $\emptyset$.

21. Substitute $y = 2x - 5$ into $y + 1 = 2(x - 2)$.

$$2x - 5 + 1 = 2(x - 2)$$

$$2x - 4 = 2x - 4$$

Since the last equation is an identity, the equations are dependent. Any ordered pair that satisfies one equation will also satisfy the other. So the solution set is $\{(x, y) \mid y = 2x - 5\}$.

23. Write $2x + y = 9$ as $y = -2x + 9$ and substitute into $2x - 5y = 15$.

$$2x - 5(-2x + 9) = 15$$
$$2x + 10x - 45 = 15$$
$$12x - 45 = 15$$
$$12x = 60$$
$$x = 5$$

Use $x = 5$ in $y = -2x + 9$ to find y.

$$y = -2(5) + 9 = -1$$

The solution set is $\{(5, -1)\}$. The equations are independent.

25. Write $x - y = 0$ as $y = x$ and substitute into $2x + 3y = 35$.

$$2x + 3x = 35$$
$$5x = 35$$
$$x = 7$$

Since $y = x$, $y = 7$ also. The solution set is $\{(7, 7)\}$ and the equations are independent.

27. Write $x + y = 40$ as $x = 40 - y$ and substitute into $0.1x + 0.08y = 3.5$.

$$0.1(40 - y) + 0.08y = 3.5$$
$$4 - 0.1y + 0.08y = 3.5$$
$$-0.02y = -0.5$$
$$y = 25$$

Since $x = 40 - y$, we get $x = 40 - 25 = 15$. The solution set is $\{(15, 25)\}$ and the equations are independent.

29. Substitute $y = 2x - 30$ into the other equation.

$$\tfrac{1}{5}x - \tfrac{1}{2}(2x - 30) = -1$$

$$\tfrac{1}{5}x - x + 15 = -1$$

$$5\left(\tfrac{1}{5}x - x + 15\right) = 5(-1)$$
$$x - 5x + 75 = -5$$
$$-4x = -80$$
$$x = 20$$

If $x = 20$, then $y = 2x - 30 = 2(20) - 30 = 10$. The solution set is $\{(20, 10)\}$ and the equations are independent.

31. Write $x - y = 5$ as $x = y + 5$ and substitute into $x + y = 4$.

$$y + 5 + y = 4$$
$$2y = -1$$
$$y = -1/2$$

If $y = -1/2$, then $x = y + 5 = -1/2 + 5 = 9/2$. The solution set is $\left\{\left(\tfrac{9}{2}, -\tfrac{1}{2}\right)\right\}$ and the equations are independent.

33. Write $2x - y = 4$ as $y = 2x - 4$ and substitute into $2x - y = 3$.

$$2x - (2x - 4) = 3$$
$$4 = 3$$

The equations are inconsistent and the solution set is $\emptyset$.

35. First simplify.

$$3(y - 1) = 2(x - 3)$$
$$3y - 3 = 2x - 6$$
$$3y - 2x = -3$$

By simplifying the first equation we see that it is the same as the second equation. There is no need to substitute now. The equations are dependent and the solution set is $\{(x, y) \mid 3y - 2x = -3\}$.

37. Write $x - y = -0.375$ as $x = y - 0.375$ and substitute into $1.5x - 3y = -2.25$.

$$1.5(\mathbf{y - 0.375}) - 3y = -2.25$$
$$1.5y - 0.5625 - 3y = -2.25$$
$$-1.5y = -1.6875$$
$$y = 1.125$$

If $y = 1.125$, then $x = 1.125 - 0.375 = 0.75$. The solution set is $\{(0.75, 1.125)\}$ and the equations are independent.

39. Let $x =$ the amount invested at 5% and $y =$ the amount invested at 10%. The interest earned on the investments is $0.05x$ and $0.10y$ respectively. We can write two equations, one expressing the total amount invested, and the other expressing the total of the interest earned.

$$x + y = 30,000$$
$$0.05x + 0.10y = 2,300$$

Write the first equation as $y = 30,000 - x$ and substitute into the second.

$$0.05x + 0.10(\mathbf{30,000 - x}) = 2,300$$
$$0.05x + 3,000 - 0.10x = 2,300$$
$$-0.05x = -700$$
$$x = 14,000$$

If $x = 14,000$, then $y = 30,000 - 14,000 = 16,000$. She invested \$14,000 at 5% and \$16,000 at 10%.

41. Let x and y be the numbers. The fact that their sum is 2 and their difference is 26 gives us a system of equations:

$$x + y = 2$$
$$x - y = 26$$

Write $x + y = 2$ as $x = 2 - y$ and substitute.

$$\mathbf{2 - y} - y = 26$$
$$-2y = 24$$
$$y = -12$$

If $y = -12$, then $x = 2 - y = 2 - (-12) = 14$. The numbers are -12 and 14.

43. Let $x =$ the number of toasters and $y =$ the number of vacation coupons. We write one equation expressing the fact that the total number of prizes is 100 and the other expressing the total bill of \$708.

$$x + y = 100$$
$$6x + 24y = 708$$

Write $x + y = 100$ as $x = 100 - y$ and substitute.

$$6(\mathbf{100 - y}) + 24y = 708$$
$$600 - 6y + 24y = 708$$
$$18y = 108$$
$$y = 6$$

If $y = 6$, then $x = 100 - y = 100 - 6 = 94$. He gave away 94 toasters and 6 vacation coupons.

45.
$$s = 0.05(100,000 - f)$$
$$f = 0.30(100,000 - s)$$

$$s = 5,000 - 0.05f$$
$$f = 30,000 - 0.30s$$

$$f = 30,000 - 0.30(5000 - 0.05f)$$
$$f = 30,000 - 1500 + 0.015f$$
$$0.985f = 28,500$$
$$f = 28,934$$
$$s = 5,000 - 0.05(28,934)$$
$$= 3553$$

State tax is \$3553 and federal tax is \$28,934.

47.
$$B = 0.20N$$
$$N = 120,000 - B$$

$$N = 120,000 - 0.20N$$
$$1.2N = 120,000$$
$$N = 100,000$$
$$B = 0.20 \cdot 100,000 = 20,000$$

The bonus is \$20,000.

49. The cost of 10,000 textbooks is \$500,000. The revenue for 10,000 textbooks is \$300,000. The cost is equal to the revenue for 20,000 textbooks. The fixed cost is \$400,000.

51. a

53. $(2.8, 2.6), (1.0, -0.2)$

10.2 WARM-UPS

1. True, because when we add the equations, the y-terms add up to 0 and one variable is eliminated. **2.** False, multiply the firs by -2 and the second by 3 to eliminate x. **3.** True, multiplying the second by 2 and adding will eliminate both x and y. **4.** True, because both ordered pairs satisfy both equations.
5. False, solution set is $\{(x, y) \mid 4x - 2y = 20\}$.
6. True, because the equations are inconsistent.
7. True. **8.** False, as long as we add the left sides and the right sides, the form does not matter. **9.** True. **10.** True.

10.2 EXERCISES

1.
$$x + y = 7$$
$$\underline{x - y = 9}$$
$$2x \quad = 16$$
$$x \quad = 8$$

Use $x = 8$ in $x + y = 7$ to find y.
$$8 + y = 7$$
$$y = -1$$

The solution set is $\{(8, -1)\}$.

3.
$$x - y = 12$$
$$\underline{2x + y = 3}$$
$$3x \quad = 15$$
$$x \quad = 5$$

Use $x = 5$ in $x - y = 12$.
$$5 - y = 12$$
$$-7 = y$$

The solution set is $\{(5, -7)\}$.

5. If we multiply $2x - y = -5$ by 2, we get $4x - 2y = -10$. Add this equation to the second equation.
$$4x - 2y = -10$$
$$\underline{3x + 2y = 3}$$
$$7x \qquad = -7$$
$$x \quad = -1$$

Use $x = -1$ in $2x - y = -5$.
$$2(-1) - y = -5$$
$$-2 - y \quad = -5$$
$$3 \quad = y$$

The solution set is $\{(-1, 3)\}$.

7. Multiply the first equation by 4 and the second by 5.
$$4(2x - 5y) = 4(13)$$
$$5(3x + 4y) = 5(-15)$$
$$8x - 20y = 52$$
$$\underline{15x + 20y = -75}$$
$$23x \qquad = -23$$
$$x \quad = -1$$

Use $x = -1$ in $2x - 5y = 13$.
$$2(-1) - 5y = 13$$
$$-2 - 5y = 13$$
$$-5y = 15$$
$$y = -3$$

The solution set is $\{(-1, -3)\}$.

9. Rewrite the first equation to match the form of the second.
$$2x - 3y = 11$$
$$7x - 4y = 6$$

Multiply the first equation by -7 and the second by 2.
$$-7(2x - 3y) = -7(11)$$
$$2(7x - 4y) = 2(6)$$

$$-14x + 21y = -77$$
$$\underline{14x - 8y = 12}$$
$$13y = -65$$
$$y = -5$$

Use $y = -5$ in $2x = 3y + 11$.
$$2x = 3(-5) + 11$$
$$2x = -4$$
$$x = -2$$

The solution set is $\{(-2, -5)\}$.

11.
$$-12(x + y) = -12(48)$$
$$12x + 14y = 628$$

$$-12x - 12y = -576$$
$$\underline{12x + 14y = 628}$$
$$2y = 52$$
$$y = 26$$
$$x + 26 = 48$$
$$x = 22$$

The solution set is $\{(22, 26)\}$.

13.
$$3x - 4y = 9$$
$$\underline{-3x + 4y = 12}$$
$$0 = 21$$

Since we obtained an incorrect equation by addition, the equations are inconsistent and the solution set is $\emptyset$.

15. Multiply $5x - y = 1$ by -2, to get $-10x + 2y = -2$. Add this equation to the second equation.
$$-10x + 2y = -2$$
$$\underline{10x - 2y = 2}$$
$$0 = 0$$

Since we obtained an identity by adding the equations, the equations are dependent and the solution set is $\{(x, y) \mid 5x - y = 1\}$.

17.
$$2x - y = 5$$
$$\underline{2x + y = 5}$$
$$4x \quad = 10$$
$$x = 10/4 = 5/2$$

Use $x = 5/2$ in $2x + y = 5$ to find y.
$$2(5/2) + y = 5$$
$$5 + y = 5$$
$$y = 0$$

The solution set is $\{(5/2, 0)\}$ and the equations are independent.

19. Multiplying the first equation by 12 to eliminate the fractions gives us the following system.
$$3x + 4y = 60$$
$$x - y = 6$$

Multiply the second equation by 4 and then add.

$$3x + 4y = 60$$
$$\underline{4x - 4y = 24}$$
$$7x \quad\quad = 84$$
$$x = 12$$

Use $x = 12$ in $x - y = 6$ to find y.
$$12 - y = 6$$
$$6 = y$$
The solution set is $\{(12, 6)\}$.

21. If we multiply the first equation by 12 and the second by 24 to eliminate fractions, we get the following system.
$$3x - 4y = -48$$
$$\underline{3x + 4y = 0}$$
$$6x \quad\quad = -48$$
$$x = -8$$
Use $x = -8$ in $3x + 4y = 0$ to find y.
$$3(-8) + 4y = 0$$
$$4y = 24$$
$$y = 6$$
The solution set is $\{(-8, 6)\}$.

23. Multiply the first equation by 8 and the second by -16 to get the following system.
$$x + 2y = 40$$
$$\underline{-x - 8y = -112}$$
$$-6y = -72$$
$$y = 12$$
$$x + 2(12) = 40$$
$$x = 16$$

The solution set is $\{(16, 12)\}$.

25. Multiply the first equation by 100 and the second by -10 to get the following system.
$$5x + 10y = 130$$
$$\underline{-10x - 10y = -190}$$
$$-5x \quad\quad = -60$$
$$x = 12$$
Use $x = 12$ in $x + y = 19$ to get $y = 7$. The solution set is $\{(12, 7)\}$.

27. Multiply the first equation by -9 and the second by 100 to get the following system.
$$-9x - 9y = -10800$$
$$\underline{12x + 9y = 12000}$$
$$3x \quad\quad = 1200$$
$$x = 400$$
Use $x = 400$ in $x + y = 1200$ to get $y = 800$. The solution set is $\{(400, 800)\}$.

29. Multiply the first equation by -2 to get the following system.
$$-3x + 4y = 0.5$$
$$\underline{3x + 1.5y = 6.375}$$
$$5.5y = 6.875$$

$$y = 1.25$$
Use $y = 1.25$ in $3x + 1.5y = 6.375$ to find x.
$$3x + 1.5(1.25) = 6.375$$
$$3x + 1.875 = 6.375$$
$$3x = 4.5$$
$$x = 1.5$$
The solution set is $\{(1.5, 1.25)\}$.

31. Let $x =$ the price of one doughnut and $y =$ the price of one cup of coffee. His bills for Monday and Tuesday give us the following system of equations.
$$3x + 2y = 2.54$$
$$2x + 3y = 2.46$$
Multiply the first equation by -2 and the second by 3.
$$-6x - 4y = -5.08$$
$$\underline{6x + 9y = 7.38}$$
$$5y = 2.30$$
$$y = 0.46$$
Use $y = 0.46$ in $3x + 2y = 2.54$ to find x.
$$3x + 2(0.46) = 2.54$$
$$3x + 0.92 = 2.54$$
$$3x = 1.62$$
$$x = 0.54$$
Doughnuts are $0.54 each and a cup of coffee is $0.46. So his total bill on Wednesday was $1.00.

33. Let $x =$ the number of boys and $y =$ the number of girls at Freemont High. We get a system of equations from the number of boys and girls who attended the dance and game.

$$\tfrac{1}{2}x + \tfrac{1}{3}y = 570$$

$$\tfrac{1}{3}x + \tfrac{1}{2}y = 580$$

Multiply the first equation by 12 and the second equation by -18.

$$6x + 4y = 6840$$
$$\underline{-6x - 9y = -10440}$$
$$-5y = -3600$$
$$y = 720$$

Use $y = 720$ in $6x + 4y = 6840$ to find x.
$$6x + 4(720) = 6840$$
$$6x + 2880 = 6840$$
$$6x = 3960$$
$$x = 660$$

There are 660 boys and 720 girls at Freemont High, or 1380 students.

35. Let x = the number of dimes and y = the number of nickels. We write one equation about the total number of the coins and the other about the value of the coins (in cents).

$$x + y = 35$$
$$10x + 5y = 330$$

Multiply the first equation by −5.

$$-5x - 5y = -175$$
$$\underline{10x + 5y = 330}$$
$$5x \qquad = 155$$
$$x = 31$$

Use x = 31 in x + y = 35 to find that y = 4. He has 31 dimes and 4 nickels.

37. Let x = the number of pounds of Chocolate fudge and y = the number of pounds of Peanut Butter fudge.

$$x + y = 50$$
$$0.35x + 0.25y = 0.29(50)$$

$$y = 50 - x$$
$$0.35x + 0.25(50 - x) = 14.5$$
$$0.10x = 2$$
$$x = 20$$
$$20 + y = 50$$
$$y = 30$$

Use 20 pounds of Chocolate fudge and 30 pounds of Peanut Butter fudge.

39. Let a = the time from Allentown to Harrisburg and h = the time from Harrisburg to Pittsburgh.

$$a + h = 6$$
$$42a + 51h = 288$$

$$-51a - 51h = -306$$
$$\underline{42a + 51h = 288}$$
$$-9a \qquad = -18$$
$$a = 2$$
$$h = 6 - 2 = 4$$

It took 4 hours to go from Harrisburg to Pittsburgh.

41. Let p = the probability of rain and q = the probability that it doesn's rain:

$$p + q = 1$$
$$p = 4q$$

$$4q + q = 1$$
$$5q = 1$$
$$q = 0.20 = 20\%$$
$$p = 80\%$$

The probability of rain is 80%.

43. Let W = the width and L = the length.

$$W = 0.75L$$
$$2W + 2L = 700$$

$$2(0.75L) + 2L = 700$$
$$1.5L + 2L = 700$$
$$3.5L = 700$$
$$L = 200$$
$$W = 0.75(200) = 150$$

The width is 150 m and the length is 200 m.

10.3 WARM-UPS

1. False, because $1 + (-2) - 3 = 4$ is incorrect. **2.** False, because there are infinitely many ordered triples that satisfy $x + y - z = 4$. **3.** True, because if x = 1, y = −1, and z = 2 then each of the equations is satisfied. **4.** False, we can use substitution and addition to eliminate variables. **5.** False, because two planes never intersect in a single point. **6.** True, because if we multiply the second one by −1 and add, then we get 0 = 2. **7.** True, because −2 times the first equation is the same as the second equation. **8.** False, because the graph is a plane. **9.** False, because the value of x nickels, y dimes, and z quarters is $5x + 10y + 25z$ cents. **10.** False, because x = −2, z = 3, and −2 + y + 3 = 6 implies y = 5.

10.3 EXERCISES

1. Add the first two equations.

$$x + y + z = 2$$
$$\underline{x + 2y - z = 6}$$
$$2x + 3y = 8 \qquad \text{A}$$

Add the first and third equations.

$$x + y + z = 2$$
$$\underline{2x + y - z = 5}$$
$$3x + 2y = 7 \qquad \text{B}$$

Multiply equation A by 3 and equation B by −2.

$$6x + 9y = 24 \qquad 3 \times \text{A}$$
$$\underline{-6x - 4y = -14} \qquad -2 \times \text{B}$$
$$5y = 10$$
$$y = 2$$

Use y = 2 in the equation $3x + 2y = 7$.

$$3x + 2(2) = 7$$
$$3x = 3$$
$$x = 1$$

Use x = 1 and y = 2 in x + y + z = 2.

$$1 + 2 + z = 2$$
$$z = -1$$

The solution set is $\{(1, 2, -1)\}$.

3. Multiply the first equation by -1 to get $-x + 2y - 4z = -3$. Now add this equation and the second, and this equation and the third.

$$-x + 2y - 4z = -3 \qquad -x + 2y - 4z = -3$$
$$\underline{x + 3y - 2z = 6} \qquad \underline{x - 4y + 3z = -5}$$
$$5y - 6z = 3 \qquad\quad -2y - z = -8$$

Multiply $-2y - z = -8$ by -6 and add the result to $5y - 6z = 3$.

$$12y + 6z = 48$$
$$\underline{5y - 6z = 3}$$
$$17y \quad\;\; = 51$$
$$y \quad\;\; = 3$$

Use $y = 3$ in $5y - 6z = 3$.

$$5(3) - 6z = 3$$
$$-6z = -12$$
$$z = 2$$

Use $y = 3$ and $z = 2$ in $x + 3y - 2z = 6$.

$$x + 3(3) - 2(2) = 6$$
$$x = 1$$

The solution set is $\{(1, 3, 2)\}$.

5. Multiply $2x - y + z = 10$ by 2 to get $4x - 2y + 2z = 20$. Add this equation and the second and this equation and the third.

$$4x - 2y + 2z = 20 \qquad 4x - 2y + 2z = 20$$
$$\underline{3x - 2y - 2z = 7} \qquad\; \underline{x - 3y - 2z = 10}$$
$$7x - 4y \quad\;\; = 27 \qquad 5x - 5y \quad\;\; = 30$$
$$x - y \quad\;\; = 6$$

Multiply $x - y = 6$ by -4 to get $-4x + 4y = -24$. Add this to $7x - 4y = 27$.

$$-4x + 4y = -24$$
$$\underline{7x - 4y = 27}$$
$$3x \quad\;\; = 3$$
$$x = 1$$

Use $x = 1$ in $7x - 4y = 27$ to find y.

$$7(1) - 4y = 27$$
$$-4y = 20$$
$$y = -5$$

Use $x = 1$ and $y = -5$ in $2x - y + z = 10$.

$$2(1) - (-5) + z = 10$$
$$z = 3$$

The solution set is $\{(1, -5, 3)\}$.

7. Multiply $x - y + 2z = -5$ by 2 to get $2x - 2y + 4z = -10$. Add this equation and the second equation, and add the first and second equations.

$$2x - 2y + 4z = -10 \qquad 2x - 3y + z = -9$$
$$\underline{-2x + y - 3z = 7} \qquad \underline{-2x + y - 3z = 7}$$
$$-y + z = -3 \qquad\quad -2y - 2z = -2$$
$$\qquad\qquad\qquad\qquad\qquad y + z = 1$$

Add the last two equations to eliminate y.

$$-y + z = -3$$
$$\underline{y + z = 1}$$
$$2z = -2$$
$$z = -1$$

Use $z = -1$ in $y + z = 1$ to get $y = 2$. Now use $z = -1$ and $y = 2$ in $x - y + 2z = -5$.

$$x - 2 + 2(-1) = -5$$
$$x = -1$$

The solution set is $\{(-1, 2, -1)\}$.

9. Multiply the first equation by -2 and add the result to the last equation.

$$-4x + 10y - 4z = -32$$
$$\underline{4x - 3y + 4z = 18}$$
$$7y \quad\;\; = -14$$
$$y \quad\;\; = -2$$

Multiply the first equation by 3 and the second by 2 and add the results.

$$6x - 15y + 6z = 48$$
$$\underline{6x + 4y - 6z = -38}$$
$$12x - 11y \quad\;\; = 10$$

Use $y = -2$ in the last equation to find x.

$$12x - 11(-2) = 10$$
$$12x = -12$$
$$x = -1$$

Use $x = -1$ and $y = -2$ in $2x - 5y + 2z = 16$.

$$2(-1) - 5(-2) + 2z = 16$$
$$2z = 8$$
$$z = 4$$

The solution set is $\{(-1, -2, 4)\}$.

11. If we add the last two equations we get $x + 2y = 7$. Multiply the first equation by -1 to get $-x - y = -4$. Add these two equations.

$$x + 2y = 7$$
$$\underline{-x - y = -4}$$
$$y = 3$$

Use $y = 3$ in $x + y = 4$ to get $x = 1$. Use $y = 3$ in $y - z = -2$.

$$3 - z = -2$$
$$5 = z$$

The solution set is $\{(1, 3, 5)\}$.

13. Multiply the first equation by -1 and add the result to the second equation.

$$-x - y \quad\;\; = -7$$
$$\underline{y - z = -1}$$
$$-x \quad\; - z = -8$$

Add this result to the last equation.

$$-x - z = -8$$
$$\underline{x + 3z = 18}$$
$$2z = 10$$
$$z = 5$$

Use $z = 5$ in $x + 3z = 18$ to get $x + 15 = 18$ or $x = 3$. Use $x = 3$ in $x + y = 7$ to get $3 + y = 7$ or $y = 4$. The solution set is $\{(3, 4, 5)\}$.

15. Add the first two equations.

$$x - y + 2z = 3$$
$$\underline{2x + y - z = 5}$$
$$3x \quad\quad + z = 8 \quad\quad A$$

Multiply the second equation by 3 and add the result to the last equation.

$$6x + 3y - 3z = 15$$
$$\underline{3x - 3y + 6z = 4}$$
$$9x \quad\quad + 3z = 19 \quad\quad B$$

Multiply A by -3 and add the result to B.

$$-9x - 3z = -24$$
$$\underline{9x + 3z = 19}$$
$$0 = -5$$

Since the last equation is false no matter what values the variables have, the solution set is $\emptyset$.

17. The second equation is the same as 3 times the first equation, and the third equation is the same as -4 times the first equation. So the system is a dependent system. All three equations are equivalent. The solution set is $\{(x, y, z) \mid 3x - y + z = 5\}$.

19. Add the first and second equation to get $x + z = 11$. Multiply this equation by -2 and then add the result to the last equation.

$$-2x - 2z = -22$$
$$\underline{2x + 2z = 7}$$
$$0 = -15$$

Since the last result is false, the solution set is $\emptyset$.

21. Multiply the first equation by 300 to get $30x + 24y - 12z = 900$. Multiply the second equation by 6 to get $30x + 24y - 12z = 900$. Multiply the third equation by 100 to get $30x + 24y - 12z = 900$. Since all three of these equations are different forms of the same equation, the system is dependent. The solution set is $\{(x, y, z) \mid 5x + 4y - 2z = 150\}$.

23. Multiply the second equation by 10 and add the result to the first equation.

$$37x - 2y + 0.5z = 4.1$$
$$\underline{3x + 2y - 0.4z = 0.1}$$
$$40x \quad\quad + 0.1z = 4.2 \quad\quad A$$

Multiply the second equation by 19 and add the result to the last equation.

$$70.3x - 3.8y + 0.95z = 7.79$$
$$\underline{-2x + 3.8y - 2.1z = -3.26}$$
$$68.3x \quad\quad - 1.15z = 4.53 \quad\quad B$$

Multiply equation A by 11.5 and add the result to equation B.

$$460x + 1.15z = 48.3$$
$$\underline{68.3x - 1.15z = 4.53}$$
$$528.3x \quad\quad = 52.83$$
$$x = 0.1$$

Use $x = 0.1$ in $40x + 0.1z = 4.2$.

$$40(0.1) + 0.1z = 4.2$$
$$0.1z = 0.2$$
$$z = 2$$

Use $x = 0.1$ and $z = 2$ in $3x + 2y - 0.4z = 0.1$.

$$3(0.1) + 2y - 0.4(2) = 0.1$$
$$0.3 + 2y - 0.8 = 0.1$$
$$2y = 0.6$$
$$y = 0.3$$

The solution set is $\{(0.1, 0.3, 2)\}$.

25. Let $x =$ her investment in stocks, $y =$ her investment in bonds, and $z =$ her investment in mutual funds. We can write 3 equations concerning x, y, and z.

$$x + y + z = 12000$$
$$0.10x + 0.08y + 0.12z = 1230$$
$$x + y = z$$

Substitute $z = x + y$ into the first equation.

$$x + y + x + y = 12000$$
$$2x + 2y = 12000$$
$$x + y = 6000 \quad\quad A$$

Substitute $z = x + y$ into the second equation.

$$0.10x + 0.08y + 0.12(x + y) = 1230$$
$$0.22x + 0.20y = 1230$$
$$-5(0.22x + 0.20y) = -5(1230)$$
$$-1.1x - y = -6150 \quad\quad B$$

Add equations A and B.

$$x + y = 6000$$
$$\underline{-1.1x - y = -6150}$$
$$-0.1x \quad\quad = -150$$
$$x = 1500$$

Use $x = 1500$ in $x + y = 6000$ to get $y = 4500$. Since $z = x + y$, we must have $z = 6000$. So Ann invested \$1500 in stocks, \$4500 in bonds, and \$6000 in a mutual fund.

27. Let $x =$ the price of one cup of coffee, $y =$ the price of one doughnut, and $z =$ the amount of the tip. We can write an equation for each day.

$$2x + y + z = 170$$
$$x + 2y + z = 165$$
$$x + y + z = 130$$

Multiply the first equation by -1 and add the result to the second and the third equation.

$$-2x - y - z = -170$$
$$\underline{x + 2y + z = 165}$$
$$-x + y \qquad = -5$$

$$-2x - y - z = -170$$
$$\underline{x + y + z = 130}$$
$$-x \qquad = -40$$
$$x = 40$$

Use $x = 40$ in the $-x + y = -5$.
$$-40 + y = -5$$
$$y = 35$$
Use $x = 40$ and $y = 35$ in $x + y + z = 130$.
$$40 + 35 + z = 130$$
$$z = 55$$

The price of coffee is 0.40, the price of a doughnut is 0.35, and the tip is always 0.55.

29. Let $x =$ the price of a banana, $y =$ the price of an apple, and $z =$ the price of an orange. We can write 3 equations.
$$3x + 2y + z = 180$$
$$4x + 3y + 3z = 305$$
$$6x + 5y + 4z = 465$$
Multiply the first equation by -3 and add the result to the second equation.
$$-9x - 6y - 3z = -540$$
$$\underline{4x + 3y + 3z = 305}$$
$$-5x - 3y \qquad = -235 \qquad \text{A}$$
Multiply the first equation by -4 and add the result to the third equation.
$$-12x - 8y - 4z = -720$$
$$\underline{6x + 5y + 4z = 465}$$
$$-6x - 3y \qquad = -255$$
$$6x + 3y \qquad = 255 \qquad \text{B}$$
Add equation B and equation A.
$$6x + 3y = 255$$
$$\underline{-5x - 3y = -235}$$
$$x \qquad = 20$$
Use $x = 20$ in $6x + 3y = 255$ to find y.
$$6(20) + 3y = 255$$
$$3y = 135$$
$$y = 45$$
Use $x = 20$ and $y = 45$ in $3x + 2y + z = 180$.
$$3(20) + 2(45) + z = 180$$
$$z = 30$$

The price of one banana is 0.20, one apple is 0.45, and one orange is 0.30. So the lunch-box -special should sell for 0.95.

31. Let $x =$ the weight of a can of soup, $y =$ the weight of a can of tuna, and $z =$ the constant error. We can write three equations.
$$x + y + z = 24$$
$$4x + 3y = 80$$
$$2y + z = 18$$
Multiply the third equation by -1 and add the result to the first equation.

$$-2y - z = -18$$
$$\underline{x + y + z = 24}$$
$$x - y \qquad = 6$$
Multiply $x - y = 6$ by 3 and add the result to the second equation.
$$3x - 3y = 18$$
$$\underline{4x + 3y = 80}$$
$$7x \qquad = 98$$
$$x = 14$$

Use $x = 14$ in $x - y = 6$ to get $y = 8$. Use $x = 14$ and $y = 8$ in $x + y + z = 24$ to get $z = 2$. So a can of soup weighs 14 ounces, a can of tuna weighs 8 ounces, and the scale has a constant error of 2 ounces.

33. Let $x =$ his income from teaching, $y =$ his income from house painting, and $z =$ his royalties.
$$x + y + z = 48000$$
$$x - y = 6000$$
$$z = \frac{1}{7}(x + y)$$

The last equation can be written as $x + y = 7z$. Replacing $x + y$ in the first equation by $7z$ gives us $7z + z = 48000$, or $z = 6000$. If $z = 6000$, then $x + y = 42000$. Add $x + y = 42000$ and the second equation.
$$x + y = 42000$$
$$\underline{x - y = 6000}$$
$$2x \qquad = 48000$$
$$x \qquad = 24000$$

Use $x = 24000$ in $x - y = 6000$ to get $y = 18000$. So he made $24,000$ teaching, $18,000$ house painting, and $6,000$ from royalties.

10.4 WARM-UPS

1. True, because any equation of the form $y = ax^2 + bx + c$ has a graph that is a parabola. **2.** False, because absolute value has a v-shaped graph. **3.** False, because $-4 = \sqrt{5(3) + 1}$ is incorrect. **4.** True, because $y = \sqrt{x}$ lies entirely in the first quadrant, and $y = -x - 2$ has no points in the first quadrant. **5.** False, because we can also use addition. **6.** True. **7.** True, because a 30-60-90 triangle is half of an equilateral triangle. **8.** True, because for any rectangular solid we have $V = L \cdot W \cdot H$. **9.** True, because the surface area consists of 6 rectangles, of which two have area LW, two have area WH, and two have area LH.

10. True, because the area of a triangle is $(bh)/2$.

10.4 EXERCISES

1. The graph of $y = x^2$ is a parabola and the graph of $x + y = 6$ is a straight line.

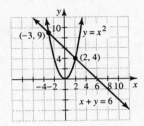

To solve the system, substitute $y = x^2$ into $x + y = 6$.

$$x + x^2 = 6$$
$$x^2 + x - 6 = 0$$
$$(x + 3)(x - 2) = 0$$
$$x = -3 \text{ or } x = 2$$

If $x = -3$, $y = (-3)^2 = 9$, and if $x = 2$, $y = 2^2 = 4$. The solution set to the system is $\{(2, 4), (-3, 9)\}$.

3. The graph of $y = |x|$ is v-shaped, and the graph of $2y - x = 6$ is a straight line. To solve the system, substitute $x = 2y - 6$ into $y = |x|$.

$$y = |2y - 6|$$
$$y = 2y - 6 \text{ or } y = -(2y - 6)$$
$$6 = y \qquad \text{ or } y = -2y + 6$$
$$3y = 6$$
$$y = 2$$

Use $y = 6$ in $x = 2y - 6$ to get $x = 6$. Use $y = 2$ in $x = 2y - 6$ to get $x = -2$. The graphs intersect at the points $(-2, 2)$ and $(6, 6)$.

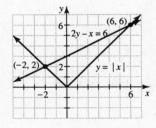

The solution set is $\{(-2, 2), (6, 6)\}$.

5. The graph of $y = \sqrt{2x}$ includes the points $(0, 0)$, $(2, 2)$, and $(4.5, 3)$. The graph of $x - y = 4$ is a straight line with slope 1 and y-intercept $(0, -4)$.

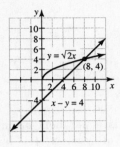

Substitute $y = \sqrt{2x}$ into $y = x - 4$.

$$\sqrt{2x} = x - 4$$
$$(\sqrt{2x})^2 = (x - 4)^2$$
$$2x = x^2 - 8x + 16$$
$$0 = x^2 - 10x + 16$$
$$0 = (x - 8)(x - 2)$$
$$x = 8 \qquad \text{or} \quad x = 2$$
$$y = \sqrt{2(8)} \quad \text{or} \quad y = \sqrt{2(2)}$$
$$= 4 \qquad \text{or} \quad = 2$$

Since we squared both sides, we must check. The pair $(8, 4)$ satisfies both equations, but $(2, 2)$ does not. The solution set to the system is $\{(8, 4)\}$.

7. The graph of $4x - 9y = 9$ is a straight line. The equation $xy = 1$ can be written as $y = 1/x$. Substitute $y = 1/x$ into $4x - 9y = 9$.

$$4x - 9\left(\tfrac{1}{x}\right) = 9$$
$$4x^2 - 9 = 9x$$
$$4x^2 - 9x - 9 = 0$$
$$(4x + 3)(x - 3) = 0$$
$$x = -\frac{3}{4} \qquad \text{or} \qquad x = 3$$
$$y = \frac{1}{-3/4} = -\frac{4}{3} \qquad \qquad y = \frac{1}{3}$$

The solution set is $\left\{\left(-\frac{3}{4}, -\frac{4}{3}\right), \left(3, \frac{1}{3}\right)\right\}$.

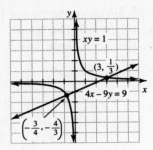

9. The graph of $y = x^2$ is a parabola opening upward, and the graph of $y = -x^2 + 1$ is a parabola opening downward.

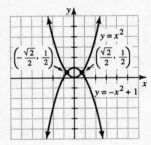

Substitute $y = x^2$ into $y = -x^2 + 1$.

$$x^2 = -x^2 + 1$$

$$2x^2 = 1$$

$$x^2 = \frac{1}{2}$$

$$x = \pm\sqrt{\frac{1}{2}} = \pm\frac{\sqrt{2}}{2}$$

Since $y = x^2$, $y = \frac{1}{2}$ for either value of x. The

solution set is $\left\{ \left(\frac{\sqrt{2}}{2}, \frac{1}{2} \right), \left(-\frac{\sqrt{2}}{2}, \frac{1}{2} \right) \right\}$.

11. Write $y = x^2 - 5$ as $x^2 = y + 5$ and substitute into $x^2 + y^2 = 25$

$$y + 5 + y^2 = 25$$

$$y^2 + y - 20 = 0$$

$$(y + 5)(y - 4) = 0$$

$$y = -5 \quad \text{or} \quad y = 4$$

Use $y = -5$ in $x^2 = y + 5$ to get $x^2 = 0$ or $x = 0$. Use $y = 4$ in $x^2 = y + 5$ to get $x^2 = 9$ or $x = \pm 3$. The solution set to the system is $\{(0, -5), (3, 4), (-3, 4)\}$.

13. Substitute $y = x + 1$ into $xy - 3x = 8$.

$$x(x + 1) - 3x = 8$$

$$x^2 + x - 3x = 8$$

$$x^2 - 2x - 8 = 0$$

$$(x - 4)(x + 2) = 0$$

$$x = 4 \quad \text{or} \quad x = -2$$

$$y = 5 \qquad\qquad y = -1 \quad \text{Since } y = x + 1$$

The solution set is $\{(4, 5), (-2, -1)\}$.

15. Write $xy - x = 8$ as $xy = x + 8$, and substitute for xy in $xy + 3x = -4$.

$$x + 8 + 3x = -4$$

$$4x = -12$$

$$x = -3$$

Use $x = -3$ in $xy = x + 8$.

$$-3y = -3 + 8$$

$$-3y = 5$$

$$y = -5/3$$

The solution set is $\left\{ \left(-3, -\frac{5}{3} \right) \right\}$.

17. If we just add the equations as they are given, then y is eliminated.

$$\frac{1}{x} - \frac{1}{y} = 5$$

$$\frac{2}{x} + \frac{1}{y} = -3$$

$$\overline{}$$

$$\frac{3}{x} = 2$$

$$3 = 2x$$

$$3/2 = x$$

Use $x = 3/2$ in $\frac{1}{x} - \frac{1}{y} = 5$.

$$\frac{1}{3/2} - \frac{1}{y} = 5$$

$$\frac{2}{3} - \frac{1}{y} = 5$$

$$2y - 3 = 15y$$

$$-3 = 13y$$

$$-3/13 = y$$

The solution set is $\left\{ \left(\frac{3}{2}, -\frac{3}{13} \right) \right\}$.

19. Substitute $y = 20/x^2$ into $xy + 2 = 6x$.

$$x \cdot \frac{20}{x^2} + 2 = 6x$$

$$\frac{20}{x} + 2 = 6x$$

$$20 + 2x = 6x^2$$

$$0 = 6x^2 - 2x - 20$$
$$3x^2 - x - 10 = 0$$
$$(3x + 5)(x - 2) = 0$$
$$x = -5/3 \quad \text{or} \quad x = 2$$
$$y = 36/5 \qquad\qquad y = 5 \quad \text{Since } y = 20/x^2$$

The solution set is $\left\{\left(-\frac{5}{3}, \frac{36}{5}\right), (2, 5)\right\}$.

21. If we add the equations we get $2x^2 = 10$.
$$x^2 = 5$$
$$x = \pm\sqrt{5}$$
Use $x = \sqrt{5}$ in $y^2 = 8 - x^2$, to get $y = \pm\sqrt{3}$. Use $x = -\sqrt{5}$ in $y^2 = 8 - x^2$, to get $y = \pm\sqrt{3}$.

The solution set contains four points,

$\{(\sqrt{5}, \sqrt{3}), (\sqrt{5}, -\sqrt{3}), (-\sqrt{5}, \sqrt{3}), (-\sqrt{5}, -\sqrt{3})\}$.

23. Substitute $y = 7 - x$ in $x^2 + xy - y^2 = -11$.
$$x^2 + x(7 - x) - (7 - x)^2 = -11$$
$$x^2 + 7x - x^2 - 49 + 14x - x^2 = -11$$
$$-x^2 + 21x - 38 = 0$$
$$x^2 - 21x + 38 = 0$$
$$(x - 2)(x - 19) = 0$$
$$x = 2 \quad \text{or} \quad x = 19$$
$$y = 5 \qquad\qquad y = -12 \quad \text{Since } y = 7 - x$$
The solution set is $\{(2, 5), (19, -12)\}$.

25. If $x^2 = y$, then $x^4 = y^2$. Substitute for x^4 in the equation $3y - 2 = x^4$.

$$3y - 2 = y^2$$
$$0 = y^2 - 3y + 2$$
$$0 = (y - 1)(y - 2)$$
$$y = 1 \quad \text{or} \quad y = 2$$

Use $y = 1$ in $x^2 = y$ to get $x^2 = 1$, or $x = \pm 1$.

Use $y = 2$ in $x^2 = y$ to get $x^2 = 2$, or $x = \pm\sqrt{2}$.

The solution set is

$\{(\sqrt{2}, 2), (-\sqrt{2}, 2), (1, 1), (-1, 1)\}$.

27. Eliminate y by substitution.
$$\log_2(x - 1) = 3 - \log_2(x + 1)$$
$$\log_2(x - 1) + \log_2(x + 1) = 3$$
$$\log_2[x^2 - 1] = 3$$
$$x^2 - 1 = 8$$
$$x^2 = 9$$
$$x = \pm 3$$
If $x = -3$, we get a logarithm of a negative number. If $x = 3$, then $y = \log_2(3 - 1) = 1$. The solution set is $\{(3, 1)\}$.

29. Use substitution to eliminate y.
$$\log_2(x - 1) = 2 + \log_2(x + 2)$$
$$\log_2(x - 1) - \log_2(x + 2) = 2$$
$$\log_2\left(\frac{x - 1}{x + 2}\right) = 2$$

$$\frac{x - 1}{x + 2} = 4$$

$$x - 1 = 4x + 8$$
$$-9 = 3x$$
$$-3 = x$$

If $x = -3$ in either of the original equations, we get a logarithm of a negative number. So the solution set is the empty set, $\emptyset$.

31. Use substitution to eliminate y.

$$2^{3x + 4} = 4^{x - 1}$$
$$2^{3x + 4} = (2^2)^{x - 1}$$
$$2^{3x + 4} = 2^{2x - 2}$$
$$3x + 4 = 2x - 2$$
$$x = -6$$

If $x = -6$, then $y = 4^{-6 - 1} = 4^{-7}$. So the solution set is $\{(-6, 4^{-7})\}$.

33. First multiply the second equation by -1 and then add the result to the first equation.
$$x^2 + xy + y^2 = 19$$
$$\underline{-x^2 \qquad - y^2 = -13}$$
$$xy \qquad = 6$$
Use $y = 6/x$ in $x^2 + y^2 = 13$.
$$x^2 + \frac{36}{x^2} = 13$$
$$x^4 + 36 = 13x^2$$
$$x^4 - 13x^2 + 36 = 0$$
$$(x^2 - 9)(x^2 - 4) = 0$$
$$x^2 = 9 \qquad \text{or} \qquad x^2 = 4$$
$$x = \pm 3 \quad \text{or} \qquad x = \pm 2$$
For each of the four values of x, we find y from the equation $y = 6/x$. The solution set is $\{(2, 3), (-2, -3), (3, 2), (-3, -2)\}$.

35. Multiply the second equation by -1 and add the result to the first equation.

$$x^2 + xy + 2y^2 = 11$$
$$\underline{-x^2 + 2xy - 2y^2 = -5}$$
$$3xy \qquad = 6$$
$$xy \qquad = 2$$
Use $y = 2/x$ in the first equation.

$$x^2 + x\left(\tfrac{2}{x}\right) + 2\left(\tfrac{2}{x}\right)^2 = 11$$
$$x^2 + 2 + \frac{8}{x^2} = 11$$

$$x^4 + 2x^2 + 8 = 11x^2$$
$$x^4 - 9x^2 + 8 = 0$$
$$(x^2 - 8)(x^2 - 1) = 0$$
$$x^2 = 8 \quad \text{or} \quad x^2 = 1$$
$$x = \pm 2\sqrt{2} \quad \text{or} \quad x = \pm 1$$

For each of the four values of x, we find the y-value using the equation $y = 2/x$. Solution set:

$$\left\{ \left(2\sqrt{2}, \tfrac{\sqrt{2}}{2}\right), \left(-2\sqrt{2}, -\tfrac{\sqrt{2}}{2}\right), (1, 2), (-1, -2) \right\}$$

37. Let x = the length of one leg and y = the length of the other. We write one equation for the area: $3 = \tfrac{1}{2}xy$. We write the other equation from the Pythagorean theorem: $x^2 + y^2 = 15$. Substitute $y = 6/x$ into $x^2 + y^2 = 15$.

$$x^2 + \left(\tfrac{6}{x}\right)^2 = 15$$
$$x^2 + \frac{36}{x^2} = 15$$
$$x^4 + 36 = 15x^2$$
$$x^4 - 15x^2 + 36 = 0$$
$$(x^2 - 3)(x^2 - 12) = 0$$
$$x = \sqrt{3} \quad \text{or} \quad x = \sqrt{12} = 2\sqrt{3}$$

If $x = \sqrt{3}$, then $y = 6/\sqrt{3} = 2\sqrt{3}$. If $x = 2\sqrt{3}$, then $y = 6/(2\sqrt{3}) = \sqrt{3}$. So the lengths of the legs are $\sqrt{3}$ feet and $2\sqrt{3}$ feet.

39. Let x = the height of each triangle, and y = the length of the base of each triangle. Since the 7 triangles are to have a total area of 3,500, the area of each must be 500: $\tfrac{1}{2}xy = 500$. The ratio of the height to base must be 1 to 4 is expressed as $\tfrac{x}{y} = \tfrac{1}{4}$. These two equations can be written as $xy = 1000$ and $y = 4x$. Use substitution.

$$x(4x) = 1000$$
$$x^2 = 250$$
$$x = \sqrt{250} = 5\sqrt{10}$$

Since $y = 4x$, $y = 20\sqrt{10}$. So the height is $5\sqrt{10}$ inches and the base is $20\sqrt{10}$ inches.

41. Let x = the number of hours for pump A to fill the tank alone, and y = the number of hours for pump B to fill the tank alone.

$$\tfrac{1}{x} + \tfrac{1}{y} = \tfrac{1}{6}$$
$$\tfrac{1}{y} - \tfrac{1}{x} = \tfrac{1}{12}$$

Adding these two equations eliminates x.

$$\tfrac{2}{y} = \tfrac{1}{6} + \tfrac{1}{12}$$
$$\tfrac{2}{y} = \tfrac{1}{4}$$
$$y = 8$$

Use $y = 8$ in the first equation.

$$\tfrac{1}{x} + \tfrac{1}{8} = \tfrac{1}{6}$$
$$\tfrac{1}{x} = \tfrac{1}{24}$$
$$x = 24$$

It would take pump A 24 hours to fill the tank alone and pump B 8 hours to fill the tank alone.

43. Let x = the time for Jan to do the job alone. Let y = the time for Beth to do the job alone. Since they do the job together in 24 minutes we have the equation

$$\tfrac{1}{x} + \tfrac{1}{y} = \tfrac{1}{24}.$$

In 50 minutes the job was completed, but not by working together. This implies that the total of their times alone is 100, $x + y = 100$. Substitute $y = 100 - x$ into the first equation.

$$\tfrac{1}{x} + \frac{1}{100 - x} = \tfrac{1}{24}$$

$$24(100 - x) + 24x = x(100 - x)$$
$$2400 - 24x + 24x = 100x - x^2$$
$$x^2 - 100x + 2400 = 0$$
$$(x - 40)(x - 60) = 0$$
$$x = 40 \quad \text{or} \quad x = 60$$
$$y = 60 \qquad\qquad y = 40 \quad \text{Since } y = 100 - x$$

Since Jan is the faster worker, it would take her 40 minutes to complete the catfish by herself.

45. Let $x =$ the length and $y =$ the width. The area is 72 is expressed as $xy = 72$. The perimeter is 34 is expressed as $2x + 2y = 34$ or $x + y = 17$. Substitute $y = 17 - x$ into $xy = 72$.

$$x(17 - x) = 72$$
$$-x^2 + 17x - 72 = 0$$
$$x^2 - 17x + 72 = 0$$
$$(x - 9)(x - 8) = 0$$
$$x = 9 \quad \text{or} \quad x = 8$$
$$y = 8 \qquad\quad y = 9$$

The rectangular area is 8 feet by 9 feet.

47. Let x and y represent the numbers.

$$x + y = 8$$
$$xy = 20$$

Substitute $y = 8 - x$ into $xy = 20$.

$$x(8 - x) = 20$$
$$-x^2 + 8x - 20 = 0$$
$$x^2 - 8x + 20 = 0$$
$$x = \frac{8 \pm \sqrt{64 - 4(1)(20)}}{2} = \frac{8 \pm \sqrt{-16}}{2}$$
$$= 4 \pm 2i$$

If $x = 4 + 2i$, then $y = 8 - (4 + 2i) = 4 - 2i$. If $x = 4 - 2i$, then $y = 8 - (4 - 2i) = 4 + 2i$. So the numbers are $4 - 2i$ and $4 + 2i$.

49. Let $x =$ the length of the side of the square, and $y =$ the height of the triangle. The total height of 10 feet means that $x + y = 10$. The total area is 72 means that $x^2 + 0.5xy = 72$. Substitute $y = 10 - x$ into the second equation.

$$x^2 + 0.5x(10 - x) = 72$$
$$x^2 + 5x - 0.5x^2 = 72$$
$$0.5x^2 + 5x - 72 = 0$$
$$x^2 + 10x - 144 = 0$$
$$(x - 8)(x + 18) = 0$$
$$x = 8 \quad \text{or} \quad x = -18$$
$$y = 2$$

The side of the square is 8 feet and the height of the triangle is 2 feet.

51. a) $(1.71, 1.55)$, $(-2.98, -3.95)$

b) $(1, 1)$, $(0.40, 0.16)$

c) $(1.17, 1.62)$, $(-1.17, -1.62)$

10.5 WARM-UPS

1. True **2.** True. **3.** True, replace R_2 of (a) by $R_1 + R_2$ to get matrix (b). **4.** False, because matrix (c) corresponds to an inconsistent system and (d) corresponds to a dependent system. **5.** True, because the last row represents the equation $0 = 7$. **6.** False, replace R_2 by $2R_1 + R_2$ to get $0 = -3$ which is inconsistent. **7.** False, the system corresponding to (d) consists of two equations equivalent to $x + 3y = 5$. **8.** False, the augmented matrix is a 2×3 matrix. **9.** True. **10.** False, it means to interchage R_1 and R_2.

10.5 EXERCISES

1. 2×2 **3.** 3×2 **5.** 3×1

7. Use the coefficients 2 and -3, and the constant 9 as the first row. Use the coefficients -3 and 1, and the constant -1 as the second row.

$$\begin{bmatrix} 2 & -3 & | & 9 \\ -3 & 1 & | & -1 \end{bmatrix}$$

9. Use the coefficients 1, -1, and 1, and the constant 1 as the first row. Use the coefficients 1, 1, and -2, and the constant 3 as the second row. Use the coefficients 0, 1, and -3, and the constant 4 as the third row.

$$\begin{bmatrix} 1 & -1 & 1 & | & 1 \\ 1 & 1 & -2 & | & 3 \\ 0 & 1 & -3 & | & 4 \end{bmatrix}$$

11. The entries in the first row $(5, 1, -1)$ represent the equation $5x + y = -1$. The entries in the second row $(2, -3, 0)$ represent the equation $2x - 3y = 0$. So the matrix represents the following system.

$$5x + y = -1$$
$$2x - 3y = 0$$

13. The entries in the first row $(1, 0, 0, 6)$ represent the equation $x = 6$. The entries in the second row $(-1, 0, 1, -3)$ represent the equation $-x + z = -3$. The entries in the third row $(1, 1, 0, 1)$ represent the equation $x + y = 1$. So the matrix represents the following system.

$$x = 6$$
$$-x + z = -3$$
$$x + y = 1$$

15. $R_1 \leftrightarrow R_2$

17. $\frac{1}{5}R_2 \rightarrow R_2$

19. $\begin{bmatrix} 1 & 1 & \vline & 3 \\ -3 & 1 & \vline & -1 \end{bmatrix}$

$\begin{bmatrix} 1 & 1 & \vline & 3 \\ 0 & 4 & \vline & 8 \end{bmatrix}$ $3R_1 + R_2 \rightarrow R_2$

$\begin{bmatrix} 1 & 1 & \vline & 3 \\ 0 & 1 & \vline & 2 \end{bmatrix}$ $R_2 \div 4 \rightarrow R_2$

$\begin{bmatrix} 1 & 0 & \vline & 1 \\ 0 & 1 & \vline & 2 \end{bmatrix}$ $-R_2 + R_1 \rightarrow R_1$

The solution set is $\{(1, 2)\}$.

21. $\begin{bmatrix} 2 & -1 & \vline & 3 \\ 1 & 1 & \vline & 9 \end{bmatrix}$

$\begin{bmatrix} 1 & 1 & \vline & 9 \\ 2 & -1 & \vline & 3 \end{bmatrix}$ $R_2 \rightarrow R_1$ and $R_1 \rightarrow R_2$

$\begin{bmatrix} 1 & 1 & \vline & 9 \\ 0 & -3 & \vline & -15 \end{bmatrix}$ $-2R_1 + R_2 \rightarrow R_2$

$\begin{bmatrix} 1 & 1 & \vline & 9 \\ 0 & 1 & \vline & 5 \end{bmatrix}$ $R_2 \div (-3) \rightarrow R_2$

$\begin{bmatrix} 1 & 0 & \vline & 4 \\ 0 & 1 & \vline & 5 \end{bmatrix}$ $-R_2 + R_1 \rightarrow R_1$

The solution set is $\{(4, 5)\}$.

23. $\begin{bmatrix} 3 & -1 & \vline & 4 \\ 2 & 1 & \vline & 1 \end{bmatrix}$

$\begin{bmatrix} 1 & -2 & \vline & 3 \\ 2 & 1 & \vline & 1 \end{bmatrix}$ $-R_2 + R_1 \rightarrow R_1$

$\begin{bmatrix} 1 & -2 & \vline & 3 \\ 0 & 5 & \vline & -5 \end{bmatrix}$ $-2R_1 + R_2 \rightarrow R_2$

$\begin{bmatrix} 1 & -2 & \vline & 3 \\ 0 & 1 & \vline & -1 \end{bmatrix}$ $R_2 \div 5 \rightarrow R_2$

$\begin{bmatrix} 1 & 0 & \vline & 1 \\ 0 & 1 & \vline & -1 \end{bmatrix}$ $2R_2 + R_1 \rightarrow R_1$

The solution set is $\{(1, -1)\}$

25. $\begin{bmatrix} 6 & -7 & \vline & 0 \\ 2 & 1 & \vline & 20 \end{bmatrix}$

$\begin{bmatrix} 0 & -10 & \vline & -60 \\ 2 & 1 & \vline & 20 \end{bmatrix}$ $-3R_2 + R_1 \rightarrow R_1$

$\begin{bmatrix} 0 & 1 & \vline & 6 \\ 2 & 1 & \vline & 20 \end{bmatrix}$ $R_1 \div (-10) \rightarrow R_1$

$\begin{bmatrix} 2 & 1 & \vline & 20 \\ 0 & 1 & \vline & 6 \end{bmatrix}$ $R_1 \leftrightarrow R_2$

$\begin{bmatrix} 2 & 0 & \vline & 14 \\ 0 & 1 & \vline & 6 \end{bmatrix}$ $-R_2 + R_1 \rightarrow R_1$

$\begin{bmatrix} 1 & 0 & \vline & 7 \\ 0 & 1 & \vline & 6 \end{bmatrix}$ $R_1 \div 2 \rightarrow R_1$

The solution set is $\{(7, 6)\}$.

27. $\begin{bmatrix} 2 & -3 & | & 4 \\ -2 & 3 & | & 5 \end{bmatrix}$

$\begin{bmatrix} 2 & -3 & | & 4 \\ 0 & 0 & | & 9 \end{bmatrix}$ $\quad R_1 + R_2 \rightarrow R_2$

Since the second row represents the equation $0 = 9$, there is no solution to the system.

29. $\begin{bmatrix} 1 & 2 & | & 1 \\ 3 & 6 & | & 3 \end{bmatrix}$

$\begin{bmatrix} 1 & 2 & | & 1 \\ 0 & 0 & | & 0 \end{bmatrix}$ $\quad -3R_1 + R_2 \rightarrow R_2$

Since the system is equivalent to the single equation $x + 2y = 1$, the equations are dependent and the solution set is $\{(x, y) \mid x + 2y = 1\}$.

31. $\begin{bmatrix} 1 & 1 & 1 & | & 6 \\ 1 & -1 & 1 & | & 2 \\ 0 & 2 & -1 & | & 1 \end{bmatrix}$

$\begin{bmatrix} 1 & 1 & 1 & | & 6 \\ 0 & -2 & 0 & | & -4 \\ 0 & 2 & -1 & | & 1 \end{bmatrix}$ $\quad -R_1 + R_2 \rightarrow R_2$

$\begin{bmatrix} 1 & 1 & 1 & | & 6 \\ 0 & 1 & 0 & | & 2 \\ 0 & 2 & -1 & | & 1 \end{bmatrix}$ $\quad R_2 \div (-2) \rightarrow R_2$

$\begin{bmatrix} 1 & 0 & 1 & | & 4 \\ 0 & 1 & 0 & | & 2 \\ 0 & 0 & -1 & | & -3 \end{bmatrix}$ $\quad \begin{array}{l} -R_2 + R_1 \rightarrow R_1 \\ -2R_2 + R_3 \rightarrow R_3 \end{array}$

$\begin{bmatrix} 1 & 0 & 1 & | & 4 \\ 0 & 1 & 0 & | & 2 \\ 0 & 0 & 1 & | & 3 \end{bmatrix}$ $\quad R_3 \div (-1) \rightarrow R_3$

$\begin{bmatrix} 1 & 0 & 0 & | & 1 \\ 0 & 1 & 0 & | & 2 \\ 0 & 0 & 1 & | & 3 \end{bmatrix}$ $\quad -R_3 + R_1 \rightarrow R_1$

The solution set is $\{(1, 2, 3)\}$.

33. $\begin{bmatrix} 2 & 1 & 1 & | & 4 \\ 1 & 1 & -1 & | & 1 \\ 1 & -1 & 2 & | & 2 \end{bmatrix}$

$\begin{bmatrix} 1 & 1 & -1 & | & 1 \\ 2 & 1 & 1 & | & 4 \\ 1 & -1 & 2 & | & 2 \end{bmatrix}$ $\quad R_1 \leftrightarrow R_2$

$\begin{bmatrix} 1 & 1 & -1 & | & 1 \\ 0 & -1 & 3 & | & 2 \\ 0 & -2 & 3 & | & 1 \end{bmatrix}$ $\quad \begin{array}{l} -2R_1 + R_2 \rightarrow R_2 \\ -R_1 + R_3 \rightarrow R_3 \end{array}$

$\begin{bmatrix} 1 & 1 & -1 & | & 1 \\ 0 & 1 & -3 & | & -2 \\ 0 & -2 & 3 & | & 1 \end{bmatrix}$ $\quad -1 \cdot R_2 \rightarrow R_2$

$\begin{bmatrix} 1 & 0 & 2 & | & 3 \\ 0 & 1 & -3 & | & -2 \\ 0 & 0 & -3 & | & -3 \end{bmatrix}$ $\quad \begin{array}{l} -1 \cdot R_2 + R_1 \rightarrow R_1 \\ 2R_2 + R_3 \rightarrow R_3 \end{array}$

$\begin{bmatrix} 1 & 0 & 2 & | & 3 \\ 0 & 1 & -3 & | & -2 \\ 0 & 0 & 1 & | & 1 \end{bmatrix}$ $\quad -1 \cdot R_3 \rightarrow R_3$

$$\begin{bmatrix} 1 & 0 & 0 & | & 1 \\ 0 & 1 & 0 & | & 1 \\ 0 & 0 & 1 & | & 1 \end{bmatrix} \qquad \begin{array}{l} -2R_3 + R_1 \to R_1 \\ 3R_3 + R_2 \to R_2 \end{array}$$

The solution set is $\{(1, 1, 1)\}$.

35. $\begin{bmatrix} 1 & 1 & -3 & | & 3 \\ 2 & -1 & 1 & | & 0 \\ 1 & -1 & 1 & | & -1 \end{bmatrix}$

$$\begin{bmatrix} 1 & 1 & -3 & | & 3 \\ 0 & -3 & 7 & | & -6 \\ 0 & -2 & 4 & | & -4 \end{bmatrix} \qquad \begin{array}{l} -2R_1 + R_2 \to R_2 \\ -R_1 + R_3 \to R_3 \end{array}$$

$$\begin{bmatrix} 1 & 1 & -3 & | & 3 \\ 0 & 1 & -1 & | & 2 \\ 0 & -2 & 4 & | & -4 \end{bmatrix} \qquad -2R_3 + R_2 \to R_2$$

$$\begin{bmatrix} 1 & 0 & -2 & | & 1 \\ 0 & 1 & -1 & | & 2 \\ 0 & 0 & 2 & | & 0 \end{bmatrix} \qquad \begin{array}{l} -R_2 + R_1 \to R_1 \\ 2R_2 + R_3 \to R_3 \end{array}$$

$$\begin{bmatrix} 1 & 0 & -2 & | & 1 \\ 0 & 1 & -1 & | & 2 \\ 0 & 0 & 1 & | & 0 \end{bmatrix} \qquad R_3 \div 2 \to R_3$$

$$\begin{bmatrix} 1 & 0 & 0 & | & 1 \\ 0 & 1 & 0 & | & 2 \\ 0 & 0 & 1 & | & 0 \end{bmatrix} \qquad \begin{array}{l} 2R_2 + R_1 \to R_1 \\ R_3 + R_2 \to R_2 \end{array}$$

The solution set is $\{(1, 2, 0)\}$.

37. $\begin{bmatrix} 1 & -1 & -4 & | & -3 \\ -1 & 3 & 1 & | & 0 \\ 1 & 1 & 2 & | & 3 \end{bmatrix}$

$$\begin{bmatrix} 1 & -1 & -4 & | & -3 \\ 0 & 2 & -3 & | & -3 \\ 0 & 2 & 6 & | & 6 \end{bmatrix} \qquad \begin{array}{l} R_1 + R_2 \to R_2 \\ -R_1 + R_3 \to R_3 \end{array}$$

$$\begin{bmatrix} 1 & -1 & -4 & | & -3 \\ 0 & 2 & -3 & | & -3 \\ 0 & 0 & 9 & | & 9 \end{bmatrix} \qquad -R_2 + R_3 \to R_3$$

$$\begin{bmatrix} 1 & -1 & -4 & | & -3 \\ 0 & 2 & -3 & | & -3 \\ 0 & 0 & 1 & | & 1 \end{bmatrix} \qquad R_3 \div 9 \to R_3$$

$$\begin{bmatrix} 1 & -1 & 0 & | & 1 \\ 0 & 2 & 0 & | & 0 \\ 0 & 0 & 1 & | & 1 \end{bmatrix} \qquad \begin{array}{l} 4R_3 + R_1 \to R_1 \\ 3R_3 + R_2 \to R_2 \end{array}$$

$$\begin{bmatrix} 1 & -1 & 0 & | & 1 \\ 0 & 1 & 0 & | & 0 \\ 0 & 0 & 1 & | & 1 \end{bmatrix} \qquad R_2 \div 2 \to R_2$$

$$\begin{bmatrix} 1 & 0 & 0 & | & 1 \\ 0 & 1 & 0 & | & 0 \\ 0 & 0 & 1 & | & 1 \end{bmatrix} \qquad R_2 + R_1 \to R_1$$

The solution set is $\{(1, 0, 1)\}$.

39. $\begin{bmatrix} 1 & -1 & 1 & | & 1 \\ 2 & -2 & 2 & | & 2 \\ -3 & 3 & -3 & | & -3 \end{bmatrix}$

$$\begin{bmatrix} 1 & -1 & 1 & | & 1 \\ 0 & 0 & 0 & | & 0 \\ 0 & 0 & 0 & | & 0 \end{bmatrix} \qquad \begin{array}{l} -2R_1 + R_2 \to R_2 \\ 3R_1 + R_3 \to R_3 \end{array}$$

Since the system of equations is equivalent to the first equation, the solution set is $\{(x, y, z) \mid x - y + z = 1\}$.

41.
$$\left[\begin{array}{rrr|r} 1 & 1 & -1 & 2 \\ 2 & -1 & 1 & 1 \\ 3 & 3 & -3 & 8 \end{array} \right]$$

$$\left[\begin{array}{rrr|r} 1 & 1 & -1 & 2 \\ 0 & -3 & 3 & -3 \\ 0 & 0 & 0 & 2 \end{array} \right] \quad \begin{array}{l} -2R_1 + R_2 \rightarrow R_2 \\ -3R_1 + R_3 \rightarrow R_3 \end{array}$$

Since the third equation is $0 = 2$, there is no solution to the system.

10.6 WARM-UPS

1. True. because the determinant is $(-1)(-5) - (2)(3) = -1$. **2.** False, because the determinant is $2 \cdot 8 - (-4)(4) = 32$. **3.** False, because if $D = 0$, then Cramer's rule fails to give us the solution. **4.** True, the determinant is the value of $ad - bc$. **5.** True, this is the case where Cramer's rule fails to give the precise solution. **6.** True. **7.** True, this is precisely when Cramer's rule works. **8.** True, because if x is the tens digit then the value is $10x + y$, and if y is the tens digit then the value is $10y + x$. **9.** False, because the total of the perimeters is $4x + 3y$. **10.** False, because if the digits are reversed then b is the tens digit and the value of the number is $10b + a$.

10.6 EXERCISES

1. $\begin{vmatrix} 2 & 5 \\ 3 & 7 \end{vmatrix} = 2 \cdot 7 - 3 \cdot 5 = -1$

3. $\begin{vmatrix} 0 & 3 \\ 1 & 5 \end{vmatrix} = 0 \cdot 5 - 1 \cdot 3 = -3$

5. $\begin{vmatrix} -3 & -2 \\ -4 & 2 \end{vmatrix} = -3 \cdot 2 - (-4)(-2) = -14$

7. $\begin{vmatrix} 0.05 & 0.06 \\ 10 & 20 \end{vmatrix} = 0.05(20) - 0.06(10) = 0.4$

9. $D = \begin{vmatrix} 2 & -1 \\ 3 & 2 \end{vmatrix} = 7 \quad D_x = \begin{vmatrix} 5 & -1 \\ -3 & 2 \end{vmatrix} = 7$

$D_y = \begin{vmatrix} 2 & 5 \\ 3 & -3 \end{vmatrix} = -21$

$x = \dfrac{D_x}{D} = \dfrac{7}{7} = 1 \qquad y = \dfrac{D_y}{D} = \dfrac{-21}{7} = -3$

The solution set is $\{(1, -3)\}$.

11. $D = \begin{vmatrix} 3 & -5 \\ 2 & 3 \end{vmatrix} = 19$

$D_x = \begin{vmatrix} -2 & -5 \\ 5 & 3 \end{vmatrix} = 19 \qquad D_y = \begin{vmatrix} 3 & -2 \\ 2 & 5 \end{vmatrix} = 19$

$x = \dfrac{D_x}{D} = \dfrac{19}{19} = 1 \qquad y = \dfrac{D_y}{D} = \dfrac{19}{19} = 1$

The solution set is $\{(1, 1)\}$.

13. $D = \begin{vmatrix} 4 & -3 \\ 2 & 5 \end{vmatrix} = 26$

$D_x = \begin{vmatrix} 5 & -3 \\ 7 & 5 \end{vmatrix} = 46 \qquad D_y = \begin{vmatrix} 4 & 5 \\ 2 & 7 \end{vmatrix} = 18$

$x = \dfrac{D_x}{D} = \dfrac{46}{26} = \dfrac{23}{13} \qquad y = \dfrac{D_y}{D} = \dfrac{18}{26} = \dfrac{9}{13}$

The solution set is $\left\{ \left(\dfrac{23}{13}, \dfrac{9}{13} \right) \right\}$.

15. $D = \begin{vmatrix} 0.5 & 0.2 \\ 0.4 & -0.6 \end{vmatrix} = -0.38$

$D_x = \begin{vmatrix} 8 & 0.2 \\ -5 & -0.6 \end{vmatrix} = -3.8 \quad D_y = \begin{vmatrix} 0.5 & 8 \\ 0.4 & -5 \end{vmatrix} = -5.7$

$x = \dfrac{D_x}{D} = \dfrac{-3.8}{-0.38} = 10 \qquad y = \dfrac{D_y}{D} = \dfrac{-5.7}{-0.38} = 15$

The solution set is $\{(10, 15)\}$.

17. Multiply the first equation by 4 and the second by 6 to eliminate the fractions.
$$2x + y = 20$$
$$2x - 3y = -6$$

$D = \begin{vmatrix} 2 & 1 \\ 2 & -3 \end{vmatrix} = -8$

$$D_x = \begin{vmatrix} 20 & 1 \\ -6 & -3 \end{vmatrix} = -54 \quad D_y = \begin{vmatrix} 2 & 20 \\ 2 & -6 \end{vmatrix} = -52$$

$$x = \frac{D_x}{D} = \frac{-54}{-8} = \frac{27}{4} \qquad y = \frac{D_y}{D} = \frac{-52}{-8} = \frac{13}{2}$$

The solution set is $\left\{\left(\frac{27}{4}, \frac{13}{2}\right)\right\}$.

19. $D = \begin{vmatrix} 2 & -3 \\ 4 & -6 \end{vmatrix} = 0$

Since Cramer's rule does not apply, multiply the first equation by -2 and add the result to the second equation.
$$\begin{array}{r} -4x + 6y = -10 \\ 4x - 6y = 8 \\ \hline 0 = -2 \end{array}$$
The solution set is $\emptyset$.

21. $D = \begin{vmatrix} 1 & -1 \\ 1 & 2 \end{vmatrix} = 3$

$$D_x = \begin{vmatrix} 4 & -1 \\ 6 & 2 \end{vmatrix} = 14 \qquad D_y = \begin{vmatrix} 1 & 4 \\ 1 & 6 \end{vmatrix} = 2$$

$$x = \frac{D_x}{D} = \frac{14}{3} \qquad y = \frac{D_y}{D} = \frac{2}{3}$$
The solution set is $\left\{\left(\frac{14}{3}, \frac{2}{3}\right)\right\}$.

23. $D = \begin{vmatrix} 4 & -1 \\ -8 & 2 \end{vmatrix} = 0$

Since Cramer's rule does not apply, multiply the first equation by 2 and add the result to the second equation.
$$\begin{array}{r} 8x - 2y = 12 \\ -8x + 2y = -12 \\ \hline 0 = 0 \end{array}$$
The solution set is $\{(x, y) \mid 4x - y = 6\}$.

25. To use Cramer's rule, we must rewrite each equation.

$$\begin{array}{ll} y = 3x - 12 & 3(x+1) - 11 = y + 4 \\ -3x + y = -12 & 3x - 8 = y + 4 \\ 3x - y = 12 & 3x - y = 12 \end{array}$$

After rewriting the equations, we see that the equations are equivalent. If we calculate D for

Cramer's rule we get $D = \begin{vmatrix} 3 & -1 \\ 3 & -1 \end{vmatrix} = 0$.

The solution set is $\{(x, y) \mid y = 3x - 12\}$.

27. Rewrite each equation.

$$\begin{array}{ll} x - 6 = y + 1 & y = -3x + 1 \\ x - y = 7 & 3x + y = 1 \end{array}$$

Apply Cramer's rule to the following system.
$$\begin{array}{l} x - y = 7 \\ 3x + y = 1 \end{array}$$

$$D = \begin{vmatrix} 1 & -1 \\ 3 & 1 \end{vmatrix} = 4$$

$$D_x = \begin{vmatrix} 7 & -1 \\ 1 & 1 \end{vmatrix} = 8 \qquad D_y = \begin{vmatrix} 1 & 7 \\ 3 & 1 \end{vmatrix} = -20$$

$$x = \frac{D_x}{D} = \frac{8}{4} = 2 \qquad y = \frac{D_y}{D} = \frac{-20}{4} = -5$$

The solution set is $\{(2, -5)\}$.

29. Rewrite the system.
$$\begin{array}{r} -0.05x + y = 0 \\ x + y = 504 \end{array}$$

$$D = \begin{vmatrix} -0.05 & 1 \\ 1 & 1 \end{vmatrix} = -1.05 \quad D_x = \begin{vmatrix} 0 & 1 \\ 504 & 1 \end{vmatrix} = -504$$

$$D_y = \begin{vmatrix} -0.05 & 0 \\ 1 & 504 \end{vmatrix} = -25.2$$

$$x = \frac{D_x}{D} = \frac{-504}{-1.05} = 480 \qquad y = \frac{D_y}{D} = \frac{-25.2}{-1.05} = 24$$

The solution set is $\{(480, 24)\}$.

31. Rewrite the system.
$$\begin{array}{r} x\sqrt{2} - y\sqrt{2} = -2 \\ x + y = \sqrt{18} \end{array}$$

$$D = \begin{vmatrix} \sqrt{2} & -\sqrt{2} \\ 1 & 1 \end{vmatrix} = 2\sqrt{2} \quad D_x = \begin{vmatrix} -2 & -\sqrt{2} \\ \sqrt{18} & 1 \end{vmatrix} = 4$$

$$D_y = \begin{vmatrix} \sqrt{2} & -2 \\ 1 & \sqrt{18} \end{vmatrix} = 8$$

$$x = \frac{D_x}{D} = \frac{4}{2\sqrt{2}} = \sqrt{2} \qquad y = \frac{D_y}{D} = \frac{8}{2\sqrt{2}} = 2\sqrt{2}$$

The solution set is $\{(\sqrt{2}, 2\sqrt{2})\}$.

33. Let $x =$ the number of servings of canned peas and $y =$ the number of servings of canned beets. In the first equation we find the total grams of protein and in the second the total grams of carbohydrates.

$$3x + y = 38$$
$$11x + 8y = 187$$

$$D = \begin{vmatrix} 3 & 1 \\ 11 & 8 \end{vmatrix} = 13 \qquad D_x = \begin{vmatrix} 38 & 1 \\ 187 & 8 \end{vmatrix} = 117$$

$$D_y = \begin{vmatrix} 3 & 38 \\ 11 & 187 \end{vmatrix} = 143$$

$$x = \frac{D_x}{D} = \frac{117}{13} = 9 \qquad y = \frac{D_y}{D} = \frac{143}{13} = 11$$

To get the required grams of protein and carbohydrates we need 9 servings of peas and 11 servings of beets.

35. Let $x =$ the tens digit and $y =$ the ones digit. The value of the number is $10x + y$, and if the digits are reversed the value is $10y + x$.
$$x + y = 10$$
$$2(10x + y) = 10y + x + 1$$
After rewriting the second equation we get the following system.

$$x + y = 10$$
$$19x - 8y = 1$$

$$D = \begin{vmatrix} 1 & 1 \\ 19 & -8 \end{vmatrix} = -27 \qquad D_x = \begin{vmatrix} 10 & 1 \\ 1 & -8 \end{vmatrix} = -81$$

$$D_y = \begin{vmatrix} 1 & 10 \\ 19 & 1 \end{vmatrix} = -189$$

$$x = \frac{D_x}{D} = \frac{-81}{-27} = 3 \qquad y = \frac{D_y}{D} = \frac{-189}{-27} = 7$$

The number is 37.

37. Let $x =$ the price of a gallon of milk and $y =$ the price of the magazine. Note that she paid $0.30 tax. The first equation expresses the total price of the goods and the second expresses the total tax.
$$x + y = 4.65$$
$$0.05x + 0.08y = 0.30$$

$$D = \begin{vmatrix} 1 & 1 \\ 0.05 & 0.08 \end{vmatrix} = 0.03$$

$$D_x = \begin{vmatrix} 4.65 & 1 \\ 0.3 & 0.08 \end{vmatrix} = 0.072$$

$$D_y = \begin{vmatrix} 1 & 4.65 \\ 0.05 & 0.3 \end{vmatrix} = 0.0675$$

$$x = \frac{D_x}{D} = \frac{0.072}{0.03} = 2.4 \qquad y = \frac{D_y}{D} = \frac{0.0675}{0.03} = 2.25$$

The price of the milk was $2.40 and the price of the magazine was $2.25.

39. Let $x =$ the number of singles and $y =$ the number of doubles. Write one equation for the patties and one equation for the tomato slices.
$$x + 2y = 32$$
$$2x + y = 34$$

$$D = \begin{vmatrix} 1 & 2 \\ 2 & 1 \end{vmatrix} = -3 \qquad D_x = \begin{vmatrix} 32 & 2 \\ 34 & 1 \end{vmatrix} = -36$$

$$D_y = \begin{vmatrix} 1 & 32 \\ 2 & 34 \end{vmatrix} = -30$$

$$x = \frac{D_x}{D} = \frac{-36}{-3} = 12 \qquad y = \frac{D_y}{D} = \frac{-30}{-3} = 10$$

He must sell 12 singles and 10 doubles.

41. Let $x =$ Gary's age and $y =$ Harry's age. Since Gary is 5 years older than Harry, $x = y + 5$. Twenty-nine years ago Gary was $x - 29$ and Harry was $y - 29$. Gary was twice as old as Harry (29 years ago) is expressed as $x - 29 = 2(y - 29)$. These equations can be rewritten as follows.
$$x - y = 5$$
$$x - 2y = -29$$

$$D = \begin{vmatrix} 1 & -1 \\ 1 & -2 \end{vmatrix} = -1 \qquad D_x = \begin{vmatrix} 5 & -1 \\ -29 & -2 \end{vmatrix} = -39$$

$$D_y = \begin{vmatrix} 1 & 5 \\ 1 & -29 \end{vmatrix} = -34$$

$$x = \frac{D_x}{D} = \frac{-39}{-1} = 39 \qquad y = \frac{D_y}{D} = \frac{-34}{-1} = 34$$

So Gary is 39 and Harry is 34.

43. Let $x =$ the length of a side of the square and $y =$ the length of a side of the equilateral triangle. Since the perimeters are to be equal, $4x = 3y$. Since the total of the two perimeters is 80, $4x + 3y = 80$. Rewrite the equations as follows.
$$4x - 3y = 0$$
$$4x + 3y = 80$$

$$D = \begin{vmatrix} 4 & -3 \\ 4 & 3 \end{vmatrix} = 24 \qquad D_x = \begin{vmatrix} 0 & -3 \\ 80 & 3 \end{vmatrix} = 240$$

$$D_y = \begin{vmatrix} 4 & 0 \\ 4 & 80 \end{vmatrix} = 320$$

$$x = \frac{D_x}{D} = \frac{240}{24} = 10 \qquad y = \frac{D_y}{D} = \frac{320}{24} = \frac{40}{3}$$

The length of the side of the square should be 10 feet and the length of the side of the triangle should be 40/3 feet.

45. Let $x =$ the number of gallons of 10% solution and $y =$ the number of gallons of 25% solution. Since the total mixture is to be 30 gallons, $x + y = 30$. The next equation comes from the fact that the chlorine in the two parts is equal to the chlorine in the 30 gallons, $0.10x + 0.25y = 0.20(30)$. Rewrite the equations as follows.

$$\begin{aligned} x + \quad y &= 30 \\ 0.10x + 0.25y &= 6 \end{aligned}$$

$$D = \begin{vmatrix} 1 & 1 \\ 0.10 & 0.25 \end{vmatrix} = 0.15$$

$$D_x = \begin{vmatrix} 30 & 1 \\ 6 & 0.25 \end{vmatrix} = 1.5 \qquad D_y = \begin{vmatrix} 1 & 30 \\ 0.10 & 6 \end{vmatrix} = 3$$

$$x = \frac{D_x}{D} = \frac{1.5}{0.15} = 10 \qquad y = \frac{D_y}{D} = \frac{3}{0.15} = 20$$

Use 10 gallons of 10% solution and 20 gallons of 25% solution.

10.7 WARM-UPS

1. True, because of the definition of minor.
2. False, the row and column of the element are deleted from the matrix.
3. True, because of the definition of determinant of a 3×3 matrix.   **4.** False, because expansion by minors is used to find the determinant.   **5.** False, because $x = D_x/D$ by Cramer's rule.   **6.** True.   **7.** True.
8. True, because we can choose to expand about a row or column that has a zero in it.
9. False, because if $D = 0$ there is either no solution or infinitely many solutions.
10. False, because Cramer's rule does not work on nonlinear equations.

10.7 EXERCISES

1. The minor for 3 is obtained by deleting the row and column containing 3, the first row and first column.
$$\begin{bmatrix} -3 & 7 \\ 1 & -6 \end{bmatrix}$$

3. The minor for 5 is obtained by deleting the first row and third column.
$$\begin{bmatrix} 4 & -3 \\ 0 & 1 \end{bmatrix}$$

5. The minor for 7 is obtained by deleting the second row and third column.
$$\begin{bmatrix} 3 & -2 \\ 0 & 1 \end{bmatrix}$$

7. The minor for 1 is obtained by deleting the third row and second column.
$$\begin{bmatrix} 3 & 5 \\ 4 & 7 \end{bmatrix}$$

9. $1\begin{vmatrix} 3 & 1 \\ 1 & 5 \end{vmatrix} - 2\begin{vmatrix} 1 & 2 \\ 1 & 5 \end{vmatrix} + 3\begin{vmatrix} 1 & 2 \\ 3 & 1 \end{vmatrix}$
$= 1(14) - 2(3) + 3(-5) = -7$

11. $2\begin{vmatrix} 0 & 1 \\ 1 & 2 \end{vmatrix} - 1\begin{vmatrix} 1 & 0 \\ 1 & 2 \end{vmatrix} + 3\begin{vmatrix} 1 & 0 \\ 0 & 1 \end{vmatrix}$
$= 2(-1) - 1(2) + 3(1) = -1$

13. $-2\begin{vmatrix} 3 & 1 \\ 4 & 0 \end{vmatrix} + 3\begin{vmatrix} 1 & 2 \\ 4 & 0 \end{vmatrix} - 5\begin{vmatrix} 1 & 2 \\ 3 & 1 \end{vmatrix}$
$= -2(-4) + 3(-8) - 5(-5) = 9$

15. $1\begin{vmatrix} 3 & 2 \\ 2 & 3 \end{vmatrix} - 0\begin{vmatrix} 1 & 5 \\ 2 & 3 \end{vmatrix} + 0\begin{vmatrix} 1 & 5 \\ 3 & 2 \end{vmatrix}$
$= 1(5) - 0(-7) + 0(-13) = 5$

17. Expand by minors about the second column because it has two zeros in it.

$-1\begin{vmatrix} 2 & 6 \\ 4 & 1 \end{vmatrix} + 0\begin{vmatrix} 3 & 5 \\ 4 & 1 \end{vmatrix} - 0\begin{vmatrix} 3 & 5 \\ 2 & 6 \end{vmatrix}$
$= -1(-22) = 22$

19. Expand by minors about the first column because it has one zero in it.

$-2\begin{vmatrix} 1 & -1 \\ -4 & -3 \end{vmatrix} - 0\begin{vmatrix} 1 & 3 \\ -4 & -3 \end{vmatrix} + 2\begin{vmatrix} 1 & 3 \\ 1 & -1 \end{vmatrix}$
$= -2(-7) + 2(-4) = 6$

21. Expand by minors about the third column because it has two zeros in it.

$$0 \begin{vmatrix} 4 & -1 \\ 0 & 3 \end{vmatrix} - 0 \begin{vmatrix} -2 & -3 \\ 0 & 3 \end{vmatrix} + 5 \begin{vmatrix} -2 & -3 \\ 4 & -1 \end{vmatrix}$$

$$= 5(14) = 70$$

23. Expand by minors about the second column.

$$-1 \begin{vmatrix} 0 & 5 \\ 5 & 4 \end{vmatrix} + 0 \begin{vmatrix} 2 & 1 \\ 5 & 4 \end{vmatrix} - 0 \begin{vmatrix} 2 & 1 \\ 0 & 5 \end{vmatrix}$$

$$= -1(-25) = 25$$

25. $D = \begin{vmatrix} 1 & 1 & 1 \\ 1 & -1 & 1 \\ 2 & 1 & 1 \end{vmatrix} = 2 \qquad D_x = \begin{vmatrix} 6 & 1 & 1 \\ 2 & -1 & 1 \\ 7 & 1 & 1 \end{vmatrix} = 2$

$D_y = \begin{vmatrix} 1 & 6 & 1 \\ 1 & 2 & 1 \\ 2 & 7 & 1 \end{vmatrix} = 4 \qquad D_z = \begin{vmatrix} 1 & 1 & 6 \\ 1 & -1 & 2 \\ 2 & 1 & 7 \end{vmatrix} = 6$

$x = \dfrac{D_x}{D} = \dfrac{2}{2} = 1, \; y = \dfrac{D_y}{D} = \dfrac{4}{2} = 2, \; z = \dfrac{D_z}{D} = \dfrac{6}{2} = 3$

The solution set is $\{(1, 2, 3)\}$.

27. $D = \begin{vmatrix} 1 & -3 & 2 \\ 1 & 1 & 1 \\ 1 & -1 & 1 \end{vmatrix} = -2 \qquad D_x = \begin{vmatrix} 0 & -3 & 2 \\ 2 & 1 & 1 \\ 0 & -1 & 1 \end{vmatrix} = 2$

$D_y = \begin{vmatrix} 1 & 0 & 2 \\ 1 & 2 & 1 \\ 1 & 0 & 1 \end{vmatrix} = -2 \qquad D_z = \begin{vmatrix} 1 & -3 & 0 \\ 1 & 1 & 2 \\ 1 & -1 & 0 \end{vmatrix} = -4$

$x = \dfrac{D_x}{D} = \dfrac{2}{-2} = -1, \quad y = \dfrac{D_y}{D} = \dfrac{-2}{-2} = 1,$

$z = \dfrac{D_z}{D} = \dfrac{-4}{-2} = 2$

The solution set is $\{(-1, 1, 2)\}$.

29. $D = \begin{vmatrix} 1 & 1 & 0 \\ 0 & 2 & -1 \\ 1 & 1 & 1 \end{vmatrix} = 2 \qquad D_x = \begin{vmatrix} -1 & 1 & 0 \\ 3 & 2 & -1 \\ 0 & 1 & 1 \end{vmatrix} = -6$

$D_y = \begin{vmatrix} 1 & -1 & 0 \\ 0 & 3 & -1 \\ 1 & 0 & 1 \end{vmatrix} = 4 \qquad D_z = \begin{vmatrix} 1 & 1 & -1 \\ 0 & 2 & 3 \\ 1 & 1 & 0 \end{vmatrix} = 2$

$x = \dfrac{D_x}{D} = \dfrac{-6}{2} = -3, \quad y = \dfrac{D_y}{D} = \dfrac{4}{2} = 2,$

$z = \dfrac{D_z}{D} = \dfrac{2}{2} = 1$

The solution set is $\{(-3, 2, 1)\}$.

31. $D = \begin{vmatrix} 1 & 1 & -1 \\ 2 & 2 & 1 \\ 1 & -3 & 0 \end{vmatrix} = 12 \qquad D_x = \begin{vmatrix} 0 & 1 & -1 \\ 6 & 2 & 1 \\ 0 & -3 & 0 \end{vmatrix} = 18$

$D_y = \begin{vmatrix} 1 & 0 & -1 \\ 2 & 6 & 1 \\ 1 & 0 & 0 \end{vmatrix} = 6 \qquad D_z = \begin{vmatrix} 1 & 1 & 0 \\ 2 & 2 & 6 \\ 1 & -3 & 0 \end{vmatrix} = 24$

$x = \dfrac{D_x}{D} = \dfrac{18}{12} = \dfrac{3}{2}, \quad y = \dfrac{D_y}{D} = \dfrac{6}{12} = \dfrac{1}{2},$

$z = \dfrac{D_z}{D} = \dfrac{24}{12} = 2$

The solution set is $\left\{ \left(\dfrac{3}{2}, \dfrac{1}{2}, 2 \right) \right\}$.

33. $D = \begin{vmatrix} 1 & 1 & 1 \\ 0 & 2 & 2 \\ 3 & -1 & 0 \end{vmatrix} = 2 \qquad D_x = \begin{vmatrix} 0 & 1 & 1 \\ 0 & 2 & 2 \\ -1 & -1 & 0 \end{vmatrix} = 0$

$D_y = \begin{vmatrix} 1 & 0 & 1 \\ 0 & 0 & 2 \\ 3 & -1 & 0 \end{vmatrix} = 2 \qquad D_z = \begin{vmatrix} 1 & 1 & 0 \\ 0 & 2 & 0 \\ 3 & -1 & -1 \end{vmatrix} = -2$

$x = \dfrac{D_x}{D} = \dfrac{0}{2} = 0, \quad y = \dfrac{D_y}{D} = \dfrac{2}{2} = 1,$

$z = \dfrac{D_z}{D} = \dfrac{-2}{2} = -1$

The solution set is $\{(0, 1, -1)\}$.

35. $D = \begin{vmatrix} 2 & -1 & 1 \\ -6 & 3 & -3 \\ 4 & -2 & 2 \end{vmatrix} = 0$

Since $D = 0$, we must use elimination of variables to solve the system. Note that if we multiply the first equation by -3, the result is the same as the second equation. If we multiply the first equation by 2, the result is the same as the third equation. So the system is dependent and the solution set is
$\{(x, y, z) \mid 2x - y + z = 1\}$.

37. $D = \begin{vmatrix} 1 & 1 & 0 \\ 0 & 1 & 2 \\ 1 & 2 & 2 \end{vmatrix} = 0$

Since $D = 0$, we must solve the system by elimination of variables. Multiply the second equation by -1 and add the result to the third equation to get $x + y = 2$. Multiply $x + y = 2$ by

−1 and add the result to the first equation.

$$-x - y = -2$$
$$\underline{x + y = 1}$$
$$0 = -1$$

The last equation is false no matter what values the variables have. The solution set is ∅.

39. $D = \begin{vmatrix} 1 & 1 & 0 \\ 0 & 1 & 1 \\ 1 & 0 & 1 \end{vmatrix} = 2$ $D_x = \begin{vmatrix} 4 & 1 & 0 \\ -3 & 1 & 1 \\ -5 & 0 & 1 \end{vmatrix} = 2$

$D_y = \begin{vmatrix} 1 & 4 & 0 \\ 0 & -3 & 1 \\ 1 & -5 & 1 \end{vmatrix} = 6$ $D_z = \begin{vmatrix} 1 & 1 & 4 \\ 0 & 1 & -3 \\ 1 & 0 & -5 \end{vmatrix} = -12$

$$x = \frac{D_x}{D} = \frac{2}{2} = 1, \qquad y = \frac{D_y}{D} = \frac{6}{2} = 3,$$

$$z = \frac{D_z}{D} = \frac{-12}{2} = -6$$

The solution set is $\{(1, 3, -6)\}$.

41. Let x = Mimi's weight, y = Mitzi's weight, and z = Cassandra's weight. We can write 3 equations.

$$x + y + z = 175$$
$$x + z = 143$$
$$y + z = 139$$

$D = \begin{vmatrix} 1 & 1 & 1 \\ 1 & 0 & 1 \\ 0 & 1 & 1 \end{vmatrix} = -1$ $D_x = \begin{vmatrix} 175 & 1 & 1 \\ 143 & 0 & 1 \\ 139 & 1 & 1 \end{vmatrix} = -36$

$D_y = \begin{vmatrix} 1 & 175 & 1 \\ 1 & 143 & 1 \\ 0 & 139 & 1 \end{vmatrix} = -32$ $D_z = \begin{vmatrix} 1 & 1 & 175 \\ 1 & 0 & 143 \\ 0 & 1 & 139 \end{vmatrix} = -107$

$$x = \frac{D_x}{D} = \frac{-36}{-1} = 36, \qquad y = \frac{D_y}{D} = \frac{-32}{-1} = 32,$$

$$z = \frac{D_z}{D} = \frac{-107}{-1} = 107$$

So Mimi weights 36 pounds, Mitzi weighs 32 pounds, and Cassandra weighs 107 pounds.

43. Let x = the number of degrees in the larger of the two acute angles, y = the number of degrees in the smaller acute angle, and z = the number of degrees in the right angle. We can write 3 equations about the sizes of the angles.

$$x + y + z = 180$$
$$x - y = 12$$
$$z = 90$$

$D = \begin{vmatrix} 1 & 1 & 1 \\ 1 & -1 & 0 \\ 0 & 0 & 1 \end{vmatrix} = -2$ $D_x = \begin{vmatrix} 180 & 1 & 1 \\ 12 & -1 & 0 \\ 90 & 0 & 1 \end{vmatrix} = -102$

$D_y = \begin{vmatrix} 1 & 180 & 1 \\ 1 & 12 & 0 \\ 0 & 90 & 1 \end{vmatrix} = -78$ $D_z = \begin{vmatrix} 1 & 1 & 180 \\ 1 & -1 & 12 \\ 0 & 0 & 90 \end{vmatrix} = -180$

$$x = \frac{D_x}{D} = \frac{-102}{-2} = 51, \qquad y = \frac{D_y}{D} = \frac{-78}{-2} = 39,$$

$$z = \frac{D_z}{D} = \frac{-180}{-2} = 90$$

The measures of the three angles of the triangle are 39°, 51°, and 90°.

CHAPTER 10   REVIEW

1. The graph of $y = 2x - 1$ has y-intercept $(0, -1)$ and slope 2. the graph of $y = -x + 2$ has y-intercept $(0, 2)$ and slope -1. The graphs appear to intersect at $(1, 1)$. Check that $(1, 1)$ satisfies both equations. The solution set is $\{(1, 1)\}$ and the system is independent.

3. The graph of $y = 3x - 4$ has y-intercept $(0, -4)$ and slope 3. The graph of $y = -2x + 1$ has y-intercept $(0, 1)$ and slope -2. The lines appear to intersect at $(1, -1)$. After checking that $(1, -1)$ satisfies both equations, we can be certain that the solution set is $\{(1, -1)\}$ and the system is independent.

5. Substitute $y = 3x + 11$ into $2x + 3y = 0$.

$$2x + 3(3x + 11) = 0$$
$$11x + 33 = 0$$
$$11x = -33$$
$$x = -3$$
$$y = 3(-3) + 11 = 2$$

The solution set is $\{(-3, 2)\}$ and the system is independent.

7. Substitute $x = y + 5$ into $2x - 2y = 12$.

$$2(y + 5) - 2y = 12$$
$$2y + 10 - 2y = 12$$
$$10 = 12$$

The solution set is ∅ and the system is inconsistent.

9. Multiply the first equation by 2 and the second by 3, and then add the resulting equations.

$$10x - 6y = -40$$
$$\underline{9x + 6y = 21}$$
$$19x \qquad = -19$$
$$x = -1$$

Use $x = -1$ in $3x + 2y = 7$.

$$3(-1) + 2y = 7$$
$$2y = 10$$
$$y = 5$$

The solution set is $\{(-1, 5)\}$ and the system is independent.

11. Rewrite the first equation.

$$2(y - 5) + 4 = 3(x - 6)$$
$$2y - 10 + 4 = 3x - 18$$
$$-3x + 2y = -12$$

Add this last equation to the original second equation.

$$-3x + 2y = -12$$
$$\underline{3x - 2y = 12}$$
$$0 = 0$$

The two equations are just different forms of an equation for the same straight line. The solution set is $\{(x, y) \mid 3x - 2y = 12\}$ and the system is dependent.

13. Add the first and second equations.

$$2x - y \ - z = 3$$
$$\underline{3x + y + 2z = 4}$$
$$5x \qquad + z = 7 \qquad \text{A}$$

Multiply the first equation by 2 and add the result to the last equation.

$$4x - 2y - 2z = 6$$
$$\underline{4x + 2y - z = -4}$$
$$8x \qquad - 3z = 2 \qquad \text{B}$$

Multiply equation A by 3 and add the result to equation B.

$$15x + 3z = 21$$
$$\underline{8x - 3z = \ 2}$$
$$23x \qquad = 23$$
$$x = 1$$

Use $x = 1$ in $8x - 3z = 2$.

$$8(1) - 3z = 2$$
$$-3z = -6$$
$$z = 2$$

Use $x = 1$ and $z = 2$ in $3x + y + 2z = 4$.

$$3(1) + y + 2(2) = 4$$
$$y = -3$$

The solution set is $\{(1, -3, 2)\}$.

15. Add the first two equations.

$$x - 3y + z = 5$$
$$\underline{2x - 4y - z = 7}$$
$$3x - 7y \qquad = 12 \qquad \text{A}$$

Multiply the second equation by 2 and add the result to the last equation.

$$4x - 8y - 2z = 14$$
$$\underline{2x - 6y + 2z = 6}$$
$$6x - 14y \qquad = 20 \qquad \text{B}$$

Multiply equation A by -2 and add the result to equation B.

$$-6x + 14y = -24$$
$$\underline{6x - 14y = 20}$$
$$0 = -4$$

The solution set is $\emptyset$.

17. The graph of $y = x^2$ is a parabola and the graph of $y = -2x + 15$ is a straight line. Use substitution to eliminate y.

$$x^2 = -2x + 15$$
$$x^2 + 2x - 15 = 0$$
$$(x + 5)(x - 3) = 0$$
$$x = -5 \quad \text{or} \quad x = 3$$
$$y = 25 \qquad \qquad y = 9 \quad \text{Since } y = x^2$$

The solution set is $\{(3, 9), (-5, 25)\}$.

19. First graph $y = 3x$ and $y = 1/x$. It appears that the graphs intersect at two points. Use substitution to eliminate y.

$$3x = \frac{1}{x}$$
$$3x^2 = 1$$
$$x^2 = \frac{1}{3}$$
$$x = \pm\sqrt{\frac{1}{3}} = \pm\frac{\sqrt{3}}{3}$$

If $x = \frac{\sqrt{3}}{3}$, then $y = \sqrt{3}$. If $x = -\frac{\sqrt{3}}{3}$, then $y = -\sqrt{3}$. The solution set is

$$\left\{ \left(\frac{\sqrt{3}}{3}, \sqrt{3} \right), \left(-\frac{\sqrt{3}}{3}, -\sqrt{3} \right) \right\}.$$

21. The second equation can be written as $x^2 = 3y$. Substitute this equation into the first equation.

$$3y + y^2 = 4$$
$$y^2 + 3y - 4 = 0$$
$$(y + 4)(y - 1) = 0$$
$$y = -4 \quad \text{or} \quad y = 1$$

If $y = -4$, then $x^2 = 3(-4)$ has no solution. If $y = 1$, then $x^2 = 3(1)$ gives us $x = \pm\sqrt{3}$. The solution set is $\{(\sqrt{3}, 1), (-\sqrt{3}, 1)\}$.

23. Use substitution to eliminate y.

$$x^2 + (x+2)^2 = 34$$
$$x^2 + x^2 + 4x + 4 = 34$$
$$2x^2 + 4x - 30 = 0$$
$$x^2 + 2x - 15 = 0$$
$$(x+5)(x-3) = 0$$
$$x = -5 \quad \text{or} \quad x = 3$$
$$y = -3 \qquad y = 5 \qquad \text{Since } y = x+2$$

The solution set is $\{(-5, -3), (3, 5)\}$.

25. Use substitution to eliminate y.

$$\log(x-3) = 1 - \log(x)$$
$$\log(x-3) + \log(x) = 1$$
$$\log(x^2 - 3x) = 1$$
$$x^2 - 3x = 10$$
$$x^2 - 3x - 10 = 0$$
$$(x-5)(x+2) = 0$$
$$x = 5 \quad \text{or} \quad x = -2$$

If $x = 5$, then $y = \log(5-3) = \log(2)$. If $x = -2$, we get a logarithm of a negative number. So the solution set is $\{(5, \log(2))\}$.

27. Use substitution to eliminate x.

$$y^2 = 2(12 - y)$$
$$y^2 = 24 - 2y$$
$$y^2 + 2y - 24 = 0$$
$$(y+6)(y-4) = 0$$
$$y = -6 \quad \text{or} \quad y = 4$$

If $y = -6$, then $-6 = x^2$ has no solution. If $y = 4$, then $x^2 = 4$ and $x = \pm 2$. The solution set is $\{(2, 4), (-2, 4)\}$.

29. $\begin{bmatrix} 1 & -3 & | & 14 \\ 2 & 1 & | & 0 \end{bmatrix}$

$\begin{bmatrix} 1 & -3 & | & 14 \\ 0 & 7 & | & -28 \end{bmatrix}$ $-2R_1 + R_2 \to R_2$

$\begin{bmatrix} 1 & -3 & | & 14 \\ 0 & 1 & | & -4 \end{bmatrix}$ $R_2 \div 7 \to R_2$

$\begin{bmatrix} 1 & 0 & | & 2 \\ 0 & 1 & | & -4 \end{bmatrix}$ $3R_2 + R_1 \to R_1$

The solution set is $\{(2, -4)\}$.

31. $\begin{bmatrix} 1 & 1 & -1 & | & 0 \\ 1 & -1 & 2 & | & 4 \\ 2 & 1 & -1 & | & 1 \end{bmatrix}$

$\begin{bmatrix} 1 & 1 & -1 & | & 0 \\ 0 & -2 & 3 & | & 4 \\ 0 & -1 & 1 & | & 1 \end{bmatrix}$ $\begin{array}{l} -R_1 + R_2 \to R_2 \\ -2R_1 + R_3 \to R_3 \end{array}$

$\begin{bmatrix} 1 & 1 & -1 & | & 0 \\ 0 & -1 & 1 & | & 1 \\ 0 & -2 & 3 & | & 4 \end{bmatrix}$ $R_2 \leftrightarrow R_3$

$\begin{bmatrix} 1 & 1 & -1 & | & 0 \\ 0 & 1 & -1 & | & -1 \\ 0 & -2 & 3 & | & 4 \end{bmatrix}$ $-R_2 \to R_2$

$\begin{bmatrix} 1 & 0 & 0 & | & 1 \\ 0 & 1 & -1 & | & -1 \\ 0 & 0 & 1 & | & 2 \end{bmatrix}$ $\begin{array}{l} -R_2 + R_1 \to R_1 \\ 2R_2 + R_3 \to R_3 \end{array}$

$\begin{bmatrix} 1 & 0 & 0 & | & 1 \\ 0 & 1 & 0 & | & 1 \\ 0 & 0 & 1 & | & 2 \end{bmatrix}$ $R_3 + R_2 \to R_2$

The solution set is $\{(1, 1, 2)\}$.

33. $\begin{vmatrix} 1 & 3 \\ 0 & 2 \end{vmatrix} = 1 \cdot 2 - 0 \cdot 3 = 2$

35. $\begin{vmatrix} 0.01 & 0.02 \\ 50 & 80 \end{vmatrix} = 0.01(80) - 0.02(50)$

$= 0.8 - 1 = -0.2$

37. $D = \begin{vmatrix} 2 & -1 \\ 3 & 1 \end{vmatrix} = 5 \quad D_x = \begin{vmatrix} 0 & -1 \\ -5 & 1 \end{vmatrix} = -5$

$D_y = \begin{vmatrix} 2 & 0 \\ 3 & -5 \end{vmatrix} = -10$

$x = \dfrac{D_x}{D} = \dfrac{-5}{5} = -1 \qquad y = \dfrac{D_y}{D} = \dfrac{-10}{5} = -2$

The solution set is $\{(-1, -2)\}$.

39. Write the system in standard form.
$$-2x + y = -3$$
$$3x - 2y = 4$$

$D = \begin{vmatrix} -2 & 1 \\ 3 & -2 \end{vmatrix} = 1 \quad D_x = \begin{vmatrix} -3 & 1 \\ 4 & -2 \end{vmatrix} = 2$

$D_y = \begin{vmatrix} -2 & -3 \\ 3 & 4 \end{vmatrix} = 1$

$x = \dfrac{D_x}{D} = \dfrac{2}{1} = 2 \qquad y = \dfrac{D_y}{D} = \dfrac{1}{1} = 1$

The solution set is $\{(2, 1)\}$.

41. Rewrite the system in standard form.
$$3x - y = -1$$
$$-6x + 2y = 5$$

$D = \begin{vmatrix} 3 & -1 \\ -6 & 2 \end{vmatrix} = 0$

Since $D = 0$, we cannot use Cramer's rule. Multiply the first equation by 2 and add the result to the second equation.

$$\begin{array}{r} 6x - 2y = -2 \\ -6x + 2y = 5 \\ \hline 0 = 3 \end{array}$$

The solution set is $\emptyset$.

43. Expand by minors using the first column.

$2\begin{vmatrix} 2 & 4 \\ 1 & 1 \end{vmatrix} - (-1)\begin{vmatrix} 3 & 1 \\ 1 & 1 \end{vmatrix} + 6\begin{vmatrix} 3 & 1 \\ 2 & 4 \end{vmatrix}$

$= 2(-2) + 2 + 6(10) = 58$

45. Expand by minors using the second column.

$-3\begin{vmatrix} 2 & 4 \\ -1 & 3 \end{vmatrix} + 0\begin{vmatrix} 2 & -2 \\ -1 & 3 \end{vmatrix} - 0\begin{vmatrix} 2 & -2 \\ 2 & 4 \end{vmatrix}$

$= -3(10) = -30$

47. $D = \begin{vmatrix} 1 & 1 & 0 \\ 1 & 1 & 1 \\ 1 & -1 & -1 \end{vmatrix} = 2 \quad D_x = \begin{vmatrix} 3 & 1 & 0 \\ 0 & 1 & 1 \\ 2 & -1 & -1 \end{vmatrix} = 2$

$D_y = \begin{vmatrix} 1 & 3 & 0 \\ 1 & 0 & 1 \\ 1 & 2 & -1 \end{vmatrix} = 4 \quad D_z = \begin{vmatrix} 1 & 1 & 3 \\ 1 & 1 & 0 \\ 1 & -1 & 2 \end{vmatrix} = -6$

$x = \dfrac{D_x}{D} = \dfrac{2}{2} = 1, \qquad y = \dfrac{D_y}{D} = \dfrac{4}{2} = 2,$

$z = \dfrac{D_z}{D} = \dfrac{-6}{2} = -3$

The solution set is $\{(1, 2, -3)\}$.

49. $D = \begin{vmatrix} 2 & -1 & 1 \\ 4 & 6 & -2 \\ 1 & -2 & -1 \end{vmatrix} = -36 \quad D_x = \begin{vmatrix} 0 & -1 & 1 \\ 0 & 6 & -2 \\ -9 & -2 & -1 \end{vmatrix} = 36$

$D_y = \begin{vmatrix} 2 & 0 & 1 \\ 4 & 0 & -2 \\ 1 & -9 & -1 \end{vmatrix} = 72 \quad D_z = \begin{vmatrix} 2 & -1 & 0 \\ 4 & 6 & 0 \\ 1 & -2 & -9 \end{vmatrix} = -144$

$x = \dfrac{D_x}{D} = \dfrac{36}{-36} = -1, \qquad y = \dfrac{D_y}{D} = \dfrac{72}{-36} = -2,$

$z = \dfrac{D_z}{D} = \dfrac{-144}{-36} = 4$

The solution set is $\{(-1, 2, 4)\}$.

51. Let $x =$ the tens digit and $y =$ the ones digit. The value of the number is $10x + y$. If the digits are reversed the value will be $10y + x$. We can write the following two equations.

$$x + y = 15$$
$$10x + y = 10y + x - 9$$

$$x + y = 15$$
$$9x - 9y = -9$$

$$\begin{array}{r} x + y = 15 \\ x - y = -1 \\ \hline 2x = 14 \\ x = 7 \\ y = 8 \qquad \text{Since } x + y = 15 \end{array}$$

The number is 78.

53. Let $x =$ the length and $y =$ the width. We can write the following 2 equations.

$$2x + 2y = 16$$
$$xy = 12$$

The first equation can be written as $y = 8 - x$. Substitute $y = 8 - x$ into the second equation.

$$x(8 - x) = 12$$
$$-x^2 + 8x - 12 = 0$$
$$x^2 - 8x + 12 = 0$$
$$(x - 6)(x - 2) = 0$$
$$x = 6 \quad \text{or} \quad x = 2$$
$$y = 2 \qquad\qquad y = 6 \quad \text{Since } y = 8 - x$$

So the length is 6 feet and the width is 2 feet.

55. Let $x =$ the larger radius and $y =$ the smaller radius.

$$\pi x^2 - \pi y^2 = 10\pi$$
$$x - y = 2$$

$$x^2 - y^2 = 10$$
$$x = y + 2$$

$$(y+2)^2 - y^2 = 10$$

$$y^2 + 4y + 4 - y^2 = 10$$

$$4y = 6$$

$$y = \frac{3}{2}$$

If $y = 3/2$ and $x = y + 2$, then $x = 7/2$. So the radius of the larger circle is 7/2 inches and the radius of the smaller circle is 3/2 inches.

57. Let $x =$ Gasper's age now and $y =$ Gasper's age when he is old enough to drive.

$$x + 4 = y$$
$$x - 3 = \frac{1}{2}(y + 2)$$

Use substitution.

$$x - 3 = \frac{1}{2}(x + 4 + 2)$$
$$2x - 6 = x + 6$$
$$x = 12$$
$$y = 12 + 4 = 16$$

Gasper is now 12 years old and he must be 16 years old to drive.

59. Let $x =$ the amount invested at 10% and $y =$ the amount invested at 8%. We can write two equations.

$$x + y = 20,000$$
$$0.10x + 0.08y = 1920$$

Substitute $y = 20,000 - x$ into the second equation.

$$0.10x + 0.08(20,000 - x) = 1920$$
$$0.10x + 1,600 - 0.08x = 1920$$
$$0.02x = 320$$
$$x = 16,000$$

If $x = 16,000$ and $x + y = 20,000$, then $y = 4,000$. She invested $16,000 at 10% and $4,000 at 8%.

Chapter 10 TEST

1. The graph of $y = -x + 4$ is a straight line with y-intercept $(0, 4)$ and slope -1. The graph of $y = 2x + 1$ is a straight line with y-intercept $(0, 1)$ and slope 2. The graphs appear to intersect at $(1, 3)$. After checking that $(1, 3)$ satisfies both equations, we can be sure that the solution set is $\{(1, 3)\}$.

2. Substitute $y = 2x - 8$ into $4x + 3y = 1$.

$$4x + 3(2x - 8) = 1$$
$$10x = 25$$
$$x = \frac{5}{2}$$

If $x = 5/2$, then $y = 2(5/2) - 8 = -3$. The solution set is $\left\{\left(\frac{5}{2}, -3\right)\right\}$.

3. Substitute $y = x - 5$ into the second equation $3x - 4(y - 2) = 28 - x$.

$$3x - 4(x - 5 - 2) = 28 - x$$
$$3x - 4(x - 7) = 28 - x$$
$$28 - x = 28 - x$$

Since the last equation is an identity, the solution set is $\{(x,y) \mid y = x - 5\}$.

4. Multiply the first equation by 3 and the second by 2 and add the results.

$$9x + 6y = 9$$
$$\underline{8x - 6y = -26}$$
$$17x \quad\quad = -17$$
$$x = -1$$

Use $x = -1$ in $3x + 2y = 3$ to find y.

$$3(-1) + 2y = 3$$
$$2y = 6$$
$$y = 3$$

The solution set is $\{(-1, 3)\}$.

5. Multiply the first equation by 2 and then add the result to the second equation.

$$6x - 2y = 10$$
$$\underline{-6x + 2y = 1}$$
$$0 = 11$$

The solution set to the system is $\emptyset$.

6. The lines both have slope 3 and different y-intercepts, so they are parallel. There is no solution to the system. The system is inconsistent.

7. If we multiply the second equation by 2, the result is the same as the first equation. These are two equations for the same straight line. The system is dependent.

8. The lines have different slopes and different y-intercepts. They will intersect in exactly one point. The system is independent.

9. The second equation can be written as $x^2 = 13 - y$. Substitute this equation into $y^2 - x^2 = 7$.

$$y^2 - (13 - y) = 7$$
$$y^2 + y - 20 = 0$$

$$(y + 5)(y - 4) = 0$$

$$y = -5 \quad \text{or} \quad y = 4$$

If $y = -5$, then $x^2 = 18$ or $x = \pm 3\sqrt{2}$. If $y = 4$, then $x^2 = 9$ or $x = \pm 3$. The solution set is $\{(3, 4), (-3, 4), (3\sqrt{2}, -5), (-3\sqrt{2}, -5)\}$.

10. Add the first two equations.

$$\begin{array}{l} x + y - z = 2 \\ \underline{2x - y + 3z = -5} \\ 3x \quad\;\; + 2z = -3 \qquad \text{A} \end{array}$$

Multiply the first equation by 3 and add the result to the last equation.

$$\begin{array}{l} 3x + 3y - 3z = 6 \\ \underline{x - 3y + z = 4} \\ 4x \quad\quad - 2z = 10 \qquad \text{B} \end{array}$$

Add equations A and B.

$$\begin{array}{l} 3x + 2z = -3 \\ \underline{4x - 2z = 10} \\ 7x \quad\quad = 7 \\ \quad\;\; x = 1 \end{array}$$

Use $x = 1$ in $3x + 2z = -3$.

$$\begin{array}{l} 3(1) + 2z = -3 \\ 2z = -6 \\ z = -3 \end{array}$$

Use $x = 1$ and $z = -3$ in $x + y - z = 2$.

$$\begin{array}{l} 1 + y - (-3) = 2 \\ y = -2 \end{array}$$

The solution set is $\{(1, -2, -3)\}$.

11.
$$\left[\begin{array}{cc|c} 1 & 2 & 12 \\ 3 & -1 & 1 \end{array}\right]$$

$$\left[\begin{array}{cc|c} 1 & 2 & 12 \\ 0 & -7 & -35 \end{array}\right] \quad -3R_1 + R_2 \rightarrow R_2$$

$$\left[\begin{array}{cc|c} 1 & 2 & 12 \\ 0 & 1 & 5 \end{array}\right] \quad R_2 \div (-7) \rightarrow R_2$$

$$\left[\begin{array}{cc|c} 1 & 0 & 2 \\ 0 & 1 & 5 \end{array}\right] \quad -2R_2 + R_1 \rightarrow R_1$$

The solution set is $\{(2, 5)\}$.

12.
$$\left[\begin{array}{ccc|c} 1 & -1 & -1 & 1 \\ -1 & -1 & 2 & -2 \\ -1 & -3 & 1 & -5 \end{array}\right]$$

$$\left[\begin{array}{ccc|c} 1 & -1 & -1 & 1 \\ 0 & -2 & 1 & -1 \\ 0 & -4 & 0 & -4 \end{array}\right] \quad \begin{array}{l} R_1 + R_2 \rightarrow R_2 \\ \\ R_1 + R_3 \rightarrow R_3 \end{array}$$

$$\left[\begin{array}{ccc|c} 1 & -1 & -1 & 1 \\ 0 & -4 & 0 & -4 \\ 0 & -2 & 1 & -1 \end{array}\right] \quad R_2 \leftrightarrow R_3$$

$$\left[\begin{array}{ccc|c} 1 & -1 & -1 & 1 \\ 0 & 1 & 0 & 1 \\ 0 & -2 & 1 & -1 \end{array}\right] \quad R_2 \div (-4) \rightarrow R_2$$

$$\left[\begin{array}{ccc|c} 1 & 0 & -1 & 2 \\ 0 & 1 & 0 & 1 \\ 0 & 0 & 1 & 1 \end{array}\right] \quad \begin{array}{l} R_2 + R_1 \rightarrow R_1 \\ \\ 2R_2 + R_3 \rightarrow R_3 \end{array}$$

$$\left[\begin{array}{ccc|c} 1 & 0 & 0 & 3 \\ 0 & 1 & 0 & 1 \\ 0 & 0 & 1 & 1 \end{array}\right] \quad R_3 + R_1 \rightarrow R_1$$

The solution set is $\{(3, 1, 1)\}$.

13. $\begin{vmatrix} 2 & 3 \\ 4 & -3 \end{vmatrix} = 2(-3) - 4(3) = -18$

14. Expand by minors using the third column.

$$-1 \begin{vmatrix} 2 & 3 \\ 1 & 1 \end{vmatrix} - 1 \begin{vmatrix} 1 & -2 \\ 1 & 1 \end{vmatrix} + 0 \begin{vmatrix} 1 & -2 \\ 2 & 3 \end{vmatrix}$$

$$= -1(-1) - 1(3) + 0(7) = -2$$

15. $D = \begin{vmatrix} 2 & -1 \\ 3 & 1 \end{vmatrix} = 5 \quad D_x = \begin{vmatrix} -4 & -1 \\ -1 & 1 \end{vmatrix} = -5$

$$D_y = \begin{vmatrix} 2 & -4 \\ 3 & -1 \end{vmatrix} = 10$$

$$x = \frac{D_x}{D} = \frac{-5}{5} = -1 \qquad y = \frac{D_y}{D} = \frac{10}{5} = 2$$

The solution set is $\{(-1, 2)\}$.

16. $D = \begin{vmatrix} 1 & 1 & 0 \\ 1 & -1 & 2 \\ 2 & 1 & -1 \end{vmatrix} = 4 \quad D_x = \begin{vmatrix} 0 & 1 & 0 \\ 6 & -1 & 2 \\ 1 & 1 & -1 \end{vmatrix} = 8$

$$D_y = \begin{vmatrix} 1 & 0 & 0 \\ 1 & 6 & 2 \\ 2 & 1 & -1 \end{vmatrix} = -8 \quad D_z = \begin{vmatrix} 1 & 1 & 0 \\ 1 & -1 & 6 \\ 2 & 1 & 1 \end{vmatrix} = 4$$

$$x = \frac{D_x}{D} = \frac{8}{4} = 2, \qquad y = \frac{D_y}{D} = \frac{-8}{4} = -2,$$

$$z = \frac{D_z}{D} = \frac{4}{4} = 1 \quad \text{The solution set is } \{(2, -2, 1)\}.$$

17. Let x = the length and y = the width. We can write the following two equations.

$$2x + 2y = 42$$
$$xy = 108$$

If we solve $2x + 2y = 42$ for y, we get $y = 21 - x$.

$$x(21 - x) = 108$$
$$-x^2 + 21x - 108 = 0$$
$$x^2 - 21x + 108 = 0$$
$$(x - 9)(x - 12) = 0$$
$$x = 9 \quad \text{or} \quad x = 12$$
$$y = 12 \qquad y = 9 \text{ Since } y = 21 - x$$

The length is 12 feet and the width is 9 feet.

18. Let x = the price of a single and y = the price of a double. We can write an equation for each night.

$$5x + 12y = 390$$
$$9x + 10y = 412$$

Multiply the first equation by -9 and the second by 5, then add the results.

$$-45x - 108y = -3510$$
$$\underline{45x + 50y = 2060}$$
$$-58y = -1450$$

$$y = 25$$

Use $y = 25$ in $5x + 12y = 390$ to find x.

$$5x + 12(25) = 390$$
$$5x + 300 = 390$$
$$5x = 90$$
$$x = 18$$

So singles rent for \$18 per night and doubles rent for \$25 per night.

19. Let x = Jill's study time, y = Karen's study time, and z = Betsy's study time. We can write three equations.

$$x + y + z = 93$$
$$x + y = \frac{1}{2}z$$
$$x = y + 3$$

Rewrite the equations as follows.

$$x + y + z = 93$$
$$2x + 2y - z = 0$$
$$x - y = 3$$

Adding the first two equations to eliminate z gives us $3x + 3y = 93$, or $x + y = 31$. Add this result to the third equation.

$$x + y = 31$$
$$\underline{x - y = 3}$$
$$2x = 34$$
$$x = 17$$

Since $x + y = 31$, we get $y = 14$. Use $x = 17$ and $y = 14$ in the first equation.

$$17 + 14 + z = 93$$
$$z = 62$$

So Jill studied 17 hours, Karen studied 14 hours, and Betsy studied 62 hours.

Tying It All Together Chapters 1-10

1.
$$2(x - 3) + 5x = 8$$
$$2x - 6 + 5x = 8$$
$$7x = 14$$
$$x = 2$$

The solution set is $\{2\}$.

2. $|2x - 3| = 6$
$$2x - 3 = 6 \text{ or } 2x - 3 = -6$$
$$2x = 9 \text{ or } \quad 2x = -3$$
$$x = 9/2 \text{ or } \quad x = -3/2$$

The solution set is $\left\{ -\frac{3}{2}, \frac{9}{2} \right\}$.

3. $\sqrt{2x+1} = 3$

$$(\sqrt{2x+1})^2 = 3^2$$
$$2x+1 = 9$$
$$2x = 8$$
$$x = 4$$

The solution set is $\{4\}$.

4. $4x - 3 = 0$
$$4x = 3$$
$$x = 3/4$$

The solution set is $\left\{\frac{3}{4}\right\}$.

5. $\dfrac{x^2}{9} = 1$

$$x^2 = 9$$
$$x = \pm 3$$

The solution set is $\{-3, 3\}$.

6. $4x^2 - 3 = 0$
$$4x^2 = 3$$
$$x^2 = \frac{3}{4}$$
$$x = \pm\sqrt{\frac{3}{4}} = \pm\frac{\sqrt{3}}{2}$$

The solution set is $\left\{\pm\dfrac{\sqrt{3}}{2}\right\}$.

7. $x^{2/3} = 4$
$$(x^{2/3})^3 = 4^3$$
$$x^2 = 64$$
$$x = \pm 8$$

The solution set is $\{-8, 8\}$.

8. $x^2 + 6x + 8 = 0$
$$(x+2)(x+4) = 0$$
$$x = -2 \quad \text{or} \quad x = -4$$

The solution set is $\{-4, -2\}$.

9. $x^2 + 8x - 2 = 0$

$$x = \frac{-8 \pm \sqrt{8^2 - 4(1)(-2)}}{2(1)} = \frac{-8 \pm \sqrt{72}}{2}$$

$$= \frac{-8 \pm 6\sqrt{2}}{2} = -4 \pm 3\sqrt{2}$$

The solution set is $\{-4 \pm 3\sqrt{2}\}$.

10. $\dfrac{1}{2} + \dfrac{1}{x} - \dfrac{4}{x^2} = 0$

$$2x^2\left(\frac{1}{2} + \frac{1}{x} - \frac{4}{x^2}\right) = 2x^2(0)$$

$$x^2 + 2x - 8 = 0$$
$$(x+4)(x-2) = 0$$
$$x = -4 \quad \text{or} \quad x = 2$$

The solution set is $\{-4, 2\}$.

11. $(x+3)^2 = x^2$

$$x^2 + 6x + 9 = x^2$$
$$6x = -9$$
$$x = -\frac{3}{2}$$

The solution set is $\left\{-\dfrac{3}{2}\right\}$.

12. $x^4 - 10x^2 + 9 = 0$
$$(x^2 - 9)(x^2 - 1) = 0$$
$$x^2 = 9 \quad \text{or} \quad x^2 = 1$$
$$x = \pm 3 \qquad x = \pm 1$$

The solution set is $\{\pm 3, \pm 1\}$.

13. $\dfrac{x+1}{x-2} = \dfrac{5}{4}$

$$4x + 4 = 5x - 10$$
$$14 = x$$

The solution set is $\{14\}$.

14. $2^{x-1} = 6$
$$x - 1 = \log_2(6)$$
$$x = 1 + \log_2(6)$$

The solution set is $\{1 + \log_2(6)\}$.

15. $\ln(x) + \ln(x+2) = 3 \cdot \ln(2)$

$$\ln(x^2 + 2x) = \ln(2^3)$$
$$x^2 + 2x = 8$$
$$x^2 + 2x - 8 = 0$$
$$(x+4)(x-2) = 0$$
$$x = -4 \quad \text{or} \quad x = 2$$

If $x = -4$ in the original equation, we get a logarithm of a negative number. So the solution set is $\{2\}$.

16. $\log_2(x) = 5$
$$x = 2^5 = 32$$

The solution set is $\{32\}$.

17. $3x - 5 \geq 10$
$3x \geq 15$
$x \geq 5$

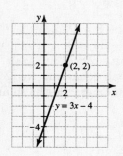

18. $3 - x < 6$ and $3x < 6$
$-x < 3$ and $x < 2$
$x > -3$ and $x < 2$

19. $|2x - 1| < 7$
$-7 < 2x - 1 < 7$
$-6 < 2x < 8$
$-3 < x < 4$

20. $x^2 + x - 6 > 0$
$(x + 3)(x - 2) > 0$

We have $x^2 + x - 6 = 0$ if $x = -3$ or $x = 2$. Test points to the left of -3, between -3 and 2, and to the right of 2: say -5, 0, and 5. We see that both -5 and 5 satisfy the original inequality. So all numbers to the left of -3 or to the right of 2 satisfy the original inequality.

21. The graph of $y = 3x - 4$ is a straight line with y-intercept $(0, -4)$ and slope 3.

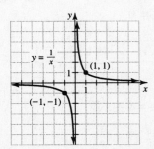

22. The graph of $f(x) = 3 - x$ is a straight line with y-intercept $(0, 3)$ and slope -1.

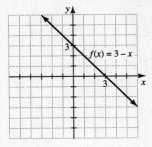

23. The graph of $h(x) = |x| + 1$ is v-shaped and includes the points $(0, 1)$, $(1, 2)$, and $(-1, 2)$.

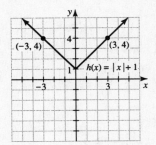

24. The graph of $y = \sqrt{x} - 1$ includes the points $(0, -1)$, $(1, 0)$, and $(4, 1)$.

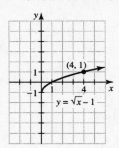

25. The graph of $y = 1/x$ includes the points $(1, 1)$, $(2, 1/2)$, and $(1/2, 2)$.

26. The graph of $y = 1 - x^2$ is a parabola opening downward. It includes the points $(0, 1)$, $(1, 0)$, and $(-1, 0)$.

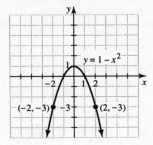

27. The graph of $y = 2^x$ includes the points $(0, 1)$, $(1, 2)$, and $(2, 4)$.

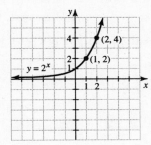

28. The graph of $f(x) = \log_3(x)$ includes the points $(1/3, -1)$, $(1, 0)$, and $(3, 1)$.

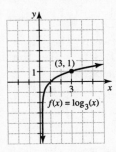

29. The graph of $f(x) = 3^{-x}$ includes the points $(0, 1)$, $(-1, 3)$, and $(1, 1/3)$.

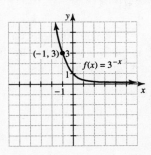

30. The graph of $g(x) = \ln(x)$ includes the points $(1, 0)$, $(e, 1)$, and $(1/e, -1)$, where e is approximately 2.7.

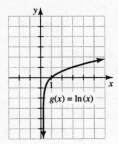

31. $\sqrt{18} + \sqrt{50} = 3\sqrt{2} + 5\sqrt{2} = 8\sqrt{2}$

32. $\sqrt{242} - \sqrt{98} = 11\sqrt{2} - 7\sqrt{2} = 4\sqrt{2}$

33. $(3 + 2\sqrt{2})(4 - 3\sqrt{2}) = 12 - \sqrt{2} - 12 = -\sqrt{2}$

34. $(5 - \sqrt{3})(5 + \sqrt{3}) = 25 - 3 = 22$

35. $(3 + \sqrt{2})^2 = 9 + 6\sqrt{2} + 2 = 11 + 6\sqrt{2}$

36. $\dfrac{6 - \sqrt{12}}{2} = \dfrac{6 - 2\sqrt{3}}{2} = 3 - \sqrt{3}$

37. $\dfrac{\sqrt{2}}{\sqrt{3} - 2} = \dfrac{\sqrt{2}(\sqrt{3} + 2)}{(\sqrt{3} - 2)(\sqrt{3} + 2)} = \dfrac{\sqrt{6} + 2\sqrt{2}}{3 - 4}$

$$= \dfrac{\sqrt{6} + 2\sqrt{2}}{-1} = -\sqrt{6} - 2\sqrt{2}$$

38. $\dfrac{3}{\sqrt{2}} - \dfrac{1}{2} + \sqrt{12} = \dfrac{3\sqrt{2}}{\sqrt{2}\sqrt{2}} - \dfrac{1}{2} + 2\sqrt{3}$

$$= \dfrac{3\sqrt{2}}{2} - \dfrac{1}{2} + \dfrac{4\sqrt{3}}{2} = \dfrac{3\sqrt{2} + 4\sqrt{3} - 1}{2}$$

39. a) From the y-intercept we see that the purchase price for A is $4000 and for B is $2000. So A has the larger purchase price.
b) Machine B makes 300,000 copies for $12,000, or $0.04 per copy. Machine A makes 300,000 copies for $9,000 or $0.03 per copy.
c) The slope of the line for machine B is 0.04 and the slope of the line for machine A is 0.03, which is the per copy cost for each machine.
d) Machine B: $y = 0.04x + 2000$
Machine A: $y = 0.03x + 4000$
e) $0.04x + 2000 = 0.03x + 4000$
$$0.01x = 2000$$
$$x = 200,000$$
The total costs are equal at 200,000 copies.

11.1 WARM-UPS

1. False, because if the focus is (2, 3) and the directrix is $y = 1$, then the vertex is (2, 2).
2. True, because the focus is 1 unit above the vertex which is (0, 1).   3. True, because in $y = a(x - h)^2 + k$, the vertex is (h, k).
4. False, because $y = 6x + 3x + 2$ is equivalent to $y = 9x + 2$, which is a straight line.
5. False, because $y = -2x^2 + 2x + 9$ opens downward.   6. True, the vertex is (0, 0) and so is the y-intercept.   7. True, because it opens upward from (2, 3) and it cannot intersect the x-axis.   8. False, because the y-intercept is (0, 0).   9. True, because (3, 5) is 1 unit from $x = 2$ and (1, 5) has the same y-coordinate and is 1 unit on the other side of $x = 2$.
10. True, because it opens upward from its vertex (5, 0).

11.1 EXERCISES

1. The vertex is (0, 0). Since $2 = 1/(4p)$, $p = 1/8$. So the focus is (0, 1/8) and the directrix is $y = -1/8$.
3. The vertex is (0, 0). Since $-1/4 = 1/(4p)$, $4p = -4$, or $p = -1$. So the focus is (0, -1) and the directrix is $y = 1$.
5. The vertex is (3, 2). Since $1/2 = 1/(4p)$, $p = 1/2$. So the focus is (3, 2.5) and the directrix is $y = 1.5$.
7. The vertex is (-1, 6). Since $-1 = 1/(4p)$, $p = -1/4$. So the focus is (-1, 5.75) and the directrix is $y = 6.25$.
9. Since the distance between the focus and directrix is 4, $p = 2$ and $a = \frac{1}{4p} = \frac{1}{4(2)} = \frac{1}{8}$. Since the vertex is half way between the focus and directrix, the vertex is (0, 0). Use the form $y = a(x - h)^2 + k$ to get the equation.
$$y = \frac{1}{8}(x - 0)^2 + 0$$
$$y = \frac{1}{8}x^2$$

11. Since the distance between the focus and directrix is 1 and the focus is below the directrix,

$p = -\frac{1}{2}$ and $a = \frac{1}{4p} = \frac{1}{4(-\frac{1}{2})} = -\frac{1}{2}$. Since the vertex is half way between the focus and directrix, the vertex is (0, 0). Use the form $y = a(x - h)^2 + k$ to get the equation.
$$y = -\frac{1}{2}(x - 0)^2 + 0$$
$$y = -\frac{1}{2}x^2$$

13. Since the distance between the focus and directrix is 1 and the parabola opens upward, $p = \frac{1}{2}$ and $a = \frac{1}{4p} = \frac{1}{4(\frac{1}{2})} = \frac{1}{2}$. Since the vertex is half way between the focus and directrix, the vertex is $\left(3, \frac{3}{2}\right)$. Use the form $y = a(x - h)^2 + k$ to get the equation.
$$y = \frac{1}{2}(x - 3)^2 + \frac{3}{2}$$
$$y = \frac{1}{2}x^2 - 3x + \frac{9}{2} + \frac{3}{2}$$
$$y = \frac{1}{2}x^2 - 3x + 6$$

15. Since the distance between the focus and directrix is 4 and the parabola opens downward, $p = -2$ and $a = \frac{1}{4p} = \frac{1}{4(-2)} = -\frac{1}{8}$. Since the vertex is half way between the focus and directrix, the vertex is (1, 0). Use the form $y = a(x - h)^2 + k$ to get the equation.
$$y = -\frac{1}{8}(x - 1)^2 + 0$$
$$y = -\frac{1}{8}x^2 + \frac{1}{4}x - \frac{1}{8}$$

17. Since the distance between the focus and directrix is $\frac{1}{2}$ and the parabola opens upward, $p = \frac{1}{4}$ and $a = \frac{1}{4p} = \frac{1}{4(0.25)} = 1$. Since the vertex is half way between the focus and directrix, the vertex is (-3, 1). Use the form $y = a(x - h)^2 + k$ to get the equation.
$$y = 1(x + 3)^2 + 1$$
$$y = x^2 + 6x + 10$$

19.
$$y = x^2 - 6x + 1$$
$$y = x^2 - 6x + 9 - 9 + 1$$
$$y = (x - 3)^2 - 8$$
The vertex is $(3, -8)$. Because $a = 1$, the parabola opens upward, and $p = 1/4$. The focus is $(3, -7.75)$ and the directrix is $y = -8.25$.

21.
$$y = 2x^2 + 12x + 5$$
$$y = 2(x^2 + 6x) + 5$$
$$y = 2(x^2 + 6x + 9) + 5 - 18$$
$$y = 2(x + 3)^2 - 13$$
The vertex is $(-3, -13)$ and the parabola opens upward. Because $a = 2$, $p = 1/8 = 0.125$ The focus is $(-3, -12.875)$ and the directrix is $y = -13.125$.

23.
$$y = -2x^2 + 16x + 1$$
$$y = -2(x^2 - 8x) + 1$$
$$y = -2(x^2 - 8x + 16) + 1 + 32$$
$$y = -2(x - 4)^2 + 33$$
The vertex is $(4, 33)$ and the parabola opens downward. Because $a = -2$,

$p = -1/8 = -0.125$. The focus is $(4, 32\frac{7}{8})$ and the directrix is $y = 33\frac{1}{8}$.

25.
$$y = 5x^2 + 40x$$
$$y = 5(x^2 + 8x)$$
$$y = 5(x^2 + 8x + 16) - 80$$
$$y = 5(x + 4)^2 - 80$$
The vertex is $(-4, -80)$ and the parabola opens upward. Because $a = 5$, $p = 1/20$. The focus is $(-4, -79\frac{19}{20})$ and the directrix is $y = -80\frac{1}{20}$.

27. The x-coordinate of the vertex is $x = \frac{-b}{2a}$.
$$x = \frac{-(-4)}{2(1)} = 2$$
$$y = 2^2 - 4(2) + 1 = -3$$
The vertex is $(2, -3)$ and the parabola opens upward. Because $a = 1$, $p = 1/4$. The focus is $(2, -2\frac{3}{4})$ and the directrix is $y = -3\frac{1}{4}$.

29. The x-coordinate of the vertex is $x = \frac{-b}{2a}$.
$$x = \frac{-(2)}{2(-1)} = 1$$
$$y = -1^2 + 2(1) - 3 = -2$$
The vertex is $(1, -2)$ and the parabola opens downward. Because $a = -1$, $p = -1/4$. The focus is $(1, -2\frac{1}{4})$ and the directrix is $y = -1\frac{3}{4}$.

31. The x-coordinate of the vertex is $x = \frac{-b}{2a}$.
$$x = \frac{-(-6)}{2(3)} = 1$$
$$y = 3(1^2) - 6(1) + 1 = -2$$
The vertex is $(1, -2)$ and the parabola opens upward. Because $a = 3$, $p = 1/12$. The focus is $(1, -1\frac{11}{12})$ and the directrix is $y = -2\frac{1}{12}$.

33. The x-coordinate of the vertex is $x = \frac{-b}{2a}$.
$$x = \frac{-(-3)}{2(-1)} = -\frac{3}{2}$$
$$y = -(-\frac{3}{2})^2 - 3(-\frac{3}{2}) + 2 = \frac{17}{4}$$
The vertex is $(-\frac{3}{2}, \frac{17}{4})$ and the parabola opens downward. Because $a = -1$, $p = -1/4$. The focus is $(-\frac{3}{2}, 4)$ and the directrix is $y = \frac{9}{2}$.

35. The x-coordinate of the vertex is $x = \frac{-b}{2a}$.
$$x = \frac{-(0)}{2(3)} = 0$$
$$y = 3(0)^2 + 5 = 5$$
The vertex is $(0, 5)$ and the parabola opens upward. Because $a = 3$, $p = 1/12$. The focus is $(0, 5\frac{1}{12})$ and the directrix is $y = 4\frac{11}{12}$.

37. Since $a = 1$, the parabola opens upward. The x-coordinate of the vertex is $x = \frac{-(-3)}{2(1)} = \frac{3}{2}$. The vertex is $(\frac{3}{2}, -\frac{1}{4})$. The axis of symmetry is the vertical line $x = \frac{3}{2}$. The y-intercept is $(0, 2)$. The x-intercepts are found by solving $x^2 - 3x + 2 = 0$. They are $(1, 0)$ and $(2, 0)$.

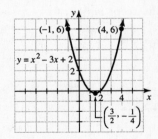

39. Since $a = -1$, the parabola opens downward. The x-coordinate of the vertex is $x = \frac{-(-2)}{2(-1)} = -1$. The vertex is $(-1, 9)$. The axis of symmetry is the vertical line $x = -1$. The y-intercept is $(0, 8)$. The x-intercepts are found by solving $-x^2 - 2x + 8 = 0$. They are $(-4, 0)$ and $(2, 0)$.

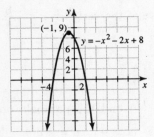

41. Since $a = 1$, the parabola opens upward. The vertex is $(-2, 1)$. The axis of symmetry is the vertical line $x = -2$. The y-intercept is $(0, 5)$. Since the graph opens upward from $(-2, 1)$, there are no x-intercepts.

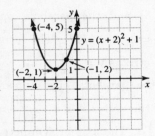

43. We can write $y = x^2 + 2x + 1$ as $y = (x + 1)^2$. The vertex is $(-1, 0)$. The axis of symmetry is $x = -1$. The y-intercept is $(0, 1)$. Since the parabola opens upward from $(-1, 0)$, the only x-intercept is $(-1, 0)$.

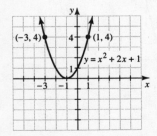

45. Since $a = -4$, the parabola opens downward. The x-coordinate of the vertex is $x = \frac{-(4)}{2(-4)} = \frac{1}{2}$. The vertex is $(\frac{1}{2}, 0)$. The axis of symmetry is the vertical line $x = \frac{1}{2}$. The intercepts are $(0, -1)$ and $(\frac{1}{2}, 0)$. Since the parabola opens downward from the x-axis, the only x-intercept is at the vertex.

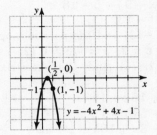

47. Since $a = 1$, the parabola opens upward. The x-coordinate of the vertex is $x = \frac{-(-5)}{2(1)} = \frac{5}{2}$. The vertex is $(\frac{5}{2}, -\frac{25}{4})$. The axis of symmetry is the vertical line $x = \frac{5}{2}$. The y-intercept is $(0, 0)$. The x-intercepts are found by solving $x^2 - 5x = 0$. They are $(0, 0)$ and $(5, 0)$.

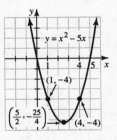

49. Since $a = 3$, the parabola opens upward. The x-coordinate of the vertex is $x = \frac{-(0)}{2(3)} = 0$. The vertex is $(0, 5)$. The axis of symmetry is the vertical line $x = 0$. The y-intercept is $(0, 5)$. Since the graph opens upward from $(0, 5)$ there are no x-intercepts.

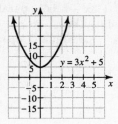

51. Since $a = 1$, the parabola opens upward. The x-coordinate of the vertex is $x = \dfrac{-(-2)}{2(1)} = 1$. The vertex is $(1, -2)$. The axis of symmetry is the vertical line $x = 1$. The y-intercept is $(0, -1)$. The x-intercepts are found by solving $x^2 - 2x - 1 = 0$. They are $(1 + \sqrt{2}, 0)$ and $(1 - \sqrt{2}, 0)$.

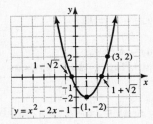

53. The vertex $y = (x - 5)^2$ is $(5, 0)$. Since $a = 1$, the parabola opens upward. The axis of symmetry is the vertical line $x = 5$. The y-intercept is $(0, 25)$. Since the parabola opens upward from $(5, 0)$, the x-intercept is $(5, 0)$.

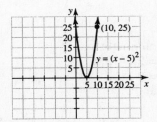

55. Two numbers that have a sum of 8 are expressed as x and $8 - x$. If y is their product, then $y = x(8 - x) = -x^2 + 8x$. The maximum value for y is obtained when $x = \dfrac{-8}{2(-1)} = 4$. If $x = 4$ then $8 - x = 4$ also. The numbers 4 and 4 have the largest product among numbers that have a sum of 8.

57. Let $x =$ the length and $y =$ the width. We have $2x + 2y = 160$, or $x + y = 80$. So $A = xy$ or $A = x(80 - x) = -x^2 + 80x$. So A is maximized if $x = \dfrac{-80}{2(-1)} = 40$. If $x = 40$, then $y = 40$. The maximum area is obtained when the rectangle is 40 ft by 40 ft.

59. Since the sum of the lengths of the legs is 6 feet, we can use x and $6 - x$ to represent the lengths of the legs. By the Pythagorean theorem, the square of the hypotenuse, c^2, is the sum of the squares of the legs.

$$c^2 = x^2 + (6 - x)^2 = 2x^2 - 12x + 36$$

The minimum value for c^2 occurs when $x = \dfrac{-(-12)}{2(2)} = 3$. If $x = 3$, then $6 - x = 3$ also. So the minimum value for c^2 occurs when each leg is 3 feet.

61. The maximum value for s in the formula $s = -16t^2 + 128t + 6$ is attained when

$$t = \dfrac{-128}{2(-16)} = 4.$$

If $t = 4$, then $s = -16(4)^2 + 128(4) + 6 = 262$. So the maximum height reached by the ball is 262 feet and it will not hit the roof.

63. The minimum value of C in the formula $C = 0.009x^2 - 1.8x + 100$ is attained when

$$x = \dfrac{-(-1.8)}{2(0.009)} = 100.$$

The cost per hour will be at a minimum when they produce 100 balls per hour.

65. The distance from the vertex to the focus is 15. So $p = 15$ and $a = 1/(4 \cdot 15) = 1/60$. the equation of the parabola is $y = \frac{1}{60}x^2$.

67. The graph of $y = -x^2 + 3$ is a parabola opening downward, and the graph of $y = x^2 + 1$ is a parabola opening upward. To find the points of intersection, eliminate y by substitution.

$$x^2 + 1 = -x^2 + 3$$
$$2x^2 = 2$$
$$x^2 = 1$$
$$x = \pm 1$$

If $x = \pm 1$, then $y = 2$ (since $y = x^2 + 1$). The solution set for the system is $\{(-1, 2), (1, 2)\}$.

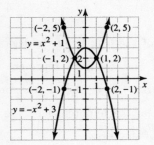

69. The graph of $y = x^2 - 2$ is a parabola opening upward, and the graph of $y = 2x - 3$ is a straight line.

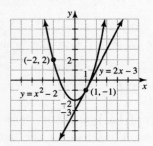

We eliminate y by substitution.

$$x^2 - 2 = 2x - 3$$
$$x^2 - 2x + 1 = 0$$
$$(x - 1)^2 = 0$$
$$x = 1$$

If $x = 1$, then $y = -1$ (because $y = 2x - 3$). The solution set for the system is $\{(1, -1)\}$.

71. The graph of $y = x^2 + 3x - 4$ is a parabola opening upward, and the graph of $y = -x^2 - 2x + 8$ is a parabola opening downward.

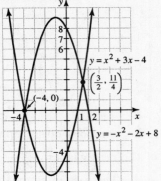

Eliminate y by substitution.

$$x^2 + 3x - 4 = -x^2 - 2x + 8$$
$$2x^2 + 5x - 12 = 0$$
$$(2x - 3)(x + 4) = 0$$
$$x = 3/2 \quad \text{or} \quad x = -4$$

If $x = 3/2$, then
$$y = (3/2)^2 + 3(3/2) - 4 = 11/4.$$
If $x = -4$, then
$$y = (-4)^2 + 3(-4) - 4 = 0.$$
The solution set for the system is $\left\{(\frac{3}{2}, \frac{11}{4}), (-4, 0)\right\}$.

73. The graph of $y = x^2 + 3x - 4$ is a parabola opening upward. The graph of $y = 2x + 2$ is a straight line with slope 2 and y-intercept $(0, 2)$.

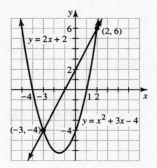

Use substitution to eliminate y.

$$x^2 + 3x - 4 = 2x + 2$$
$$x^2 + x - 6 = 0$$
$$(x + 3)(x - 2) = 0$$
$$x = -3 \quad \text{or} \quad x = 2$$
$$y = -4 \qquad\qquad y = 6 \quad \text{Since } y = 2x + 2$$

The solution set for the system is $\{(-3, -4), (2, 6)\}$.

220

75. Since vertex is (0, 4), the equation is $y = a(x - 0)^2 + 4$, or $y = ax^2 + 4$. Since (2, 0) is on the parabola, $0 = a \cdot 2^2 + 4$, or $a = -1$. So the equation is $y = -x^2 + 4$.

77. The vertex is (2, 2). So $y = a(x - 2)^2 + 2$. Since (0, 0) is on the parabola, $0 = a(0 - 2)^2 + 2$, or $a = -\frac{1}{2}$. The equation is $y = -\frac{1}{2}(x - 2)^2 + 2$, or $y = -\frac{1}{2}x^2 + 2x$.

11.2 WARM-UPS

1. False, the radius of a circle is a positive real number. **2.** False, the coordinates of the center do not satisfy the equation of the circle, only points on the circle satisfy the equation. **3.** True, the center is (0, 0). **4.** False, the equation of the circle centered at the origin with radius 9 is $x^2 + y^2 = 81$. **5.** False, because $(x - 2)^2 + (y - 3)^2 = 4$ has radius 2. **6.** False, because $(x - 3)^2 + (y + 5)^2 = 9$ is a circle of radius 3 centered at (3, −5) **7.** True, because the distance from (−3, −1) to (0, 0) is the radius. **8.** False, the center is (3, 4). **9.** True, because there is only an x^2 term and no x-term. **10.** False, because if we complete the square for x then the right side will no longer be 4.

11.2 EXERCISES

1. Use $h = 0$, $k = 3$, and $r = 5$ in the standard equation $(x - h)^2 + (y - k)^2 = r^2$ to get the equation $x^2 + (y - 3)^2 = 25$.

3. Use $h = 1$, $k = -2$, and $r = 9$ in the standard equation $(x - h)^2 + (y - k)^2 = r^2$ to get the equation $(x - 1)^2 + (y + 2)^2 = 81$.

5. Use $h = 0$, $k = 0$, and $r = \sqrt{3}$ in the standard equation $(x - h)^2 + (y - k)^2 = r^2$ to get the equation $x^2 + y^2 = 3$.

7. Use $h = -6$, $k = -3$, and $r = 1/2$ in the standard equation $(x - h)^2 + (y - k)^2 = r^2$ to get the equation $(x + 6)^2 + (y + 3)^2 = \frac{1}{4}$.

9. Use $h = 1/2$, $k = 1/3$, and $r = 0.1$ in the standard equation $(x - h)^2 + (y - k)^2 = r^2$ to get the equation $(x - \frac{1}{2})^2 + (y - \frac{1}{3})^2 = 0.01$.

11. Compare $(x - 3)^2 + (y - 5)^2 = 2$ with $(x - h)^2 + (y - k)^2 = r^2$ (*where the center is (h, k) and radius is r*), to get a center at (3, 5) and radius $\sqrt{2}$.

13. Compare $x^2 + (y - \frac{1}{2})^2 = \frac{1}{2}$ with $(x - h)^2 + (y - k)^2 = r^2$ (*where the center is (h, k) and radius is r*), to get a center at $(0, \frac{1}{2})$ and radius $\frac{\sqrt{2}}{2}$.

15. Divide each side by 4 to get $x^2 + y^2 = \frac{9}{4}$. Compare with $(x - h)^2 + (y - k)^2 = r^2$ (*where the center is (h, k) and radius is r*), to get a center at (0, 0) and radius $\sqrt{\frac{9}{4}} = \frac{3}{2}$.

17. Rewrite the equation as $(x - 2)^2 + y^2 = 3$. Compare with $(x - h)^2 + (y - k)^2 = r^2$ (*where the center is (h, k) and radius is r*), to get a center at (2, 0) and radius $\sqrt{3}$.

19. The graph of $x^2 + y^2 = 9$ is a circle with radius 3, centered at (0, 0).

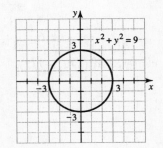

21. The graph of $x^2 + (y-3)^2 = 9$ is a circle with radius 3, centered at $(0, 3)$.

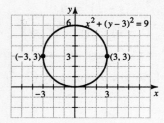

23. The graph of $(x+1)^2 + (y-1)^2 = 2$ is a circle with radius $\sqrt{2}$, centered at $(-1, 1)$.

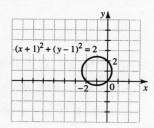

25. The graph of $(x-4)^2 + (y+3)^2 = 16$ is a circle of radius 4, centered at $(4, -3)$.

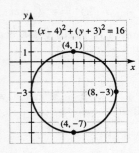

27. The graph of $(x-\frac{1}{2})^2 + (y+\frac{1}{2})^2 = \frac{1}{4}$ is a circle of radius 1/2, centered at $(1/2, -1/2)$.

29.

$$x^2 + 4x + \quad y^2 + 6y \quad = 0$$
$$x^2 + 4x + 4 + y^2 + 6y + 9 = 0 + 4 + 9$$
$$(x+2)^2 + (y+3)^2 = 13$$

This circle has radius $\sqrt{13}$ and center $(-2, -3)$.

31.

$$x^2 - 2x + \quad y^2 - 4y - 3 = 0$$
$$x^2 - 2x + 1 + y^2 - 4y + 4 = 3 + 1 + 4$$
$$(x-1)^2 + (y-2)^2 = 8$$

This circle has radius $\sqrt{8} = 2\sqrt{2}$ and center $(1, 2)$.

33.

$$x^2 - 10x + y^2 - 8y = -32$$
$$x^2 - 10x + 25 + y^2 - 8y + 16 = -32 + 25 + 16$$
$$(x-5)^2 + (y-4)^2 = 9$$

This circle has radius 3 and center $(5, 4)$.

35.

$$x^2 - x + \quad y^2 + y \quad = 0$$
$$x^2 - x + \frac{1}{4} + y^2 + y + \frac{1}{4} = \frac{1}{4} + \frac{1}{4}$$
$$(x-\tfrac{1}{2})^2 + (y+\tfrac{1}{2})^2 = \frac{1}{2}$$

Center is $(1/2, -1/2)$ and radius is $\sqrt{\frac{1}{2}} = \frac{\sqrt{2}}{2}$.

37.

$$x^2 - 3x + \quad y^2 - y \quad = 1$$
$$x^2 - 3x + \frac{9}{4} + y^2 - y + \frac{1}{4} = 1 + \frac{9}{4} + \frac{1}{4}$$
$$(x-\tfrac{3}{2})^2 + (y-\tfrac{1}{2})^2 = \frac{7}{2}$$

Center is $(3/2, 1/2)$ and radius $\sqrt{\frac{7}{2}} = \frac{\sqrt{14}}{2}$.

39.

$$x^2 - \tfrac{2}{3}x + \quad y^2 + \tfrac{3}{2}y \quad = 0$$
$$x^2 - \tfrac{2}{3}x + \frac{1}{9} + y^2 + \tfrac{3}{2}y + \frac{9}{16} = 0 + \frac{1}{9} + \frac{9}{16}$$
$$(x-\tfrac{1}{3})^2 + (y+\tfrac{3}{4})^2 = \frac{97}{144}$$

Center is $(1/3, -3/4)$ and radius is $\sqrt{\frac{97}{144}} = \frac{\sqrt{97}}{12}$.

41. The graph of $x^2 + y^2 = 10$ is a circle centered at $(0, 0)$ with radius $\sqrt{10}$. The graph of $y = 3x$ is a straight line through $(0, 0)$ with slope 3. Use substitution to eliminate y.

$$x^2 + (3x)^2 = 10$$
$$10x^2 = 10$$
$$x^2 = 1$$
$$x = 1 \quad \text{or} \quad x = -1$$
$$y = 3 \quad \text{or} \quad y = -3 \quad \text{Since } y = 3x$$

The solution set is $\{(1, 3), (-1, -3)\}$.

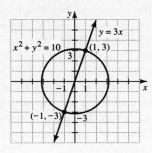

43. The graph of $x^2 + y^2 = 9$ is a circle centered at $(0, 0)$ with radius 3. The graph of $y = x^2 - 3$ is a parabola opening upward.

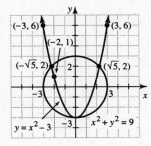

Use $x^2 = y + 3$ to eliminate x.
$$y + 3 + y^2 = 9$$
$$y^2 + y - 6 = 0$$
$$(y + 3)(y - 2) = 0$$
$$y = -3 \quad \text{or} \quad y = 2$$
If $y = -3$, then $x^2 = -3 + 3 = 0$ or $x = 0$. If $y = 2$, then $x^2 = 2 + 3 = 5$ or $x = \pm\sqrt{5}$. The solution set is $\{(0, -3), (\sqrt{5}, 2), (-\sqrt{5}, 2)\}$.

45. The graph of $(x - 2)^2 + (y + 3)^2 = 4$ is a circle centered at $(2, -3)$ with radius 2. The graph of $y = x - 3$ is a line through $(0, -3)$ with slope 1.

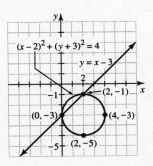

Use substitution to eliminate y.
$$(x - 2)^2 + (x - 3 + 3)^2 = 4$$
$$x^2 - 4x + 4 + x^2 = 4$$
$$2x^2 - 4x = 0$$
$$x^2 - 2x = 0$$
$$x(x - 2) = 0$$
$$x = 0 \quad \text{or} \quad x = 2$$
$$y = -3 \qquad\qquad y = -1 \quad \text{Since } y = x - 3$$

The solution set is $\{(0, -3), (2, -1)\}$.

47. To find the y-intercepts, use $x = 0$ in the equation of the circle.
$$(0 - 1)^2 + (y - 2)^2 = 4$$
$$1 + y^2 - 4y + 4 = 4$$
$$y^2 - 4y + 1 = 0$$

$$y = \frac{4 \pm \sqrt{16 - 4(1)(1)}}{2(1)} = \frac{4 \pm 2\sqrt{3}}{2} = 2 \pm \sqrt{3}$$

The y-intercepts are $(0, 2 + \sqrt{3})$ and $(0, 2 - \sqrt{3})$.

49. The radius is the distance from $(2, -5)$ to $(0, 0)$.
$$r = \sqrt{(2 - 0)^2 + (-5 - 0)^2} = \sqrt{4 + 25} = \sqrt{29}$$

51. The radius is the distance from $(2, 3)$ to $(-2, -1)$.

$$r = \sqrt{(-2 - 2)^2 + (-1 - 3)^2} = \sqrt{16 + 16} = \sqrt{32}$$

The equation of a circle centered at $(2, 3)$ with radius $\sqrt{32}$ is $(x - 2)^2 + (y - 3)^2 = 32$.

53. Substitute $y^2 = 9 - x^2$ into $(x - 5)^2 + y^2 = 9$ to eliminate y.
$$(x - 5)^2 + 9 - x^2 = 9$$
$$x^2 - 10x + 25 + 9 - x^2 = 9$$
$$-10x = -25$$
$$x = 5/2$$
Use $x = 5/2$ in $y^2 = 9 - x^2$ to find y.

$$y^2 = 9 - \left(\frac{5}{2}\right)^2 = \frac{11}{4}$$

$$y = \pm\sqrt{\frac{11}{4}} = \pm\frac{\sqrt{11}}{2}$$

The circles intersect at $\left(\frac{5}{2}, -\frac{\sqrt{11}}{2}\right)$ and $\left(\frac{5}{2}, \frac{\sqrt{11}}{2}\right)$.

55. From the equation of the bore we see that the radius is $\sqrt{83.72}$. The volume is given by

$$V = \pi r^2 h = \pi \cdot 83.72 \cdot 2874 = 755{,}903 \text{ mm}^3.$$

57. Since both x^2 and y^2 are nonnegative for real values of x and y, the only way their sum could be zero is to choose both x and y equal to zero. So the graph consists of $(0, 0)$ only.

59. If we square both sides of $y = \sqrt{1 - x^2}$ we get $x^2 + y^2 = 1$, which is a circle with center $(0, 0)$ and radius 1. In the equation $y = \sqrt{1 - x^2}$, y must be nonnegative. So the graph of $y = \sqrt{1 - x^2}$ is the top half of the graph of $x^2 + y^2 = 1$.

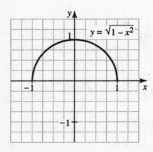

61. B and D can be any real numbers, but A must equal C. So that the radius us positive, we must also have

$$\frac{E}{A} + \left(\frac{B}{2A}\right)^2 + \left(\frac{D}{2A}\right)^2 > 0, \text{ or } 4AE + B^2 + D^2 > 0.$$

63. $y = \pm\sqrt{4 - x^2}$ **65.** $y = \pm\sqrt{x}$

67. $x = (y + 1)^2$

$y + 1 = \pm\sqrt{x}$

$y = -1 \pm\sqrt{x}$

11.3 WARM-UPS

1. False, the x-intercepts are $(-6, 0)$ and $(6, 0)$.
2. False, because both x^2 and y^2 must appear in the equation for an ellipse.   **3.** True, because if the foci coincide then every point on the ellipse will be a fixed distance from one fixed point.   **4.** True, because if we divide each side of the equation by 2, then it fits the standard equation for an ellipse.   **5.** True, because if we use $x = 0$ in the equation, we get $y = \pm\sqrt{3}$.

6. False, because both x^2 and y^2 must appear in the equation of a hyperbola.   **7.** False, because in this hyperbola there are no y-intercepts.   **8.** True, because it has y-intercepts at $(0, -3)$ and $(0, 3)$.   **9.** True, because if we divide each side of the equation by 4, then it fits the standard form for the equation of a hyperbola.   **10.** True, the asymptotes are the extended diagonals of the fundamental rectangle.

11.3 EXERCISES

1. The graph of $\dfrac{x^2}{9} + \dfrac{y^2}{4} = 1$ is an ellipse with x-intercepts $(-3, 0)$ and $(3, 0)$, and y-intercepts $(0, 2)$ and $(0, -2)$.

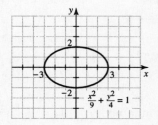

3. The graph of $\dfrac{x^2}{9} + y^2 = 1$ is an ellipse with x-intercepts $(-3, 0)$ and $(3, 0)$, and y-intercepts $(0, -1)$ and $(0, 1)$.

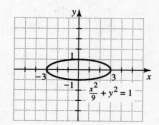

5. The graph of $\frac{x^2}{36} + \frac{y^2}{25} = 1$ is an ellipse with x-intercepts $(-6, 0)$ and $(6, 0)$, and y-intercepts $(0, -5)$ and $(0, 5)$.

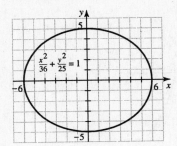

7. The graph of $\frac{x^2}{24} + \frac{y^2}{5} = 1$ is an ellipse with x-intercepts $(-\sqrt{24}, 0)$ and $(\sqrt{24}, 0)$, and y-intercepts $(0, -\sqrt{5})$ and $(0, \sqrt{5})$.

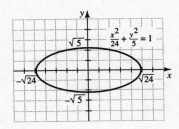

9. The graph $9x^2 + 16y^2 = 144$ is an ellipse with x-intercepts $(-4, 0)$ and $(4, 0)$, and y-intercepts $(0, -3)$ and $(0, 3)$.

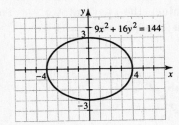

11. The graph of $25x^2 + y^2 = 25$ is an ellipse with x-intercepts $(-1, 0)$ and $(1, 0)$, and y-intercepts $(0, -5)$ and $(0, 5)$.

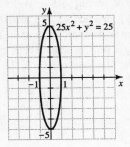

13. The graph of $4x^2 + 9y^2 = 1$ is an ellipse with x-intercepts $(1/2, 0)$ and $(-1/2, 0)$, and y-intercepts $(0, 1/3)$ and $(0, -1/3)$.

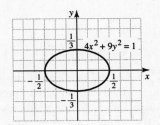

15. The graph of $\frac{(x-3)^2}{4} + \frac{(y-1)^2}{9} = 1$ is an ellipse centered at $(3, 1)$. The 4 in the denominator indicates that the ellipse passes through points that are 2 units to the right and 2 units to the left of the center: $(5, 1)$ and $(1, 1)$. The 9 in the denominator indicates that the ellipse passes through points that are 3 units above and 3 units below the center: $(3, 4)$ and $(3, -2)$.

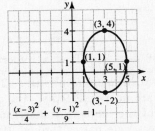

17. The graph of $\frac{(x+1)^2}{16} + \frac{(y-2)^2}{25} = 1$ is an ellipse centered at $(-1, 2)$. The 16 in the denominator indicates that the ellipse passes through points that are 4 units to the right and 4 units to the left of the center: $(3, 2)$ and $(-5, 2)$. The 25 in the denominator indicates that the ellipse passes through points that are 5 units above and 5 units below the center: $(-1, 7)$ and $(-1, -3)$.

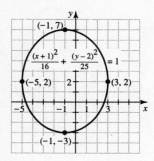

19. The graph of $(x-2)^2 + \frac{(y+1)^2}{36} = 1$ is an ellipse centered at $(2, -1)$. The 1 in the denominator indicates that the ellipse passes through points that are 1 unit to the right and 1 unit to the left of the center: $(3, -1)$ and $(1, -1)$. The 36 in the denominator indicates that the ellipse passes through points that are 6 units above and 6 units below the center: $(2, 5)$ and $(2, -7)$.

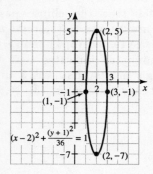

21. The graph of $\frac{x^2}{4} - \frac{y^2}{9} = 1$ is a hyperbola centered at $(0, 0)$ with x-intercepts at $(-2, 0)$ and $(2, 0)$. There are no y-intercepts. Use 9 to determine the size of the fundamental rectangle. The fundamental rectangle passes through $(0, 3)$ and $(0, -3)$. Extend the diagonals of the rectangle to determine the asymptotes. The

hyperbola opens to the left and right. The equations of the asymptotes are $y = \pm\frac{3}{2}x$.

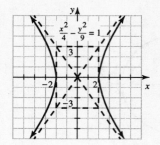

23. The graph of $\frac{y^2}{4} - \frac{x^2}{25} = 1$ is a hyperbola with y-intercepts $(0, -2)$ and $(0, 2)$. The fundamental rectangle passes through the y-intercepts, $(-5, 0)$, and $(5, 0)$. The extended diagonals determine the asymptotes.

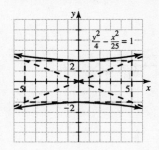

25. The graph of $\frac{x^2}{25} - y^2 = 1$ is a hyperbola with x-intercepts $(-5, 0)$ and $(5, 0)$. The fundamental rectangle passes through the intercepts, $(0, -1)$, and $(0, 1)$. The extended diagonals determine the asymptotes. The equations of the asymptotes are $y = \pm\frac{1}{5}x$.

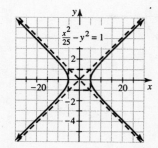

27. The graph of $x^2 - \dfrac{y^2}{25} = 1$ is a hyperbola with x-intercepts $(-1, 0)$ and $(1, 0)$. Plot $(0, 5)$ and $(0, -5)$ for the fundamental rectangle and extend the diagonals. The equations of the asymptotes are $y = \pm 5x$.

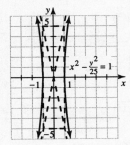

29. Divide each side of $9x^2 - 16y^2 = 144$ by 144 to get $\dfrac{x^2}{16} - \dfrac{y^2}{9} = 1$. The graph is a hyperbola with x-intercepts at $(-4, 0)$ and $(4, 0)$. The fundamental rectangle passes through the y-intercepts, $(0, -3)$, and $(0, 3)$. The extended diagonals determine the asymptotes.

The equations of the asymptotes are $y = \pm \dfrac{3}{4}x$.

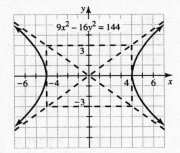

31. The graph of $x^2 - y^2 = 1$ is a hyperbola with x-intercepts $(-1, 0)$ and $(1, 0)$. The fundamental rectangle passes through the y-intercepts, $(0, -1)$, and $(0, 1)$. The extended diagonals determine the asymptotes, $y = \pm x$.

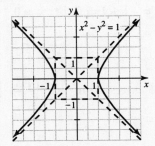

33. First graph the hyperbola and the ellipse.

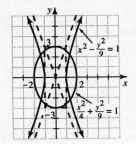

If we add the equations the variable y will be eliminated and we get the following equation.

$$\frac{x^2}{4} + x^2 = 2$$
$$x^2 + 4x^2 = 8$$
$$5x^2 = 8$$
$$x^2 = \frac{8}{5}$$
$$x = \pm \sqrt{\frac{8}{5}} = \pm \frac{2\sqrt{10}}{5}$$

Use $x^2 = 8/5$ in the second equation.

$$\frac{8}{5} - \frac{y^2}{9} = 1$$
$$-\frac{y^2}{9} = -\frac{3}{5}$$
$$y^2 = \frac{27}{5}$$
$$y = \pm \sqrt{\frac{27}{5}} = \pm \frac{3\sqrt{15}}{5}$$

The graphs intersect at $\left(\dfrac{2\sqrt{10}}{5}, \dfrac{3\sqrt{15}}{5} \right)$, $\left(\dfrac{2\sqrt{10}}{5}, -\dfrac{3\sqrt{15}}{5} \right)$, $\left(-\dfrac{2\sqrt{10}}{5}, \dfrac{3\sqrt{15}}{5} \right)$, and $\left(-\dfrac{2\sqrt{10}}{5}, -\dfrac{3\sqrt{15}}{5} \right)$.

35. The graphs are an ellipse and a circle.

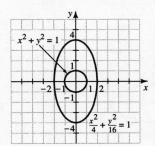

227

From the graph we can see that there are no points of intersection. If we eliminate y by substituting $y^2 = 1 - x^2$ into the first equation, we get the following equation.

$$\frac{x^2}{4} + \frac{1 - x^2}{16} = 1$$
$$4x^2 + 1 - x^2 = 16$$
$$3x^2 = 15$$
$$x^2 = 5$$

From $y^2 = 1 - x^2$, we get $y^2 = -4$, which has no real solution. There are no points of intersection.

37. The graphs are a circle and a hyperbola.

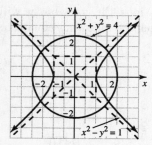

It appears that there are 4 points of intersection. To find them, add the equations to get the following equation.

$$2x^2 = 5$$
$$x^2 = \frac{5}{2}$$
$$x = \pm\sqrt{\frac{5}{2}} = \pm\frac{\sqrt{10}}{2}$$

From $y^2 = 4 - x^2$, we get $y^2 = 4 - \frac{5}{2} = \frac{3}{2}$, or

$$y = \pm\sqrt{\frac{3}{2}} = \pm\frac{\sqrt{6}}{2}.$$

The graphs intersect at $\left(\frac{\sqrt{10}}{2}, \frac{\sqrt{6}}{2}\right)$, $\left(\frac{\sqrt{10}}{2}, -\frac{\sqrt{6}}{2}\right)$, $\left(-\frac{\sqrt{10}}{2}, \frac{\sqrt{6}}{2}\right)$, and $\left(-\frac{\sqrt{10}}{2}, -\frac{\sqrt{6}}{2}\right)$.

39. The graphs are an ellipse and a circle.

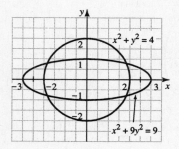

The graphs appear to intersect at 4 points. To find the points, substitute $y^2 = 4 - x^2$ into the first equation.

$$x^2 + 9(4 - x^2) = 9$$
$$-8x^2 + 36 = 9$$
$$-8x^2 = -27$$
$$x^2 = \frac{27}{8}$$
$$x = \pm\sqrt{\frac{27}{8}} = \pm\frac{3\sqrt{6}}{4}$$

Use $x^2 = 27/8$ in $y^2 = 4 - x^2$.

$$y^2 = 4 - \frac{27}{8} = \frac{5}{8}$$
$$y = \pm\sqrt{\frac{5}{8}} = \pm\frac{\sqrt{10}}{4}$$

The graphs intersect at $\left(\frac{3\sqrt{6}}{4}, \frac{\sqrt{10}}{4}\right)$,

$\left(\frac{3\sqrt{6}}{4}, -\frac{\sqrt{10}}{4}\right)$, $\left(-\frac{3\sqrt{6}}{4}, \frac{\sqrt{10}}{4}\right)$, and $\left(-\frac{3\sqrt{6}}{4}, -\frac{\sqrt{10}}{4}\right)$.

41. Graph the parabola and the ellipse.

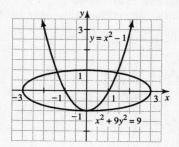

The graphs appear to intersect at 3 points. To find them, substitute $x^2 = y + 1$ into the first equation.

$$y + 1 + 9y^2 = 9$$
$$9y^2 + y - 8 = 0$$
$$(9y - 8)(y + 1) = 0$$
$$y = \frac{8}{9} \quad \text{or} \quad y = -1$$

If $y = -1$, the $x^2 = -1 + 1 = 0$ or $x = 0$. If $y = 8/9$, then

$$x^2 = \frac{8}{9} + 1 = \frac{17}{9} \text{ or } x = \pm\frac{\sqrt{17}}{3}.$$

The three points of intersection are $\left(\frac{\sqrt{17}}{3}, \frac{8}{9}\right)$, $\left(-\frac{\sqrt{17}}{3}, \frac{8}{9}\right)$, and $(0, -1)$.

43. Graph the hyperbola and the line.

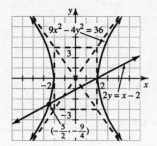

To find the points of intersection, substitute $x = 2y + 2$ into the first equation.

$$9(2y+2)^2 - 4y^2 = 36$$
$$9(4y^2 + 8y + 4) - 4y^2 = 36$$
$$36y^2 + 72y + 36 - 4y^2 = 36$$
$$32y^2 + 72y = 0$$
$$4y^2 + 9y = 0$$
$$y(4y + 9) = 0$$
$$y = 0 \quad \text{or} \quad y = -9/4$$

If $y = 0$, then $x = 2(0) + 2 = 2$. If $y = -9/4$, then $x = 2(-9/4) + 2 = -5/2$. The two points of intersection are $(2, 0)$ and $(-5/2, -9/4)$.

45. Substitute $x^2 = 1 + 3y^2$ into $4y^2 - x^2 = 1$.

$$4y^2 - (1 + 3y^2) = 1$$
$$y^2 = 2$$
$$y = \pm\sqrt{2}$$

$x^2 = 1 + 3(\pm\sqrt{2})^2 = 1 \pm 3(2) = 1 \pm 6 = 7 \text{ or } -5$

If $x^2 = 7$, then $x = \pm\sqrt{7}$. If $x^2 = -5$, there is no real solution. Since the boat is in the first quadrant, the boat is at $(\sqrt{7}, \sqrt{2})$.

11.4 WARM-UPS

1. False, because $x^2 + y = 4$ is the equation of a parabola. **2.** True, because if we divide each side by 9 we get the standard form for an ellipse. **3.** True, because it can be written as $y^2 - x^2 = 1$. **4.** True, because $2(0)^2 - 0 < 3$ is correct. **5.** False, because $0 > 0^2 - 3(0) + 2$ is incorrect. **6.** False, because the origin is on the parabola $x^2 = y$. **7.** False, because $(0, 0)$ does not satisfy the inequality. **8.** True, because $x^2 + y^2 = 4$ is a circle of radius 2 centered at $(0, 0)$, and the points inside this circle satisfy the inequality. **9.** True, because both $0^2 - 4^2 < 1$ and $4 > 0^2 - 2(0) + 3$ are correct. **10.** True, because both $0^2 + 0^2 < 1$ and $0 < 0^2 + 1$ are correct.

11.4 EXERCISES

1. Graph the parabola $y = x^2$. Since $(0, 4)$ satisfies $y > x^2$, shade the region containing $(0, 4)$.

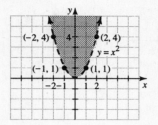

3. First graph the parabola $y = x^2 - x$. Since $(5, 0)$ satisfies $y < x^2 - x$, we shade the region containing $(5, 0)$.

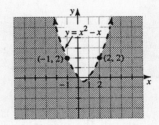

5. First graph the parabola $y = x^2 - x - 2$. Since $(0, 5)$ satisfies $y > x^2 - x - 2$, shade the region containing $(0, 5)$.

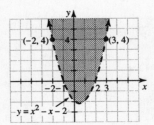

7. First graph the circle $x^2 + y^2 = 9$ using a solid curve. Since $(0, 0)$ satisfies $x^2 + y^2 \leq 9$, shade the inside of the circle.

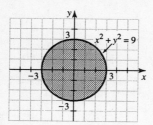

9. First graph the ellipse $x^2 + 4y^2 = 4$ using a dashed curve. Since $(0, 0)$ does not satisfy the inequality $x^2 + 4y^2 > 4$, we shade the region outside the ellipse.

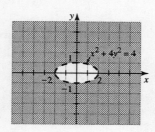

11. First divide both sides of the inequality by 36.

$$\frac{x^2}{9} - \frac{y^2}{4} < 1$$

Now graph the hyperbola $\frac{x^2}{9} - \frac{y^2}{4} = 1$. Testing a point in each of the three regions, we find that only points in the region containing the origin satisfy $4x^2 - 9y^2 < 36$.

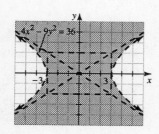

13. First graph the circle centered at $(2, 3)$ with radius 2, using a dashed curve. Since $(2, 3)$ satisfies $(x-2)^2 + (y-3)^2 < 4$, shade the region inside the circle.

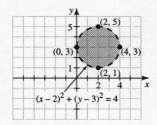

15. First graph the circle centered at $(0, 0)$ with radius 1. Since $(0, 0)$ does not satisfy $x^2 + y^2 > 1$, shade the region outside the circle.

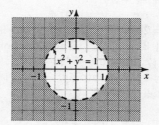

17. Graph the hyperbola $4x^2 - y^2 = 4$, using a dashed curve. After testing a point in each of the three regions, we see that only points in the region containing $(0, 0)$ fail to satisfy $4x^2 - y^2 > 4$.

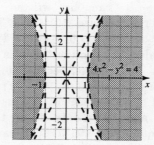

19. Graph the hyperbola $y^2 - x^2 = 1$. After testing a point in each of the three regions, we see that only points in the region containing $(0, 0)$ satisfy $y^2 - x^2 \leq 1$.

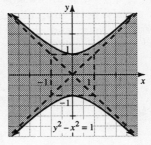

21. The graph of $y = x$ is a line with slope 1 and y-intercept $(0, 0)$. Since $(5, -5)$ satisfies $x > y$, we shade the region below the line.

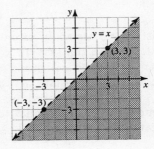

23. Graph the circle $x^2 + y^2 = 9$ and the line $y = x$. The graph of $x^2 + y^2 < 9$ is the region inside the circle. the graph of $y > x$ is the region above the line. The intersection of these two regions is shown as follows.

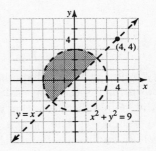

We could have used test points to determine which of the 4 regions determined by the circle and the line satisfies both inequalities.

25. The graph of $x^2 - y^2 = 1$ is a hyperbola. The graph of $x^2 + y^2 = 4$ is a circle of radius 2. Points that satisfy $x^2 + y^2 < 4$ are inside the circle. The hyperbola divides the plane into 3 regions. Points in the 2 regions not containing $(0, 0)$ are the points that satisfy $x^2 - y^2 > 1$. The intersection of these regions is shown in the following graph.

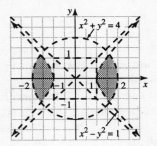

27. The graph of $y = x^2 + x$ is a parabola opening upward, with x-intercepts at $(0, 0)$ and $(-1, 0)$. The graph of $y > x^2 + x$ is the region inside this parabola. The graph of $y = 5$ is a horizontal line through $(0, 5)$. The graph of $y < 5$ is the region below this horizontal line. The points that satisfy both inequalities are the points inside the parabola and below $y = 5$, as shown in the following graph.

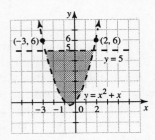

29. The graph of $y = x + 2$ is a line with y-intercept $(0, 2)$ and slope 1. The graph of $y \geq x + 2$ is the region above and including the line. The graph of $y = 2 - x$ is a line with y-intercept $(0, 2)$ and slope -1. The graph of $y \leq 2 - x$ is the region below and including this line. The region above the first line and below the second is shown in the following graph. We could have tested a point in each of the 4 regions determined by the two lines. We would find that only points in the region shown satisfy both inequalities.

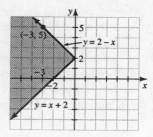

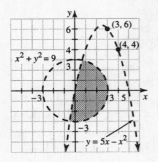

35. The graph of $y = 5x - x^2$ is a parabola opening downward. Points that satisfy $y < 5x - x^2$ are the points below the parabola. The graph of $x^2 + y^2 = 9$ is a circle of radius 3. Points that satisfy $x^2 + y^2 < 9$ are inside the circle. Points that satisfy both inequalities are the points inside the circle and below the parabola.

31. The graph of $4x^2 - y^2 = 4$ is a hyperbola opening to the left and right. Points that satisfy $4x^2 - y^2 < 4$ are in the region between the two branches of the hyperbola. The graph of $x^2 + 4y^2 = 4$ is an ellipse passing through $(0, 1)$, $(0, -1)$, $(2, 0)$, and $(-2, 0)$. Points that satisfy $x^2 + 4y^2 > 4$ are outside the ellipse. Points that satisfy both inequalities are outside the ellipse and between the two branches of the hyperbola.

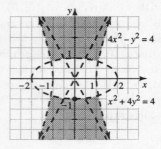

37. Points that satisfy $y \geq 3$ are above or on the horizontal line $y = 3$. Points that satisfy $x \leq 1$ are on or to the left of the vertical line $x = 1$. Points that satisfy both inequalities are shown in the following graph. The points graphed are the points with x-coordinate less than or equal to 1, and y-coordinate greater than or equal to 3.

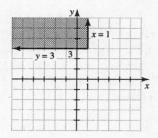

33. The graph of $x - y = 0$ is the same as the line $y = x$ (through $(0, 0)$ with slope 1). Points that satisfy $x - y < 0$ are above the line $y = x$. The graph of $y + x^2 = 1$ is the same as the parabola $y = -x^2 + 1$, which opens downward. Points that satisfy $y + x^2 < 1$ are below the parabola. The points that satisfy both inequalities, below the parabola and above the line, are shown in the following graph.

39. The graph of $4y^2 - 9x^2 = 36$ is a hyperbola opening up and down. Points that satisfy $4y^2 - 9x^2 < 36$ are the points between the two branches of the hyperbola. The graph of $x^2 + y^2 = 16$ is a circle of radius 4. Points that satisfy $x^2 + y^2 < 16$ are inside the circle. The points that satisfy both inequalities are indicated as follows.

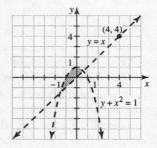

232

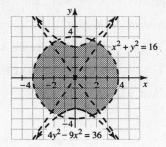

41. The graph of $y = x^2$ is a parabola opening upward. Points that satisfy $y < x^2$ are below this parabola. The graph of $x^2 + y^2 = 1$ is a circle of radius 1. Points that satisfy $x^2 + y^2 < 1$ are inside the circle. The graph of the system of inequalities consists of points that are below the parabola and inside the circle.

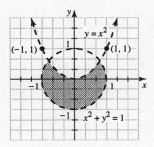

43. Let x = the number of paces he walked to the east and y = the number of paces he walked to the north. So $x + y \geq 50$, $x^2 + y^2 \leq 50^2$, and $y > x$. The graph of this system follows.

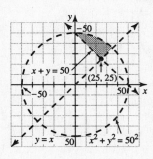

45. No solution

CHAPTER 11   REVIEW

1. The y-intercept is $(0, -18)$. To find the x-intercepts, solve $x^2 + 3x - 18 = 0$.
$$(x + 6)(x - 3) = 0$$
$$x = -6 \text{ or } x = 3$$
The x-intercepts are $(-6, 0)$ and $(3, 0)$. Use $x = -b/(2a)$ to find the x-coordinate of the vertex. The vertex is $(-\frac{3}{2}, -\frac{81}{4})$ and the axis of symmetry is $x = -3/2$. Since $a = 1$, $p = 1/4$. The focus is $(-\frac{3}{2}, -20)$ and the directrix is $y = -\frac{41}{2}$.

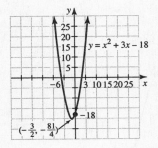

3. The y-intercept is $(0, 2)$. To find the x-intercepts, solve $x^2 + 3x + 2 = 0$.
$$(x + 2)(x + 1) = 0$$
$$x = -2 \text{ or } x = -1$$
The x-intercepts are $(-2, 0)$ and $(-1, 0)$. Use $x = -b/(2a)$ to find the x-coordinate of the vertex. The vertex is $(-\frac{3}{2}, -\frac{1}{4})$ and the axis of symmetry is $x = -3/2$. Since $a = 1$, $p = 1/4$. The focus is $(-\frac{3}{2}, 0)$ and the directrix is $y = -\frac{1}{2}$.

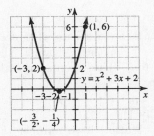

5. The y-intercept is (0, 1). To find the x-intercepts solve

$$-\frac{1}{2}(x-2)^2 + 3 = 0$$
$$(x-2)^2 = 6$$
$$x - 2 = \pm\sqrt{6}$$
$$x = 2 \pm \sqrt{6}$$

The x-intercepts are $(2 \pm \sqrt{6}, 0)$. The vertex is (2, 3) and the parabola opens down. The axis of symmetry is x = 2. Since a = −1/2, p = −1/2. The focus is $(2, 2\frac{1}{2})$ and the directrix is $y = 3\frac{1}{2}$.

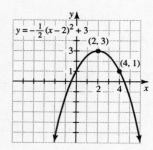

7. $y = 2(x^2 - 4x) + 1$
  $y = 2(x^2 - 4x + 4 - 4) + 1$
  $y = 2(x^2 - 4x + 4) - 8 + 1$
  $y = 2(x-2)^2 - 7$
Vertex (2, −7)

9. $y = -\frac{1}{2}(x^2 + 2x) + \frac{1}{2}$

  $y = -\frac{1}{2}(x^2 + 2x + 1 - 1) + \frac{1}{2}$

  $y = -\frac{1}{2}(x+1)^2 + \frac{1}{2} + \frac{1}{2}$

  $y = -\frac{1}{2}(x+1)^2 + 1$

Vertex (−1, 1)

11. Two numbers that have a sum of 6 are x and 6 − x. If we let y be their product, then $y = x(6-x) = -x^2 + 6x$. The maximum value for y is attained when

$$x = \frac{-b}{(2a)} = \frac{-6}{2(-1)} = 3.$$

If x = 3, then 6 − x = 3 also. Two number that have the largest product among numbers with a sum of 6 are 3 and 3.

13. Since the perimeter is 90 feet, the length and width have a sum of 45 feet. Let x = the length and 45 − x = the width. If y is the area, then $y = x(45 - x) = -x^2 + 45x$. The maximum value for y is attained when

$$x = \frac{-b}{2a} = \frac{-45}{2(-1)} = 22.5.$$

If x = 22.5, then 45 − x = 22.5 also. A square 22.5 feet by 22.5 feet will be the rectangular shape with the maximum area for the amount of fencing available.

15. The graph of $x^2 + y^2 = 100$ is a circle centered at (0, 0) with radius 10.

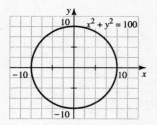

17. The graph of $(x-2)^2 + (y+3)^2 = 81$ is a circle centered at (2, −3) with radius 9.

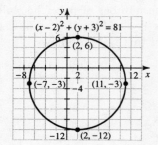

19. $9y^2 + 9x^2 = 4$
  $x^2 + y^2 = \frac{4}{9}$

The graph is a circle with center (0, 0) and radius 2/3.

234

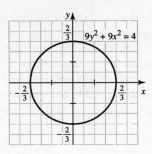

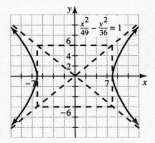

29. The graph of $\frac{x^2}{49} - \frac{y^2}{36} = 1$ is a hyperbola with x-intercepts $(-7,\ 0)$ and $(7,\ 0)$. The fundamental rectangle passes through the x-intercepts and the points $(0,\ -6)$ and $(0,\ 6)$. Extend the diagonals of the fundamental rectangle to get the asymptotes and draw a hyperbola opening to the left and right.

21. The equation of a circle with center (h, k) and radius r is $(x-h)^2 + (y-k)^2 = r^2$. The equation of a circle with center $(0,\ 3)$ and radius 6 is $x^2 + (y-3)^2 = 36$.

23. The equation of a circle with center $(2,\ -7)$ and radius 5 is $(x-2)^2 + (y+7)^2 = 25$.

25. The graph of $\frac{x^2}{36} + \frac{y^2}{49} = 1$ is an ellipse with x-intercepts at $(-6,\ 0)$ and $(6,\ 0)$, y-intercepts at $(0,\ -7)$ and $(0,\ 7)$, and centered at the origin.

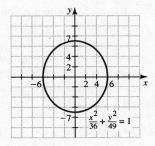

27. The graph of $25x^2 + 4y^2 = 100$ is an ellipse with x-intercepts $(-2,\ 0)$ and $(2,\ 0)$, y-intercepts $(0,\ -5)$ and $(0,\ 5)$, and centered at the origin.

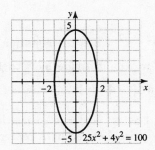

31. Write $4x^2 - 25y^2 = 100$ as $\frac{x^2}{25} - \frac{y^2}{4} = 1$.
The graph has x-intercepts at $(-5,\ 0)$ and $(5,\ 0)$. The fundamental rectangle passes through the x-intercepts and the points $(0,\ -2)$ and $(0,\ 2)$. Extend the diagonals of the rectangle to get the asymptotes and draw a hyperbola opening to the left and right.

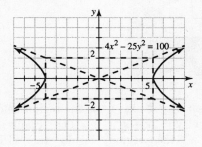

33. First graph the line $4x - 2y = 3$. It goes through $(0,\ -3/2)$ and $(3/4,\ 0)$. Since $(0,\ 0)$ fails to satisfy $4x - 2y > 3$, we shade the region not containing $(0,\ 0)$.

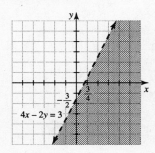

35. Write $y^2 < x^2 - 1$ as $x^2 - y^2 > 1$. Graph the hyperbola $x^2 - y^2 = 1$ and test a point in each of the three regions: $(-5, 0)$, $(0, 0)$, and $(5, 0)$. Since $(-5, 0)$ and $(5, 0)$ satisfy the inequality, we shade the regions containing those points.

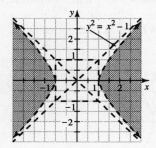

37. Write $4x^2 + 9y^2 > 36$ as $\frac{x^2}{9} + \frac{y^2}{4} > 1$. First graph the ellipse $\frac{x^2}{9} + \frac{y^2}{4} = 1$ through $(-3, 0)$, $(3, 0)$, $(0, -2)$, and $(0, 2)$. Since $(0, 0)$ fails to satisfy the inequality, we shade the region outside of the ellipse.

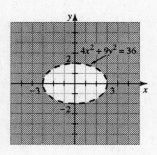

39. The graph of $y < 3x - x^2$ is the region below the parabola $y = 3x - x^2$. The graph of $x^2 + y^2 < 9$ is the region inside the circle $x^2 + y^2 = 9$. Points that are inside the circle and below the parabola are shown in the following graph.

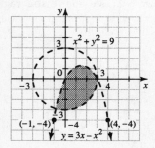

41. The set of points that satisfy $4x^2 + 9y^2 > 36$ is the set of points outside the ellipse $4x^2 + 9y^2 = 36$. The set of points that satisfy $x^2 + y^2 < 9$ is the set of points inside the circle $x^2 + y^2 = 9$. The solution set to the system consists of points that are outside the ellipse and inside the circle, as shown in the following graph.

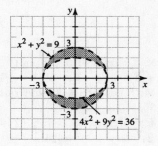

43. The equation $x^2 = y^2 + 1$ is the equation of a hyperbola because it could be written as $x^2 - y^2 = 1$.

45. The equation $x^2 = 1 - y^2$ is the equation of a circle because it could be written as $x^2 + y^2 = 1$.

47. The equation $x^2 + x = 1 - y^2$ is the equation of a circle because we could write $x^2 + x + y^2 = 1$, and then complete the square to get the standard equation for a circle.

49. The equation $x^2 + 4x = 6y - y^2$ is the equation of a circle because we could write it as $x^2 + 4x + y^2 - 6y = 0$, and then complete the squares for both x and y to get the standard equation of a circle.

51. The equation is the equation of a hyperbola in standard form.

53. The equation $4y^2 - x^2 = 8$ is the equation of a hyperbola because we could divide by 8 to get the standard equation $\frac{y^2}{2} - \frac{x^2}{8} = 1$.

55. Write $x^2 = 4 - y^2$ as $x^2 + y^2 = 4$ to see that it is the equation of a circle of radius 2 centered at the origin.

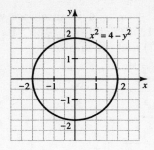

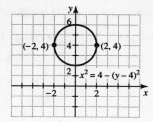

57. Write $x^2 = 4y + 4$ as $y = \frac{1}{4}x^2 - 1$ to see that it is the equation of a parabola opening upward with vertex at $(0, -1)$.

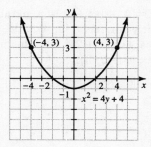

59. Write $x^2 = 4 - 4y^2$ as $\frac{x^2}{4} + y^2 = 1$ to see that it is the equation of an ellipse centered at $(0, 0)$ and passing through $(0, -1)$, $(0, 1)$, $(-2, 0)$, and $(2, 0)$.

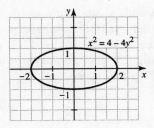

61. Write $x^2 = 4 - (y - 4)^2$ as $x^2 + (y - 4)^2 = 4$ to see that it is the equation of a circle of radius 2 centered at $(0, 4)$.

63. The radius of the circle is the distance between $(0, 0)$ and $(3, 4)$.

$$r = \sqrt{(3 - 0)^2 + (4 - 0)^2} = \sqrt{9 + 16} = \sqrt{25} = 5$$

The equation of the circle centered at $(0, 0)$ with radius 5 is $x^2 + y^2 = 25$.

65. Use the center $(-1, 5)$ and radius 6 in the form $(x - h)^2 + (y - k)^2 = r^2$ to get the equation $(x + 1)^2 + (y - 5)^2 = 36$.

67. The vertex is half way between the focus and directrix at $(1, 3)$. Since the distance from $(1, 3)$ to $(1, 4)$ is 1, we have $p = 1$, and $a = 1/4$. So the equation is
$$y = \frac{1}{4}(x - 1)^2 + 3.$$

69. Since the vertex is below the focus, $p = 1/4$ and $a = 1$. The equation is
$y = 1(x - 0)^2 + 0$ of $y = x^2$.

71. Since the vertex is $(0, 0)$ the parabola has equation $y = ax^2$. Since $(3, 2)$ is on the parabola, $2 = a \cdot 3^2$, or $a = 2/9$. The equation is $y = \frac{2}{9}x^2$.

73. Substitute $y = -x + 1$ into $x^2 + y^2 = 25$ to eliminate y.

$$x^2 + (-x + 1)^2 = 25$$
$$x^2 + x^2 - 2x + 1 = 25$$
$$2x^2 - 2x - 24 = 0$$
$$x^2 - x - 12 = 0$$
$$(x - 4)(x + 3) = 0$$

$x = 4$ or $x = -3$
$y = -3$ $y = 4$ Since $y = -x + 1$

The solution set is $\{(4, -3), (-3, 4)\}$

237

75. If we add the two equations to eliminate y, we get the equation $5x^2 = 25$. Solving this equation gives $x^2 = 5$ or $x = \pm\sqrt{5}$. Use $x^2 = 5$ in the first equation.

$$4(5) + y^2 = 4$$
$$y^2 = -16$$

There are no real numbers that satisfy the system of equations. The solution set is $\emptyset$.

CHAPTER 11 TEST

1. The graph of $x^2 + y^2 = 25$ is a circle of radius 5 centered at $(0, 0)$.

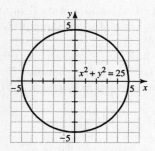

2. The graph of $\dfrac{x^2}{16} - \dfrac{y^2}{25} = 1$ is a hyperbola centered at the origin. the x-intercepts are $(-4, 0)$ and $(4, 0)$. The fundamental rectangle passes through the x-intercepts, $(0, -5)$, and $(0, 5)$. Extend the diagonals of the fundamental rectangle to obtain the asymptotes. The hyperbola opens to the left and right.

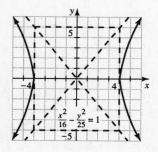

3. The graph of $y^2 + 4x^2 = 4$ is an ellipse with x-intercepts at $(-1, 0)$ and $(1, 0)$. Its y-intercepts are $(0, -2)$ and $(0, 2)$.

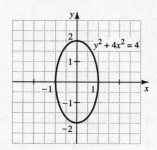

4. The graph of $y = x^2 + 4x + 4$ is a parabola opening upward with y-intercept $(0, 4)$. We can write this equation as $y = (x + 2)^2$. In this form we see that the vertex is $(-2, 0)$.

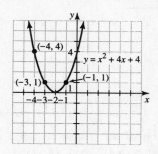

5. Write $y^2 - 4x^2 = 4$ as $\dfrac{y^2}{4} - x^2 = 1$. The graph is a hyperbola with y-intercepts at $(0, -2)$ and $(0, 2)$. The fundamental rectangle passes through the y-intercepts, $(-1, 0)$, and $(1, 0)$. Extend the diagonals to get the asymptotes. The hyperbola opens up and down.

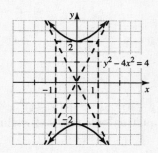

6. The graph of $y = -x^2 - 2x + 3$ is a parabola opening downward. The vertex is at $(-1, 4)$. The y-intercept is $(0, 3)$. The x-intercepts are $(-3, 0)$ and $(1, 0)$.

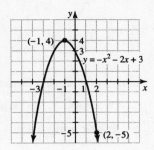

7. First graph the hyperbola $\frac{x^2}{9} - \frac{y^2}{9} = 1$. Since $(0, 0)$ satisfies the inequality, the region containing $(0, 0)$ is shaded.

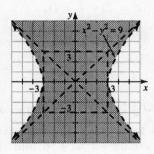

8. First graph the circle $x^2 + y^2 = 9$, centered at $(0, 0)$ with radius 3. Since $(0, 0)$ fails to satisfy $x^2 + y^2 > 9$, we shade the region outside the circle.

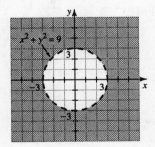

9. The graph of $y = x^2 - 9$ is a parabola opening upward. Its vertex is $(0, -9)$. To find its x-intercepts solve $x^2 - 9 = 0$. The x-intercepts are $(-3, 0)$ and $(3, 0)$. The graph of $y > x^2 - 9$ is the region containing the origin because $(0, 0)$ satisfies the inequality.

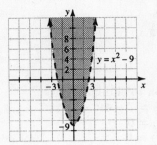

10. The graph of $x^2 + y^2 < 9$ is the region inside the circle $x^2 + y^2 = 9$. To find the graph of $x^2 - y^2 > 1$ first graph the hyperbola $x^2 - y^2 = 1$. By testing points, we can see that the two regions not containing the origin satisfy $x^2 - y^2 > 1$. The points in these two regions that are also inside the circle are the points that satisfy both inequalities of the system.

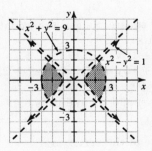

11. The graph of $y = -x^2 + x$ is a parabola opening downward. Points below this parabola satisfy $y < -x^2 + x$. The graph of $y = x - 4$ is a line through $(0, -4)$ with slope 1. Points below this line satisfy $y < x - 4$. The solution set to the system of inequalities consists of points below the parabola and below the line.

239

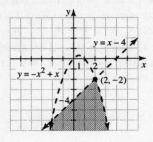

12. Use substitution to eliminate y.
$$x^2 - 2x - 8 = 7 - 4x$$
$$x^2 + 2x - 15 = 0$$
$$(x + 5)(x - 3) = 0$$
$$x = -5 \quad \text{or} \quad x = 3$$
$$y = 27 \qquad y = -5 \quad \text{Since } y = 7 - 4x$$
The solution set is $\{(-5, 27), (3, -5)\}$.

13. Substitute $x^2 = y$ into $x^2 + y^2 = 12$.
$$y + y^2 = 12$$
$$y^2 + y - 12 = 0$$
$$(y + 4)(y - 3) = 0$$
$$y = -4 \quad \text{or} \quad y = 3$$
If $y = -4$, the $x^2 = -4$ has no solution. If $y = 3$, then $x^2 = 3$ or $x = \pm\sqrt{3}$. The solution set is $\{(\sqrt{3}, 3), (-\sqrt{3}, 3)\}$.

14. Complete the square to get the equation into the standard form.
$$x^2 + 2x + \qquad y^2 + 10y \qquad = 10$$
$$x^2 + 2x + 1 + y^2 + 10y + 25 = 10 + 1 + 25$$
$$(x + 1)^2 + (y + 5)^2 = 36$$
The center is $(-1, -5)$ and the radius is 6.

15. The x-coordinate of the vertex is
$$x = \frac{-b}{2a} = \frac{-1}{2(1)} = -\frac{1}{2}.$$

Use $x = -1/2$ in $y = x^2 + x + 3$, to get the vertex $(-\frac{1}{2}, \frac{11}{4})$. Since $a = 1$, we have $p = 1/4$. The focus is $(-\frac{1}{2}, 3)$ and the directrix is $y = \frac{5}{2}$. The axis of symmetry is $x = -\frac{1}{2}$ and the parabola opens up.

16. $y = \frac{1}{2}x^2 - 3x - \frac{1}{2}$
$$y = \frac{1}{2}(x^2 - 6x) - \frac{1}{2}$$
$$y = \frac{1}{2}(x^2 - 6x + 9 - 9) - \frac{1}{2}$$
$$y = \frac{1}{2}(x^2 - 6x + 9) - \frac{9}{2} - \frac{1}{2}$$
$$y = \frac{1}{2}(x - 3)^2 - 5$$

17. The maximum value of s in the formula $s = -16t^2 + 64t + 20$ occurs when $t = \frac{-64}{2(-16)} = 2.$ The maximum height is $s = -16(2)^2 + 64(2) + 20 = 84$ feet.

18. The radius of the circle is the distance between $(-1, 3)$ and $(2, 5)$.
$$r = \sqrt{(-1 - 2)^2 + (3 - 5)^2} = \sqrt{9 + 4} = \sqrt{13}$$
The equation of the circle with center $(-1, 3)$ and radius $\sqrt{13}$ is $(x + 1)^2 + (y - 3)^2 = 13$.

Tying It All Together Chapters 1-11

1. The graph of $y = 9x - x^2$ is a parabola that opens downward. Solve $9x - x^2 = 0$ to find the x-intercepts.
$$x(9 - x) = 0$$
$$x = 0 \quad \text{or} \quad x = 9$$
The x-intercepts are $(0, 0)$ and $(9, 0)$.

The x-coordinate of the vertex is $x = \frac{-(9)}{2(-1)} = \frac{9}{2}.$ The vertex is $(9/2, 81/4)$.

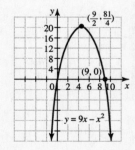

2. The graph of $y = 9x$ is a line through $(0, 0)$ with slope 9.

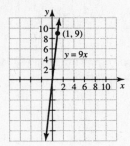

3. The graph of $y = (x - 9)^2$ is a parabola opening upward, with vertex at $(9, 0)$.

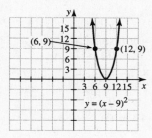

4. Write $y^2 = 9 - x^2$ as $x^2 + y^2 = 9$. It is the equation of a circle of radius 3 centered at $(0, 0)$.

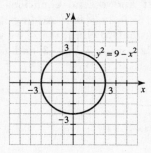

5. The graph of $y = 9x^2$ is a parabola opening upward with vertex at $(0, 0)$.

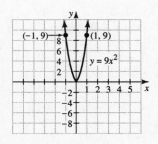

6. The graph of $y = |9x|$ is v-shaped. It contains the points $(0, 0)$, $(-1, 9)$, and $(1, 9)$.

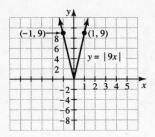

7. Write $4x^2 + 9y^2 = 36$ as $\frac{x^2}{9} + \frac{y^2}{4} = 1$ to see that it is the equation of an ellipse through $(-3, 0)$, $(3, 0)$, $(0, 2)$, and $(0, -2)$.

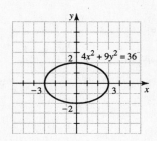

8. Write $4x^2 - 9y^2 = 36$ as $\frac{x^2}{9} - \frac{y^2}{4} = 1$ to see that it is the equation of a hyperbola with x-intercepts $(-3, 0)$ and $(3, 0)$. The fundamental rectangle goes through the x-intercepts, $(0, -2)$, and $(0, 2)$. Extend the diagonals of the rectangle to get the asymptotes.

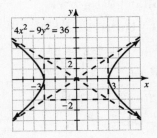

241

9. The graph of $y = 9 - x$ is a line through $(0, 9)$ with slope -1.

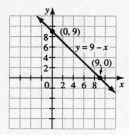

10. The graph of $y = 9^x$ goes through $(0, 1)$, $(1, 9)$, and $(-1, 1/9)$.

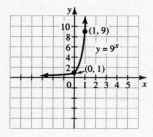

11. $(x + 2y)^2 = x^2 + 2(x)(2y) + (2y)^2$
$$= x^2 + 4xy + 4y^2$$

12. $(x + y)(x^2 + 2xy + y^2)$
$$= x(x^2 + 2xy + y^2) + y(x^2 + 2xy + y^2)$$
$$= x^3 + 2x^2y + xy^2 + x^2y + 2xy^2 + y^3$$
$$= x^3 + 3x^2y + 3xy^2 + y^3$$

13. $(a + b)^3 = (a + b)(a^2 + 2ab + b^2)$
$$= a(a^2 + 2ab + b^2) + b(a^2 + 2ab + b^2)$$
$$= a^3 + 2a^2b = ab^2 + a^2b + 2ab^2 + b^3$$
$$= a^3 + 3a^2b + 3ab^2 + b^3$$

14. $(a - 3b)^2 = a^2 - 2(a)(3b) + (3b)^2$
$$= a^2 - 6ab + 9b^2$$

15. $(2a + 1)(3a - 5) = 6a^2 + 3a - 10a - 5$
$$= 6a^2 - 7a - 5$$

16. $(x - y)(x^2 + xy + y^2)$
$$= x(x^2 + xy + y^2) - y(x^2 + xy + y^2)$$
$$= x^3 + x^2y + xy^2 - x^2y - xy^2 - y^3$$
$$= x^3 - y^3$$

17. Multiply the second equation by -2 and add the result to the first equation.
$$2x - 3y = -4$$
$$\underline{-2x - 4y = -10}$$
$$-7y = -14$$
$$y = 2$$

Use $y = 2$ in $x + 2y = 5$ to find x.
$$x + 2(2) = 5$$
$$x = 1$$
The solution set is $\{(1, 2)\}$.

18. Substitute $y = 7 - x$ into $x^2 + y^2 = 25$.
$$x^2 + (7 - x)^2 = 25$$
$$x^2 + 49 - 14x + x^2 = 25$$
$$2x^2 - 14x + 24 = 0$$
$$x^2 - 7x + 12 = 0$$
$$(x - 3)(x - 4) = 0$$

$x = 3$ or $x = 4$
$y = 4$ $y = 3$ Since $y = 7 - x$
The solution set is $\{(3, 4), (4, 3)\}$.

19. Adding the first and second equations to eliminate z, we get $3x - 3y = 9$. Adding the second and third equations to eliminate z, we get $2x - y = 4$. Divide $3x - 3y = 9$ by -3 and add the result to $2x - y = 4$.
$$-x + y = -3$$
$$\underline{2x - y = 4}$$
$$x \qquad = 1$$
Use $x = 1$ in $-x + y = -3$.
$$-1 + y = -3$$
$$y = -2$$
Use $x = 1$ and $y = -2$ in $x + y + z = 2$.
$$1 + (-2) + z = 2$$
$$z = 3$$
The solution set is $\{(1, -2, 3)\}$.

20. Substitute $y = x^2$ into $y - 2x = 3$.
$$x^2 - 2x = 3$$
$$x^2 - 2x - 3 = 0$$
$$(x - 3)(x + 1) = 0$$

$x = 3$ or $x = -1$
$y = 9$ $y = 1$ Since $y = x^2$

The solution set is $\{(-1, 1), (3, 9)\}$.

21.
$$ax + b = 0$$
$$ax = -b$$
$$x = -\frac{b}{a}$$

22. Use the quadratic formula with $a = w$, $b = d$, and $c = m$.
$$x = \frac{-d \pm \sqrt{d^2 - 4wm}}{2w}$$

23.
$$A = \frac{1}{2}h(B + b)$$
$$2A = h(B + b)$$
$$2A = hB + hb$$
$$2A - bh = hB$$
$$B = \frac{2A - bh}{h}$$

24.
$$\frac{1}{x} + \frac{1}{y} = \frac{1}{2}$$
$$2xy\left(\frac{1}{x} + \frac{1}{y}\right) = 2xy\left(\frac{1}{2}\right)$$
$$2y + 2x = xy$$
$$2y = xy - 2x$$
$$2y = x(y - 2)$$
$$x = \frac{2y}{y - 2}$$

25.
$$L = m + mxt$$
$$L = m(1 + xt)$$
$$m = \frac{L}{1 + xt}$$

26.
$$y = 3a\sqrt{t}$$
$$(y)^2 = (3a\sqrt{t})^2$$
$$y^2 = 9a^2t$$
$$\frac{y^2}{9a^2} = t$$
$$t = \frac{y^2}{9a^2}$$

27. First find the slope.
$$m = \frac{-3 - 1}{2 - (-4)} = \frac{-4}{6} = -\frac{2}{3}$$
Use point-slope form with $(2, -3)$.
$$y - (-3) = -\frac{2}{3}(x - 2)$$
$$y + 3 = -\frac{2}{3}x + \frac{4}{3}$$
$$y = -\frac{2}{3}x - \frac{5}{3}$$

28. Write $2x - 4y = 5$ as $y = \frac{1}{2}x - \frac{5}{4}$. The slope of any line perpendicular to this line is -2. The line through $(0, 0)$ with slope -2 is $y = -2x$.

29. The radius is the distance between $(2, 5)$ and $(-1, -1)$.

$$r = \sqrt{(-1 - 2)^2 + (-1 - 5)^2} = \sqrt{9 + 36} = \sqrt{45}$$

The equation of circle with center $(2, 5)$ and radius $\sqrt{45}$ is $(x - 2)^2 + (y - 5)^2 = 45$.

30. Use completing the square to get the equation into standard form for a circle.
$$x^2 + 3x + y^2 - 6y = 0$$
$$x^2 + 3x + \frac{9}{4} + y^2 - 6y + 9 = 0 + \frac{9}{4} + 9$$
$$\left(x + \frac{3}{2}\right)^2 + (y - 3)^2 = \frac{45}{4}$$

The center is $\left(-\frac{3}{2}, 3\right)$ and radius is $\sqrt{\frac{45}{4}} = \frac{3\sqrt{5}}{2}$.

31. $2i(3 + 5i) = 6i + 10i^2 = -10 + 6i$

32. $i^6 = i^4 \cdot i^2 = 1(-1) = -1$

33. $(2i - 3) + (6 - 7i) = 3 - 5i$

34. $(3 + i\sqrt{2})^2 = 9 + 6i\sqrt{2} + 2i^2 = 7 + 6i\sqrt{2}$

35.
$$(2 - 3i)(5 - 6i) = 10 - 15i - 12i + 18i^2$$
$$= -8 - 27i$$

36. $(3 - i) + (-6 + 4i) = -3 + 3i$

37. $(5 - 2i)(5 + 2i) = 25 - 4i^2 = 29$

38. $\dfrac{2 - 3i}{2i} = \dfrac{(2 - 3i)(-i)}{2i(-i)} = \dfrac{-2i + 3i^2}{2(1)} = -\dfrac{3}{2} - i$

39. $\dfrac{(4 + 5i)(1 + i)}{(1 - i)(1 + i)} = \dfrac{4 + 9i + 5i^2}{2}$
$$= \frac{-1 + 9i}{2} = -\frac{1}{2} + \frac{9}{2}i$$

40. $\dfrac{4 - \sqrt{-8}}{2} = \dfrac{4 - 2i\sqrt{2}}{2} = 2 - i\sqrt{2}$

41. a) $m = \dfrac{250 - 200}{0.30 - 0.40} = \dfrac{50}{-0.1} = -500$
$$q - 250 = -500(x - 0.30)$$
$$q - 250 = -500x + 150$$
$$q = -500x + 400$$

b) $R = qx = (-500x + 400)x$
$$R = -500x^2 + 400x$$

c) $x = \dfrac{-400}{2(-500)} = 0.40$
$$R = -500(0.40)^2 + 400(0.40)$$
$$R = 80$$

The maximum revenue of $80 occurs when the bananas are $0.40 per pound.

12.1 WARM-UPS

1. True, because the formula $a_n = 2n$ will produce even numbers when n is a positive integer. **2.** True, because the formula $a_n = 2n - 1$ will produce odd numbers when n is a positive integer. **3.** True, by the definition of sequence. **4.** False, because the domain of a finite sequence is the set of positive integers less than or equal to some fixed positive integer. **5.** False, because if $n = 1$, $a_1 = (-1)^2 \cdot 1^2 = 1$. **6.** False, because the independent variable is n. **7.** True. **8.** False, because the 6th term is $a_6 = (-1)^7 2^6 = -64$. **9.** True. **10.** True, because each term starting with the third is the sum of the two terms preceding it.

12.1 EXERCISES

1. The terms of the sequence a_n are found by squaring the integers from 1 through 8, because of the formula $a_n = n^2$. So the terms are 1, 4, 9, 16, 25, 36, 49, and 64.

3. $b_1 = \dfrac{(-1)^1}{1} = -1$, $\quad b_2 = \dfrac{(-1)^2}{2} = \dfrac{1}{2}$,

$b_3 = \dfrac{(-1)^3}{3} = -\dfrac{1}{3}$, $\quad b_4 = \dfrac{(-1)^4}{4} = \dfrac{1}{4}$,

$b_5 = \dfrac{(-1)^5}{5} = -\dfrac{1}{5}$, etc.

The 10 terms of the sequence are

$-1, \dfrac{1}{2}, -\dfrac{1}{3}, \dfrac{1}{4}, -\dfrac{1}{5}, \dfrac{1}{6}, -\dfrac{1}{7}, \dfrac{1}{8}, -\dfrac{1}{9}$, and $\dfrac{1}{10}$.

5. $c_1 = (-2)^{1-1} = 1$, $c_2 = (-2)^{2-1} = -2$,

$c_3 = (-2)^{3-1} = 4$, $c_4 = (-2)^{4-1} = -8$,

$c_5 = (-2)^{5-1} = 16$

The five terms of the sequence are 1, -2, 4, -8, and 16.

7. $a_1 = 2^{-1} = \dfrac{1}{2}$, $\quad a_2 = 2^{-2} = \dfrac{1}{4}$,

$a_3 = 2^{-3} = \dfrac{1}{8}$, $\quad a_4 = 2^{-4} = \dfrac{1}{16}$, etc.

The six terms of the sequence are $\dfrac{1}{2}, \dfrac{1}{4}, \dfrac{1}{8}, \dfrac{1}{16}, \dfrac{1}{32}$, and $\dfrac{1}{64}$.

9. $b_1 = 2(1) - 3 = -1$, $\quad b_2 = 2(2) - 3 = 1$,

$b_3 = 2(3) - 3 = 3$, $\quad b_4 = 2(4) - 3 = 5$,

$b_5 = 2(5) - 3 = 7$, $\quad b_6 = 2(6) - 3 = 9$,

$b_7 = 2(7) - 3 = 11$

The seven terms of the sequence are -1, 1, 3, 5, 7, 9, and 11.

11. $c_1 = 1^{-1/2} = 1$, $\quad c_2 = 2^{-1/2} = \dfrac{1}{\sqrt{2}}$,

$c_3 = 3^{-1/2} = \dfrac{1}{\sqrt{3}} = \dfrac{\sqrt{3}}{3}$, $\quad c_4 = 4^{-1/2} = \dfrac{1}{2}$,

$c_5 = 5^{-1/2} = \dfrac{1}{\sqrt{5}} = \dfrac{\sqrt{5}}{5}$

The five terms are 1, $\dfrac{\sqrt{2}}{2}, \dfrac{\sqrt{3}}{3}, \dfrac{1}{2}$, and $\dfrac{\sqrt{5}}{5}$.

13. $a_1 = \dfrac{1}{1^2 + 1} = \dfrac{1}{2}$, $\quad a_2 = \dfrac{1}{2^2 + 2} = \dfrac{1}{6}$,

$a_3 = \dfrac{1}{3^2 + 3} = \dfrac{1}{12}$, $\quad a_4 = \dfrac{1}{4^2 + 4} = \dfrac{1}{20}$

The first four terms are $\dfrac{1}{2}, \dfrac{1}{6}, \dfrac{1}{12}$, and $\dfrac{1}{20}$.

15. $b_1 = \dfrac{1}{2(1) - 5} = -\dfrac{1}{3}$, $\quad b_2 = \dfrac{1}{2(2) - 5} = -1$,

$b_3 = \dfrac{1}{2(3) - 5} = 1$, $\quad b_4 = \dfrac{1}{2(4) - 5} = \dfrac{1}{3}$

The first four terms are $-\dfrac{1}{3}$, -1, 1, and $\dfrac{1}{3}$.

17. $c_1 = (-1)^1 (1 - 2)^2 = -1$,

$c_2 = (-1)^2 (2 - 2)^2 = 0$, $\quad c_3 = (-1)^3 (3 - 2)^2 = -1$,

$c_4 = (-1)^4 (4 - 2)^2 = 4$

The first four terms are -1, 0, -1, and 4.

19. $a_1 = \dfrac{(-1)^{2(1)}}{1^2} = 1$, $\quad a_2 = \dfrac{(-1)^{2(2)}}{2^2} = \dfrac{1}{4}$,

$a_3 = \dfrac{(-1)^{2(3)}}{3^2} = \dfrac{1}{9}$, $\quad a_4 = \dfrac{(-1)^{2(4)}}{4^2} = \dfrac{1}{16}$

The first four terms are 1, $\dfrac{1}{4}, \dfrac{1}{9}$, and $\dfrac{1}{16}$.

21. This sequence is a sequence of odd integers starting at 1. Even integers are represented by 2n and odd integers are all one less than an even integer. Try the formula $a_n = 2n - 1$. To see that it is correct, find a_1 through a_5.

23. A sequence of alternating ones and negative ones is obtained by using a power of -1. Since we want a_1 to be positive when $n = 1$, we use the $n + 1$ power on -1. A formula for the general term is $a_n = (-1)^{n+1}$.

25. This sequence is a sequence of even integers starting at 0. Even integers can be generated by 2n, but to make the first one 0 (when $n = 1$), we use the formula $a_n = 2n - 2$.

27. This sequence is a sequence consisting of the positive integral multiples of 3. Try the formula $a_n = 3n$ to see that it generates the appropriate numbers.

29. Each term in this sequence is one larger than the corresponding term in the sequence of exercise 27. Try the formula $a_n = 3n + 1$ to see that it generates the given sequence.

31. To get the alternating signs on the terms of the sequence, we use a power of -1. Next we observe that the numbers 1, 2, 4, 8, 16 are powers of 2. The formula $a_n = (-1)^n 2^{n-1}$ will produce the given sequence. Remember that n always starts at 1, so in this case we use $n - 1$ as the power of 2 to get $2^0 = 1$.

33. Notice that the numbers 0, 1, 4, 9, 16, . . . are the squares of the nonnegative integers: 0^2, 1^2, 2^2, 3^2, 4^2,... . We would use n^2 to generate squares, but since n starts at 1, we use $a_n = (n-1)^2$ to get the given sequence.

35. After the first penalty the ball is on the 4-yard line. After the second penalty the ball is on the 2-yard line. After the third penalty the ball is on the 1-yard line, and so on. The sequence of five terms is 4, 2, 1, $\frac{1}{2}$, $\frac{1}{4}$.

37. To find the amount of increase, we take 5% of $16,441 to get $822 increase. We could just multiply $16,441 by 1.05 to obtain the new price. In either case the price the next year is $17,263. To find the price for the next year we multiply the last year's price by 1.05: 1.05($17,263) = $18,126. Repeating this process gives us the prices $17,263, $18,126, $19,033, $19,984, $20,983 as the price of the car for the next five years.

39. Of the $1 million, 80% is respent in the community. So $800,000 is respent. Of the $800,000 we have 80% respent in the community. So $640,000 is respent in the community, and so on. The first four terms of the sequence are $1,000,000, $800,000, $640,000, $512,000.

41. Possible vertical repeats are 27 in., 13.5 in., 9 in., 6.75 in., and 5.4 in.

43. If we put the word great in front of the word grandparents 35 times, then we have 2^{37} of this type of relative. Use a calculator to find that $2^{37} = 137,438,953,500$. This is certainly larger than the present population of the earth.

47. Use a calculator to find

$$a_{100} = (0.999)^{100} = 0.9048,$$

$$a_{1000} = (0.999)^{1000} = 0.3677,$$

and $\quad a_{10000} = (0.999)^{10000} = 0.00004517.$

12.2 WARM-UPS

1. True, because of the definition of series.
2. False, the sum of a series can be any real number. **3.** False, there are 9 terms in 2^3 through 10^3. **4.** False, because the terms in the first series are opposites of the terms in the second series. **5.** False,

the ninth term is $\dfrac{(-1)^9}{(9+1)(9+2)} = -\dfrac{1}{110}$.

6. True, because the terms are -2 and 4 and the sum is 2.

7. True, because of the distributive property.

8. True, because the notation indicates to add the number 4 five times. **9.** True, because $2i + 7i = 9i$. **10.** False, because in the series on the left side the 1 is added in three times and in the series on the right side the 1 is only added in once.

12.2 EXERCISES

1. $\displaystyle\sum_{i=1}^{4} i^2 = 1^2 + 2^2 + 3^2 + 4^2 = 1 + 4 + 9 + 16$
$$= 30$$

3. $\displaystyle\sum_{j=0}^{5} (2j-1) = (2 \cdot 0 - 1) + (2 \cdot 1 - 1)$

$$+ (2 \cdot 2 - 1) + (2 \cdot 3 - 1) + (2 \cdot 4 - 1) + (2 \cdot 5 - 1)$$
$$= -1 + 1 + 3 + 5 + 7 + 9 = 24$$

5. $\displaystyle\sum_{i=1}^{5} 2^{-i} = 2^{-1} + 2^{-2} + 2^{-3} + 2^{-4} + 2^{-5}$
$$= \frac{1}{2} + \frac{1}{4} + \frac{1}{8} + \frac{1}{16} + \frac{1}{32} = \frac{31}{32}$$

7. $\displaystyle\sum_{i=1}^{10} 5i^0 = 5(1)^0 + 5(2)^0 + 5(3)^0 + 5(4)^0$

$$+ 5(5)^0 + 5(6)^0 + 5(7)^0 + 5(8)^0 + 5(9)^0 + 5(10)^0$$
$$= 5 + 5 + 5 + 5 + 5 + 5 + 5 + 5 + 5 + 5 = 50$$

9. $\displaystyle\sum_{i=1}^{3} (i-3)(i+1) = (1-3)((1+1)$

$$+ (2-3)(2+1) + (3-3)(3+1)$$
$$= -4 + (-3) + 0 = -7$$

11. $\displaystyle\sum_{j=1}^{10} (-1)^j = (-1)^1 + (-1)^2 + ... + (-1)^{10}$

$$= -1 + 1 - 1 + 1 - 1 + 1 - 1 + 1 - 1 + 1 = 0$$

13. The sum of the first six positive integers is written as $\displaystyle\sum_{i=1}^{6} i$. There are other ways to indicate this series in summation notation but this is the simplest.

15. To get the signs of the terms to alternate from positive to negative we use a power of -1. To get odd integers we use the formula $2i - 1$.

So this series is written $\displaystyle\sum_{i=1}^{6} (-1)^i (2i-1)$.

17. This series consists of the squares of the first six positive integers. It is written in summation notation as $\displaystyle\sum_{i=1}^{6} i^2$.

19. This series consists of the reciprocals of the positive integers. If the index i goes from 1 to 4, we must use $2 + i$ to get the numbers 3 through 6. So the series is written as $\displaystyle\sum_{i=1}^{4} \frac{1}{2+i}$.

21. The terms of this series are logarithms of positive integers. If the index i goes from 1 to 3, we must use $i + 1$ to get the numbers 2 through 4. The series is written as $\displaystyle\sum_{i=1}^{3} \ln(i+1)$.

23. Since the subscripts range from 1 through 4, we let i range from 1 through 4: $\displaystyle\sum_{i=1}^{4} a_i$.

25. The subscripts on x range from 3 through 50, so i ranges from 1 through 48: $\displaystyle\sum_{i=1}^{48} x_{i+2}$.

27. The subscripts on w range from 1 through n, so i ranges from 1 through n: $\displaystyle\sum_{i=1}^{n} w_i$.

29. If $j = 0$ when $i = 1$, then $i = j + 1$ and j ranges from 0 through 4: $\displaystyle\sum_{j=0}^{4} (j+1)^2$.

31. If $j = 1$ when $i = 0$, then $i = j - 1$ and j ranges from 1 through 13. If we substitute $i = j - 1$ into $2i - 1$ we get $2i - 1 = 2(j-1) - 1$ $= 2j - 3$. So the series is written $\displaystyle\sum_{j=1}^{13} (2j-3)$.

33. If $j = 1$ when $i = 4$, then $i = j + 3$ and j ranges from 1 through 5. Substitute $j + 3$ for i to get $\displaystyle\sum_{j=1}^{5} \frac{1}{j+3}$.

35. If $j = 0$ when $i = 1$, then $i = j + 1$ and j ranges from 0 through 3. The exponent $2i + 3$ becomes $2(j+1) + 3 = 2j + 5$. The series is written as $\displaystyle\sum_{j=0}^{3} x^{2j+5}$.

37. If $j = 0$ when $i = 1$, then $i = j + 1$ and j ranges from 0 through $n - 1$. Replacing i by $j + 1$ gives us the series $\displaystyle\sum_{j=0}^{n-1} x^{j+1}$.

39. $\displaystyle\sum_{i=1}^{6} x^i = x + x^2 + x^3 + x^4 + x^5 + x^6$

41. $\displaystyle\sum_{j=0}^{3} (-1)^j x_j$
$= (-1)^0 x_0 + (-1)^1 x_1 + (-1)^2 x_2 + (-1)^3 x_3$
$= x_0 - x_1 + x_2 - x_3$

43. $\displaystyle\sum_{i=1}^{3} i x^i = 1x^1 + 2x^2 + 3x^3 = x + 2x^2 + 3x^3$

45. On the first jump he has moved $\frac{1}{2}$ yard. On the second jump he moves $\frac{1}{4}$ yard. On the third jump he moves $\frac{1}{8}$ yard, and so on. To express the reciprocals of the powers of 2, we use 2^{-i}. His total movement after nine jumps is expressed as the series $\displaystyle\sum_{i=1}^{9} 2^{-i}$.

47. $\displaystyle\sum_{i=1}^{4} 1{,}000{,}000(0.8)^{i-1}$

12.3 WARM-UPS

1. False, because the common difference is -2 $(1 - 3 = -2)$.   **2.** False, because the difference between consecutive terms is not constant. **3.** False, because the difference between the consecutive terms is sometimes 2 and sometimes -2.   **4.** False, because the nth term is given by $a_n = a_1 + (n-1)d$.   **5.** False, because the second must be 7.5, making $d = 2.5$ and the fourth term 12.5.   **6.** False, because if the first is 6 and the third is 2, the second must be 4 in order to have a common difference.   **7.** True, because that is the definition of arithmetic sequence.   **8.** True, because the series is $5 + 7 + 9 + 11 + 13$ and the common difference is 2.   **9.** True, because of the formula for the sum of an arithmetic series.   **10.** False, because there are 11 even integers from 8 through 28 inclusive and the sum is $\frac{11}{2}(8 + 28)$.

12.3 EXERCISES

1. The common difference is $d = 6$ and the first term is $a_1 = 0$. Use the formula $a_n = a_1 + (n-1)d$ to find that the nth term is $a_n = 0 + (n-1)6 = 6n - 6$.
3. The common difference is $d = 5$ and the first term is 7. The nth term is $a_n = 7 + (n-1)5 = 5n + 2$.
5. The common difference is $d = 2$ and the first term is $a_1 = -4$. Use the formula $a_n = a_1 + (n-1)d$ to find that the nth term is $a_n = -4 + (n-1)2 = 2n - 6$.
7. The common difference is $d = 1 - 5 = -4$ and the first term is $a_1 = 5$. The nth term is $a_n = 5 + (n-1)(-4) = -4n + 9$.
9. The common difference is $d = -9 - (-2) = -7$ and the first term is -2. The nth term is $a_n = -2 + (n-1)(-7) = -7n + 5$.
11. The common difference is $d = -2.5 - (-3) = 0.5$ and the first term is -3. The nth term is $a_n = -3 + (n-1)(0.5) = 0.5n - 3.5$.
13. The common difference is $d = -6.5 - (-6) = -0.5$ and the first term is -6. The nth term is $a_n = -6 + (n-1)(-0.5) = -0.5n - 5.5$.
15. $a_1 = 9 + (1-1)4 = 9$,
$a_2 = 9 + (2-1)4 = 13$, $a_3 = 9 + (3-1)4 = 17$,
$a_4 = 9 + (4-1)4 = 21$, $a_5 = 9 + (5-1)4 = 25$
The first five terms of the arithmetic sequence are 9, 13, 17, 21, and 25.
17. $a_1 = 7 + (1-1)(-2) = 7$,
$a_2 = 7 + (2-1)(-2) = 5$,
$a_3 = 7 + (3-1)(-2) = 3$,
$a_4 = 7 + (4-1)(-2) = 1$,
$a_5 = 7 + (5-1)(-2) = -1$
The first five terms are 7, 5, 3, 1, and -1.
19. $a_1 = -4 + (1-1)3 = -4$,
$a_2 = -4 + (2-1)3 = -1$,
$a_3 = -4 + (3-1)3 = 2$,
$a_4 = -4 + (4-1)3 = 5$,
$a_5 = -4 + (5-1)3 = 8$
The first five terms of the arithmetic sequence are -4, -1, 2, 5, and 8.
21. $a_1 = -2 + (1-1)(-3) = -2$,
$a_2 = -2 + (2-1)(-3) = -5$,
$a_3 = -2 + (3-1)(-3) = -8$,
$a_4 = -2 + (4-1)(-3) = -11$,
$a_5 = -2 + (5-1)(-3) = -14$
The first five terms are -2, -5, -8, -11, and -14.

23. $a_1 = -4(1) - 3 = -7$,
$a_2 = -4(2) - 3 = -11$, $a_3 = -4(3) - 3 = -15$,
$a_4 = -4(4) - 3 = -19$, $a_5 = -4(5) - 3 = -23$

The first five terms of the arithmetic sequence
are -7, -11, -15, -19, and -23.

25. $a_1 = 0.5(1) + 4 = 4.5$,
$a_2 = 0.5(2) + 4 = 5$, $a_3 = 0.5(3) + 4 = 5.5$,
$a_4 = 0.5(4) + 4 = 6$, $a_5 = 0.5(5) + 4 = 6.5$
The first five terms of the arithmetic sequence
are 4.5, 5, 5.5, 6, and 6.5.

27. $a_1 = 20(1) + 1000 = 1020$,
$a_2 = 20(2) + 1000 = 1040$,
$a_3 = 20(3) + 1000 = 1060$,
$a_4 = 20(4) + 1000 = 1080$,
$a_5 = 20(5) + 1000 = 1100$

The first five terms of the arithmetic sequence
are 1020, 1040, 1060, 1080, and 1100.

29. Use $a_1 = 9$, $n = 8$, and $d = 6$ in the formula
$a_n = a_1 + (n - 1)d$.
$$a_8 = 9 + (8 - 1)6 = 51$$

31. Use $a_1 = 6$, $a_{20} = 82$, and $n = 20$ in the
formula $a_n = a_1 + (n - 1)d$.
$$82 = 6 + (20 - 1)d$$
$$82 = 6 + 19d$$
$$76 = 19d$$
$$4 = d$$

33. Use $a_7 = 14$, $d = -2$, and $n = 7$ in the
formula $a_n = a_1 + (n - 1)d$.
$$14 = a_1 + (7 - 1)(-2)$$
$$14 = a_1 - 12$$
$$26 = a_1$$

35. From the fact that the fifth term is 13 and
the first term is -3, we can find the common
difference.
$$13 = -3 + (5 - 1)d$$
$$13 = -3 + 4d$$
$$16 = 4d$$
$$4 = d$$
Use $a_1 = -3$, $n = 6$, and $d = 4$ in the formula
$a_n = a_1 + (n - 1)d$.
$$a_6 = -3 + (6 - 1)4 = 17$$

37. Use $a_1 = 1$, $a_{48} = 48$, and $n = 48$ in the
formula $S_n = \frac{n}{2}(a_1 + a_n)$.
$$S_{48} = \frac{48}{2}(1 + 48) = 1176$$

39. To find n, use $a_1 = 8$, $d = 2$, and $a_n = 36$ in
the formula $a_n = a_1 + (n - 1)d$.
$$36 = 8 + (n - 1)2$$
$$36 = 8 + 2n - 2$$
$$30 = 2n$$
$$15 = n$$
Use $a_1 = 8$, $a_{15} = 36$, and $n = 15$ in the
formula $S_n = \frac{n}{2}(a_1 + a_n)$.
$$S_{15} = \frac{15}{2}(8 + 36) = 330$$

41. To find n, use $a_1 = -1$, $d = -6$, and
$a_n = -73$ in the formula $a_n = a_1 + (n - 1)d$.
$$-73 = -1 + (n - 1)(-6)$$
$$-78 = -1 - 6n + 6$$
$$-78 = -6n$$
$$13 = n$$
Use $a_1 = -1$, $a_{13} = -73$, and $n = 13$ in the
formula $S_n = \frac{n}{2}(a_1 + a_n)$.
$$S_{13} = \frac{13}{2}(-1 + (-73)) = -481$$

43. To find n, use $a_1 = -6$, $d = 5$, and $a_n = 64$
in the formula $a_n = a_1 + (n - 1)d$.
$$64 = -6 + (n - 1)5$$
$$64 = -6 + 5n - 5$$
$$75 = 5n$$
$$15 = n$$
Use $a_1 = -6$, $a_{15} = 64$, and $n = 15$ in the
formula $S_n = \frac{n}{2}(a_1 + a_n)$.
$$S_{15} = \frac{15}{2}(-6 + 64) = 435$$

45. To find n, use $a_1 = 20$, $d = -8$, and
$a_n = -92$ in the formula $a_n = a_1 + (n - 1)d$.
$$-92 = 20 + (n - 1)(-8)$$
$$-92 = 20 - 8n + 8$$
$$-120 = -8n$$
$$15 = n$$
Use $a_1 = 20$, $a_{15} = -92$, and $n = 15$ in the
formula $S_n = \frac{n}{2}(a_1 + a_n)$.
$$S_{15} = \frac{15}{2}(20 + (-92)) = -540$$

47. $\sum_{i=1}^{12} (3i - 7) = -4 + (-1) + \ldots + 29$

Use $a_1 = -4$, $a_{12} = 29$, and $n = 12$ in the
formula $S_n = \frac{n}{2}(a_1 + a_n)$.
$$S_{12} = \frac{12}{2}(-4 + 29) = 150$$

49. $\sum_{i=1}^{11}(-5i+2) = -3 + (-8) + \ldots + (-53)$

Use $a_1 = -3$, $a_{11} = -53$, and $n = 11$ in the formula $S_n = \frac{n}{2}(a_1 + a_n)$.

$$S_{11} = \frac{11}{2}(-3 + (-53)) = -308$$

51. Use $a_1 = \$22,000$, $n = 7$, and $d = \$500$ in the formula $a_n = a_1 + (n-1)d$.

$$a_7 = 22,000 + (7-1)500 = \$25,000$$

53. The students read 5 pages the first day, 7 pages the second day, 9 pages the third day, and so on. To find the number they read on the 31st day, let $n = 31$, $d = 2$, and $a = 5$ in the formula $a_n = a_1 + (n-1)d$.

$$a_{31} = 5 + (31-1)2 = 65$$

To find the sum $5 + 7 + 9 + \ldots + 65$, use $n = 31$, $a_1 = 5$, and $a_{31} = 65$ in the formula $S_n = \frac{n}{2}(a_1 + a_n)$.

$$S_{31} = \frac{31}{2}(5 + 65) = 1085$$

55. b

12.4 WARM-UPS

1. False, because the ratio of two consecutive terms is not constant. **2.** False, there is a common ratio of 2 between adjacent terms. **3.** True, because the general form for a geometric sequence is $a_n = a_1 r^{n-1}$.

4. True, because if $n = 1$, then $3(2)^{-1+3} = 12$. **5.** True, because $a_1 = 12$ and $a_2 = 6$ gives $r = 1/2$. **6.** True, because of the definition of geometric series. **7.** False, because we have a formula for the sum of a finite geometric series. **8.** False, because $a_1 = 6$. **9.** True, because this is the correct formula for the sum of all of the terms of an infinite geometric series with first term 10 and ratio 1/2. **10.** False, because there is no sum for an infinite geometric series with a ratio of 2.

12.4 EXERCISES

1. Since the first term is 1/3 and the common ratio is 3, the nth term is $a_n = \frac{1}{3}(3)^{n-1}$.

3. Since the first term is 64 and the common ratio is 1/8, the nth term is $a_n = 64\left(\frac{1}{8}\right)^{n-1}$.

5. Since the first term is 8 and the common ratio is $-1/2$, the nth term is $a_n = 8\left(-\frac{1}{2}\right)^{n-1}$.

7. Since the first term is 2 and the common ratio is $-4/2 = -2$, the nth term is $a_n = 2(-2)^{n-1}$.

9. Since the first term is $-1/3$ and the common ratio is $(-1/4)/(-1/3) = 3/4$, the nth term is $a_n = -\frac{1}{3}\left(\frac{3}{4}\right)^{n-1}$.

11. $a_1 = 2(1/3)^{1-1} = 2$,

$a_2 = 2(1/3)^{2-1} = 2/3$, $a_3 = 2(1/3)^{3-1} = 2/9$,

$a_4 = 2(1/3)^{4-1} = 2/27$, $a_5 = 2(1/3)^{5-1} = 2/81$

The first 5 terms are 2, $\frac{2}{3}$, $\frac{2}{9}$, $\frac{2}{27}$, and $\frac{2}{81}$.

13. $a_1 = (-2)^{1-1} = 1$,

$a_2 = (-2)^{2-1} = -2$, $a_3 = (-2)^{3-1} = 4$,

$a_4 = (-2)^{4-1} = -8$, $a_5 = (-2)^{5-1} = 16$

The first 5 terms are 1, -2, 4, -8, and 16.

15. $a_1 = 2^{-1} = 1/2$,

$a_2 = 2^{-2} = 1/4$, $a_3 = 2^{-3} = 1/8$,

$a_4 = 2^{-4} = 1/16$, $a_5 = 2^{-5} = 1/32$

The first 5 terms are $\frac{1}{2}$, $\frac{1}{4}$, $\frac{1}{8}$, $\frac{1}{16}$, and $\frac{1}{32}$.

17. $a_1 = (0.78)^1 = 0.78$,

$a_2 = (0.78)^2 = 0.6084$, $a_3 = (0.78)^3 = 0.4746$,

$a_4 = (0.78)^4 = 0.3702$, $a_5 = (0.78)^5 = 0.2887$

The first 5 terms are 0.78, 0.6084, 0.4746, 0.3702, and 0.2887.

19. Use $a_4 = 40$, $n = 4$, and $r = 2$ in the formula $a_n = a_1 r^{n-1}$.

$$40 = a_1(2)^{4-1}$$

$$40 = 8a_1$$

$$5 = a_1$$

21. Use $a_4 = 2/9$, $n = 4$, and $a_1 = 6$ in the formula $a_n = a_1 r^{n-1}$.

$$\frac{2}{9} = 6r^{4-1}$$

$$\frac{1}{27} = r^3$$

$$\frac{1}{3} = r$$

23. Use $r = 1/3$, $n = 4$, and $a_1 = -3$ in the formula $a_n = a_1 r^{n-1}$.

$$a_4 = -3\left(\frac{1}{3}\right)^{4-1} = -3\left(\frac{1}{27}\right) = -\frac{1}{9}$$

25. Use $r = 1/2$, $a_1 = 1/2$, and $a_n = 1/512$ in the formula $a_n = a_1 r^{n-1}$ to find n.

$$\frac{1}{512} = \frac{1}{2}\left(\frac{1}{2}\right)^{n-1}$$

$$\frac{1}{2^9} = \left(\frac{1}{2}\right)^n$$

$$n = 9$$

Use $n = 9$, $a_1 = 1/2$, and $r = 1/2$ in the formula $S_n = \dfrac{a_1(1-r^n)}{1-r}$.

$$S_9 = \frac{\frac{1}{2}\left(1-\left(\frac{1}{2}\right)^9\right)}{1-\frac{1}{2}} = 1 - \frac{1}{512} = \frac{511}{512}$$

27. Use $n = 5$, $a_1 = 1/2$, and $r = -1/2$ in the formula $S_n = \dfrac{a_1(1-r^n)}{1-r}$.

$$S_5 = \frac{\frac{1}{2}\left(1-\left(-\frac{1}{2}\right)^5\right)}{1-\left(-\frac{1}{2}\right)} = \frac{\frac{1}{2}\left(\frac{33}{32}\right)}{\frac{3}{2}} = \frac{11}{32}$$

29. First determine the number of terms. Since $r = 2/3$, the nth term is $30\left(\frac{2}{3}\right)^{n-1}$. Solve

$$30\left(\frac{2}{3}\right)^{n-1} = \frac{1280}{729}$$

$$\left(\frac{2}{3}\right)^{n-1} = \frac{128}{2187} = \left(\frac{2}{3}\right)^7$$

$$n - 1 = 7$$
$$n = 8$$

Use $n = 8$, $a_1 = 30$, and $r = 2/3$ in the formula $S_n = \dfrac{a_1(1-r^n)}{1-r}$.

$$S_8 = \frac{30\left(1-\left(\frac{2}{3}\right)^8\right)}{1-\left(\frac{2}{3}\right)} = \frac{30\left(\frac{6305}{6561}\right)}{\frac{1}{3}} \approx 86.4883$$

31. $\displaystyle\sum_{i=1}^{10} 5(2)^{i-1} = S_{10} = \dfrac{5(1-(2)^{10})}{1-(2)}$

$$= \frac{5(-1023)}{-1} = 5115$$

33. $\displaystyle\sum_{i=1}^{6} (0.1)^i = S_6 = \dfrac{0.1(1-(0.1)^6)}{1-(0.1)}$

$$= \frac{0.1(0.999999)}{0.9} = 0.111111$$

35. $\displaystyle\sum_{i=1}^{6} 100(0.3)^i = S_6 = \dfrac{100(0.3)(1-(0.3)^6)}{1-(0.3)}$

$$= \frac{100(0.3)(1-(0.3)^6)}{0.7} = 42.8259$$

37. Use $a_1 = 1/8$ and $r = 1/2$ in the formula for the sum of an infinite geometric series $S = \dfrac{a_1}{1-r}$.

$$S = \frac{\frac{1}{8}}{1-\frac{1}{2}} = \frac{\frac{1}{8}}{\frac{1}{2}} = \frac{1}{4}$$

39. Use $a_1 = 3$ and $r = 2/3$ in $S = \dfrac{a_1}{1-r}$.
$$S = \frac{3}{1-\frac{2}{3}} = \frac{3}{\frac{1}{3}} = 9$$

41. Use $a_1 = 4$ and $r = -1/2$ in $S = \dfrac{a_1}{1-r}$.
$$S = \frac{4}{1-\left(-\frac{1}{2}\right)} = \frac{4}{\frac{3}{2}} = \frac{8}{3}$$

43. Use $a_1 = 0.3$ and $r = 0.3$ in $S = \dfrac{a_1}{1-r}$.
$$S = \frac{0.3}{1-0.3} = \frac{0.3}{0.7} = \frac{3}{7}$$

45. Use $a_1 = 3$ and $r = 0.5$ in $S = \dfrac{a_1}{1-r}$.
$$S = \frac{3}{1-0.5} = \frac{3}{0.5} = 6$$

47. Use $a_1 = 3$ and $r = 0.1$ in $S = \dfrac{a_1}{1-r}$.

$$S = \frac{0.3}{1-0.1} = \frac{0.3}{0.9} = \frac{1}{3}$$

49. Use $a_1 = 0.12$ and $r = 0.01$ in $S = \dfrac{a_1}{1-r}$.

$$S = \frac{0.12}{1-0.01} = \frac{0.12}{0.99} = \frac{12}{99} = \frac{4}{33}$$

51. We want the sum of the geometric series

$$2000(1.12)^{45} + 2000(1.12)^{44} + \ldots + 2000(1.12).$$

Note that the last deposit is made at the beginning of the 45th year and earns interest for only one year. Rewrite the series as

$$2000(1.12) + 2000(1.12)^2 + \ldots + 2000(1.12)^{45},$$

where the $a_1 = 2000(1.12)$, $n = 45$, and $r = 1.12$.

$$S_{45} = \frac{2000(1.12)\left(1 - (1.12)^{45}\right)}{1-1.12} = \$3,042,435.27$$

53. We want the sum of the finite geometric series $1 + 2 + 4 + 8 + 16 + \ldots + 2^{30}$, which has 31 terms and a ratio of 2.

$$S_{31} = \frac{1(1 - 2^{31})}{1-2} = 2^{31} - 1 = 2,147,483,647 \text{ cents}$$

$$= \$21,474,836.47$$

55. Use $r = 0.80$, $a_1 = 1,000,000$ in $S = \dfrac{a_1}{1-r}$.

$$S = \frac{1,000,000}{1-0.80} = \$5,000,000$$

57. d

59. Use $a_1 = 24/100 = 0.24$ and

$r = 1/100 = 0.01$ in the formula for $S = \dfrac{a_1}{1-r}$.

$$S = \frac{0.24}{1-0.01} = \frac{0.24}{0.99} = \frac{8}{33}$$

12.5 WARM-UPS

1. False, because there are 13 terms in a binomial to the 12th power. **2.** False, because the 7th term has variable part $a^6 b^6$. **3.** False, because if $x = 1$ the equation is incorrect. **4.** True, because the signs alternate in any expansion of a difference. **5.** True, because we can obtain it from the 7th line. **6.** True, because $1 + 4 + 6 + 4 + 1 = 2^4$. **7.** True, because of the binomial theorem. **8.** True,

because $2^n = (1+1)^n = \displaystyle\sum_{i=0}^{n} \frac{n!}{(n-i)!\,i!} 1^{n-i} 1^i$

$$= \sum_{i=0}^{n} \frac{n!}{(n-i)!\,i!},$$

and the last sum is the sum of the coefficients in the nth row.
9. True, by definition of 0! and 1!.

10. True, because $\dfrac{7 \cdot 6 \cdot 5 \cdot 4 \cdot 3 \cdot 2 \cdot 1}{5 \cdot 4 \cdot 3 \cdot 2 \cdot 1 \cdot 2 \cdot 1} = 21$

12.5 EXERCISES

1. $\dfrac{5!}{2!3!} = \dfrac{5 \cdot 4}{2} = 10$

3. $\dfrac{8!}{5!3!} = \dfrac{8 \cdot 7 \cdot 6}{3 \cdot 2 \cdot 1} = 56$

5. The coefficients in the 5th row of Pascal's triangle are 1, 5, 10, 10, 5, 1. Use these coefficients with the pattern for the exponents.

$$(r + t)^5 = r^5 + 5r^4 t + 10r^3 t^2 + 10r^2 t^3 + 5rt^4 + t^5$$

7. The coefficients in the 3rd row are 1, 3, 3, 1. Use these coefficients with the pattern for the exponents, and alternate the signs.

$$(m - n)^3 = m^3 - 3m^2 n + 3mn^2 - n^3$$

9. Use the coefficients 1, 3, 3, 1 and let $y = 2a$ in the binomial theorem.

$$(x + 2a)^3 = 1x^3 (2a)^0 + 3x^2 (2a)^1 + 3x(2a)^2 + (2a)^3$$

$$= x^3 + 6ax^2 + 12a^2 x + 8a^3$$

11. Use the coefficients 1, 4, 6, 4, 1 in the binomial theorem.

$$(x^2 - 2)^4 = (x^2)^4 - 4(x^2)^3 2 + 6(x^2)^2 2^2$$

$$- 4x^2 2^3 + 1(x^2)^0 2^4$$

$$= x^8 - 8x^6 + 24x^4 - 32x^2 + 16$$

13. Use the coefficients from the 7th line 1, 7, 21, 35, 35, 21, 7, 1 and alternate the signs of the terms.

$$(x - 1)^7 = x^7 - 7x^6 + 21x^5 - 35x^4 + 35x^3$$

$$- 21x^2 + 7x - 1$$

15. Use the binomial theorem to write the first 4 terms of $(a - 3b)^{12}$.

$$\frac{12!}{12!0!}a^{12}b^0 - \frac{12!}{11!1!}a^{11}b^1 + \frac{12!}{10!2!}a^{10}b^2 - \frac{12!}{9!3!}a^9b^3$$

$$= a^{12} - 36a^{11}b + 594a^{10}b^2 - 5940a^9b^3$$

17. Use the binomial theorem to write the first 4 terms of $(x^2 + 5)^9$.

$$\frac{9!}{9!0!}(x^2)^9 5^0 + \frac{9!}{8!1!}(x^2)^8 5^1 + \frac{9!}{7!2!}(x^2)^7 5^2$$
$$+ \frac{9!}{6!3!}(x^2)^6 5^3$$

$$= x^{18} + 45x^{16} + 900x^{14} + 10500x^{12}$$

19. Use the binomial theorem to write the first 4 terms of $(x - 1)^{22}$.

$$\frac{22!}{22!0!}x^{22}1^0 - \frac{22!}{21!1!}x^{21}1^1 + \frac{22!}{20!2!}x^{20}1^2$$
$$- \frac{22!}{19!3!}x^{19}1^3$$

$$= x^{22} - 22x^{21} + 231x^{20} - 1540x^{19}$$

21. Use the binomial theorem to write the first 4 terms of $\left(\frac{x}{2} + \frac{y}{3}\right)^{10}$.

$$\frac{10!}{10!0!}\left(\frac{x}{2}\right)^{10}\left(\frac{y}{3}\right)^0 + \frac{10!}{9!1!}\left(\frac{x}{2}\right)^9\left(\frac{y}{3}\right)^1 + \frac{10!}{8!2!}\left(\frac{x}{2}\right)^8\left(\frac{y}{3}\right)^2$$
$$+ \frac{10!}{7!3!}\left(\frac{x}{2}\right)^7\left(\frac{y}{3}\right)^3$$

$$= \frac{x^{10}}{1024} + \frac{5x^9y}{768} + \frac{5x^8y^2}{256} + \frac{5x^7y^3}{144}$$

23. Use the formula for the kth term of $(x + y)^n$ with $k = 6$ and $n = 13$.

$$\frac{13!}{(13 - 6 + 1)!(6 - 1)!}a^{13 - 6 + 1}w^{6 - 1}$$

$$= \frac{13!}{8!5!}a^8w^5 = 1287a^8w^5$$

25. Use the formula for the kth term with $k = 8$ and $n = 16$.

$$\frac{16!}{(16 - 8 + 1)!(8 - 1)!}m^{16 - 8 + 1}(-n)^{8 - 1}$$

$$= \frac{16!}{9!7!}m^9(-n)^7 = -11440m^9n^7$$

27. Use the formula for the kth term with $k = 4$ and $n = 8$.

$$\frac{8!}{(8 - 4 + 1)!(4 - 1)!}x^{8 - 4 + 1}(2y)^{4 - 1}$$

$$= \frac{8!}{5!3!}x^5(2y)^3 = 56x^5 8y^3 = 448x^5y^3$$

29. Use the formula for the kth term with $k = 7$ and $n = 20$.

$$\frac{20!}{(20 - 7 + 1)!(7 - 1)!}(2a^2)^{20 - 7 + 1}b^{7 - 1}$$

$$= \frac{20!}{14!6!}(2a^2)^{14}b^6 = 635043840a^{28}b^6$$

31. Use $n = 8$, $x = a$, and $y = b$ in the binomial theorem with summation notation.

$$(a + m)^8 = \sum_{i = 0}^{8} \frac{8!}{(8 - i)!i!}a^{8 - i}m^i$$

33. Use $n = 5$, $x = a$, and $y = -2x$ in the binomial theorem with summation notation.

$$(a + (-2x))^5 = \sum_{i = 0}^{5} \frac{5!}{(5 - i)!i!}a^{5 - i}(-2x)^i$$

$$= \sum_{i = 0}^{5} \frac{5!(-2)^i}{(5 - i)!i!}a^{5 - i}x^i$$

35. $(a + (b + c))^3$

$$= a^3 + 3a^2(b + c) + 3a(b + c)^2 + (b + c)^3$$
$$= a^3 + 3a^2b + 3a^2c + 3ab^2 + 6abc + 3ac^2$$
$$+ b^3 + 3b^2c + 3bc^2 + c^3$$
$$= a^3 + b^3 + c^3 + 3a^2b + 3a^2c + 3ab^2 + 3ac^2$$
$$+ 3b^2c + 3bc^2 + 6abc$$

CHAPTER 12 REVIEW

1. $a_1 = 1^3$, $a_2 = 2^3$, $a_3 = 3^3$, $a_4 = 4^3$, $a_5 = 5^3$
The terms of the sequence are 1, 8, 27, 64, 125.

3. $c_1 = (-1)^1(2 \cdot 1 - 3) = 1$,
$c_2 = (-1)^2(2 \cdot 2 - 3) = 1$,

$c_3 = (-1)^3(2 \cdot 3 - 3) = -3,$

$c_4 = (-1)^4(2 \cdot 4 - 3) = 5,$

$c_5 = (-1)^5(2 \cdot 5 - 3) = -7$

$c_6 = (-1)^6(2 \cdot 6 - 3) = 9$

The terms of the sequence are 1, 1, −3, 5, −7, 9.

5. $a_1 = -\frac{1}{1} = -1, a_2 = -\frac{1}{2}, a_3 = -\frac{1}{3}$

The first three terms are $-1, -\frac{1}{2}, -\frac{1}{3}$.

7. $b_1 = \frac{(-1)^{2 \cdot 1}}{2 \cdot 1 + 1} = \frac{1}{3}, \quad b_2 = \frac{(-1)^{2 \cdot 2}}{2 \cdot 2 + 1} = \frac{1}{5}$

$b_3 = \frac{(-1)^{2 \cdot 3}}{2 \cdot 3 + 1} = \frac{1}{7}$

The first three terms are $\frac{1}{3}, \frac{1}{5}$, and $\frac{1}{7}$.

9. $c_1 = \log_2(2^1 + 3) = \log_2(2^4) = 4$

$c_2 = \log_2(2^2 + 3) = \log_2(2^5) = 5$

$c_3 = \log_2(2^3 + 3) = \log_2(2^6) = 6$

The first three terms are 4, 5, and 6.

11. $\displaystyle\sum_{i=1}^{3} i^3 = 1^3 + 2^3 + 3^3 = 36$

13. $\displaystyle\sum_{n=1}^{5} n(n-1) = 1(1-1) + 2(2-1) + 3(3-1)$
$$+ 4(4-1) + 5(5-1)$$
$$= 0 + 2 + 6 + 12 + 20 = 40$$

15. The terms in the series are reciprocals of even integers. Even integers are usually represent as 2i, but to get 4 in the denominator when i = 1, we use 2(i + 1).

$$\sum_{i=1}^{\infty} \frac{1}{2(i+1)}$$

17. The terms in this series are the squares of integers. Squares are usually represented as i^2, but to get the first term 0 when i = 1, we use $(i - 1)^2$.

$$\sum_{i=1}^{\infty} (i-1)^2$$

19. To get alternating signs for the terms, we use a power of −1. If we use $(-1)^i$, then i = 1

makes the first term negative. So we use $(-1)^{i+1}$.

$$\sum_{i=1}^{\infty} (-1)^{i+1} x_i$$

21. $a_1 = 6 + (1-1)5 = 6$

Since the common difference is 5, the first four terms are 6, 11, 16, and 21.

23. $a_1 = -20 + (1-1)(-2) = -20$

Since the common difference is −2, the first four terms are −20, −22, −24, and −26.

25. $a_1 = 1000(1) + 2000 = 3000$

Since the common difference is 1000, the first four terms are 3000, 4000, 5000, and 6000.

27. Use $a_1 = 1/3$, $d = 1/3$, and the formula $a_n = a_1 + (n-1)d$.

$$a_n = \frac{1}{3} + (n-1)\frac{1}{3} = \frac{n}{3}$$

29. Use $a_1 = 2$, $d = 2$, and the formula $a_n = a_1 + (n-1)d$.

$$a_n = 2 + (n-1)(2) = 2n$$

31. Use $a_1 = 1$, $a_{24} = 24$, $n = 24$, and the formula $S_n = \frac{n}{2}(a_1 + a_n)$.

$$S_{24} = \frac{24}{2}(1 + 24) = 300$$

33. Use $a_1 = 1/6$, $d = 1/3$, $a_n = 11/2$, and the formula $a_n = a_1 + (n-1)d$ to find n.

$$\frac{11}{2} = \frac{1}{6} + (n-1)\frac{1}{3}$$
$$33 = 1 + (n-1)2$$
$$32 = 2n - 2$$
$$34 = 2n$$
$$17 = n$$

Now use $n = 17$, $a_1 = 1/6$, $a_{17} = 11/2$, and the formula $S_n = \frac{n}{2}(a_1 + a_n)$ to find the sum.

$$S_{17} = \frac{17}{2}\left(\frac{1}{6} + \frac{11}{2}\right) = \frac{17}{2}\left(\frac{34}{6}\right) = \frac{289}{6}$$

35. Use $a_1 = -1$, $a_7 = 11$, $n = 7$, and the formula $S_n = \frac{n}{2}(a_1 + a_n)$ to find the sum.

$$S_7 = \frac{7}{2}(-1 + 11) = 35$$

37. $a_1 = 3\left(\frac{1}{2}\right)^{1-1} = 3$, $\quad a_2 = 3\left(\frac{1}{2}\right)^{2-1} = \frac{3}{2}$,

$$a_3 = 3\left(\frac{1}{2}\right)^{3-1} = \frac{3}{4}, \quad a_4 = 3\left(\frac{1}{2}\right)^{4-1} = \frac{3}{8}.$$

The first four terms are 3, $\frac{3}{2}$, $\frac{3}{4}$, and $\frac{3}{8}$.

39. $a_1 = 2^{1-1} = 1$, $a_2 = 2^{1-2} = \frac{1}{2}$

$$a_3 = 2^{1-3} = \frac{1}{4}, \quad a_4 = 2^{1-4} = \frac{1}{8}$$

The first four terms are 1, $\frac{1}{2}$, $\frac{1}{4}$, and $\frac{1}{8}$.

41. $\quad a_1 = 23(10)^{-2(1)} = 0.23$,

$a_2 = 23(10)^{-2(2)} = 0.0023$,

$a_3 = 23(10)^{-2(3)} = 0.000023$,

$a_4 = 23(10)^{-2(4)} = 0.00000023$

The first four terms of the geometric sequence are 0.23, 0.0023, 0.000023, and 0.00000023.

43. Use $a_1 = 1/2$, $r = 6$, and the formula $a_n = a_1 r^{n-1}$.

$$a_n = \frac{1}{2}(6)^{n-1}$$

45. Use $a_1 = 7/10$, $r = 1/10$, and the formula $a_n = a_1 r^{n-1}$.

$$a_n = 0.7(0.1)^{n-1}$$

47. Use $\quad a_1 = 1/3$, $r = 1/3$, $n = 4$, and the formula $S_n = \frac{a_1(1-r^n)}{1-r}$.

$$S_4 = \frac{\frac{1}{3}\left(1 - \left(\frac{1}{3}\right)^4\right)}{1 - \frac{1}{3}} = \frac{\frac{1}{3}\left(\frac{80}{81}\right)}{\frac{2}{3}} = \frac{40}{81}$$

49. Use $\quad a_1 = 0.3$, $r = 0.1$, $n = 10$, and the formula $S_n = \frac{a_1(1-r^n)}{1-r}$.

$$S_{10} = \frac{0.3\left(1 - (0.1)^{10}\right)}{1 - 0.1} = \frac{0.3(0.9999999999)}{0.9}$$

$$= 0.3333333333$$

Your calculator may not give ten 3's after the decimal point, but doing this computation without a calculator does give ten 3's and this is the exact answer.

51. Use $a_1 = 1/4$, $r = 1/3$, and the formula for the sum of an infinite geometric series $S = \frac{a_1}{1-r}$.

$$S = \frac{\frac{1}{4}}{1 - \frac{1}{3}} = \frac{\frac{1}{4}}{\frac{2}{3}} = \frac{3}{8}$$

53. Use $a_1 = 18$, $r = 2/3$, and the formula for the sum of an infinite geometric series $S = \frac{a_1}{1-r}$.

$$S = \frac{18}{1 - \frac{2}{3}} = \frac{18}{\frac{1}{3}} = 54$$

55. The coefficients for the fifth power of a binomial are 1, 5, 10, 10, 5, and 1. $(m + n)^5 = m^5 + 5m^4n + 10m^3n^2 + 10m^2n^3 + 5mn^4 + n^5$

57. The coefficients for the third power of a binomial are 1, 3, 3, and 1. Alternate the signs because it is a difference to a power.

$$(a^2 - 3b)^3 = 1(a^2)^3(3b)^0 - 3(a^2)^2(3b)^1$$
$$+ 3(a^2)^1(3b)^2 - 1(a^2)^0(3b)^3$$
$$= a^6 - 9a^4b + 27a^2b^2 - 27b^3$$

59. Use $n = 12$ and $k = 5$ in the formula for the kth term.

$$\frac{12!}{(12-5+1)!(5-1)!}x^{12-5+1}y^{5-1} = \frac{12!}{8!4!}x^8y^4$$

$$= 495x^8y^4$$

61. Use $n = 14$ and $k = 3$ in the formula for the kth term.

$$\frac{14!}{(14-3+1)!(3-1)!}(2a)^{14-3+1}(-b)^{3-1}$$

$$= \frac{14!}{12!2!}(2a)^{12}(-b)^2 = 372,736a^{12}b^2$$

63. Use the binomial theorem expressed in summation notation, with n = 7.

$$(a + w)^7 = \sum_{i=0}^{7} \frac{7!}{(7-i)!\,i!} \, a^{7-i} w^i$$

65. The sequence has neither a constant difference nor a constant ratio. So it is neither arithmetic nor geometric.

67. There is a constant difference of 3. So the sequence is an arithmetic sequence.

69. There is a constant difference of 2. So the sequence is an arithmetic sequence.

71. Use $a_1 = 6$, $n = 4$, $a_4 = 1/30$, and the formula $a_n = a_1 r^{n-1}$.

$$\frac{1}{30} = 6r^{4-1}$$

$$\frac{1}{180} = r^3$$

$$r = \sqrt[3]{\frac{1}{180}} = \frac{1}{\sqrt[3]{180}}$$

73.
$$\sum_{i=1}^{5} \frac{(-1)^i}{i!} = \frac{(-1)^1}{1!} + \frac{(-1)^2}{2!} + \frac{(-1)^3}{3!}$$
$$+ \frac{(-1)^4}{4!} + \frac{(-1)^5}{5!}$$
$$= -1 + \frac{1}{2} - \frac{1}{6} + \frac{1}{24} - \frac{1}{120}$$

75. This is the summation notation for the binomial expansion of $(a + b)^5$.

$$\frac{5!}{5!0!}a^5b^0 + \frac{5!}{4!1!}a^4b^1 + \frac{5!}{3!2!}a^3b^2 + \frac{5!}{2!3!}a^2b^3$$
$$+ \frac{5!}{1!4!}a^1b^4 + \frac{5!}{0!5!}a^0b^5$$
$$= a^5 + 5a^4b + 10a^3b^2 + 10a^2b^3 + 5ab^4 + b^5$$

77. There are 26 terms because in the expansion of $(x + y)^n$ there are $n + 1$ terms.

79. The first $3000 earns interest for 16 years. The second $3000 earns interest for 15 years, and so on. The last $3000 earns interest for 1 year. The total in the account at the end of 16 years is the sum of the following series.

$$3000(1.1) + 3000(1.1)^2 + ... + 3000(1.1)^{16}$$

This is a geometric series with n = 16, $a_1 = 3000(1.1)$ and r = 1.1.

$$S_{16} = \frac{3000(1.1)\left(1 - (1.1)^{16}\right)}{1 - 1.1} = \$118,634.11$$

81. We compute a new balance 16 times by multiplying by $1 + 0.10$ each time.

$$3000(1.10)^{16} = \$13,784.92$$

CHAPTER 12   TEST

1. $a_1 = -10 + (1 - 1)6 = -10$
$a_2 = -10 + (2 - 1)6 = -4$
$a_3 = -10 + (3 - 1)6 = 2$
$a_4 = -10 + (4 - 1)6 = 8$
The first four terms are −10, −4, 2, and 8.

2. $a_1 = 5(0.1)^{1-1} = 5$

$$a_2 = 5(0.1)^{2-1} = 0.5$$

$$a_3 = 5(0.1)^{3-1} = 0.05$$

$$a_4 = 5(0.1)^{4-1} = 0.005$$

The first four terms are 5, 0.5, 0.05, and 0.005.

3. $a_1 = \frac{(-1)^1}{1!} = -1$, $a_2 = \frac{(-1)^2}{2!} = \frac{1}{2}$,

$$a_3 = \frac{(-1)^3}{3!} = -\frac{1}{6}, \quad a_4 = \frac{(-1)^4}{4!} = \frac{1}{24}$$

The first four terms are -1, $\frac{1}{2}$, $-\frac{1}{6}$, and $\frac{1}{24}$.

4. $a_1 = \frac{2(1) - 1}{(1)^2} = 1$, $a_2 = \frac{2(2) - 1}{(2)^2} = \frac{3}{4}$,

$$a_3 = \frac{2(3) - 1}{(3)^2} = \frac{5}{9}, \quad a_4 = \frac{2(4) - 1}{(4)^2} = \frac{7}{16}$$

The first four terms are 1, $\frac{3}{4}$, $\frac{5}{9}$, and $\frac{7}{16}$.

5. The sequence is an arithmetic sequence with $a_1 = 7$ and $d = -3$. So the general term is $a_n = 7 + (n-1)(-3) = 7 - 3n + 3 = 10 - 3n$.

6. This sequence is a geometric sequence with $a_1 = -25$, and $r = -1/5$. So the general term is $a_n = -25\left(-\frac{1}{5}\right)^{n-1}$.

7. This sequence is a sequence of even integers, which we can represent as 2n. To get the signs to alternate, we use a power of -1. So the general term is $a_n = (-1)^{n-1}2n$.

8. This sequence is a sequence of squares of the positive integers. So the general term is $a_n = n^2$.

9. $\displaystyle\sum_{i=1}^{5}(2i+3) = 2(1) + 3 + 2(2) + 3 + 2(3) + 3$
$$+ 2(4) + 3 + 2(5) + 3$$
$$= 5 + 7 + 9 + 11 + 13$$

10. $\displaystyle\sum_{i=1}^{6}5(2)^{i-1} = 5(2)^{1-1} + 5(2)^{2-1}$
$$+ 5(2)^{3-1} + 5(2)^{4-1} + 5(2)^{5-1} + 5(2)^{6-1}$$
$$= 5 + 10 + 20 + 40 + 80 + 160$$

11. $\displaystyle\sum_{i=0}^{4}\frac{4!}{(4-i)!i!}m^{4-i}q^i = \frac{4!}{4!0!}m^4q^0$
$$+ \frac{4!}{3!1!}m^3q^1 + \frac{4!}{2!2!}m^2q^2 + \frac{4!}{1!3!}m^1q^3 + \frac{4!}{0!4!}m^0q^4$$
$$= m^4 + 4m^3q + 6m^2q^2 + 4mq^3 + q^4$$

12. Use $a_1 = 9$, $a_{20} = 66$, and $n = 20$ in the formula for the sum of an arithmetic series.
$$S_{20} = \frac{20}{2}(9 + 66) = 10(75) = 750$$

13. Use $a_1 = 10$, $n = 5$, and $r = 1/2$ in the formula for the sum of a finite geometric series.
$$S_5 = \frac{10\left(1 - \left(\frac{1}{2}\right)^5\right)}{1 - \frac{1}{2}} = \frac{10\left(\frac{31}{32}\right)}{\frac{1}{2}} = \frac{155}{8}$$

14. Use $a_1 = 0.35$ and $r = 0.93$ in the formula for the sum of an infinite geometric series.
$$S = \frac{0.35}{1 - 0.93} = \frac{0.35}{0.07} = 5$$

15. Use $a_1 = 2$, $a_{100} = 200$, and $n = 100$ in the formula for the sum of a finite arithmetic series.
$$S_{100} = \frac{100}{2}(2 + 200) = 50(202) = 10,100$$

16. Use $a_1 = 1/4$ and $r = 1/2$ in the formula for the sum of an infinite geometric series.
$$S = \frac{1/4}{1 - 1/2} = \frac{1/4}{1/2} = \frac{1}{2}$$

17. Use $a_1 = 2$, $r = 1/2$ and $a_n = 1/128$ to find n:
$$2\left(\frac{1}{2}\right)^{n-1} = \frac{1}{128}$$
$$\left(\frac{1}{2}\right)^{n-1} = \frac{1}{256} = \frac{1}{2^8}$$
$$n - 1 = 8$$
$$n = 9$$
Use $a_1 = 2$, $n = 9$, and $r = 1/2$ to find the sum of the 9 terms.
$$S_9 = \frac{2(1 - (1/2)^9)}{1 - 1/2} = \frac{511}{128} \approx 3.9922$$

18. Use $a_1 = 3$, $a_5 = 48$, $n = 5$ in the formula for the general term of a geometric sequence.
$$48 = 3r^{5-1}$$
$$16 = r^4$$
$$\pm 2 = r$$

19. Use $a_1 = 1$, $a_{12} = 122$, $n = 12$, and the formula for the general term of an arithmetic sequence.
$$122 = 1 + (12 - 1)d$$
$$121 = 11d$$
$$11 = d$$

20. Use $n = 15$ and $k = 5$ in the formula for the kth term of a binomial expansion.
$$\frac{15!}{(15-5+1)!(5-1)!}r^{15-5+1}(-t)^{5-1}$$
$$= \frac{15!}{11!4!}r^{11}t^4 = 1365r^{11}t^4$$

21. Use $n = 8$ and $k = 4$ in the formula for the kth term of a binomial expansion.

$$\frac{8!}{(8 - 4 + 1)!(4 - 1)!}(a^2)^{8 - 4 + 1}(-2b)^{4 - 1}$$

$$= \frac{8!}{5!3!}a^{10}(-2)^3 b^3 = -448a^{10}b^3$$

22. $800(1.10)^1 + 800(1.10)^2 + ... + 800(1.10)^{25}$

$$= \frac{800(1.10)(1 - 1.10^{25})}{1 - 1.10} = \$86,545.41$$

Tying It All Together Chapters 1-12

1. $f(3) = 3^2 - 3 = 9 - 3 = 6$

2. $f(n) = n^2 - 3$

3. $f(x + h) = (x + h)^2 - 3 = x^2 + 2xh + h^2 - 3$

4. $f(x) - g(x) = x^2 - 3 - (2x - 1) = x^2 - 2x - 2$

5. $g(f(3)) = g(6) = 2(6) - 1 = 11$

6. $(f \circ g)(2) = f(g(2)) = f(3) = 3^2 - 3 = 6$

7. $m(16) = \log_2(16) = 4$

8. $(h \circ m)(32) = h(m(32)) = h(5) = 2^5 = 32$

9. $h(-1) = 2^{-1} = 1/2$

10. $h^{-1}(8) = \log_2(8) = 3$

11. $m^{-1}(0) = \log_2(0) = 2^0 = 1$

12. $(m \circ h)(x) = m(h(x)) = m(2^x) = \log_2(2^x)$
$= x$
So $(m \circ h)(x) = x$.

13. If y varies directly as x, then $y = kx$. Since $y = -6$ when $x = 4$, we have $-6 = 4k$, or $k = -3/2$. When $x = 9$, we can use the original formula with $k = -3/2$.

$$y = -\frac{3}{2}(9) = -\frac{27}{2}$$

14. If a varies inversely as b, then $a = k/b$. If $a = 2$ when $b = -4$, we can find k.

$$2 = \frac{k}{-4}$$

$$-8 = k$$

To find a, use $k = -8$ and $b = 3$.

$$a = \frac{-8}{3} = -\frac{8}{3}$$

15. If y varies directly as w and inversely as t, then $y = (kw)/t$. Use $y = 16$, $w = 3$, and $t = -4$ to find k.

$$16 = \frac{k(3)}{-4}$$

$$-64 = 3k$$

$$\frac{-64}{3} = k$$

To find y, use $k = -64/3$, $w = 2$, and $t = 3$.

$$y = -\frac{64(2)}{3(3)} = -\frac{128}{9}$$

16. If y varies jointly as h and the square of r, then $y = khr^2$. Use $y = 12$, $h = 2$, and $r = 3$ to find k.

$$12 = k(2)(3)^2$$
$$12 = 18k$$

$$\frac{2}{3} = k$$

To find y, use $k = 2/3$, $h = 6$, and $r = 2$.

$$y = \frac{2}{3}(6)(2)^2 = 16$$

17. The graph of $x > 3$ is the region to the right of the vertical line $x = 3$. The graph of $x + y < 0$ is the region below the line $y = -x$. The region to the right of $x = 3$ and below $y = -x$ is shown in the following graph.

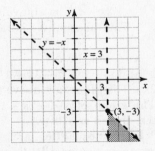

18. The inequality $|x - y| \geq 2$ is equivalent to the compound inequality $x - y \geq 2$ or $x - y \leq -2$. The graph of $x - y \geq 2$ is the region below the line $y = x - 2$. The graph of $x - y \leq -2$ is the region above the line $y = x + 2$. Since the word or is used, the graph of the compound inequality is the union of these two regions as shown in the following diagram.

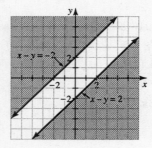

19. The graph of $y < -2x + 3$ is the region below the line $y = -2x + 3$. The graph of $y > 2^x$ is the region above the curve $y = 2^x$. Since the word and is used, the graph of the compound inequality is the intersection of these two regions, the points that lie above $y = 2^x$ and below $y = -2x + 3$. We could have used a test point in each of the four regions to see which region satisfies both inequalities.

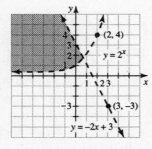

20. The inequality $|y + 2x| < 1$ is equivalent to $-1 < y + 2x < 1$. This inequality is also written as

$$y + 2x > -1 \quad \text{and } y + 2x < 1$$
$$y > -2x - 1 \text{ and } \quad y < -2x + 1$$

The graph of $y > -2x - 1$ is the region above the line $y = -2x - 1$. The graph of $y < -2x + 1$ is the region below the line $y = -2x + 1$. The points that satisfy both inequalities are the points that lie between these two parallel lines.

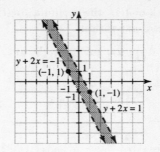

21. The graph of $x^2 + y^2 = 4$ is a circle of radius 2 centered at the origin. Since $(0, 0)$ satisfies the inequality $x^2 + y^2 < 4$, we shade the region inside the circle.

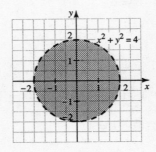

22. The graph of $x^2 - y^2 = 1$ is a hyperbola with x-intercepts $(-1, 0)$ and $(1, 0)$. The fundamental rectangle passes through the x-intercepts and $(0, 1)$ and $(0, -1)$. Extend the diagonals for the asymptotes. The hyperbola opens to the left and right. Test a point in each region to see that only points in the region containing the origin satisfy the inequality.

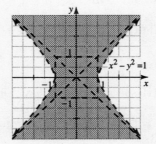

23. Graph the curve $y = \log_2(x)$ and shade the region below the curve to show the graph of $y < \log_2(x)$.

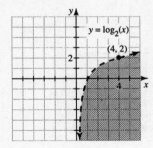

24. Write $x^2 + 2y < 4$ as $y < -\frac{1}{2}x^2 + 2$, to see that the boundary is a parabola opening downward with vertex at $(0, 2)$.

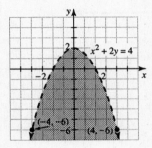

25. The graph of $\frac{x^2}{4} + \frac{y^2}{9} < 1$ is the region inside the ellipse $\frac{x^2}{4} + \frac{y^2}{9} = 1$. The graph of $y > x^2$ is the region above the parabola $y = x^2$. The points that satisfy the compound inequality are inside the ellipse and above the parabola as shown in the diagram.

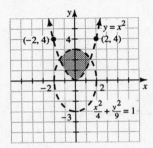

26. $\dfrac{a}{b} + \dfrac{b}{a} = \dfrac{a(a)}{b(a)} + \dfrac{b(b)}{a(b)} = \dfrac{a^2 + b^2}{ab}$

27. $1 - \dfrac{3}{y} = \dfrac{y}{y} - \dfrac{3}{y} = \dfrac{y - 3}{y}$

28. $\dfrac{x - 2}{x^2 - 9} - \dfrac{x - 4}{x^2 - 2x - 3}$

$= \dfrac{(x - 2)(x + 1)}{(x - 3)(x + 3)(x + 1)} - \dfrac{(x - 4)(x + 3)}{(x - 3)(x + 1)(x + 3)}$

$= \dfrac{x^2 - x - 2}{(x - 3)(x + 3)(x + 1)} - \dfrac{x^2 - x - 12}{(x - 3)(x + 3)(x + 1)}$

$= \dfrac{10}{(x - 3)(x + 3)(x + 1)}$

29. $\dfrac{(x - 4)(x + 4)}{2(x + 4)} \cdot \dfrac{4(x^2 + 4x + 16)}{x^3 - 16}$

$= \dfrac{2(x^3 - 64)}{x^3 - 16}$

30. $\dfrac{(a^2 b)^3}{(ab^2)^4} \cdot \dfrac{ab^3}{a^{-4} b^2} = \dfrac{a^6 b^3}{a^4 b^8} \cdot \dfrac{ab^3}{a^{-4} b^2} = \dfrac{a^7 b^6}{b^{10}} = \dfrac{a^7}{b^4}$

31. $\dfrac{x^2 y}{(xy)^3} \div \dfrac{xy^2}{x^2 y^4} = \dfrac{x^2 y}{x^3 y^3} \cdot \dfrac{x^2 y^4}{xy^2} = \dfrac{x^4 y^5}{x^4 y^5} = 1$

32. $8^{2/3} = (\sqrt[3]{8})^2 = 2^2 = 4$

33. $16^{-5/4} = \dfrac{1}{(\sqrt[4]{16})^5} = \dfrac{1}{(2)^5} = \dfrac{1}{32}$

34. $-4^{1/2} = -\sqrt{4} = -2$

35. $27^{-2/3} = \dfrac{1}{(\sqrt[3]{27})^2} = \dfrac{1}{3^2} = \dfrac{1}{9}$

36. $-2^{-3} = -\dfrac{1}{2^3} = -\dfrac{1}{8}$

37. $2^{-3/5} \cdot 2^{-7/5} = 2^{-10/5} = 2^{-2} = \dfrac{1}{2^2} = \dfrac{1}{4}$

38. $5^{-2/3} \div 5^{1/3} = 5^{-\frac{2}{3} - \frac{1}{3}} = 5^{-1} = \dfrac{1}{5}$

39. $(9^{1/2} + 4^{1/2})^2 = (3 + 2)^2 = 5^2 = 25$

40. a) Age 4 years 3 months is 4.25 years.

$h(4.25) = 79.041 + 6.39(4.25) - e^{3.261 - 0.993(4.25)}$

$= 105.8$ cm

105.8 cm $\cdot \dfrac{1 \text{ in}}{2.54 \text{ cm}} = 41.7$ in.

b) 1.3 years

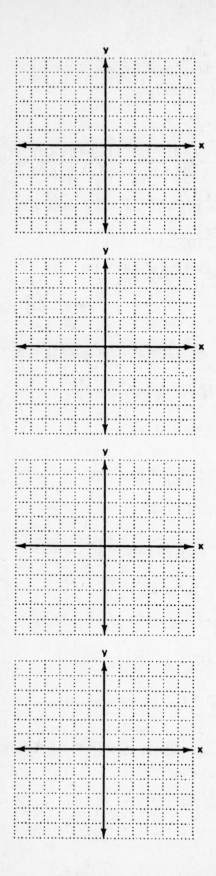
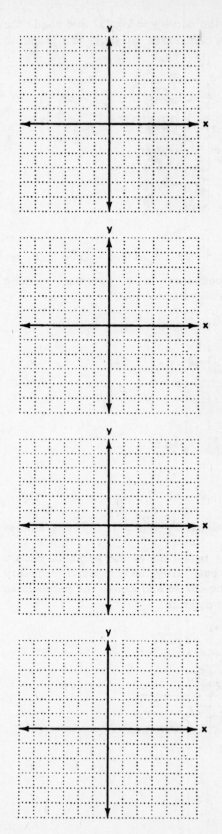

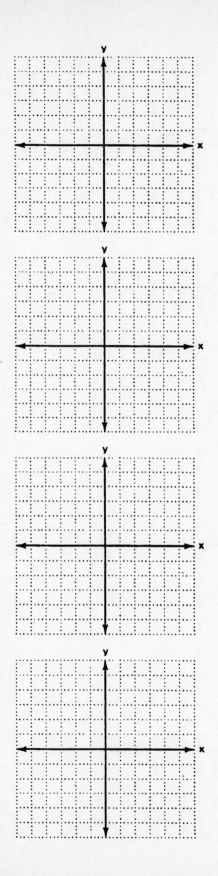

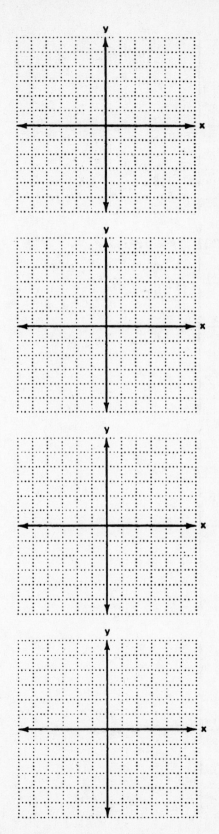

261

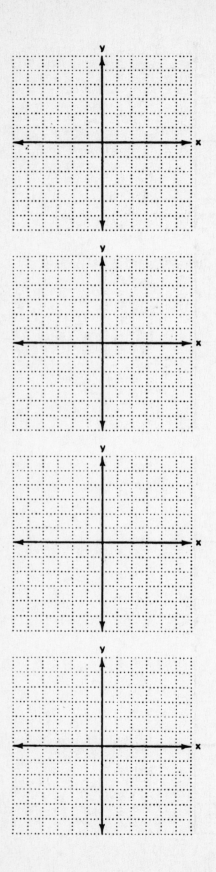
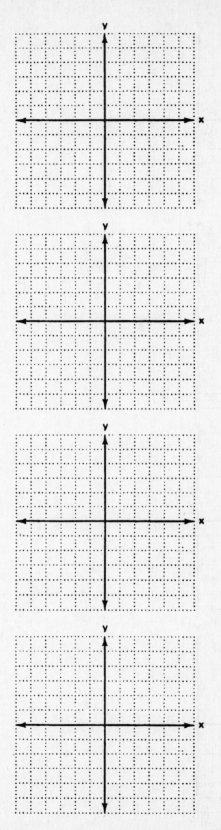

262

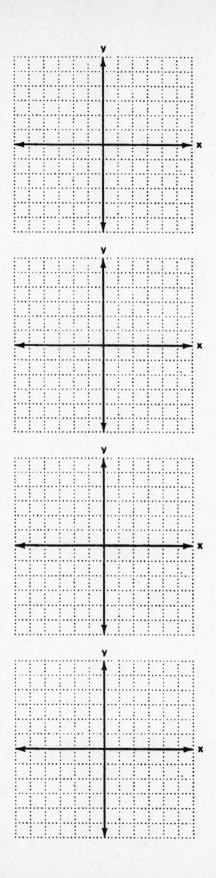

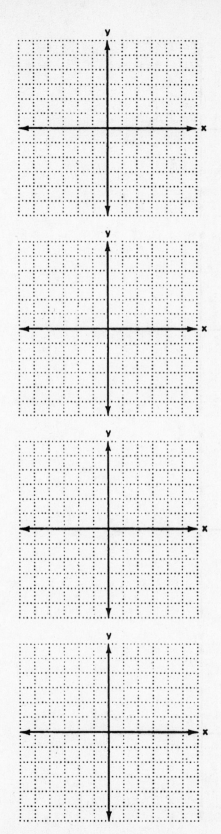

263

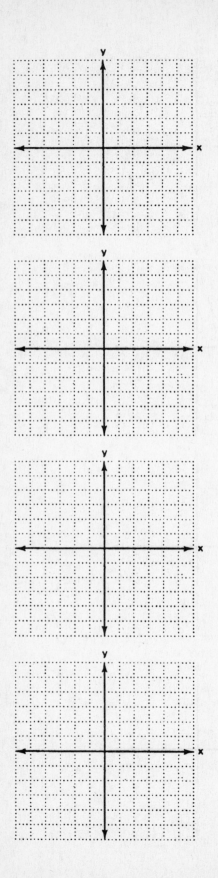

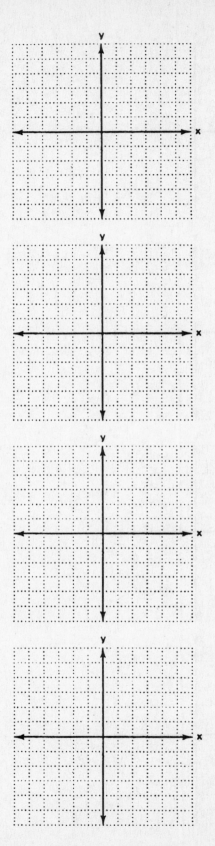

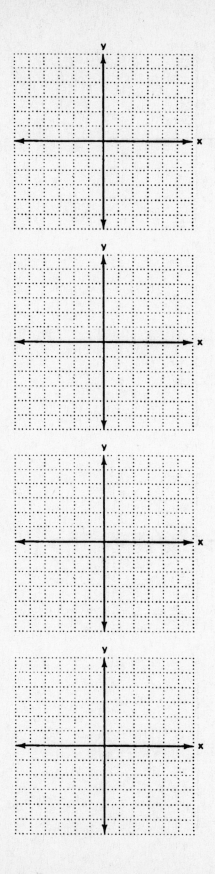

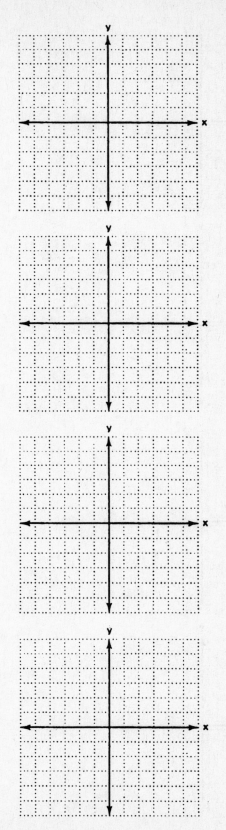

265

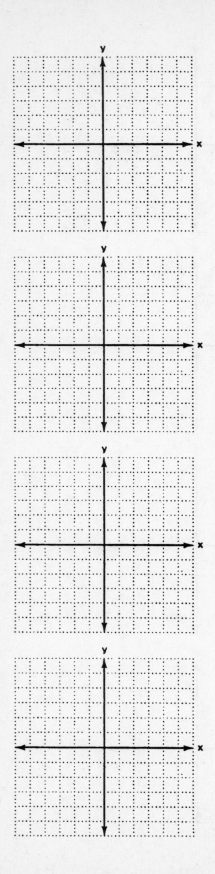

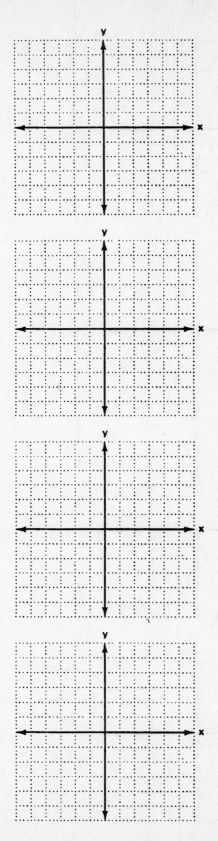

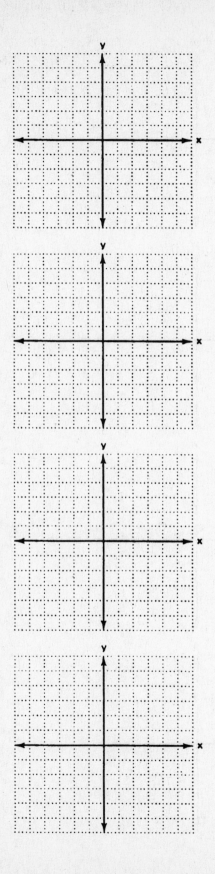

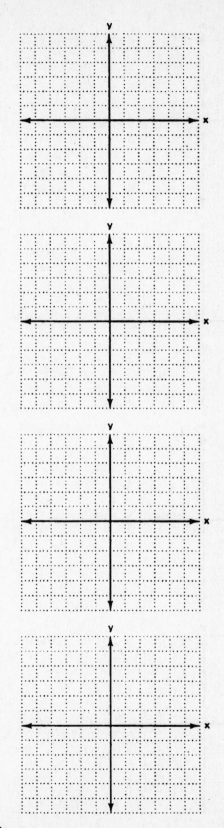

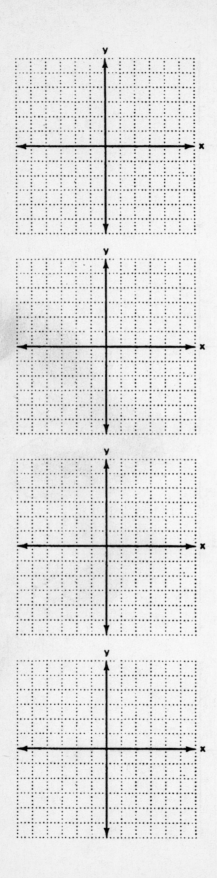
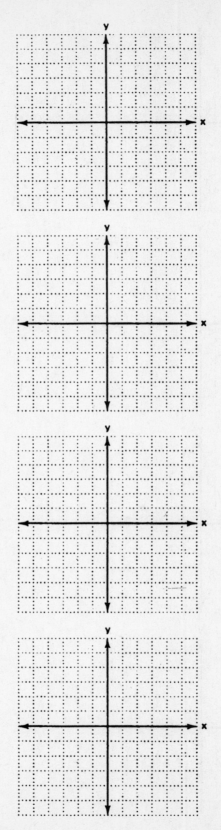

268